Medical Virology

MEDICAL VIROLOGY

Third Edition

DAVID O. WHITE

Department of Microbiology
University of Melbourne
Parkville, Victoria, Australia

FRANK FENNER

John Curtin School of Medical Research
Australian National University
Canberra, A.C.T., Australia

ACADEMIC PRESS, INC.

Harcourt Brace Jovanovich, Publishers

Orlando San Diego New York Austin
London Montreal Sydney Tokyo Toronto

ACADEMIC PRESS, INC.
Orlando, Florida 32887

United Kingdom Edition published by
ACADEMIC PRESS INC. (LONDON) LTD.
24–28 Oval Road, London NW1 7DX

Library of Congress Cataloging in Publication Data

Fenner, Frank, Date
Medical virology.

Rev. ed. of: Medical virology / Frank Fenner, David O. White. 2nd ed. c1976
Includes bibliographies and index.
1. Medical virology. I. White, David O. II. Title. [DNLM: 1. Virus Diseases. 2. Viruses. QZ 65 F336m]
QR360.F43 1986 616'.0194 85-15674
ISBN 0–12–746640–1 (alk. paper)

PRINTED IN THE UNITED STATES OF AMERICA

86 87 88 89 9 8 7 6 5 4 3 2 1

Contents

Preface *vii*

Part I. **PRINCIPLES OF ANIMAL VIROLOGY**

1. **Structure and Classification of Viruses** 3
2. **Cultivation and Assay of Viruses** 35
3. **Viral Multiplication** 53
4. **Viral Genetics and Evolution** 91
5. **Pathogenesis and Pathology of Viral Infections** 119
6. **Host Responses to Viral Infections** 147
7. **Persistent Infections** 193
8. **Oncogenic Viruses** 217
9. **Epidemiology and Control of Viral Infections** 247
10. **Immunization against Viral Diseases** 283
11. **Chemotherapy of Viral Diseases** 303
12. **Laboratory Diagnosis of Viral Diseases** 325

Part II. VIRUSES OF MAN

13. Hepadnaviruses 365

14. Papovaviruses 381

15. Adenoviruses 389

16. Herpesviruses 401

17. Poxviruses 433

18. Parvoviruses 445

19. Picornaviruses and Caliciviruses 451

20. Togaviruses and Flaviviruses 479

21. Orthomyxoviruses 509

22. Paramyxoviruses 521

23. Coronaviruses 541

24. Arenaviruses 547

25. Bunyaviruses 555

26. Rhabdoviruses and Filoviruses 567

27. Reoviruses 575

28. Other Viral Diseases 585

29. Viral Syndromes 607

Index 635

Preface

A decade has elapsed since the Second Edition of "Medical Virology" was published. Advances in the subject during that time have been such that we have elected to rewrite the book completely, retaining only the plan and layout of the 1976 edition with certain of the illustrations. In the process, the text has been expanded by about 150 pages. For the Third Edition we have assumed that today students of virology have a more sophisticated molecular and cellular biology background; thus it is pitched at a somewhat higher level.

"Medical Virology" is intended primarily as a textbook for medical students and advanced science students, but we trust it will also serve as a useful reference work for teachers of virology and for postgraduate students and others undertaking research in this discipline.

Part I of the book has been written as a self-contained synopsis of the principles of virology, which are central to the interests of each of these groups. Part II, arranged by viral family, is oriented toward the needs of medical students.

Although many of the tables and figures contain important factual information, the text has been stripped of minutiae so far as possible in order to focus on fundamental principles and basic facts. Statements have not been individually supported by references, but lists of recent authoritative reviews are provided at the end of each chapter to simplify the reader's entry into the scientific literature.

We thank the following colleagues in Melbourne and Canberra for their comments on particular chapters related to their fields of interest: Drs. A. J. Bellett, R. V. Blanden, T. Gonda, I. D. Gust, K. Hayes, J. C. Hierholzer, I. H. Holmes, I. Jack, I. D. Marshall, E. G. Westaway and A. Yung. We owe a special debt to Dr. Fred Murphy and his colleagues at the Centers for Disease Control, Atlanta, for providing several of the electron micrographs and to Ian Jack of the Royal Children's Hospital, Melbourne, for numerous excellent photographs. Professor C. A. Mims and Blackwell Scientific Publications graciously allowed us to use certain illustrations and other material from "Viral Pathogenesis and Immunology" by C. A. Mims and D. O. White (1984). We are also indebted to those scientists and publishers, too numerous to mention here, who responded generously to our requests for illustrative material; appropriate acknowledgments accompany the legends to figures and plates. Librarian Janet Cook provided sterling assistance with reference research and Belinda Lightfoot with index preparation. Our secretaries, Mrs. Judy Feary and Mrs. Marj Lee, have earned our gratitude and admiration for their devotion and skill in preparing the manuscript. The staff of Academic Press has once again cooperated to ensure rapid production of this edition.

This book is dedicated to the memory of the great Australian virologist and immunologist, Sir Macfarlane Burnet, who died at the age of 85 on August 31, 1985.

David O. White
Frank Fenner

PART I

Principles of Animal Virology

CHAPTER 1

Structure and Classification of Viruses

Introduction 3
Morphology of Viruses 4
Chemical Composition of Viruses 9
Inactivation of Viruses 17
Classification of Viruses 20
The Families of DNA Viruses 23
The Families of RNA Viruses 27
Unclassified Viruses 32
Further Reading 33

INTRODUCTION

The unicellular microorganisms can be arranged in order of decreasing size and complexity: protozoa, yeasts and certain fungi, bacteria, mycoplasmas, rickettsiae, and chlamydiae. Microorganisms, however small and simple, are cells. They always contain DNA as the repository of their genetic information, and they also have their own machinery for producing energy and macromolecules. Microorganisms grow by synthesizing their own macromolecular constituents (nucleic acid, protein, carbohydrate, and lipid), and they multiply by binary fission.

Viruses, on the other hand, cannot really be regarded as microorganisms at all for they are not cells. They contain only one type of nucleic acid, either DNA or RNA, but not both. Furthermore, since viruses have no ribosomes, mitochondria, or other organelles, they are completely dependent upon their cellular hosts for the machinery of

TABLE 1-1
Properties of Microorganisms and Viruses

PROPERTY	BACTERIA	MYCOPLASMAS	RICKETTSIAE	CHLAMYDIAE	VIRUSES
Growth on nonliving media	+	+	−	−	−
Binary fission	+	+	+	+	−
DNA and RNA	+	+	+	+	−
Ribosomes	+	+	+	+	−
Metabolism	+	+	+	±	−
Sensitivity to antibiotics	+	+	+	+	−
Sensitivity to interferon	−	−	−	+	+

protein synthesis and energy production. They are metabolically inert and can multiply only within the cells of the host that they parasitize. Indeed, unlike any of the microorganisms, many viruses can, in suitable cells, reproduce themselves from their genome, a single nucleic acid molecule. The key differences between viruses and microorganisms are listed in Table 1-1.

Several important practical consequences flow from these differences. For example, some viruses (but no microorganisms) may persist in cells by the integration of their DNA (or a DNA copy of their RNA) with that of the host cell. Moreover, being metabolically inert, viruses are not susceptible to antibiotics that act against specific steps in the metabolic pathways of microorganisms.

MORPHOLOGY OF VIRUSES

Physical Methods for Studying Viruses

It has been known for many years that viruses are smaller than microorganisms. The first unequivocal demonstration of this for an animal virus occurred in 1898, when Loeffler and Frosch demonstrated that foot-and-mouth disease, an important infectious disease of cattle, could be transferred by material that could pass through a filter of average pore diameter too small to allow passage of bacteria. The new group of "organisms" became known as the "filterable viruses." For a time they were also called "ultramicroscopic," since most viruses are beyond the limit of resolution of light microscopes [200 nanometers (nm) = 2000 angstroms (Å)]. Only with the advent of the electron microscope was it possible to study the morphology of viruses properly. It then became apparent that they range in size from about the size of the smallest microorganisms down to little bigger than the largest protein molecules. In 1959, our knowledge of viral ultrastructure was transformed when Brenner and Horne applied negative staining to the electron microscopy of viruses. Potassium phosphotungstate, which is electron dense, fills the interstices of the viral surface, giving the resulting electron micrograph a degree of fine detail not previously possible. The technique of X-ray diffraction had already provided evidence that the protein molecules of the viral coat are packed around the nucleic acid molecule in a symmetrical arrangement; negative staining confirmed and extended this finding in a most revealing way. Electron micrographs of negatively stained preparations of the virions of all families of viruses that include human pathogens are shown in the relevant chapters of Part II of this volume.

Viral Structure

In the simpler viruses the mature virus particle, the *virion,* consists of a single molecule of nucleic acid surrounded by a protein coat, the *capsid;* the capsid and its enclosed nucleic acid together constitute the *nucleocapsid* (Fig. 1-1). In some of the more complex viruses the capsid surrounds a protein *core* and in other viruses the capsid is surrounded by a cell membrane-derived lipoprotein *envelope.* In some RNA viruses the genome comprises several separate nucleic acid molecules. The viral nucleic acid is not arranged randomly inside the virion; it has a specific relationship with the polypeptides of either capsid or core. The capsid is composed of a large number of *capsomers,* which are held together by noncovalent bonds and are usually visible by electron microscopy. Each capsomer is in turn made up of one or more polypeptide chains. X-Ray crystallography reveals the symmetry of arrangement of these molecules.

For reasons of genetic economy, capsids are composed of repeating units of one or a small number of different kinds of polypeptides. In order to form a complete shell to protect the viral nucleic acid from nucleases, molecules of these polypeptides must pack together symmetrically. The virions of all viruses that infect humans are either isometric and have icosahedral symmetry, or their tubular nucleocapsids have helical symmetry but are enclosed within a rather pleomorphic envelope (Fig. 1-1).

Icosahedral Symmetry. An icosahedron has 12 vertices and 20 faces, each an equilateral triangle. It has axes of two-, three-, and fivefold rotational symmetry, passing through the edges, faces, and vertices,

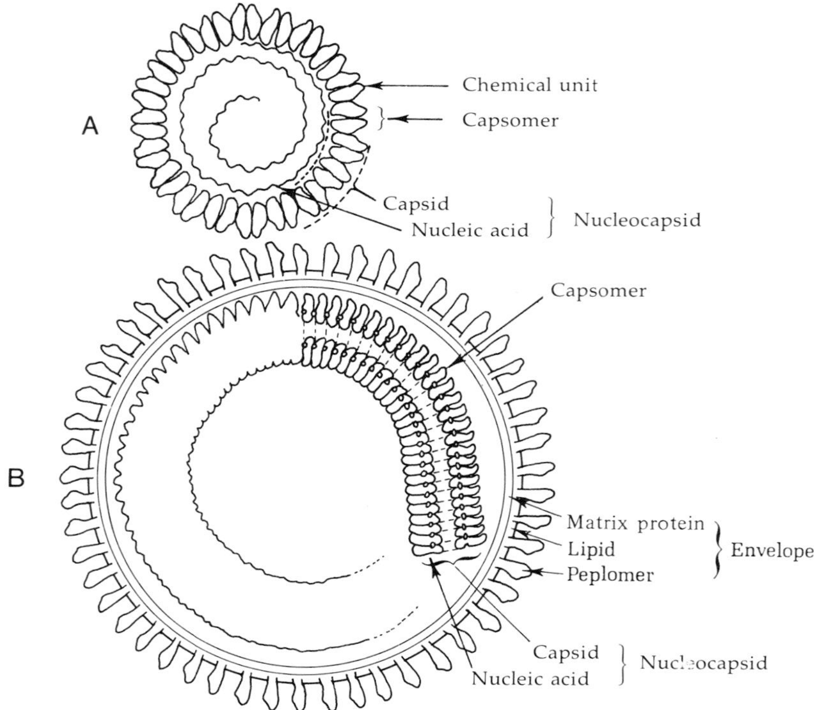

FIG. 1-1. *Schematic diagrams of the structure of a simple nonenveloped virion with an icosahedral capsid (A) and an enveloped virion with a tubular nucleocapsid with helical symmetry (B). The capsids are composed of capsomers, which are in turn made up of one or more polypeptides. Many icosahedral viruses have a "core" (not illustrated), which consists of protein(s) directly associated with the nucleic acid, inside the icosahedral capsid. In viruses of type B the envelope is a complex structure consisting of an inner virus-specified protein shell (M, for matrix protein) and a lipid layer derived from cellular lipids, in which are embedded virus-specified glycoprotein subunits (peplomers), each of which, like a capsomer, consists of one or more polypeptide chains. [Modified from D. L. D. Caspar* et al., Cold Spring Harbor Symp. Quant. Biol. **27,** *49 (1962).]*

respectively (Fig. 1-2). Only certain arrangements of the repeating subunits, the capsomers, can fit into the faces, edges, and vertices. Capsomers on the faces and edges that relate to six neighboring capsomers are called *hexamers;* those at the vertices relate to five neighbors and are called *pentamers* (Plate 1-1A). In some viruses different polypeptide subunits are used to build hexamers and pentamers. The different arrangements of capsomers have been systematically codified by electron micro-

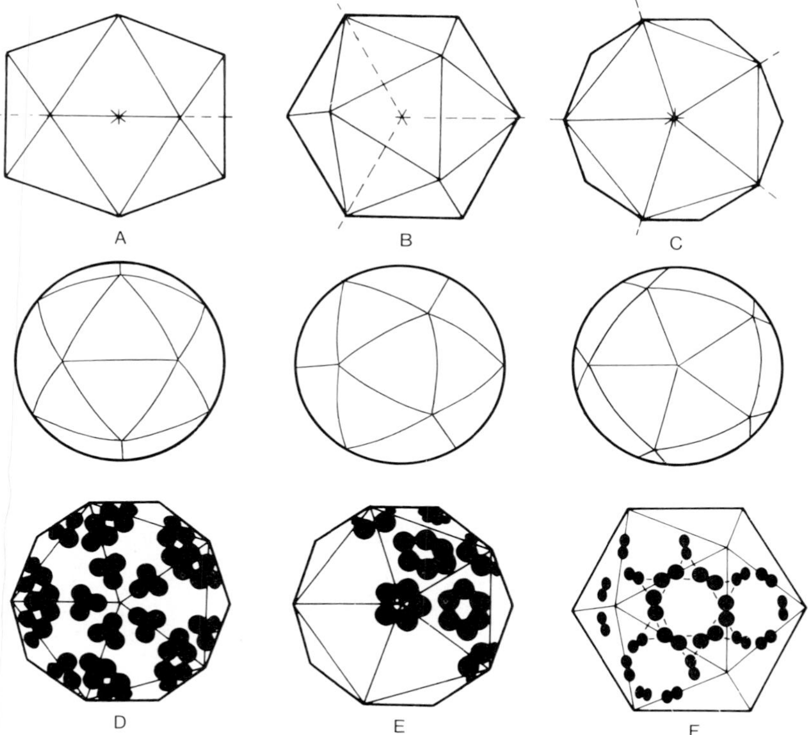

FIG. 1-2. *Features of icosahedral structure. Regular icosahedron viewed along twofold (A), threefold (B), and fivefold (C) axes. Virions may be icosahedral (upper row) or apparently spherical (middle row). Various clusterings of capsid polypeptides give characteristic appearances of the capsomers in electron micrographs (lower row). For example, they may be arranged as 60 trimers (D), capsomers being then difficult to define, as in poliovirus; or they may be grouped as 12 pentamers and 20 hexamers (E), which form bulky capsomers as in parvoviruses, or as dimers on the faces and edges of the triangular facets (F), producing an appearance of a bulky capsomer on each face, as in caliciviruses.*

scopists interested in viral structure, but the details are beyond the scope of this volume. Some possible arrangements of capsomers are shown in Fig. 1-2D, E, and F.

In a practical sense, the examination of negatively stained virus particles in the electron microscope, and analysis of their capsomer arrangement, can provide important information for the identification of a virus as a member of a known family or, in rare instances, as a candidate prototype for a new family. For example, the visualization of a nonen-

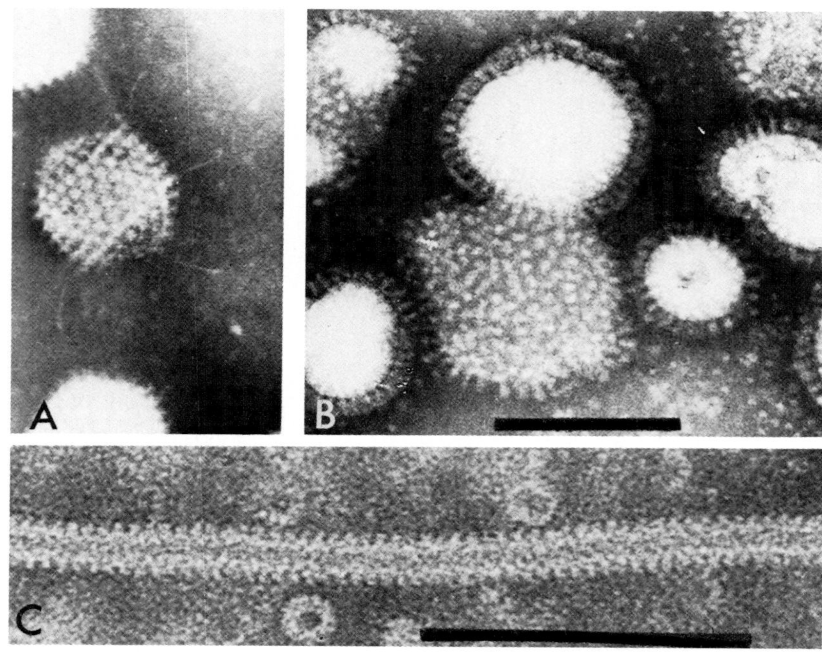

PLATE 1-1. *Morphological features of viral structure revealed by negative staining and electron microscopy (bars = 100 nm). (A) Icosahedral structure of adenovirus capsid. At each of the 12 vertices there is a penton base capsomer from which projects a fiber with a small terminal knob; each of the 20 triangular facets contains 12 identical hexon capsomers. The capsid encloses a protein core with which the DNA is associated. (Courtesy Dr. N. G. Wrigley.) (B) Envelope of influenza virus (family:* Orthomyxoviridae*). The projections (peplomers) are of two morphological types: the hemagglutinin subunits which are trimers and the neuraminidase subunits which are mushroom-shaped tetramers. Both are embedded in lipid, and this lipoprotein envelope encloses a coiled ribonucleoprotein tube. (Courtesy Dr. N. G. Wrigley.) (C) Nucleocapsid of parainfluenza virus (family:* Paramyxoviridae*). The RNA is wound within and protected by a helix composed of identical protein subunits. The complete nucleocapsid is 1 μm long, but in the intact particle is coiled up within a roughly spherical envelope 0.15 μm in diameter. (Courtesy Dr. A. J. Gibbs.)*

veloped virion with a row of four hexamers in line between vertex pentamers would identify a virus as an adenovirus (Plate 1-1A).

Helical Symmetry. The nucleocapsids of many RNA viruses show a different type of symmetry: the capsomers and nucleic acid molecule are wound together in a helix or spiral (Plate 1-1C). Each capsomer consists of a single polypeptide molecule. The rod-shaped plant viruses consist solely of such a naked nucleocapsid. However, in viruses of vertebrates this tubular nucleocapsid is wound into a coil, and surrounded by a lipoprotein envelope, possibly to give the very long nucleocapsids stability (Fig. 1-1 and see Plate 22-1).

Viral Envelopes. Viral envelopes are derived directly from the cell's plasma membrane or cytoplasmic organelle membranes, and are formed during the release of the nucleocapsid from the infected cell by a process known as "budding." The lipids of the envelope are characteristic of the cell of origin, but the cellular proteins in the original plasma membrane have been displaced by virus-specified glycoproteins called *peplomers* (*peplos* = envelope). These peplomers can be clearly seen in electron micrographs as projections from the envelope (Plate 1-1B). On the inside of the envelope of many viruses a protein, called matrix protein, provides added rigidity. Arenaviruses, bunyaviruses, and coronaviruses have no matrix protein, and consequently are rather pleomorphic and fragile.

Envelopes are not restricted to viruses of helical symmetry; some icosahedral viruses (herpesviruses, togaviruses, and flaviviruses) have envelopes. The envelope of rhabdoviruses is closely applied to a bullet-shaped shell that encloses a tubular nucleocapsid with helical symmetry. Poxviruses have a rather atypical envelope, which is not necessary for infectivity.

CHEMICAL COMPOSITION OF VIRUSES

Methods of Purification

An essential prerequisite for the chemical analysis of viruses has been the development of adequate methods of purification. Special problems are created by the close association of viruses with the cells they parasitize; it is not an easy matter to free virions of associated cell debris, or even from viral proteins synthesized in excess in the infected cell. Furthermore, the infectivity of virions is very sensitive to inactivation by heat, acid, alkali, and sometimes lipid solvents or osmotic shock. Accordingly, throughout all purification protocols the virus is maintained at near neutral pH and 4°C.

Liberation of Virus from Cells. The first step in the purification process consists of obtaining virions free from the cells in which they were grown. The supernatant fluid bathing an infected cell culture provides the cleanest starting source of virions, but some viruses must be released from the cells by methods such as sonic vibration, homogenization, or repeated freeze-thawing.

Chemical Methods of Purification. The surface of the virion consists of protein, sometimes together with associated lipid, so that chemical methods of purifying virions must avoid conditions leading to the denaturation of proteins. Virions can be adsorbed to columns of DEAE-cellulose, calcium phosphate, aluminum hydroxide gel, ion-exchange resins, or Sephadex, or to erythrocytes, and then eluted with buffers of different pH and ionic strength, or at a different temperature. Alternatively, nonenveloped viruses can be separated from lipids and soluble proteins by partitioning them into the aqueous phase of a fluorocarbon–water emulsion.

Physical Methods of Purification. After partial purification and concentration by chemical methods, or even without any preliminary treatment, virus particles can be separated from soluble contaminants by centrifugation. Differential centrifugation consists of alternate cycles of low- and high-speed centrifugation to deposit first large contaminating particles, then virions. Rate zonal centrifugation through a preformed gradient of a dense solute such as sucrose forces virions to sediment through the gradient at a rate determined by their sedimentation coefficient (a function principally of their size and shape). Equilibrium (isopycnic) gradient centrifugation in dense solutes such as cesium chloride or potassium tartrate (or sucrose in the case of enveloped viruses of low density), on the other hand, separates virions from contaminants according to their buoyant density. After prolonged ultracentrifugation at very high gravitational forces the virions will come to rest in a sharp band in that part of the tube where the solution has the same density as the virions, usually within the range 1.15–1.4.

Concentration of Viruses. Following or preceding purification, viruses can be concentrated into a small volume by ultracentrifugation, freeze-drying, pervaporation, or dialysis against a hydrophilic agent such as polyethylene glycol.

Nucleic Acid

Viruses contain only a single species of nucleic acid. This may be DNA or RNA; indeed, the RNA viruses provide the only instance in nature in which RNA is the repository of genetic information. All viral genomes are haploid, i.e., they contain only one copy of each gene, except for some retrovirus genomes, which are diploid (see Fig. 8-1). Viral DNA or RNA can be *double-stranded* (ds) or *single-stranded* (ss). Since 1978 the genomes of many of the smaller animal viruses have been sequenced and there are now no insuperable technical impediments to the sequencing of any viral genome. For example, it is already commonplace to sequence dozens of mutants of influenza virus selected by growth in the presence of particular monoclonal antibodies, in order to locate the position of amino acid substitutions in the HA protein. We would anticipate that the complete sequences of the genomes of most of the medically important viruses will be known by the end of the 1980s.

When carefully extracted from the virus particle by treatment with detergents or phenol, the nucleic acid of viruses of certain families of both DNA and RNA viruses is infectious, i.e., when introduced into a cell it can initiate the production of complete virions. In these cases, messenger RNA (mRNA) is transcribed from the viral DNA by a nuclear transcriptase, or with RNA viruses the isolated RNA functions as mRNA. In other cases, the isolated nucleic acid is not infectious even though it contains all the necessary genetic information. Among DNA viruses this occurs because transcription depends upon a viral rather than a cellular transcriptase; among RNA viruses this occurs because the viral RNA occurs as the negative (complementary) strand and its transcription depends on a virion-associated transcriptase without which multiplication cannot proceed (see Table 1-6).

DNA. The genome of all DNA viruses consists of a single molecule, which is double-stranded except in the case of the parvoviruses, and may be linear or circular.

The DNA of papovaviruses and hepadnaviruses is circular. This conformation (1) confers protection against exonucleases, (2) presents advantages for replication of the molecule, and (3) is necessary for integration into cellular DNA. The circular DNA of the papovaviruses, like that of bacterial plasmids, plant chloroplasts, and mammalian mitochondria, normally exists as a twisted circle, known as a superhelix (Plate 1-2A). When an untwisting enzyme relieves the tension by introducing a nick into one strand the molecule becomes a relaxed circle (Plate 1-2B). The circular DNA of hepadnaviruses normally contains a nick in one strand; the other strand is incomplete.

Most of the linear DNAs from other families of viruses have special characteristics that enable them to adopt a circular configuration temporarily, presumably during replication. The two strands of poxvirus DNA are covalently cross-linked at each end, so that on denaturation, the molecule becomes a large ss circle (Fig. 1-3C). The ds linear DNA of herpesviruses (and the ss linear RNA of retroviruses) contains *reiterated sequences* at the ends of the molecule (*terminal redundancy*). Following

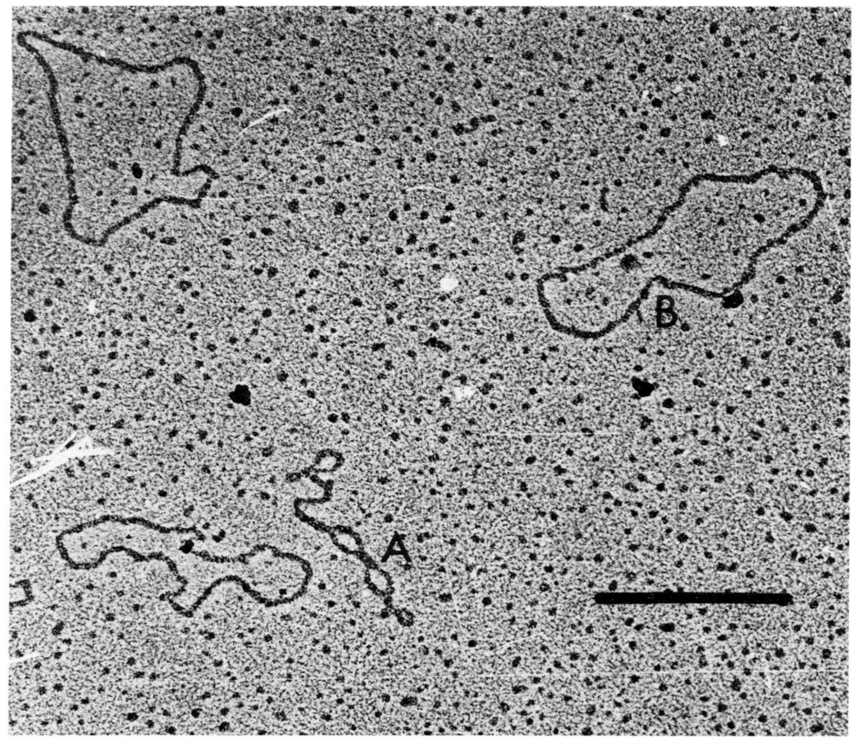

Plate 1-2. *DNA molecules extracted from the papovavirus SV40 (bar = 0.5 μm). Molecules of SV40 DNA exist in two major forms. When it is isolated from the virus particles, most of the DNA occurs in the configuration shown in (A) as double-stranded closed-circular molecules containing superhelical twists. If one of the DNA strands is broken, the superhelical twists are relieved and the molecule assumes a relaxed circular configuration (B). (Courtesy Dr. P. Sharp.)*

partial digestion of both DNA strands from their 5′ ends by an exonuclease, the exposed single-stranded ends are complementary in their nucleotide sequences ("cohesive" or "sticky" ends) so that, if the molecule is then melted, it will reanneal as a circular dsDNA. In the case of the adenoviruses, these terminal repeats are inverted; hence even without enzymatic digestion, denatured molecules self-anneal to form single-stranded circles (Fig. 1-3A). *Inverted terminal repeats* are also a feature of the ssDNA parvoviruses.

Another type of terminal structure occurs in adenoviruses, hepadnaviruses (and the ssRNA picornaviruses and caliciviruses), in all of

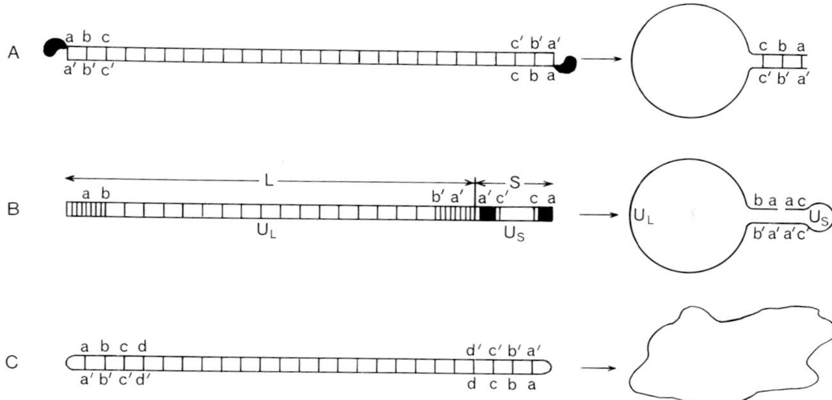

FIG. 1-3. *Diagram representing the specialized structures at the ends of the DNAs of viruses whose genome is a single linear molecule of DNA. (A) Adenovirus DNA has inverted terminal repeats, with a covalently linked protein located at each end of the molecule. Intact single strands self-anneal as shown on the right. (B) Herpes simplex virus DNA consists of two covalently linked components, L and S, each of which consists of a largely unique sequence (U_L and U_S) flanked by inverted repeats. It exists in four isomeric forms differing in the location of the unique regions relative to the terminus. Intact single strands self-anneal as shown on the right. (C) Vaccinia virus DNA has inverted terminal repeats and both of its ends are covalently closed, so that on denaturation it forms a very large cyclic single-stranded molecule.*

which a protein is covalently linked to the 5′ terminus. This is believed to have a vital function in replication of the genome.

The molecular weight (MW) of viral DNA ranges from 1.5 million (parvoviruses) to 150 million (poxviruses). This corresponds to about 4000 nucleotides (4 kilobases, kb) for the small ssDNA parvoviruses to over 200 kb pairs for the large dsDNA poxviruses. As 1 kb or kb pair contains enough genetic information to code for about one average-sized protein, we might conclude that viral DNAs contain from about 4 to 200 genes and code for 4 to 200 proteins. However, the relationship between any particular nucleotide sequence and its protein product is not as straightforward as this (Table 1-2).

First, the DNA of most of the larger viruses—like that of cells—contains what appears to be redundant information in the form of (1) *reiterated sequences* and (2) *introns* which are spliced out and discarded from the RNA transcript. On the other hand, a single such RNA transcript may be spliced in several different ways to yield several distinct mRNAs which become translated into distinct proteins with only certain amino acid sequences in common. Furthermore, a given DNA sequence may

TABLE 1-2
Characteristics of Viral Nucleic Acids

Type	DNA or RNA
MW ($\times 10^6$)	DNA: ss 1.5; ds 1.8–150 RNA: ss 2.5–7.5; ds 15
"Genes"	DNA 4–200; RNA 4–12
Configuration	Linear or circular
Segmentation	Continuous (1 molecule) or segmented (2–12 molecules)
Strandedness	ss or ds
Polarity (ssNA)	+, −, or ambisense
Infectivity	+ or −
Other features of dsDNA	Nicks, gaps in one strand Covalent cross-linking of ends Reiterated sequences Terminal repeats Inverted terminal repeats Overlapping genes Introns Consensus sequences Covalently linked protein at 5′ end
Other features of ssRNA	Secondary structure Polyadenylation at 3′ end Methylated cap at 5′ end Covalently linked protein at 5′ end

be read in two different frames (theoretically, up to three, because each codon is a triplet), giving rise to two (or three) proteins with different amino acid sequences. This fascinating example of genetic economy is well illustrated by the papovaviruses (see Fig. 3-5) and will be discussed in Chapter 3. Suffice it to say at this point that nowadays we cannot always talk in terms of a direct one-to-one relationship between a "gene" and its "gene product," although such a relationship does sometimes occur in viral nucleic acids.

Viral DNAs also contain several kinds of noncoding sequences. We have described introns and various types of terminal reiterated sequences. In addition there are *consensus sequences* which tend to be conserved through evolution because they serve vital functions, including those of RNA splice sites, polyadenylation sites, RNA polymerase recognition sites and promoters, initiation codons for translation, terminator codons, and so on.

RNA. The genome of RNA viruses may also be ss or ds. Furthermore, while some occur as a single molecule, others are *segmented.*

Arenavirus RNA consists of two segments, bunyavirus 3, orthomyxovirus 7–8, and reovirus 10–12. Each of these molecules is unique (often a single "gene"). All viral RNAs are linear; none is a covalently closed circle. However, the ssRNA of arenaviruses and bunyaviruses has sticky ends, hence these molecules exist as circles (in circular nucleocapsids). The linear RNA molecules of RNA viruses have considerable secondary structure, regions of base pairing probably serving as signals controlling transcription, translation, and/or packaging into the virion.

Single-stranded viral nucleic acid, which is generally RNA, can also be defined according to its *polarity* (or *sense*). If it serves as mRNA, it is said to have *positive polarity* (+ *sense*). This is the case with the picornaviruses, caliciviruses, togaviruses, flaviviruses, coronaviruses, and retroviruses. If, on the other hand, its nucleotide sequence is *complementary* to that of mRNA, it is said to have *negative polarity* (− *sense*). Such is the case with the paramyxoviruses, orthomyxoviruses, rhabdoviruses, arenaviruses, and bunyaviruses, all of which therefore need to carry an RNA-dependent RNA polymerase (transcriptase) in the virion, so that mRNA can be transcribed. An extraordinary situation exists with the arenaviruses and at least one genus of bunyaviruses, in which one of the RNA segments is *ambisense,* i.e., part +, part − polarity.

Because + strand RNA has mRNA activity, the RNA of positive sense RNA viruses is usually polyadenylated at its 3′ end (picornaviruses, caliciviruses, togaviruses, coronaviruses) and capped at its 5′ end (togaviruses, flaviviruses, coronaviruses). The picornaviruses and caliciviruses have a protein attached to the 5′ end of their vRNA.

The molecular weight of ssRNA viral genomes varies from 2.5–7.5 million—a much smaller range than seen with the dsDNA viruses. Accordingly they code, in general, for less than a dozen proteins. In the case of the segmented RNA genomes of orthomyxoviruses and reoviruses, one can consider most of the segments to be individual genes, each coding for one unique protein. No such simple relationship applies to the other RNA viruses. For example, the picornavirus genome [ss(+)RNA] is directly translated into a single "polyprotein" precursor, which is subsequently cleaved into smaller functional proteins.

The structure of the genomes of viruses of vertebrates is summarized in a simplified form in Table 1-3. Their remarkable variety is reflected in the diverse ways in which the information encoded in the viral genome is translated into proteins, and the ways in which the viral nucleic acid is replicated (see Chapter 3).

Virus preparations often contain some particles with an atypical content of nucleic acid (see Chapter 4). Host cell DNA is found in some papovavirus particles, and cellular ribosomes are incorporated in

TABLE 1-3
Structure of the Genome in Viruses of Different Families[a]

FAMILY	STRUCTURE OF NUCLEIC ACID
Hepadnaviridae	ds circular DNA with ss tail
Papovaviridae	ds cyclic superhelical DNA (see Plate 1-2)
Adenoviridae	ds linear DNA with inverted terminal repeats and a covalently bound protein (see Fig. 1-3)
Herpesviridae	ds linear DNA; two unique sequences flanked by reiterated sequences; isomeric configurations occur (see Fig. 1-3)
Poxviridae	ds linear DNA; both ends covalently closed, with inverted terminal repeats (see Fig. 1-3)
Parvoviridae	ss linear DNA, − sense; with repeated sequences and a mirror sequence at ends
Picornaviridae, *Caliciviridae*, *Togaviridae*, *Flaviviridae*, *Coronaviridae*	ss linear RNA, + sense; serves as mRNA; 3′ end poly-adenylated (except *Flaviviridae*); 5′ end capped, or protein covalently bound (*Picornaviridae, Caliciviridae*)
Paramyxoviridae, *Rhabdoviridae*	ss linear RNA, − sense
Orthomyxoviridae	Segmented genome; 7–8 molecules of ss linear RNA, − sense
Arenaviridae	Segmented genome; 2 molecules of ssRNA, − sense or "ambisense"[b]; "sticky ends" allow circularization
Bunyaviridae	Segmented genome; 3 molecules of ssRNA, generally − sense, sometimes ambisense; "sticky ends" allow circularization
Retroviridae	Diploid genome, dimer of ss linear RNA, + sense; hydrogen bonded at 5′ ends; terminal redundancy; both 3′ termini polyadenylated, both 5′ ends capped; may carry oncogene
Reoviridae	Segmented genome; 10–12 molecules of ds linear RNA

[a] There is considerable variation within some families, e.g., *Herpesviridae*.
[b] Ambisense, part of molecule is + and part − sense.

arenaviruses. Several copies of the complete viral genome may be enclosed within a single particle, or viral particles may be formed that contain no nucleic acid (empty particles) or that have an incomplete genome, lacking part of the nucleic acid that is needed for infectivity.

Protein

The major constituent of all viruses is protein. Its primary role is to provide the viral nucleic acid with a protective coat. The virions of all viruses of vertebrates contain several different polypeptides separable by polyacrylamide gel electrophoresis; the number ranges from two or three in the case of the simplest viruses to well over 100 in the case of the highly complex poxviruses. In isometric viruses, the proteins form an icosahedral capsid which may enclose a second (inner) capsid or core of histone-like polypeptides intimately associated with the nucleic acid.

The capsid proteins are assembled in the virion in groups to form the capsomers visible in electron micrographs. Each capsomer is composed of one to six molecules of polypeptide, usually of the same kind (homopolymers) but sometimes different (heteropolymers). Capsomers from different regions of the virion, e.g., the inner and outer capsids, or the vertex and the face of the outer capsid, are usually composed of different polypeptides. Other polypeptides again, usually glycoproteins, make up the peplomers projecting from the envelope of enveloped viruses. One or more of the proteins on the surface of the virion has a special affinity for complementary receptors present on the surface of susceptible cells; the same viral proteins contain the antigenic determinants against which the body's immune system makes neutralizing antibodies. Virions of several families carry a limited number of enzymes for specialized purposes, transcriptases being the most important.

INACTIVATION OF VIRUSES

In general, viruses are more sensitive than bacteria or fungi to inactivation by physical and chemical agents. A knowledge of their sensitivity to environmental conditions is therefore important for ensuring the preservation of viability of viruses in clinical specimens, as well as for their deliberate inactivation for such practical ends as disinfection, the provision of safe drinking water, and the production of inactivated vaccines (Table 1–4).

Preservation of Infectivity

The principal environmental condition that may adversely affect the infectivity of viruses in clinical specimens is temperature.

TABLE 1-4
Methods of Inactivating Viruses for Various Purposes

STERILIZATION	DISINFECTION	SKIN DISINFECTION	VACCINE PRODUCTION
Steam under pressure	Sodium hypochlorite	Chlorhexidine	Formaldehyde
Dry heat	Glutaraldehyde	70% ethanol	β-Propiolactone
Ethylene oxide	Formaldehyde	Iodophores	Psoralen + UV irradiation
γ-Irradiation	Peracetic acid		Detergents (for subunit vaccines)

Temperature. Most viruses are heat labile. Within a few minutes at temperatures of 55°–60°C the capsid protein is denatured, with the result that the virion is no longer infectious, presumably because it is no longer capable of normal cellular attachment and/or uncoating. At ambient temperatures the rate of decay of viral infectivity is slow but significant, mainly due to changes in the nucleic acid. Viral preparations must therefore be stored at low temperature; 4°C is usually satisfactory for a day or so, but longer term preservation requires temperatures well below zero. Two convenient temperatures are −70°C, the temperature of frozen CO_2 ("dry ice") and of some commercially available freezers, or −196°C, the temperature of liquid nitrogen. As a rule of thumb, the half-life of most viruses can be measured in seconds at 60°C, minutes at 37°C, hours at 20°C, days at 4°C, months at −70°C, and years at −196°C. The enveloped viruses are more heat labile than naked icosahedral viruses. Some enveloped viruses, notably respiratory syncytial virus, tend to be inactivated by the actual process of freezing and subsequent thawing, probably as a result of disruption of the virion by ice crystals. This poses problems in the collection and transportation of clinical specimens. Perhaps the most practical way of avoiding such problems is to deliver specimens to the laboratory as rapidly as practicable, packed without freezing, on wet ice.

In the laboratory, it is often necessary to preserve stocks of viable virus for years. This is achieved in one of two ways: (1) rapid freezing of small aliquots of virus suspended in media containing protective protein and/or dimethyl sulfoxide, followed by storage at −70° or −196°C; (2) freeze-drying (lyophilization), i.e., dehydration of a frozen virus suspension under vacuum followed by storage at 4° or −20°C. Freeze-drying prolongs viability significantly even at ambient temperatures, and is important in enabling live viral vaccines to be used in tropical countries. The fact that freeze-dried smallpox vaccine is exceptionally heat resistant, and can withstand temperatures of 37°C for over a month, was a vital factor in the achievement of the global eradication of smallpox.

Ionic Environment and pH. On the whole viruses prefer an isotonic environment at a physiological pH, but some virions tolerate a wide ionic and pH range. Enteric viruses survive the acidic pH of the stomach, but certain enveloped respiratory viruses are inactivated at pH 5.

Lipid Solvents. The infectivity of enveloped viruses is readily destroyed by lipid solvents, such as ether or chloroform, or detergents such as sodium deoxycholate, so that these agents must be avoided in laboratory procedures concerned with maintaining the viability of viruses. On the other hand, detergents are commonly used by virologists to solubilize viral envelopes and liberate proteins for use as vaccines or for chemical analysis. Sensitivity to lipid solvents is also employed as a preliminary screening test in the identification of new viral isolates, especially by arbovirologists.

Disinfection and Sterilization

The effective use of proper disinfectants and sterilization procedures is an important factor in preventing nosocomial infections. In hospitals, steam under pressure (autoclaving), dry heat, or, for heat-sensitive equipment, ethylene oxide gas, is used for sterilization. The biocidal action of chemical disinfectants is limited to a narrower range of susceptible viruses and bacteria and thus cannot be relied upon for sterilization. Sodium hypochlorite, peracetic acid, and glutaraldehyde are useful for cleaning up spills, swabbing laboratory benches, etc., but are too irritating to be applied to the skin or any other part of the body as antiseptics. For the latter purpose, e.g., disinfection of hands, thorough washing followed by the application of 0.5–1% chlorhexidine, or an iodophore (1% available iodine), or 70% ethanol, destroys enveloped viruses and some others.

The effectiveness of sterilization processes depends upon exposure of the material to sterilizing conditions for an adequate time. The conditions for sterilization in steam sterilizers and hot air ovens are defined by temperature and time. Relative humidity and the concentration of the chemical vapor are additional parameters in gas sterilization. Liquid chemical disinfectants must, of course, be used at the concentration recommended for each particular product. The Centers for Disease Control provides all hospitals in the United States with guidelines for the prevention of nosocomial infections, which provide, among other things, up-to-date information on currently available disinfectants and antiseptics.

Ultraviolet irradiation is used for sterilization of exposed surfaces and air within closed spaces, such as laboratory hoods. It has also been used for "air sanitation" within wards in hospitals, with limited success. Ultraviolet irradiation produces its damaging effects on DNA by causing adjacent pyrimidine bases on the same polynucleotide chain to form dimers; uracil dimers are responsible for most of the UV damage to RNA viruses.

Ionizing radiation, such as γ-rays from a ^{60}Co source, can be used for the sterilization of products that might be damaged by heating. The efficiency of inactivation of viruses with ss nucleic acids by ionizing radiation is relatively high, for almost every ionization causes a lethal break in the polynucleotide chain. With ds nucleic acids only a proportion of the one-strand breaks is lethal. However, the "target size" of a viral genome is very small compared with that of bacteria or higher organisms, so that higher doses of irradiation, e.g., 5–10 Mrad, cf. 2.5 Mrad, are required to inactivate viruses than to kill organisms.

Inactivated Vaccines

The traditional method of inactivating viruses for use as vaccines has been treatment with formaldehyde. Experience with the Salk poliovirus vaccine (see Chapters 10 and 19) illustrates the importance of ensuring that the viral suspension is free of clumped virions prior to treatment. Chemicals that inactivate the nucleic acid with minimal effect on the viral proteins and therefore on immunogenicity, include β-propiolactone, and psoralen followed by UV irradiation (see Chapter 10).

CLASSIFICATION OF VIRUSES

Virtually every kind of organism can be parasitized by viruses: vertebrate and invertebrate animals—including mammals, birds, reptiles, amphibia, helminths, and arthropods—as well as plants, algae, fungi, protozoa, bacteria, and mycoplasmas. Indeed, there is good reason to believe that the number of distinct species of viruses on earth is greater than the number of species of all other living things, since every species of organism that has been intensively searched, e.g., man, monkey, mouse, and *Escherichia coli,* has yielded dozens of different viruses belonging to several viral families. There are even some "satellite" viruses that are, in a sense, "parasites" on other viruses, although in such cases both agents are parasitic on their host cells.

All viruses, whatever their cellular hosts, share the features summarized in Table 1-1, hence taxonomists concerned with viral classification have now accepted a scheme of classification that embraces all viruses. From the operational point of view, however, we can divide viruses into those that affect vertebrate animals, insects, plants, and bacteria. In this volume we shall be concerned solely with the viruses of vertebrate animals and primarily with those that affect man. Some of these viruses, the arthropod-borne viruses (arboviruses), also multiply in insects or other arthropods.

Several hundred species of viruses have been recovered from man, which is by far the best studied vertebrate host, and new ones are still being discovered. In order to simplify their study we need to sort them into groups that share certain common properties. The most important criteria for classification are the physical and chemical characteristics of the virion and the mode of replication of the virus. Before discussing viral taxonomy based on these criteria, a simpler, clinically useful classification based on modes of transmission from one host to another will be described.

Classification Based on Epidemiological Criteria

Viruses of man can be transmitted by ingestion, inhalation, injection, or contact, including coitus. In discussing the epidemiology and pathogenesis of viral infections we shall frequently make use of the following terminology as a useful subsidiary way of classifying viruses.

Enteric viruses enter the body through, and multiply primarily in, the gastrointestinal tract. Human enteric viruses include the acid- and bile-resistant enteroviruses (including hepatitis A virus), caliciviruses, rotaviruses, and some adenoviruses.

Respiratory viruses enter the body through the respiratory tract, either by inhalation of small-particle aerosols or by contact. The term is usually restricted to those viruses that remain localized to the respiratory tract, and includes all orthomyxoviruses, paramyxoviruses, coronaviruses, and rhinoviruses, most adenoviruses, and some enteroviruses.

Arboviruses (*ar*thropod-*bo*rne viruses) infect arthropods that ingest vertebrate blood; they multiply in the arthropod's tissues and can then be transmitted by bite to susceptible vertebrates. Viruses belonging to four families that infect man are included: togaviruses, flaviviruses, bunyaviruses, and orbiviruses (a genus of the reovirus family).

Classification Based on Physicochemical Criteria

Although classification on epidemiological criteria is convenient for certain purposes, each group contains many viruses with very disparate physicochemical and biological properties. As knowledge of the properties of viruses increased it became clear that the physicochemical properties of the virus particle afforded a more fundamental basis for virus classification. The classification of viruses into major groups called *families*, and the subdivision of families into *genera*, has now reached a position of substantial international agreement. The subdivision of genera into *species* is much more controversial. Among higher organisms a species is defined by its inability to interbreed with another species, but sexual reproduction does not occur with viruses. Some viruses do undergo molecular recombination or genetic reassortment (see Chapter 4) and the extent to which these processes occur between different viruses, under optimal conditions, can be used as a criterion of genetic relatedness. The degree of genetic homology can be measured more directly by nucleic acid hybridization, and ultimately by sequencing. Most virologists would agree that to be designated as distinct species, two viruses should differ substantially in nucleic acid sequence. However, there is as yet no consensus on how such differences should be quantitated, and perhaps weighted. Some take the view that antigenic differences are of such paramount importance that viruses distinguishable by neutralization should be regarded as separate species, even if their genomes display more than 90% homology. Others believe that such closely related but serologically distinct viruses should be designated as serotypes. The effects of this ongoing debate on the classification of adenoviruses, picornaviruses, and bunyaviruses are discussed in Chapters 15, 19, and 25; in this chapter we will be concerned mainly with classification into families and genera.

The primary criteria for delineation of families are (1) the morphology of the virion and (2) the nature of the genome and the strategy of its replication. The morphology of the virion, including its size, shape, symmetry of the nucleocapsid, and the presence or absence of an envelope, is readily determined by electron microscopy.

Subdivision of families into genera depends upon criteria that differ in different families. The most commonly used criterion is antigenic cross-reactivity, but in some families other properties may be used. Genera contain from one to over a hundred different species, identification of which is usually dependent on serological differences. Monoclonal antibodies are proving of great value in the differentiation of viruses representing *subtypes, strains,* and *variants.* At the research rather than the routine diagnostic level, the composition of the virion nucleic acid, as revealed by molecular hybridization, oligonucleotide fingerprinting or restriction endonuclease mapping, electrophoresis in gels (especially useful for RNA viruses with segmented genomes), and nucleotide sequence analysis, is used to identify minor differences between viral strains and mutants.

Nomenclature

Since 1966 the classification and nomenclature of viruses, at the higher taxonomic levels (families and genera), has been systematically organized by an International Committee for the Taxonomy of Viruses, the decisions of which are now widely accepted. The highest taxonomic group among viruses is the family; families are named with a suffix -*viridae.* Subfamilies have the suffix *-virinae;* genera the suffix *-virus.* The prefix may be another latin word or an acronym, i.e., an abbreviation derived from some initial letters. Latinized family, subfamily, and generic names are usually written in italics; vernacular terms derived from them are written in roman letters. For example, the term poxviruses is used to designate members of the family *Poxviridae*. It is still customary to use vernacular terms for viral species and strains, e.g., measles virus, rubella virus.

Cryptograms

Cryptograms provide a kind of shorthand for encapsulating the properties of a virus, viral genus, or viral family. They consist of four sets of symbols. As used in this volume, they refer to members of the various viral families that produce disease in man. The symbols have the following meanings.

The *first set* describes the kind of nucleic acid: (*R*)NA or (*D*)NA/single-stranded (ss) or double-stranded (ds). For example: *Reovirus* = R/ds; *Flavivirus* = R/ss.

The *second set* describes the molecular weight and arrangement of the genome: molecular weight in millions/(C)ircular or (*L*)inear with superscripts (+) or (−) indicating whether single-stranded nucleic acid is positive or negative sense. The symbol Σ and a subscripted number indicate the total molecular weight and number of molecules in the genome of virions with segmented genomes. For example: *Reovirus* = Σ_{10}15/L; *Flavivirus* = 4/L^+.

The *third set* describes the virion shape: outline shape of virion/outline of nucleocapsid: (*S*)pherical, (*E*)longated with flat ends or ro(*U*)nded ends, or if none of these then (*X*), and also, whether the virion is (*e*)nveloped. For example: *Reovirus* = S/S; *Flavivirus* = Se/S.

The *fourth set* summarizes some biological characters of the virus: kind(s) of host/mode of transmission/kind of vector (omitted if not relevant or known): (*I*)nvertebrate, or (*V*)ertebrate/(C)ongenitally, by (*R*)espiration, c(*O*)ntact, (*I*)ngestion, or (*Ve*)ctor/(*Ac*)arina, (*Di*)ptera, (*Si*)phonaptera. The letter (C) after *Ve*, for arboviruses, indicates that congenital transmission may occur in the arthropod host. For example: *Reovirus* = V/R,O,I; *Flavivirus* = I,V/Ve(C)/Ac,Di.

THE FAMILIES OF DNA VIRUSES

A brief description of each major family that includes viruses pathogenic for man is given below, in order to orient the reader before embarking upon Part I of this volume. Each family is allocated a separate

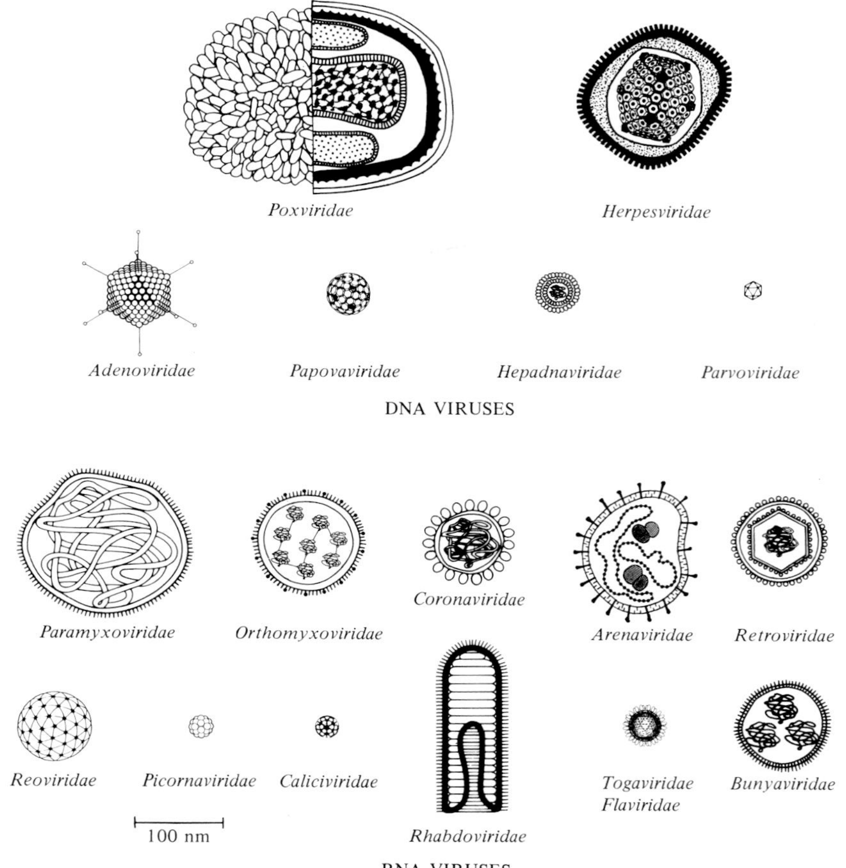

FIG. 1-4. *Diagram illustrating the shapes and sizes of animal viruses of families that include human pathogens. The virions are drawn to scale, but artistic license has been used in representing their structure. In some, the cross-sectional structures of capsid and envelope are shown, with a diagrammatic representation of the genome; with the very small virions, only their size and symmetry; in the largest, both.*

chapter in Part II, which deals with the role of the viruses in human disease. We begin with the DNA viruses (Fig. 1-4 and Table 1-5).

Hepadnaviridae [D/ds:1.8/C:S/S:V/C,O]

Human hepatitis B virus and related viruses of other animals form the family *Hepadnaviridae* (*hepa* = liver; *dna* = deoxyribonucleic acid). The

TABLE 1-5
Properties of the Virions of the Families of DNA Viruses

	GENOME[a]		VIRION		
FAMILY	MW (× 10⁶)	NATURE[b]	SHAPE	SIZE (nm)	TRANSCRIPTASE
Hepadnaviridae	1.8	ds, circular[c]	Spherical	42	+[e]
Papovaviridae	3–5	ds, circular	Icosahedral (72)[d]	45–55	−
Adenoviridae	20–25	ds, linear	Icosahedral (252)	70–90	−
Herpesviridae	100–150	ds, linear	Icosahedral (162) enveloped	Envelope 120–200; capsid 100–110	−
Poxviridae	85–140	ds, linear	Brick-shaped, sometimes enveloped	300 × 240 × 100	+
Parvoviridae	1.5–2.0	ss, linear, −	Icosahedral (32)	18–26	−

[a] Genome invariably a single molecule.
[b] ds, double-stranded; ss, single-stranded; polarity of ss nucleic acid (+) or (−).
[c] Circular molecule is ds for most of its length but contains an ss region.
[d] Values in parentheses indicate number of capsomers in icosahedral capsid.
[e] Possibly a reverse transcriptase present.

virion is a spherical particle 42 nm in diameter, consisting of a 27-nm icosahedral core within a closely adherent "envelope" that contains cellular lipids, glycoproteins, and a virus-specific surface antigen (HBsAg). The core contains two additional antigens, HBcAg and HBeAg, plus a DNA polymerase and a small circular dsDNA molecule which includes a short ss region.

The hepadnaviruses multiply in the nucleus of hepatocytes and cause hepatitis, which may progress to cirrhosis and primary hepatocellular carcinoma. Human hepatitis B virus, of which there are several subtypes, may cause persistent infections in which HbsAg and often virions circulate continuously in the bloodstream.

Papovaviridae [D/ds:3–5/C:S/S:V/O]

The papovaviruses (*pa* = papilloma; *po* = polyoma; *va* = vacuolating agent) are small nonenveloped icosahedral viruses which replicate in the nucleus and often transform infected cells. In the virion their nucleic acid occurs as a cyclic ds molecule, which is infectious. There are two genera: *Papillomavirus* (wart viruses) (55 nm diameter) have a larger genome (MW 5 million) which may persist in episomal form in the cell; *Polyomavirus* (45 nm in diameter) has a smaller genome (MW 3 million) which may persist in cells in an integrated form.

Papillomaviruses cause human warts and some species are oncogenic;

the human polyomaviruses commonly produce inapparent infection and may be reactivated by immunosuppression. Murine polyoma virus and simian virus 40 (SV40) have been extensively used in laboratory studies of viral oncogenesis.

Adenoviridae [D/ds:20–25/L:S/S:V/I,O,R]

The adenoviruses (*adeno* = gland) are nonenveloped icosahedral viruses whose genome consists of a single linear molecule of dsDNA and which multiply in the nucleus. About 40 serologically distinct types (species) of human adenovirus are currently recognized, all of which share the adenovirus group antigen with adenovirus serotypes infecting other mammals (the genus *Mastadenovirus*).

Human adenoviruses are associated with infections of the respiratory tract, and occasionally the eye or genital tract. Many are characterized by prolonged latency. Several species multiply in the intestinal tract and are visualized by electron microscopy in feces but are difficult to grow in cultured cells; some cause gastroenteritis. Some human adenoviruses produce malignant tumors when inoculated into newborn hamsters.

Herpesviridae [D/ds:100–150/L:Se/S:V/C,O]

The herpesviruses (*herpes* = creeping) have large icosahedral nucleocapsids and their genome is a single linear molecule of dsDNA. They multiply in the nucleus and mature by budding through the nuclear membrane, thus acquiring an envelope. The family, which includes several important human pathogens, has been subdivided into three subfamilies: *Alphaherpesvirinae* includes herpes simplex types 1 and 2 and varicella-zoster viruses, *Betaherpesvirinae* comprises the cytomegaloviruses, and *Gammaherpesvirinae* includes Epstein–Barr (EB) virus, which causes infectious mononucleosis.

A feature of all herpesvirus infections is prolonged persistence of the virus in the body, usually in a latent form. Episodes of recurrent clinical disease of endogenous origin may occur years after the initial infection, e.g., zoster after chickenpox and reactivation of cytomegalovirus by immunosuppression. There is suggestive evidence that some herpesviruses may have a role in human cancers, notably EB virus in nasopharyngeal cancer and Burkitt's lymphoma (see Chapter 8).

Poxviridae [D/ds:85–140/L:X or Xe/X:V/O,R]

The poxviruses (*pock* = pustule) are the largest and most complex viruses of vertebrates. The virion consists of a brick-shaped core which contains a single linear molecule of dsDNA, surrounded by a complex series of membranes of viral origin. Unlike other DNA viruses of mammals, poxviruses multiply in the cytoplasm; mRNA is transcribed by a virion-associated transcriptase.

Viruses of several different poxvirus genera may infect man; smallpox and molluscum contagiosum are specifically human diseases, whereas man is only occasionally infected with the viruses of orf, cowpox, and milker's nodes, which normally affect other animals, including livestock. All produce skin lesions, and smallpox was a very serious systemic disease, but is now extinct.

Parvoviridae [D/ss:1.5–2.0/L$^-$:S/S:V/O,R]

These tiny viruses (*parvus* = small) are only 20 nm in diameter, have icosahedral symmetry, and a genome of ssDNA. Members of the genus *Parvovirus* infect several species of rodents and domestic animals and one has recently been identified in man. A second genus, *Dependovirus*, comprises the five types of "adeno-associated viruses" which are defective viruses incapable of replication except in the presence of a "helper" adenovirus or herpesvirus.

THE FAMILIES OF RNA VIRUSES

Table 1-6 lists the properties of the virions of the families of RNA viruses.

Picornaviridae [R/ss:2.5/L$^+$:S/S:V/I,O,R]

The *Picornaviridae* (*pico* = small; *rna* = ribonucleic acid) comprise small nonenveloped icosahedral viruses which contain a single molecule of positive-sense ssRNA, and multiply in the cytoplasm.

Two genera, *Enterovirus* and *Rhinovirus*, each contain large numbers of human pathogens. The genus *Enterovirus* includes 3 polioviruses, 32 human echoviruses, 29 coxsackieviruses, and 5 other human enteroviruses. Most of these viruses usually produce inapparent enteric infections, but the polioviruses may also cause paralysis, while the other enteroviruses are sometimes associated with meningitis, rashes, myocarditis, myositis, conjunctivitis, mild upper respiratory tract infections, or hepatitis (*Enterovirus* 72). The genus *Rhinovirus* includes well over 100 serotypes that affect man, and are the most common viruses causing the common cold.

TABLE 1-6

Properties of the Virions of the Families of RNA Viruses

	GENOME		VIRION				
FAMILY	MW ($\times 10^6$)	NATURE[a]	ENVELOPE	SHAPE[b]	SIZE (nm)	TRANS-CRIPTASE	SYMMETRY OF NUCLEOCAPSID[c]
Picornaviridae	2.3	ss,1, +	−	Icosahedral	25–30	−	Icosahedral (25–30)
Caliciviridae	2.6	ss,1, +	−	Icosahedral	35–40	−	Icosahedral (35–40)
Togaviridae	4	ss,1, +	+	Spherical	60–70	−	Icosahedral (28–35)
Flaviviridae	4	ss,1, +	+	Spherical	40–50	−	Icosahedral (25–30)
Orthomyxoviridae	5	ss,8, −	+	Spherical	80–120	+	Helical (9–15)
Paramyxoviridae	5–7	ss,1, −	+	Spherical	150–300	+	Helical (12–17)
Coronaviridae	6	ss,1, +	+[d]	Spherical	60–220	−	Helical (11–13)
Arenaviridae	3–5	ss,2, −	+[d]	Spherical	50–300	+	Helical (?)
Bunyaviridae	4–7	ss,3, −	+[d]	Spherical	90–100	+	Helical (2–2.5)
Retroviridae	2×(2–3)[e]	ss,1, +	+	Spherical	80–110	+[f]	Helical (?)
Rhabdoviridae	4	ss,1, −	+	Bullet-shaped	180×75	+	Helical (5)
Filoviridae	4	ss,1, −	+	Filamentous	800×80	+(?)	Helical (?)
Reoviridae	11–15	ds,10–12	−	Icosahedral	60–80	+	Icosahedral (45–60)

[a] All molecules linear; ss, single-stranded; ds, double-stranded; 1 to 12, number of molecules in genome; + or −, polarity of ss nucleic acid.

[b] Some enveloped viruses are very pleomorphic (sometimes filamentous).

[c] Values in parentheses indicate diameter (nm) of nucleocapsid.

[d] No matrix protein.

[e] Genome is diploid, two identical molecules being held together by hydrogen bonds at their 5′ ends.

[f] Reverse transcriptase.

Caliciviridae [R/ss:2.6/L^{+}:S/S:V/I]

The caliciviruses (*calix* = cup) used to be classified as a genus of *Picornaviridae,* but differ enough in their morphology and the mode of viral replication to warrant classification as a family. Several are of veterinary importance; the only ones of medical significance are the Norwalk agent and related viruses which are important causes of gastroenteritis.

Togaviridae [R/ss:3.4–4/L^{+}:Se/S:I,V/C,R,Ve/Di]

The togaviruses (*toga* = cloak) are small spherical enveloped viruses containing positive-sense ssRNA enclosed within an icosahedral core of cubic symmetry. They multiply in the cytoplasm and mature by budding from plasma membrane. Two genera contain species that cause disease in humans: *Alphavirus* and *Rubivirus. Alphavirus* is a genus containing many species, all of which are arthropod transmitted. Important human pathogens include eastern, western, and Venezuelan equine encephalitis viruses and chikungunya virus. In nature the alphaviruses usually produce inapparent viremic infections of birds, mammals, or reptiles, being transmitted by arthropod vectors, in which they also multiply. When man or his domestic animals are bitten an inapparent infection is the usual consequence, but generalized disease or encephalitis can result.

The virus of rubella (genus *Rubivirus*), although not arthropod transmitted, is in its physicochemical properties a typical togavirus.

Flaviviridae [R/ss:4/L^{+}:Se/S:I,V/Ve(C)/Ac,Di]

Formerly classified with the togaviruses, *Flaviviridae* (*flavi* = yellow) has now been given family status, principally because of differences in the strategy of replication. In most other respects the flaviviruses resemble the togaviruses (see above). There is only one genus, *Flavivirus,* which contains several important human pathogens including the viruses of yellow fever, dengue, and St. Louis, Japanese, Murray Valley, and Russian tick-borne encephalitis.

Orthomyxoviridae [R/ss:Σ_{7-8}5/L^{-}:Se/E:V/R]

The orthomyxoviruses (*myxo* = mucus) occur as spherical or filamentous RNA viruses with a tubular nucleocapsid enclosed within an envelope acquired as they bud from the cytoplasmic membrane. Their genome consists of seven or eight pieces of complementary (negative sense) ssRNA, and is associated with a viral transcriptase. The envelope is studded with spikes (peplomers) which are of two kinds, one being the hemagglutinin of the virus, the other the neuraminidase.

The family includes two important species: influenza virus A, strains of which have been recovered from birds, horses, and swine, as well as man, and influenza virus B, which is a specifically human pathogen. Influenza virus C is an uncommon human pathogen. Influenza A virus in particular undergoes sequential point mutations which cause "antigenic drift," as well as occasional genetic reassortment with influenza A viruses of animals to cause "antigenic shift" and major pandemics of influenza.

Paramyxoviridae [R/ss:5–7/L^-:Se/E:V/O,R]

The paramyxoviruses occur as large, roughly spherical enveloped viruses or as filaments containing a tubular nucleocapsid. Their genome consists of a single linear molecule of ss complementary RNA and the virion contains a transcriptase. The envelope contains two glycoproteins, a hemagglutinin (often with neuraminidase activity also) and fusion protein.

The family *Paramyxoviridae* is subdivided into three genera: *Paramyxovirus, Morbillivirus* (*morbilli* = measles), and *Pneumovirus* (*pneumo* = lung). Viruses of the genera *Paramyxovirus* and *Pneumovirus* cause respiratory infections (parainfluenza and respiratory syncytial viruses) or generalized infections (mumps); viruses of the genus *Morbillivirus* usually cause generalized infections, some of which (measles in man) are associated with skin rashes.

Coronaviridae [R/ss:6/L^+:Se/E:V/I,R]

The coronaviruses (*corona* = crown) are medium-sized somewhat pleomorphic viruses, with widely spaced, club-shaped peplomers in their lipoprotein envelope, which lacks a matrix protein but encloses a core of undetermined symmetry and a single linear molecule of positive-sense ssRNA. The family includes several agents that cause common colds.

Arenaviridae [R/ss:$\Sigma_2$3–5/L^-:Se/*:V/C,O]

Arenaviruses (*arena* = sand) acquired their name because of the presence of ribosomes (resembling "grains of sand") incorporated within the rather pleomorphic enveloped virion, in which there is no matrix protein. Their genome consists of two pieces of negative sense or ambisense (+ and − sense, in different halves of one molecule) ssRNA, each held in a circular configuration, with an associated transcriptase. All cause natural inapparent infections of rodents. Man may occasionally contract from rodents a serious generalized disease, e.g., Lassa fever or lymphocytic choriomeningitis. Lassa, Machupo, and Junin viruses are "class-4" pathogens and may be worked with in the laboratory only under maximum biocontainment conditions.

Bunyaviridae [R/ss:$\Sigma_3$4–7/C$^-$:Se/E:I,V/e(C)/Ac,Di]

Over 100 bunyaviruses (Bunyamwera = a locality in Africa) comprise the largest single group of arboviruses. They are medium-sized enveloped RNA viruses with a nucleocapsid of helical symmetry, whose genome consists of three pieces of negative sense or ambisense ssRNA, each held in a circular configuration by "sticky ends," with an associated transcriptase. They multiply in the cytoplasm and bud into Golgi vesicles. Bunyaviruses readily undergo genetic reassortment, which may produce "antigenic shift," and some strains have been shown to undergo antigenic drift as well. Many species are transmitted transovarially in their arthropod vectors.

The *Bunyavirus* genus, mostly mosquito transmitted, includes the California group arboviruses some of which occasionally cause encephalitis in man. The genus *Phlebovirus* includes the important human pathogens of sandfly fever (transmitted by *Phlebotomus*) and Rift Valley fever (transmitted by mosquitoes). Members of the genus *Nairovirus* are tick borne and include the virus of Crimean-Congo hemorrhagic fever. The widespread human disease "hemorrhagic fever with renal syndrome" is caused by the Hantaan virus of rodents, and does not appear to be arthropod transmitted.

Retroviridae [R/ss:2x(2–3)/L$^+$:Se/*:V/C,O]

Retroviridae (*re* = reverse; *tr* = transcriptase) is a large family of medium-sized enveloped viruses with a complex structure and a remarkable enzyme, reverse transcriptase. Uniquely among viruses, the genome is diploid and consists of an inverted dimer of ss positive-sense RNA. The DNA copy of the viral genome transcribed by the viral reverse transcriptase is integrated into the cellular DNA before multiplication occurs. Such proviral DNA is found in the DNA of all normal cells of many species of animals and may under certain circumstances be induced to produce virus. These are known as *endogenous retroviruses*.

The family *Retroviridae* is subdivided into three subfamilies, only two of which are important in human medicine. The subfamily *Oncovirinae* (*onkos* = tumor) includes the human T cell leukemia viruses while the agent associated with acquired immune deficiency syndrome, discussed in Chapter 28, is a member of the *Lentivirinae* subfamily. As the oncovirus subfamily is of central importance in understanding viral oncogenesis, it will be described in Chapter 8 rather than in Part II.

Rhabdoviridae [R/ss:4/L$^-$:Ue/E:I,V/O]

The rhabdoviruses (*rhabdos* = rod) are large, bullet-shaped viruses containing a single molecule of complementary ssRNA, which is associated with a transcriptase, and a helical capsid enclosed within a shell to which is closely applied an envelope with peplomers. The virion matures at the cytoplasmic membrane. The only typical rhabdovirus infecting man is the *Lyssavirus* (*lyssa* = rabies) of rabies, a fatal disease of the CNS, transmitted to man by the bite of an infected animal. Morphologically similar viruses occur in mammals, fish, insects, and plants.

Reoviridae [R/ds:Σ_{10-12}11–15/L:S/S:I,V/R,O,I,Ve(C)/Ac,Di]

The family name is an acronym, Respiratory Enteric Orphan virus, reflecting the fact that members of the genus first discovered, *Reovirus*, are found in both the respiratory and enteric tract of man and many animals, but have yet to be associated with any disease. The distinctive feature of the family is that the virions contain dsRNA which occurs in 10–12 pieces. The virion is a medium-sized nonenveloped icosahedron, which multiplies in the cytoplasm and may undergo genetic reassortment. Species are found in animals, insects, and plants. The genera that can infect man are *Reovirus*, *Orbivirus* (*orbi* = ring), which are arboviruses that may include Colorado tick fever, and *Rotavirus* (*rota* = wheel) which is a very important cause of infantile diarrhea.

UNCLASSIFIED VIRUSES

Certain viruses, such as the non-A, non-B hepatitis viruses (Chapter 28), are known to exist by inference, but have yet to be discovered. The delta agent is a defective RNA virus, the coat of which is coded by a helper virus, hepatitis B; in this respect it resembles certain "satellite" plant viruses (see Chapter 28). Other viruses have been isolated and studied but are not yet well enough understood to be allocated unequivocally to a particular family.

The "class-4" pathogens, Ebola and Marburg viruses, that cause African hemorrhagic fever, have been tentatively allocated to a new family, *Filoviridae* [R/ss:4/L$^-$:Ee/E:V/O] discussed in Chapter 26. In many respects they closely resemble rhabdoviruses, but the virions are pleomorphic and sometimes very long (*filum* = thread), maximum infectivity being associated with a particle 790 (Marburg) or 970 (Ebola) nm long and 80 nm wide.

The causative agents of the subacute spongiform encephalopathies, which include the sheep disease scrapie and the human diseases kuru and Creutzfeldt–Jakob syndrome, remain enigmatic. Their infectivity seems to be associated with thin fibrils like those seen in amyloid plaques, but they appear to be completely nonimmunogenic. They may be small and unusual viruses, or they may not be viruses at all, as the term is currently used. All produce slow infections with incubation periods measured in months or even years leading inexorably to death from a degenerative condition of the brain characterized by a "spongiform" appearance. They will be discussed in Chapters 7 and 28.

FURTHER READING

Books

Andrewes, C. H., Periera, H. G., and Wildy, P. (1978). "Viruses of Vertebrates," 4th ed. Bailliere, London.

Fraenkel-Conrat, H., and Wagner, R. R., eds. (1982–1985). "The Viruses," Vols. 1 – (a continuing series). Plenum, New York.

Knight, C. A. (1975). "Chemistry of Viruses," 2nd ed. Springer-Verlag, Berlin and New York.

Mahy, B. W. J., and Pattison, J. R., eds. (1984). "The Microbe 1984. I: Viruses," Symp. Soc. Gen. Microbiol., Vol. 36. Cambridge Univ. Press, London and New York.

Matthews, R. E. F., ed. (1983). "A Critical Appraisal of Viral Taxonomy." CRC Press, Boca Raton, Florida.

Palmer, E. L., and Martin, M. L. (1982). "An Atlas of Mammalian Viruses." CRC Press, Boca Raton, Florida.

Reviews

Almeida, J. D. (1984). Morphology: Virus structure. *In* "Topley & Wilson's Principles of Bacteriology Virology and Immunity" (F. Brown and G. Wilson, eds.), 7th ed., Vol. 4, pp. 14–48. Williams & Wilkins, Baltimore, Maryland.

Bishop, D. H. L. (1977). Virion polymerases. *In* "Comprehensive Virology" (H. Fraenkel-Conrat and R. R. Wagner, eds.), Vol. 10, p. 117. Plenum, New York.

Caspar, D. L. D. (1965). Design principles in virus particle construction. *In* "Viral and Rickettsial Infections of Man" (F. L. Horsfall, Jr. and I. Tamm, eds.), 4th ed., pp. 51–93. Lippincott, Philadelphia, Pennsylvania.

Centers for Disease Control (1981). "Guidelines for the Prevention and Control of Nosocomial Infections." C. D. C., Atlanta, Georgia.

Compans, R. W., and Klenk, H.-D. (1979). Viral membranes. *In* "Comprehensive Virology" (H. Fraenkel-Conrat and R. R. Wagner, eds.), Vol. 12, p. 293. Plenum, New York.

Diener, T. O. (1983). Viroids. *Adv. Virus Res.* **28,** 241.

Favero, M. S. (1983). Chemical disinfection of medical and surgical materials. *In* "Disinfection, Sterilization and Preservation" (S. S. Block, ed.), pp. 469. Lea & Febiger, Philadelphia, Pennsylvania.

Fenner, F., and Gibbs, A. (1983). Cryptograms 1982. *Intervirology* **19,** 121.

Harrison, S. C. (1984). Structures of viruses. *Symp. Soc. Gen. Microbiol.* **36,** 29–74.

McGeogh, D. J. (1981). Structural analysis of animal virus genomes. *J. Gen. Virol.* **55,** 1.

Mattern, C. F. T. (1977). Symmetry in virus architecture. *In* "Molecular Biology of Animal Viruses" (D. P. Nayak, ed.), Vol. 1, p. 1. Dekker, New York.

Matthews, R. E. F. (1982). Classification and nomenclature of viruses: Fourth report of the International Committee on Taxonomy of Viruses. *Intervirology* **17,** 1–199.

"Portraits of Viruses," a series of historical reviews edited by F. Fenner and A. J. Gibbs and published in *Intervirology* (1979) **11,** 137, 201; (1980) **13,** 133, 257; (1981) **15,** 57, 121, 177; (1982) **18,** 1; (1983) **20,** 61; (1984) **22,** 1, 121.

CHAPTER 2

Cultivation and Assay of Viruses

Introduction 35
Cell Culture 35
Embryonated Eggs 43
Laboratory Animals 44
Assay of Viral Infectivity 45
Assays of Other Properties of Virions 48
Further Reading 51

INTRODUCTION

Viruses replicate only within living cells and cannot multiply in any cell-free medium, no matter how complex. Some viruses are restricted in the kinds of cells in which they replicate, and a few have not yet been cultivated at all under laboratory conditions. Fortunately, however, most viruses can be grown in cultured cells, embryonated eggs, or laboratory animals. The cultivation of viruses in cultured cells is essential for the study of their mode of replication (see Chapter 3), almost essential for the characterization of the virion (Chapter 1), and basic to classical diagnostic virology (Chapter 12), although, as will be seen in those chapters and in Chapter 4, recent developments in technology have enabled molecular biologists to discover an impressive amount about viruses such as hepatitis B even without the advantage of a cell culture system for their cultivation.

CELL CULTURE

Although over 70 years has elapsed since mammalian cells were first grown *in vitro*, it is only since the advent of antibiotics that cell culture became a matter of simple routine. Aseptic precautions are still essential, but the problems of contamination with bacteria, mycoplasmas, fungi, and yeasts are no longer insurmountable. Today, most kinds of animal cells can be cultivated *in vitro* for at least a few generations, and numerous immortal cell lines have been derived. Since 1949, when Enders, Weller, and Robbins reported that poliovirus could be grown in cultured nonneural cells with the production of recognizable cytopathic changes, hundreds of previously unknown viruses have been isolated and identified in cultured cells. The discovery of the adenoviruses, echoviruses, rhinoviruses, coronaviruses, and many others during the 1950s and 1960s was directly attributable to the use of cultured cells, as was the consequent revolution in the diagnosis of viral disease, the development of poliomyelitis, measles and rubella vaccines, and the dramatic advances in knowledge of the molecular biology of viruses of vertebrates.

Methods of Cell Culture

Cells may be grown *in vitro* as explants of tissue, such as respiratory or intestinal epithelium (*organ culture*), or as cultures of dispersed cells (*cell culture*). Organ cultures are used only for research; cell cultures form the basis of diagnostic virology, vaccine production, and most virus research. Tissue is cut into small pieces using scissors or scalpel and placed in a medium containing proteolytic enzymes such as trypsin. After the cells have been dispersed into a single-cell suspension, they are washed, counted, diluted in a growth medium, and permitted to settle on the flat surface of a glass or plastic container. Most types of cells adhere quickly, and under optimal conditions they divide about once a day until the surface is covered with a confluent monolayer.

Media

Cell culture has been greatly aided by the development of chemically defined media containing almost all the nutrients required for cell growth. The best known of these media, developed by Eagle, is an isotonic solution of simple salts, glucose, vitamins, and amino acids, buffered at pH 7.4, and containing antibiotics to inhibit the growth of bacteria and fungi. Serum must be added to Eagle's medium, and to most others, to supply additional growth factors, without which most cells will not multiply satisfactorily. In recent years several growth factors have been identified, and certain cell lines can now be grown in media that are totally defined chemically. For instance, the Madin–Darby canine kidney (MDCK) cell line has been grown on fibronectin or in polylysine- or collagen-coated dishes, in serum-free medium supplemented with hormones including insulin, binding proteins, e.g., transferrin and albumin, and attachment factors. Such serum-free media are particularly useful for the cultivation of "hybridoma" cells used for the production of monoclonal antibodies, where there is a need to ensure that all the immunoglobulin in the medium is antibody of a single specificity. Defined media also present advantages for the isolation of viruses that are likely to be neutralized by antibody present in "normal" human or animal serum, but this is generally not a problem if fetal calf serum is employed in media used for the isolation of human viruses. Fetal calf serum (5–10%) is therefore incorporated in the media used for the initial growth of cells in culture. Once the monolayer is established the "growth medium" is removed, virus is inoculated, and "maintenance medium" containing little or no serum is added to the culture.

Types of Cultured Cells

Many types of cells are capable of undergoing only a few divisions *in vitro* before dying out, whereas others will survive for up to a hundred cell generations, and some can be propagated indefinitely. These differences, the nature of which are not fully understood, give us three main types of cultured cells.

Primary Cell Cultures. When cultures are established initially from tissue taken freshly from animals they contain several cell types, most of which are capable of only limited growth *in vitro*—perhaps 5 or 10 divisions at most. This greatly restricts their value, whether for routine diagnostic work or vaccine production, because of the high cost and inconvenience of having to obtain fresh tissue each time, as well as lack of consistency from batch to batch. Furthermore, the donor animals often harbor latent viruses which can confuse diagnosis or contaminate vaccines. Nevertheless, the presence of a diverse range of differentiated cell types in such primary cultures, e.g., derived from monkey kidney, means that they tend to be very sensitive to most human viruses.

Diploid Cell Strains. These are cells of a single type that are capable of undergoing up to about 100 divisions *in vitro* before dying. They retain their original diploid chromosome number throughout. Diploid strains of fibroblasts established from human fetuses or embryos (HDF or HEF) are widely used in diagnostic virology and vaccine production. Some low-passage strains, such as WI-38 and MRC-5, having been certified diploid, nonmalignant, and virus free, are stored frozen in banks and are available to diagnostic laboratories or vaccine manufacturers.

Because diploid cells do not grow in suspension but only on surfaces, various methods have been devised to maximize the surface area to which they can attach, while keeping the overall size of the vessel and the volume of medium within reasonable bounds. Round bottles can be continuously rolled, or may be filled with glass tubes, glass beads, or spiral plastic film. Various types of rotary column or perfusion systems have also been used. Perhaps the most useful method for growing cells on an industrial scale for interferon or vaccine production is on plastic or Sephadex beads (*microcarriers*) maintained in suspension in large fermentation tanks.

Continuous Cell Lines. These are cells of a single type that are capable of indefinite propagation *in vitro.* Such immortal cell lines originate from cancers, or by spontaneous transformation of a diploid cell strain. Often they no longer bear close resemblance to their cell of origin, as they undergo many sequential mutations during their long history in culture, even in a given laboratory. The most usual indication of these changes is that the cells are "dedifferentiated," i.e., they have lost the specialized morphology and biochemical abilities that they possessed as differentiated cells *in vivo.* For example, it is no longer possible to distinguish microscopically between the various epithelial cell lines arising from cells of ectodermal or endodermal origin, or between the fibroblastic cell lines arising from cells of mesodermal origin (Plate 2-1). Cells of continuous cell lines are often aneuploid in chromosome number, especially if they are of malignant origin.

Continuous cell lines such as HEp-2 or HeLa, derived from human carcinomas, support the growth of certain viruses. These lines, and others derived from other animals such as monkeys (Vero), dogs (MDCK), mice (L929, 3T3), hamsters (BHK-21), rabbits (RK-13), etc., are widely used in experimental virology, and to a certain extent in diagnostic virology.

The great advantage of continuous cell lines over primary cell cultures is that they can be propagated indefinitely by subculturing the cells at regular intervals. Furthermore, they retain viability for many years when frozen (in serum-containing medium with added dimethyl sulfoxide) and stored at very low temperature, e.g., in liquid nitrogen (−196°C) or an electric deep freezer at −70° to −80°C. Good laboratories follow the general microbiological precept that the surest way of faithfully maintaining the credentials of a cultured cell line is to replace it periodically from frozen stocks.

Some continuous cell lines have been adapted to grow as a suspension of single cells, continuously stirred. Such suspension cultures are partic-

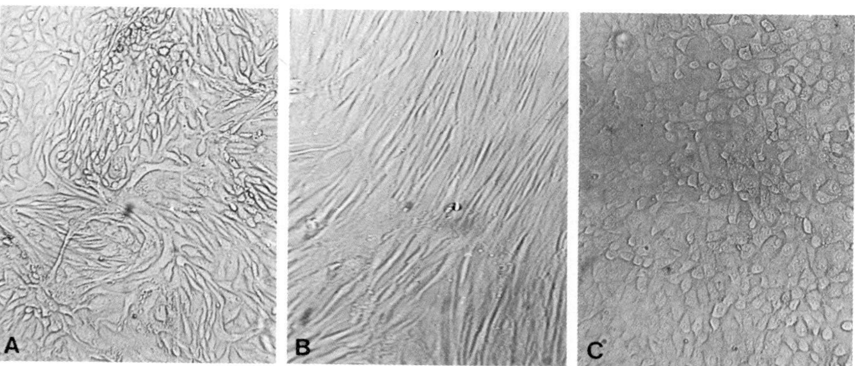

PLATE 2-1. *Cultured cells. Confluent monolayers of the three main types of cultured cell are shown as they normally appear in the living state, viewed by conventional low-power light microscopy through the wall of the tissue culture bottle.* ×60. *(A) Primary monkey kidney epithelium. (B) Diploid strain of human fetal fibroblasts. (C) Continuous line of epithelial cells. (Courtesy I. Jack.)*

ularly useful for biochemical studies of viral multiplication, because large numbers of identical cells are continuously available for regular sampling and processing.

Applications of Cell Culture

Of paramount importance for virologists is the selection of cell lines that will allow optimal growth of the virus under study. Some viruses multiply in almost any cell line and some cell lines support the replication of many different types of viruses. On the other hand, many viruses are quite restricted in the kinds of cell in which they can be isolated from human specimens. On serial passage of the isolate, however, mutants with somewhat greater growth potential for a given cell line can be selected.

Cultured cells serve three main purposes: (1) isolation of viruses from clinical specimens (see Chapter 12), for which cells of high sensitivity and readily recognized cytopathic effects are required, (2) production of vaccines, and antigens for diagnostic serology, in which the emphasis is placed on high yield of virus and freedom from contaminating agents (see Chapter 10), and (3) basic biochemical research (see Chapter 3), for which continuous cell lines, preferably growing as suspension cultures, are usually chosen for convenience of sampling.

Recognition of Viral Growth in Cell Culture

The growth of viruses in cell culture can be monitored by a number of biochemical procedures indicative of the intracellular increase in viral macromolecules and virions, as described in Chapter 3. In addition, there are simpler methods that are more commonly used for diagnostic work.

Cytopathic Effects (CPE). Many viruses kill the cells in which they multiply, so that infected cell monolayers gradually develop visible evidence of cell damage, as newly formed virions spread to involve more and more cells in the culture. These changes are known as *cytopathic effects (CPE)* and the responsible virus is said to be *cytopathogenic*. Most CPE can be readily observed in unfixed, unstained cell cultures, even in test tubes, under low power of the light microscope with the condenser racked down and the iris diaphragm partly closed to obtain the required contrast.

Detection of CPE is an important first step in diagnostic virology. Some viruses multiply readily in cell culture on primary isolation; the time at which cytopathic changes first become detectable depends to some extent on the number of virions that the specimen contained, but, far more importantly, on the growth rate of the virus in question. Enteroviruses, for example, which grow rapidly, often show detectable CPE after 24 to 48 hours, destroying the monolayer completely by 3 days. On the other hand, some of the more slowly growing viruses may not produce detectable CPE for weeks, during which time the condition of even uninfected control monolayers may deteriorate unless the medium is changed periodically.

A trained virologist can distinguish several types of CPE, even in unstained, living cultures (Plate 2-2, Table 2-1). Fixation and staining of the cell monolayer reveals further diagnostic details, notably *inclusion bodies* (Plate 5-1, Fig. 5-1, Table 12-3) and *syncytia* (Plate 2-2C, Table 2-2).

PLATE 2-2. *Cytopathic effects (CPE) produced in monolayers of cultured cells by different viruses. The cultures are shown as they would normally be viewed in the laboratory, unfixed and unstained. ×60. (A) Enterovirus—rapid, rounding of cells progressing to complete cell destruction. (B) Herpesvirus—focal areas of swollen rounded cells. (C) Paramyxovirus—focal areas of fused cells (syncytia). (D) Hemadsorption. Erythrocytes adhere to those cells in the monolayer that are infected. The technique is applicable to any virus that causes a hemagglutinin to be incorporated into the plasma membrane. Most of the enveloped viruses that mature by budding from cytoplasmic membranes produce hemadsorption. (Courtesy I. Jack.)*

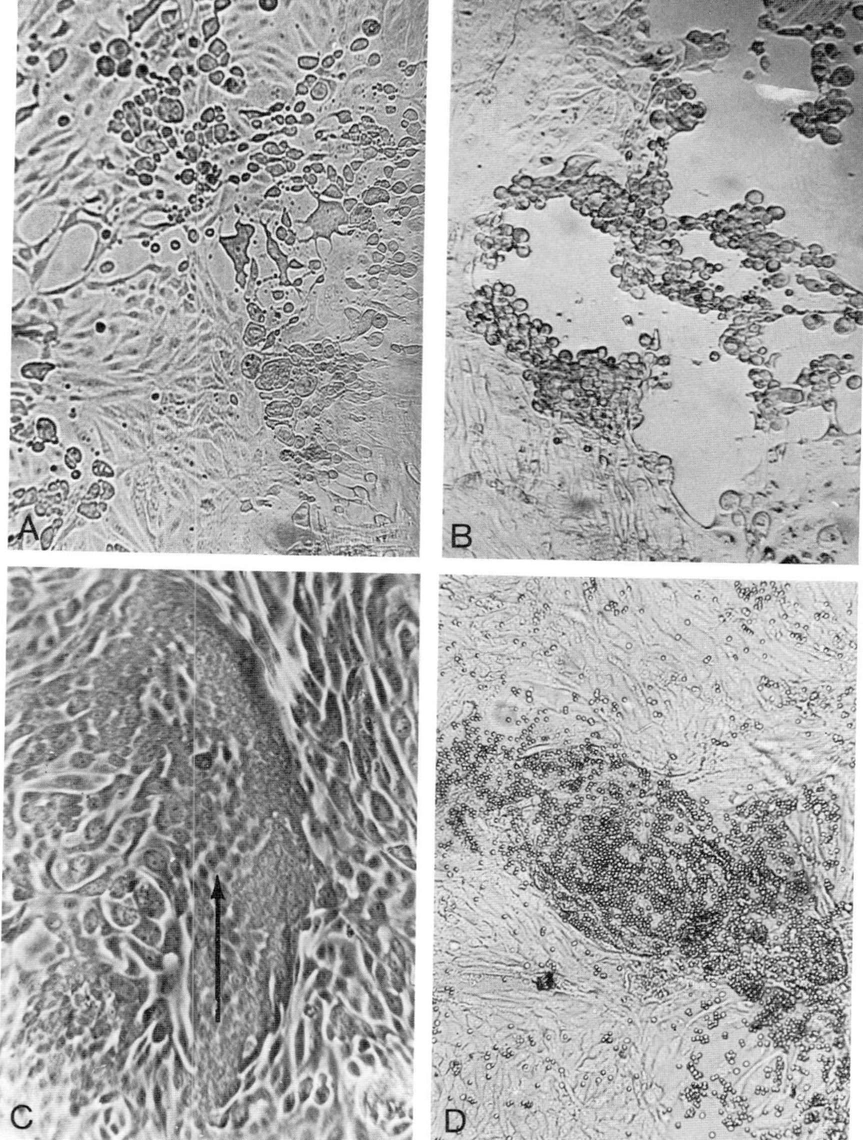
A
B
C
D

TABLE 2-1
Cytopathic Effects of Viruses in Cell Culture

CYTOPATHIC EFFECT	VIRUS
Cell destruction[a]	Enteroviruses
	Adenoviruses[b]
	Poxviruses
	Reoviruses
	Herpes simplex
	Togaviruses[c]
	Rhinoviruses[c]
Cell fusion (syncytium)	Paramyxoviruses
	Herpesviruses
Focal[d]	Cytomegalovirus
	Varicella-zoster
Minimal	Retroviruses
	Rhabdoviruses
	Arenaviruses
	Coronaviruses
	Orthomyxoviruses[e]
	Rubella[e]
	Hepatitis A[e]

[a] Rounding, pyknosis, then sloughing.
[b] Some types cause aggregation of cells, and some cause foci of CPE.
[c] Often produce incomplete cytopathic effect.
[d] Slowly expanding focus of affected cells.
[e] In most cell types.

The basis of these morphological consequences of viral infection is discussed in Chapter 5.

Hemadsorption and Hemagglutination. Cultured cells infected with orthomyxoviruses, paramyxoviruses, togaviruses, and flaviviruses which bud from cytoplasmic membranes, acquire the ability to adsorb erythrocytes. This phenomenon, known as *hemadsorption,* is due to the incorporation into the plasma membrane of newly synthesized viral protein that binds red blood cells (Plate 2-2D). Hemadsorption can be used to demonstrate infection with noncytopathogenic viruses, as well as cytocidal viruses, and can be demonstrated very early, e.g., after 24 hours, when only a small number of cells in the culture are infected.

Hemagglutination is a different, though related phenomenon, in which erythrocytes are agglutinated by free virus (see below). Virions or hemagglutinin may be demonstrated in the supernatant fluid of an infected culture.

Immunofluorescence. An even more sensitive technique for recognizing infection in a small number of cells, well before CPE is detectable, is immunofluorescence. Newly synthesized intracellular viral antigen is detected by staining the fixed cell monolayer with specific antiviral antibody that has been labeled with a fluorescent dye, or with an enzyme such as peroxidase. Full details of these important techniques are given in Chapter 12.

Interference. The multiplication of one virus in a cell usually inhibits the multiplication of another virus entering subsequently (see Chapters 4 and 6). The viruses of rubella and of the common cold were first discovered by showing that infected cell cultures, which showed no CPE, were nevertheless resistant to challenge with an unrelated echovirus. Cell lines have now become available in which these viruses produce CPE, but interference is a useful technique when searching for new noncytopathogenic viruses.

EMBRYONATED EGGS

Prior to the 1950s, when cell culture began to be widely adopted for the cultivation of viruses, the standard host for the cultivation of many viruses was the embryonated hen's egg (developing chick embryo). The technique was devised by Goodpasture in 1930 and was extensively developed by Burnet over the ensuing years. Nearly all of the viruses that were known at that time could be grown in the cells of one or another of the embryonic membranes, namely the amnion (plus the lung of the chick within the amniotic sac), allantois, chorion, or yolk sac (Fig. 2-1).

Fertilized eggs are inoculated after incubation for 5–14 days, depending on the state of development of the membrane it is proposed to infect. A hole is drilled in the shell, and virus is injected into the fluid bathing the appropriate membrane. Following incubation for an additional 2–5 days, viral growth can be recognized by one or more of the criteria listed in Table 2-2. In the case of the chorioallantoic membrane, the chorionic side of which is not bathed by fluid, clearly visible pocks develop after a few days (Plate 2-4).

Embryonated eggs are rarely employed now for viral isolation, but the allantois produces such high yields of certain viruses, notably influenza,

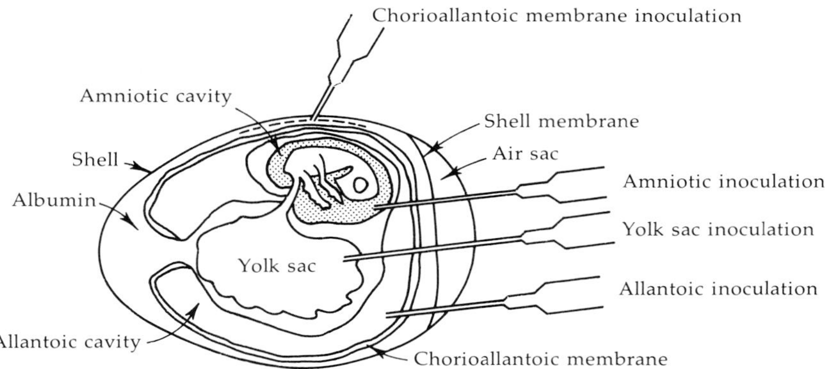

FIG. 2-1. *Routes of inoculation of the embryonated egg. Yolk sac inoculation is usually carried out with 5-day-old embryos; amniotic and allantoic inoculation with 10-day-old embryos; chorioallantoic inoculation with 11- or 12-day-old embryos. (Modified from B. D. Davis, R. Dulbecco, H. N. Eisen, H. S. Ginsberg, and W. B. Wood, "Microbiology." Harper, New York, 1967.)*

that this system is used by research laboratories and for vaccine production (Chapter 21).

LABORATORY ANIMALS

Like embryonated eggs, laboratory animals have almost disappeared now from diagnostic laboratories, since cell cultures are so much simpler to handle and much more versatile. Infant mice are still used for the isolation of some coxsackieviruses (see Chapter 19) and many arboviruses (Chapters 20 and 25).

TABLE 2-2
Growth of Viruses in Embryonated Eggs

MEMBRANE	VIRUSES	SIGNS OF GROWTH
Yolk sac	Herpes simplex	Death
Chorion	Herpes simplex	Pocks
	Poxviruses	
	Togaviruses[a]	
Allantois	Influenza	Hemagglutination
Amnion	Mumps	Hemagglutination
	Influenza	

[a] Some only.

However, laboratory animals are still essential for many kinds of virological research. Experiments on pathogenic mechanisms and the immune response are commonly carried out in inbred strains of mice, and sometimes other rodents. Primates are used to study human viruses such as the subacute spongiform encephalopathy agents and hepatitis viruses that will not grow in other laboratory animals or in cultured cells. Chimpanzees or smaller primates are also employed to evaluate experimental human viral vaccines for immunogenicity and safety prior to instigation of trials in man. Hamsters and other rodents are widely used in tumor virology, because they are highly susceptible to cancer production by oncogenic viruses. Finally, since serology looms large in much virological research, laboratory animals, usually rabbits, are extensively used for producing antisera. Mice are also commonly used for intraperitoneal implantation of hybridomas that secrete monoclonal antibody into the ascitic fluid.

ASSAY OF VIRAL INFECTIVITY

All scientific research depends upon reliable methods of measurement, and with viruses the property we are most obviously concerned with measuring is infectivity. The content of infectious virions in a given suspension can be titrated by infecting cell cultures (or, more rarely, chick embryos or laboratory animals) with dilutions of viral suspensions and then observing them for evidence of viral multiplication. Two types of infectivity assay should be distinguished: quantitative and quantal.

Quantitative Assays

A familiar example of this type of assay is the bacterial colony count on an agar plate. Each viable organism multiplies to produce a discrete clone; the colony count therefore represents a direct estimate of the number of organisms originally plated. The parallel in virology is the plaque assay, using monolayers of cultured cells.

Plaque Assays. Dulbecco introduced to animal virology a modification of the bacteriophage plaque assay which is now used very widely for the quantitation of animal viruses. A viral suspension is added to a monolayer of cultured cells for an hour or so to allow the virions to absorb to the cells. The infected cells are then overlayed with medium in an agar or methylcellulose gel, to ensure that the spread of progeny is restricted to the immediate vicinity of the originally infected cell. Hence, each infective particle gives rise to a localized focus of infected cells that becomes, after a few days, large enough to see with the naked eye as an

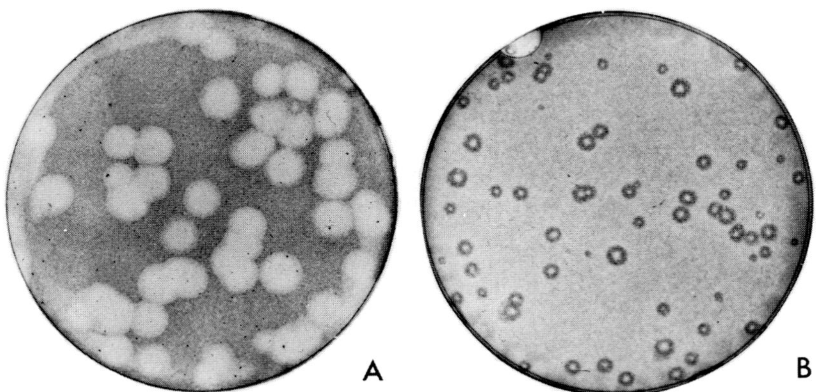

PLATE 2-3. *Plaques produced by influenza virus in monolayers of a continuous cell line derived from human conjunctival cells (Chang). Each plaque is initiated by a single virion and yields a clone. (A) Normal plaques, seen as clear areas in monolayer stained with neutral red. (B) "Red" plaques, characteristic of certain strains of influenza virus, and some other viruses. [From E. D. Kilbourne,* Bull. WHO **41,** *643 (1969); courtesy Dr. E. D. Kilbourne.]*

area of CPE. To render the plaques more readily visible, the cell monolayers are usually stained with a vital dye such as neutral red; the living (uninfected) cells take up the stain and the plaques appear as clear areas against a red background (Plate 2-3). Some viruses, e.g., herpesviruses and poxviruses, will produce plaques even in cell monolayers maintained in liquid medium, because most of the newly formed virus remains cell associated and plaques form by direct spread of virus to adjacent cells through intercellular bridges. Infection with a single virus particle is sufficient to form a plaque eventually; the infectivity titer of the original viral suspension is expressed in terms of *plaque-forming units (pfu)* per milliliter.

Transformation Assays. Some viruses, notably the oncogenic viruses, do not kill cells but "transform" them (see Chapter 5 and Plate 5-2). Compared with noninfected cells, transformed cells display reduced contact inhibition, so that they grow in an unrestrained fashion to produce a heaped-up "microtumor" that stands out conspicuously against the background of normal cells in the monolayer. Like malignant cells excised from a tumor, the transformed cells have also acquired the ability to grow in semisolid ("sloppy") agar or methylcellulose. Both properties have been exploited by tumor virologists to assay oncogenic viruses (see Chapter 8).

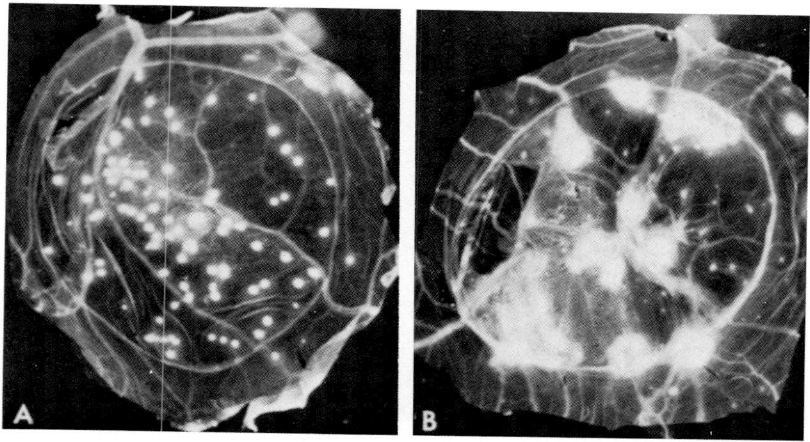

PLATE 2-4. *Pocks produced on the chorioallantoic membrane of embryonated eggs 3 days after inoculation with variola virus (A) and vaccinia virus (B). Small secondary pocks of vaccinia are visible. (Courtesy Professor A. W. Downie.)*

Pock Assays. A much older assay, still occasionally used for the poxviruses, is the titration of viruses on the chorioallantoic membrane of the chick embryo. Newly synthesized virus escaping from infected cells spreads mainly to adjacent cells, so that each infecting particle eventually gives rise to a localized lesion, known as a *pock* (Plate 2-4). The morphology and color of the pock are often characteristic of a particular group of viruses or even a particular mutant.

Quantal Assays

The second type of infectivity assay is not quantitative but quantal, i.e., it does not measure the number of infectious virus particles in the inoculum, but only whether there are any at all. Being an all-or-none assay, it is not nearly as precise as a quantitative assay. Serial, e.g., 10-fold, dilutions of virus are inoculated into several replicate cell cultures, eggs, or animals. Adequate time is allowed for the virus to multiply and spread to destroy the whole cell culture, or kill the animal, as the case may be. Hence, each host yields only a single piece of information, namely, whether or not it was infected by that particular dilution of virus. A more economical procedure can be used with viruses such as vaccinia that produce localized skin lesions in an animal such as the rabbit. Twenty or thirty skin sites can be separately inoculated, each giving an all-or-none answer equivalent to the death or survival of a cell culture or a mouse.

TABLE 2-3
Comparison of Quantitative and Quantal Infectivity Titrations

VIRUS DILUTION	QUANTITATIVE ASSAY (PLAQUE COUNT)[a]	QUANTAL ASSAY (CPE)
10^{-2}	C, C, C, C, C	+ + + + + + + + + +
10^{-3}	50, 42, 54, 59, 45	+ + + + + + + + + +
10^{-4}	5, 7, 3, 6, 4	+ + − + + + + + + +
10^{-5}	0, 0, 1, 0, 1	− + + − + − − + − +
10^{-6}	0, 0, 0, 0, 0	− − − − − − − + − − −
10^{-7}	0, 0, 0, 0, 0	− − − − − − − − − − −

[a] C, Confluent (uncountable).

Statistical Considerations. The results of typical quantal and quantitative infectivity titrations are given in Table 2-3. It will be seen that at high dilutions of the inoculum all the hosts remained uninfected because they failed to receive even a single infectious unit. The end point of a quantal titration is taken to be the dilution of virus that infects 50% of the inoculated hosts; the infectivity titer of the original virus suspension is then expressed in terms of *50% infectious doses (ID_{50})* per milliliter ($TCID_{50}$ in cell cultures). At first sight, it may be thought that Table 2-3 shows some anomalous results, in that some hosts have become infected following inoculation with higher dilutions of virus than those that failed to infect others. This sort of finding is quite normal and explicable in terms of the Poisson distribution. At any given multiplicity (m) of virus there is a finite probability (e^{-m}) that a given aliquot of the inoculum will contain no virus particle. In the example shown, each 0.1 ml sample of the 10^{-5} dilution contained an average of 1 ID_{50}, i.e., 0.5 of an infectious unit. Hence, each 0.1 ml sample of the 10^{-6} dilution contained an average of 0.05 infectious units, i.e., about 1 host in 20 received one infectious particle, whereas each 0.1 ml sample of the 10^{-4} dilution contained an average of five infectious units, i.e., e^{-5} (about one in a hundred) of these hosts failed to receive an infectious particle at all. In practice, quantal assays rarely produce such a nicely balanced result as the example presented in Table 2-3, and statistical procedures must be used to calculate the end point of the titration.

ASSAYS OF OTHER PROPERTIES OF VIRIONS

Hemagglutination

Many viruses contain, in their outer coat, virus-coded proteins capable of binding to erythrocytes (Table 2-4). Such viruses can, therefore,

TABLE 2-4
Hemagglutination by Viruses

VIRUS		
FAMILY	GENERA OR SPECIES	ERYTHROCYTES
Adenoviridae	Most species	Monkey and/or rat, 37°C
Poxviridae	Most species	Chicken, 37°C
Parvoviridae	Adeno-associated virus 4	Human, guinea pig, 4°C
Picornaviridae	*Enterovirus* (some types)	Human
	Rhinovirus (some types)	Sheep, 4°C
Togaviridae	*Alphavirus*, rubella	Goose, pigeon, chick; pH and temperature critical
Flaviviridae	*Flavivirus*	Goose, pigeon, chick; pH and temperature critical
Orthomyxoviridae	Influenza A and B	Chick, human, guinea pig, 4°C
Paramyxoviridae	Parainfluenza, mumps	Chick, human, guinea pig, 4°C
	Measles	Monkey, 37°C
Rhabdoviridae	Rabies	Goose, 4°C
Coronaviridae	OC43	Rat, mouse, chick
Bunyaviridae	Several species	Goose, pH critical
Reoviridae	*Reovirus*	Human
	Rotavirus	Chick

bridge red blood cells to form a lattice. This phenomenon, known as *hemagglutination*, was first described in 1941 by Hirst, who then went on to analyze the mechanism of hemagglutination by influenza virus. The *hemagglutinin* of influenza is a glycoprotein, which projects from the envelope of the virion (Plate 21-1). The virus will attach to any species of erythrocyte carrying complementary receptors, which are glycoproteins of a different sort. Hemagglutination by influenza virus (and the paramyxoviruses) is complicated by the fact that the virion also carries an enzyme, neuraminidase, which destroys the glycoprotein receptors on the erythrocyte surface (by removing terminal neuraminic acid) and allows the virus to elute, unless the test is carried out at a temperature too low for the enzyme to act (4°C). About 10^7 influenza virions are needed to cause agglutination of sufficient erythrocytes to permit the test to be read with the naked eye. Thus, hemagglutination is not a sensitive indicator of the presence of small numbers of virions, but because of its simplicity it provides a very convenient assay if large amounts of virus are available (Plate 12-6).

Counting Virions by Electron Microscopy

Negative staining with potassium phosphotungstate is such a simple technique that virus suspensions can be quite readily counted by electron microscopy. There are two main approaches. The virus can be mixed with a known concentration of polystyrene latex particles (or a morphologically distinct virus), to provide an easily recognizable marker; the ratio of "unknown" virus to latex beads enables its concentration to be determined. Alternatively, a known volume of virus suspension is deposited by ultracentrifugation (e.g., in an "Airfuge"). Or water and salts may be removed from a drop of virus suspension hanging from the underside of an ultrathin carbon-coated plastic ("Formvar") film mounted on a copper grid, by diffusion downward into agar (*agar filtration*); thus the number of particles in a known volume may be counted.

Comparison of Different Assays

If a given preparation of virus particles were to be assayed by all of the methods described above, the "titer" would be different in every case. For example, an influenza virus suspension may provide the data set out in Table 2-5.

The difference between the EM and HA titers reflects merely the difference in sensitivity between the two assays. On the other hand, the ratio between the infectivity titer and the EM particle count, known as the *infectivity-to-particle ratio*, requires deeper analysis. In part, it is explained by the fact that most of the virus particles visible by electron microscopy are noninfectious, having being inactivated by heat (even at 37°C) or by other mechanisms during extraction and purification, or having been put together unsatisfactorily in the first place (typically having a defective genome, hence known as "defective" or "incomplete" virus—see Chapter 4). However, the situation is complicated by another factor, namely the susceptibility of the cell system in which viral infectivity is assayed; here one speaks of the *plating efficiency* of the virus. Even a fully infectious virus particle has only a certain chance of successfully negotiating all the barriers it may encounter in the course of

TABLE 2-5
Comparison of Assays for Virus

METHOD	TITER (per ml)
Electron microscope count	10^{10} virions
Quantal infectivity assay in eggs	10^{9} egg ID_{50}
Quantal infectivity assay in cultured cells[a]	$10^{8.3}$ $TCID_{50}$
Quantitative infectivity assay by plaque formation[a]	10^{8} pfu
Hemagglutination assay	10^{3} HA units

[a] Assume same cell line.

entering and infecting a cell. The susceptibility of one cell system may be much lower (i.e., resistance much higher) than that of another (see Chapter 6), hence a given preparation of virus may have a lower efficiency of plating in one system than in the other (note 0.7 $\log_{10}$, i.e., fivefold, difference between egg ID_{50} and $TCID_{50}$ in Table 2-5).

FURTHER READING

Books

Bachmann, P. A., ed. (1983). "New Developments in Diagnostic Virology," Curr. Top. Microbiol. Immunol., Vol. 104. Springer-Verlag, Berlin and New York.

Freshney, R. I. (1983). "Culture of Animal Cells: A Manual of Basic Technique." Alan R. Liss/Wiley, New York.

Habel, K., and Salzman, N. P., eds.(1969). "Fundamental Techniques in Virology." Academic Press, New York.

Howard, C. R., ed. (1982). "New Developments in Practical Virology," Lab. Res. Methods Biol. Med., Vol. 5. Alan R. Liss, Inc., New York.

Maramorosch, K., and Koprowski, H., eds. (1967–1984). "Methods in Virology," Vols. 1–8. Academic Press, New York.

Reviews

Cooper, P. D. (1967). The plaque assay of animal viruses. *In* "Methods in Virology" (K. Maramorosch and H. Koprowski, eds.), Vol. 3, p. 243. Academic Press, New York.

Howe, C., and Lee, L. T. (1972). Virus-erythrocyte interactions. *Adv. Virus Res.* **17,** 1.

Nermut, M. V. (1982). Advanced methods in electron microscopy of viruses. *In* "New Developments in Practical Virology" (C. R. Howard, ed.), pp. 2–58. Alan R. Liss, Inc., New York.

Schmidt, N. J. (1972). Tissue culture in the laboratory diagnosis of viral infections. *Am. J. Clin. Pathol.* **57,** 820.

Schmidt, N. J. (1979). Cell culture techniques for diagnostic virology. *In* "Diagnostic Procedures for Viral and Rickettsial Infections" (E. H. Lennette and N. J. Schmidt, eds.), 5th ed., p. 65. Am. Public Health Assoc., New York.

Schrom, M., and Bablanian, R. (1981). Altered cellular morphology resulting from cytocidal virus infection. *Arch. Virol.* **70,** 173.

CHAPTER 3

Viral Multiplication

Introduction 53
Methods of Investigation of Viral Multiplication 54
The Viral Multiplication Cycle 60
Strategy of Expression of the Viral Genome 61
Attachment (Adsorption) 65
Uptake (Penetration) 65
Uncoating 69
Transcription 70
Translation 77
Replication of Viral Nucleic Acid 79
Assembly (Morphogenesis) 82
Inhibition of Synthesis of Cellular Macromolecules 87
Epilogue 88
Further Reading 88

INTRODUCTION

Viral multiplication is the central focus not only of much experimental virology but a significant part of molecular biology as well. The explosion of knowledge in the area of gene expression and its regulation owes much to viruses and to virologists. In the 1940s Delbruck, recognizing the advantage of viruses as models for study of the expression of genetic information, founded the discipline of bacteriophage genetics—and molecular biology was born. The field attracted some brilliant experimentalists and several won Nobel Prizes in the ensuing years. As techniques for handling mammalian cells became more sophisticated, the focus of attention began to switch to animal viruses in the 1960s and 1970s. Progress has been such that the complete genome of several mammalian viruses has now been sequenced and in a few cases every gene product has been identified. The basic mechanisms of transcription, translation, and nucleic acid replication have been characterized for all of the major families of animal viruses and the strategy of gene expression and its regulation clarified. Many of the biochemical processes elucidated by virologists, such as splicing and other types of posttranscriptional processing of messenger RNA, posttranslational cleavage and glycosylation of proteins, replication of RNA, reverse transcription, and integration and translocation of viral genes and cellular oncogenes, have already turned out to have general application right across the spectrum of cell biology.

Molecular virology has a literature amounting to thousands of papers a year, not to mention several multivolume treatises, a few of which are listed at the end of this chapter. Obviously, it is neither possible nor appropriate to devote more than a chapter or two to this highly specialized and rapidly moving subject in a general text of this type. Nevertheless, every student of virology, including students of medicine, must understand the fundamentals of viral multiplication, if only to appreciate the basis of antiviral chemotherapy, pathogenesis of viral disease, and the role of viruses in cancer.

Every one of the families of mammalian viruses boasts its own unique strategy of replication, each teaching us something new about gene expression. To run through them all seriatim would be the easiest way of disposing of this difficult subject, but also the most repetitious and mindboggling for the reader. Accordingly, we have decided to attempt a synthetic description of the multiplication of mammalian viruses as a whole. We shall highlight the key common features of the multiplication of all viruses and draw attention to the major differences in replication strategy between families. Where an important phenomenon has been thoroughly worked out using a particular virus as a model, that example may be used to illuminate the picture as a whole. In general, however, detailed descriptions of the multiplication of viruses of each of the 18 families will be held over to the appropriate chapters of Part II of this volume. Readers particularly interested in the replication strategy of any given family of viruses are referred to Chapters 13–27. The special case of the retroviruses that induce cancer will be dealt with in Chapter 8.

METHODS OF INVESTIGATION OF VIRAL MULTIPLICATION

Following the pattern established in experiments with bacteriophages, studies of the multiplication of animal viruses are based on the *one-step (single-cycle)* growth experiment. In such experiments events in

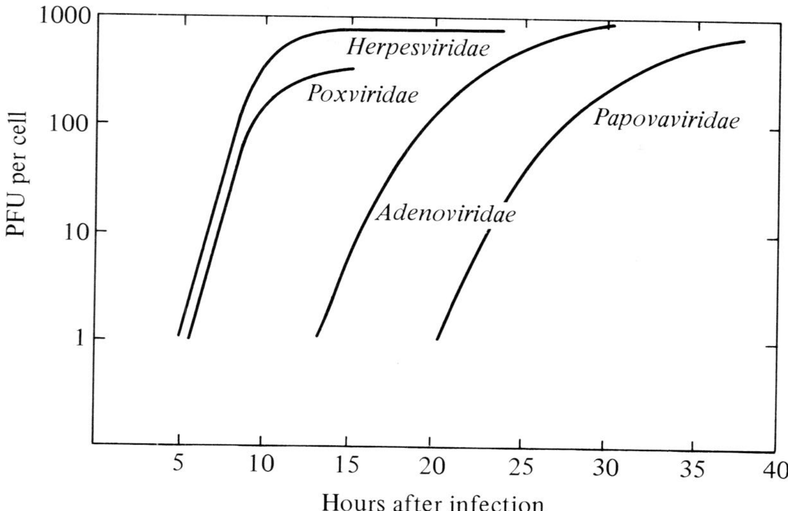

FIG. 3-1. *One-step (single-cycle) growth curves of representatives of the major families of DNA viruses. Yields are expressed as plaque-forming units (PFU), i.e., infectious particles scoring as plaques when titrated in cultured cells; the total number of particles would be 1–2 log_{10} higher. These curves are approximate only, as the latent period, multiplication rate, and total yield are all affected by the particular strain of virus employed, as well as by the cell strain and culture conditions, while the latent period and multiplication rate are also influenced by the multiplicity of infection.*

individual cells are synchronized by infecting all cells in a culture simultaneously, i.e., at high *multiplicity of infection (m.o.i.)*. Excess input virus is removed or neutralized, and the subsequent increase in numbers of infective virus particles is followed in the cells and the surrounding medium. Shortly after infection the inoculated virus disappears; it cannot be demonstrated, even intracellularly. This *eclipse period* or *latent period* continues until the first progeny become detectable some hours later. The eclipse period in such one-step growth cycles ranges from 5 to 20 hours for the various DNA viruses (Fig. 3-1) and from 2 to 10 hours for RNA viruses (Fig. 3-2).

Early studies, relying on quantitative electron microscopy and assay of infectious virions, provided a limited amount of information about the very early and the late events in the multiplication cycle but could not tell us anything about the all-important molecular events that occurred during the eclipse period. Investigation of the expression and replication of the viral genome became possible only with the development of several biochemical techniques some of which are outlined very briefly below.

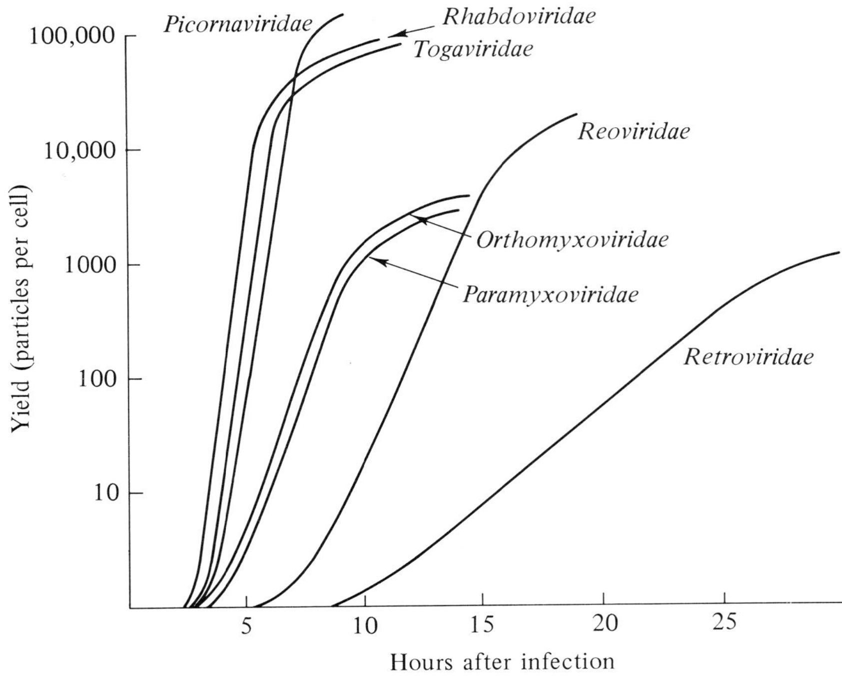

FIG. 3-2. *One-step (single-cycle) growth curves of representatives of the major families of RNA viruses. These curves are approximate only, as the latent period, multiplication rate, and total yield are all affected by the particular strain of virus employed, as well as by the cell strain and culture conditions, while the latent period and multiplication rate are also influenced by the multiplicity of infection.*

Infected cells, usually labeled with an appropriate radioactive isotope, can be broken open by freeze-thawing, physical shearing, homogenization, sonic vibration, or detergent solubilization. If desired, individual organelles or membranes can then be separated from soluble material by density gradient centrifugation. These in turn can be solubilized by appropriate detergents, and macromolecules such as nucleic acids and proteins separated by standard biochemical procedures.

Viral nucleic acids can be separated from those of the cell, initially on the basis of molecular weight or density, by techniques such as polyacrylamide gel electrophoresis, column chromatography, or rate zonal or equilibrium gradient centrifugation. Single-stranded nonsegmented viral RNA of positive polarity is infectious in its own right, as is DNA from all but the largest viruses, and may be identified and assayed by plaquing, providing appropriate measures are taken to protect the naked molecule against degradation by nucleases and to facilitate its entry into cells. Nucleic acid of other viruses is not infectious and cannot be assayed in this way, nor of course can complementary RNA (cRNA) transcribed from vRNA. However, these will anneal to nucleic acid extracted from virions. A range of such hybridization techniques has been developed to recover the various types of virus-specific DNA and RNA molecules from infected cells, to identify them, to determine their polarity, and to pinpoint which part of the genome they represent (for details see Chapter 12). The latter can be determined visually by electron microscopy (heteroduplex mapping, denaturation mapping, or R-loop mapping), or chemically by oligonucleotide mapping or restriction endonuclease mapping, Southern or Northern blotting, or ultimately by nucleotide sequencing (see Chapter 12). Complementary RNA can be transcribed *in vitro,* and mRNA can be translated *in vitro* into proteins which can themselves be identified. Indeed, for certain viruses with segmented genomes such as influenza and reovirus, it can now be said that every gene has been separately isolated and the function of the protein coded by each gene has been identified (Plate 3-1 and Table 21-2).

Viral proteins can be released from the cell by the techniques described above. Specific, perhaps monoclonal, antibodies may be employed to precipitate particular viral proteins which can be separated by polyacrylamide gel electrophoresis or to identify proteins following such separation (Western blotting). The activity of viral enzymes may be assayed directly, while other viral proteins are identifiable by serology, peptide mapping, or, ultimately, amino acid sequencing. The three-dimensional structure of a number of viral proteins has now been determined by X-ray crystallography (Fig. 4-5).

The intracellular location of viral macromolecules can be visualized by autoradiography, immunofluorescence, or immunoelectron microscopy. Intracellular movement and processing of viral macromolecules are followed by the pulse-chase technique, in which a short pulse of radioactively labeled precursor (e.g., thymidine for DNA, uridine for RNA, amino acid for protein) is chased by a large excess of cold precursor, and its intracellular migration and modification are followed over time.

Chemical inhibitors are widely used to block, reversibly or irreversibly, particular intracellular events such as transcription (e.g., actinomycin D, α-amanitin), translation (e.g., puromycin, pactamycin, cycloheximide, hypertonic salt), glycosylation (e.g., tunicamycin, monensin), or DNA synthesis (e.g., fluorodeoxyuridine, cytosine arabinoside). For instance,

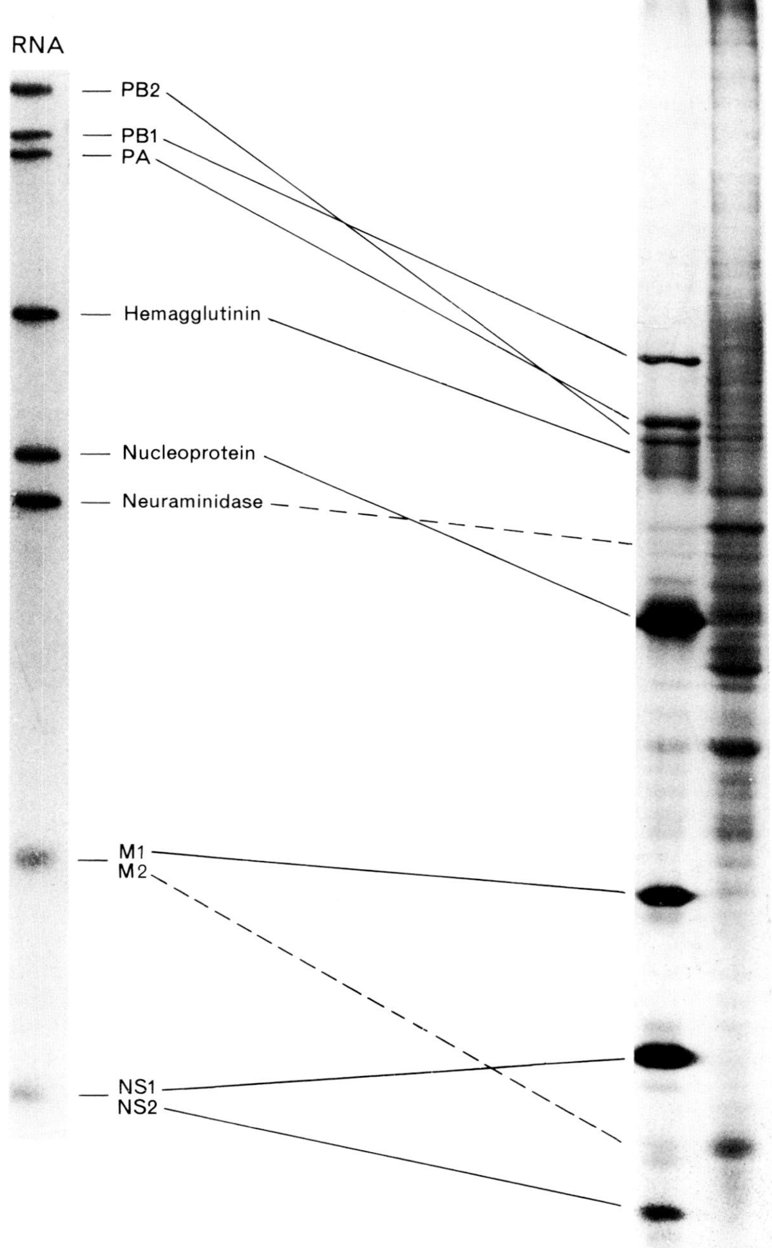

Protein
I U
RNA
PB2
PB1
PA
Hemagglutinin
Nucleoprotein
Neuraminidase
M1
M2
NS1
NS2

the fact that some viruses, such as poliovirus, multiply perfectly satisfactorily in cells whose RNA and protein synthesis has been blocked by actinomycin D has proved to be a most important aid to the study of their multiplication.

Transcription and translation mapping are techniques that make it possible to determine the linear order of the genes along an RNA molecule, or of the proteins for which these genes code. For example, ultraviolet irradiation creates uracil dimers in RNA, which block its transcription or translation; as these will occur at random positions in a polycistronic mRNA molecule, the probability that a particular gene product will be affected at any given UV dosage will be proportional to the distance of its gene downstream from the point of initiation of transcription or translation.

Viral variants with point mutations or deletions in known positions can be exploited to identify the function of the affected gene. Even before the advent of nucleotide sequencing, complementation and recombination tests with conditional lethal mutants provided virologists with much information about the number, location, and function of viral genes. These and other aspects of viral genetics will be discussed in the next chapter.

Finally, the development of rapid and reliable techniques for the cloning and sequencing of nucleic acids has revolutionized molecular virology. Comparison of the sequences of viral genomes with that of the mRNAs and proteins for which they code has already told us a great deal about the transcription and translation strategies of the different families. Even the conserved sequences, once despised as "noncoding" regions, are being identified as essential recognition signals, such as initiation codons, termination codons, splice sites, polyadenylation

---

PLATE 3-1. *The structure of the influenza A virus genome.* ^{32}*P-labeled RNA extracted from purified influenza virus A/PR/8/34 was separated into its eight segments following electrophoresis on 2.6% polyacrylamide gels containing 6* M *urea (left track).* [^{35}*S]Methionine-labeled proteins in extracts from virus-infected cells (I) or uninfected control cells (U) were separated on a 5–13% gradient polyacrylamide gel (right track). The hemagglutinin, nucleoprotein, and neuraminidase genes each code for a single protein. The large RNAs code for the polymerase proteins PB1, PB2, and PA. The M and NS genes each code for two different polypeptides. Broken lines indicate the approximate positions of the neuraminidase and M2 proteins (not seen here when [*^{35}*S]methionine is used as label—visualization of these polypeptides is improved by using labeled cysteine and isoleucine, respectively.) The molecular weights of the RNAs range from 0.28 to 0.7* $\times 10^6$*, and they encode polypeptides of 121 to 759 amino acids in length. (Courtesy Dr. P. Palese.)*

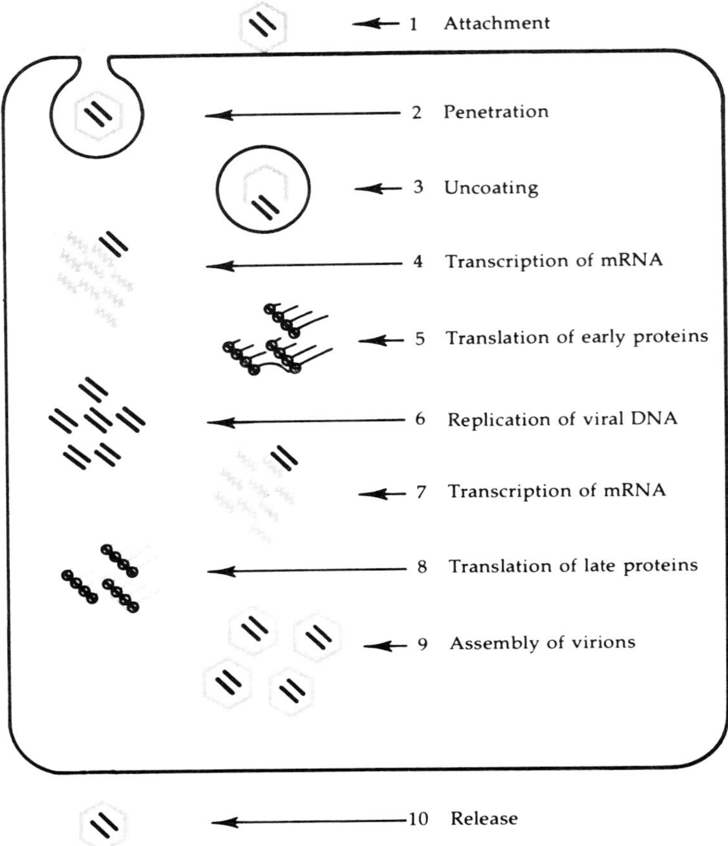

FIG. 3-3. *The viral multiplication cycle. Highly stylized diagram, using an icosahedral DNA virus as a model. No topographical location for each event is implied. Only the main steps are shown; further details and variations are described in the text. Of course, one step grades into the next such that, as the cycle progresses, several of these processes are proceeding simultaneously.*

sites, and binding sites for polymerases, ribosomes, and various regulatory viral proteins.

THE VIRAL MULTIPLICATION CYCLE

Figure 3-3 illustrates in a stylized and greatly simplified diagram the major steps in the multiplication of a "typical" mammalian DNA virus.

TABLE 3-1
Proteins Coded by the Viral Genome

Structural proteins of the virion
Enzymes located in the virion
Enzymes, nonstructural, mainly involved in transcription and replication of viral nucleic acid
Regulatory proteins affecting cellular transcription or translation
Regulatory proteins controlling expression of viral genes

Following attachment, the virion is taken up by its host cell and is partially uncoated to expose the viral genome. Certain *early viral genes* are transcribed into mRNA which may then be processed in a number of ways, including splicing. The early gene products translated from this mRNA are of two main types: proteins that regulate the expression of the viral and cellular genomes, and enzymes required for the replication of viral nucleic acid (Table 3-1). Following viral nucleic acid replication, *late viral genes* become available for transcription. The late proteins are principally structural proteins for incorporation into new virions; some of these are subject to posttranslational glycosylation and/or cleavage. Assembly of icosahedral virions occurs in the nucleus or cytoplasm, depending on the particular family. Enveloped viruses, on the other hand, are completed only as they "bud" through cellular membranes. Each infected cell yields several thousand new virions over a period of several hours.

Some of the more obvious differences between the families of mammalian viruses in respect of their multiplication are set out in Table 3-2. It will be noted that with some families many of the crucial events take place in the cell nucleus, while others multiply exclusively within the cytoplasm (and, in some cases, can multiply successfully in enucleated cells). Some viruses acquire an envelope by budding through plasma membrane, others through nuclear membrane, and still others into Golgi vesicles or endoplasmic reticulum. Some viruses shut down (repress) the synthesis of cellular macromolecules very effectively whereas others do not. Indeed, some viruses are noncytocidal, and others actually induce the cell to divide, or even transform it to the malignant state.

STRATEGY OF EXPRESSION OF THE VIRAL GENOME

The story of viral multiplication is the story of the expression of the information contained within the viral genome. While Fig. 3-3 adequately summarizes the way in which the genome of a DNA virus is

TABLE 3-2
Characteristics of Multiplication of Selected Viruses[a]

FAMILY	EXAMPLE	SITE OF NA REPLICATION	ECLIPSE PERIOD (HOURS)[b]	BUDDING	CELL SHUTDOWN[b,c]
Parvoviridae	AAV	N	?	—	–
Papovaviridae	SV40	N	15	—	+[d]
Adenoviridae	Human adenovirus 2	N	12	—	+
Herpesviridae	Herpes simplex	N	5	N	+
Poxviridae	Vaccinia	C	5	GP	+
Picornaviridae	Poliovirus	C	2.5	—	+
Caliciviridae	Feline calicivirus	C	2.5	—	+
Togaviridae	Sindbis	C	2.5	P	+
Flaviviridae	Kunjin	C	2.5	?	+
Coronaviridae	Murine hepatitis	C	5	G	+
Paramyxoviridae	NDV	C	4	P	+
Rhabdoviridae	VSV	C	2.5	P	+
Arenaviridae	Pichinde	C	5	P	–
Bunyaviridae	Snowshoe hare	C	4	G	+
Orthomyxoviridae	Influenza A	N	3	P	+
Retroviridae	Avian leukosis	N	10	P	–
Reoviridae	Reovirus 3	C	5	—	+

[a] N, Nucleus; C, cytoplasm; P, plasma membrane; G, Golgi apparatus.

[b] Differs with multiplicity of infection, strain of virus, cell type, and physiological conditions.

[c] Differs markedly in degree and in rapidity, from early and profound, to late and partial.

[d] No shutdown in transformation.

transcribed, translated, and replicated, quite different events are involved in the case of RNA viruses. Moreover, there are very important differences between viruses containing single-stranded rather then double-stranded nucleic acid, and between viruses with RNA of positive or negative polarity. Evolution has left us with a quite remarkable diversity of types of viral genome, each with its own unique strategy of replication. With characteristic insight Baltimore identified the nub of the matter. To make protein one must first have mRNA; hence, either viral RNA must itself be mRNA, or it must be transcribed into mRNA by a transcriptase carried in the virion. Figure 3-4 illustrates the strategies employed by the 6 basic classes of viral genome to transcribe mRNA.

1. The single-stranded (ss) viral RNA (vRNA) of the *Picornaviridae, Caliciviridae, Togaviridae, Flaviviridae,* and *Coronaviridae* is of positive (+) polarity (+ sense), hence is immediately available as mRNA for translation into protein.

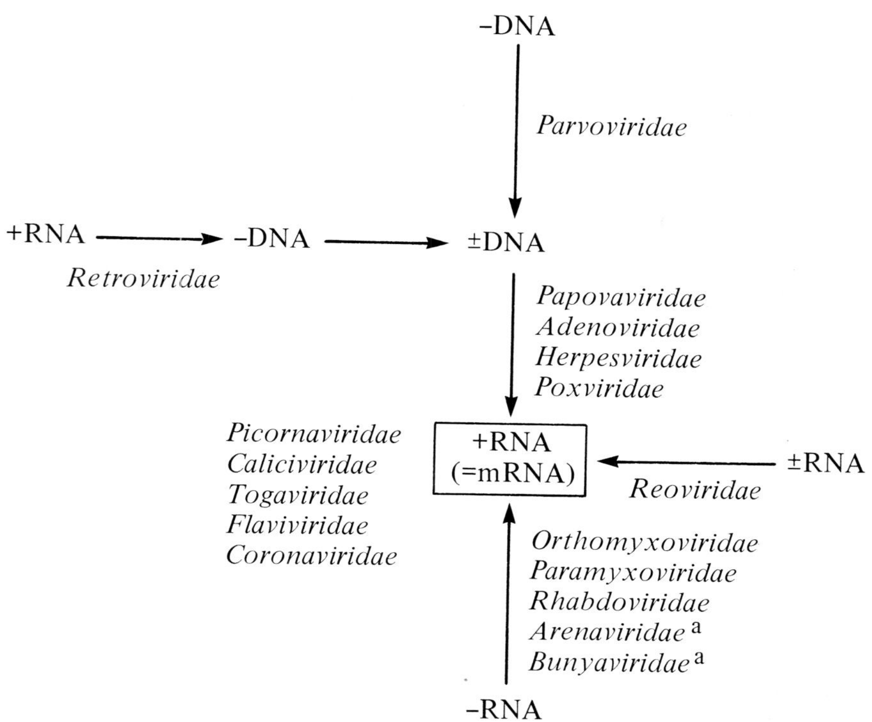

FIG. 3-4. *Six basic strategies for transcribing messenger RNA from different types of viral genome. [Based on a scheme originally conceived by D. Baltimore,* Bacteriol. Rev. **35,** *235(1971).]* [a]*At least some gene segments of the segmented RNA genomes of at least some arenaviruses and of the* Phlebovirus *genus of the* Bunyaviridae *family are ambisense, i.e., part +, part − polarity.*

2. The ssvRNA of the *Orthomyxoviridae, Paramyxoviridae, Rhabdoviridae, Arenaviridae,* and *Bunyaviridae* is minus-stranded, i.e., of negative polarity (− sense), hence must be transcribed by a virion-associated RNA-dependent RNA polymerase into complementary (+ sense) mRNA before any proteins can be made. (The vRNA of certain members of the *Arenaviridae* and *Bunyaviridae* is ambisense, i.e., part +, part − polarity.)

3. The dsvRNA of the *Reoviridae* is transcribed by a virion-associated RNA polymerase into ss(+)mRNA.

4. The dsDNA of the *Papovaviridae, Adenoviridae, Herpesviridae,* and *Poxviridae* is transcribed into mRNA by a DNA-dependent RNA polymerase.

5. The ssDNA of the *Parvoviridae* is converted into dsDNA which is in turn transcribed into mRNA.

6. The ss(+)vRNA of the *Retroviridae* is transcribed by a unique virion-associated reverse transcriptase into complementary ssDNA, which in turn is converted to dsDNA, which, following circularization, becomes integrated into the cellular genome, and may subsequently be transcribed into mRNA.

While Fig. 3-4 indicates the alternative routes leading to the production of mRNA from viral nucleic acid, it does not address the further question of how individual functional proteins are obtained from this mRNA. This brings us to the matter of posttranscriptional cleavage of mRNA and posttranslational cleavage of proteins. Presumably because eukaryotic cells, unlike prokaryotes, lack the machinery to initiate translation at internal sites part-way along a polycistronic mRNA molecule, viruses have developed various *alternative strategies for converting the information contained in the polycistronic viral genome into the several discrete proteins* for which it codes:

1. Viral genes may occur as separate molecules. For example, each segment of the segmented RNA genome of members of the *Reoviridae* and the *Orthomyxoviridae* can be thought of as a single gene. With few exceptions, the mRNA transcribed from each segment is translated into a single species of protein.

2. Monocistronic mRNA molecules, each encoding a unique protein, may be transcribed from polycistronic viral genomes, e.g., of *Paramyxoviridae* and *Rhabdoviridae.*

3. Polycistronic mRNA transcribed from the whole or part of the viral genome may subsequently be cleaved enzymatically into monocistronic mRNA species. In the case of most families of DNA viruses as well as certain RNA viruses, e.g., *Retroviridae,* this involves the removal of introns by splicing.

4. The + sense vRNA of *Picornaviridae,* or part of the vRNA of *Togaviridae* or *Caliciviridae,* is translated directly into a single polyprotein, which is subsequently cleaved enzymatically into individual proteins.

Furthermore, in the interests of genetic economy, the genome of some viruses may be read in different ways to code for different proteins; there are examples of functional mRNA being transcribed from both strands of dsDNA, from overlapping regions of one strand, or from the same sequence but translated in different reading frames. One can only marvel at the astonishing variety of alternative strategies that have evolved to make the most economical use of viral nucleic acid by maximizing the number of proteins for which they code, while minimizing replication time and constraints on packaging (Table 3-3).

The replication strategies of different viruses may influence the expression of the viral genome at any of several levels: transcription, posttranscriptional processing of mRNA, translation, posttranslational processing of proteins, replication of viral nucleic acid, and assembly and maturation of the virion. Key examples of the various alternatives at each level will be discussed where they are most readily comprehended, namely in the context of the individual steps in the viral multiplication cycle—to which we now turn.

ATTACHMENT (ADSORPTION)

Because virions and cells are both negatively charged at physiological pH, they tend to repel one another, but random collisions do occur and initial (reversible) attachment may be facilitated by cations. Firm binding, however, requires the presence of *receptors* for the virus on the plasma membrane. For instance, orthomyxo- and paramyxoviruses bind via an envelope protein known as the hemagglutinin (HA) to glycoproteins or glycolipids with oligosaccharide side chains terminating in *N*-acetylneuraminic acid (sialic acid). Polioviruses infect primate, not nonprimate, cells because only the former carry a receptor for the relevant viral capsid protein. While there is a degree of specificity about the recognition of particular receptors by particular viruses, many quite different viruses may utilize the same receptor. No doubt viruses have evolved to make opportunistic use of a limited number of membrane glycoproteins, the primary function of which would be quite different and presumably of benefit to the cell.

UPTAKE (PENETRATION)

Long-standing controversy surrounds the question of how the virion enters the cell. Electron microscopic and other data clearly show that virions can enter by at least three different mechanisms: (1) *endocytosis,* (2) *fusion,* and (3) *translocation.* What is unclear, however, is which of these leads to successful infection. The majority of virions entering a cell fail to initiate infection. The conventional wisdom is that most of the particles taken up by endocytosis are degraded by lysosomal enzymes. Nevertheless, recent evidence suggests that this may be the normal route to successful infection by some viruses at least.

TABLE 3-3

Strategy of Expression of Viral Genome

FAMILY	VIRAL NUCLEIC ACID				TRANSCRIPTASE IN VIRION	STRATEGY[a]		
	TYPE	STRANDS	POLARITY	SEGMENTS		MONOCISTRONIC mRNA DIRECTLY TRANSCRIBED[b]	SUBGENOMIC mRNA TRANSCRIBED THEN CLEAVED[c]	GENOME TRANSLATED DIRECTLY INTO POLYPROTEIN[d]
Parvoviridae	DNA	1	−[e]	1	−		+	
Papovaviridae	DNA	2	±	1	−		+[f]	
Adenoviridae	DNA	2	±	1	−		+[f]	
Herpesviridae	DNA	2	±	1	−		+[f]	
Poxviridae	DNA	2	±	1	+		+	
Picornaviridae	RNA	1	+	1	−			+
Caliciviridae	RNA	1	+	1	−			+[g]
Togaviridae	RNA	1	+	1	−			+[g]
Flaviviridae	RNA	1	+	1	−			[h]
Coronaviridae	RNA	1	+	1	−	+[i]		
Paramyxoviridae	RNA	1	−	1	+	+		
Rhabdoviridae	RNA	1	−	1	+	+		
Arenaviridae	RNA	1	−	2	+	?+[j]		
Bunyaviridae	RNA	1	−	3	+	+[j]		
Orthomyxoviridae	RNA	1	−	8	+	+[k]		
Retroviridae	RNA	1	+	2(diploid)	+			+[g]
Reoviridae	RNA	2	±	10–12	+	+		

[a] Only the principal feature of the strategy of transcription/translation is given here. Important ancillary features are mentioned in footnotes and/or described in the text.

[b] From monocistronic vRNA in the case of the separate vRNA molecules of certain viruses with segmented genomes (orthomyxoviruses, reoviruses); from the single polycistronic vRNA molecule of paramyxoviruses and rhabdoviruses; an intermediate situation obtains with the segmented genomes of the arenaviruses and bunyaviruses (see footnote *j*).

[c] Polycistronic mRNAs are transcribed from large tracts of the DNA, then cleaved to yield monocistronic mRNA species.

[d] The polyprotein is then cleaved enzymatically into separate functional proteins.

[e] − or + in different virions of members of the *Dependovirus* genus.

[f] Splicing is required to generate monocistronic mRNA from the subgenomic transcript.

[g] One end of viral (+)RNA is translated into a polyprotein; subgenomic (+)mRNA corresponding to the other end is transcribed from full-length complementary (−)RNA (or from integrated cDNA in the case of retroviruses) and translated into another polyprotein.

[h] Neither monocistronic mRNA nor subgenomic mRNA nor polyprotein has been found.

[i] A nested set of subgenomic mRNA molecules is transcribed from complementary (−)RNA. However the 3′ sequence shared with the next smallest mRNA in the nested set is not translated. Only the unique 5′ end of each mRNA is translated, each yielding a distinct protein.

[j] But some of the vRNA segments are polycistronic and some are ambisense.

[k] The transcripts of the two smallest influenza vRNA segments are spliced, producing an additional mRNA from each.

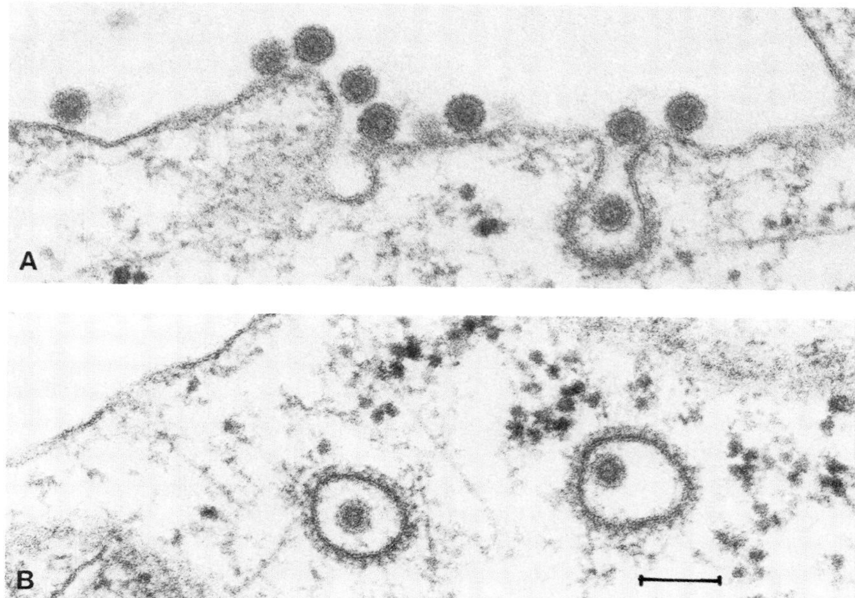

PLATE 3-2. *Entry of a togavirus. (A) Attachment and movement into a coated pit. (B) Endocytosis, coated vesicle (bar = 100 nm). [A from E. Fries and A. Helenius,* Eur. J. Biochem. **97,** *213 (1979); B from K. Simons* et al., Sci. Am. **246,** *46 (1982), Courtesy Dr. A. Helenius.]*

Endocytosis

Virtually all mammalian cells, not only those known as phagocytes, are continuously engaged in phagocytosis, pinocytosis, and receptor-mediated (adsorptive) endocytosis, a specific process for the uptake of essential macromolecules. Viruses may use this normal cellular function to initiate infection (Plate 3-2). Following multipoint attachment to receptors, virions move down into *coated pits.* These pits, coated with clathrin, fold inward to produce *coated vesicles* that enter the cytoplasm and fuse with a lysosome to form what is sometimes called a *lysosomal vesicle* or *secondary lysosome* or *phagosome.* At the acid pH (5) of the lysosome the envelope of endocytosed virus fuses with the lysosomal membrane, expelling the viral nucleocapsid into the cytoplasm. This explains how a virion can be uncoated by a lysosome yet escape total degradation by the lysosome's hydrolytic enzymes. The whole process seems to resemble closely that whereby many large molecules, such as hormones, are taken up by cells yet avoid degradation in lysosomes.

Recent studies with influenza virus have identified a pH 5-mediated conformational change in the cleaved form of the HA molecule which enables fusion to occur between the viral envelope and the membrane of the secondary lysosome.

Fusion with Plasma Membrane

The F (fusion) glycoprotein of paramyxoviruses, in its cleaved form, enables the envelope of these viruses to fuse directly with the plasma membrane of cells, even at pH 7. This may allow the nucleocapsid to be released directly into the cytoplasm. Though a number of enveloped viruses display a capacity to fuse mammalian cells or to lyse erythrocytes it is not clear whether this is the normal way in which they infect cells.

Translocation

Some nonenveloped icosahedral viruses appear to be capable of passing directly through the plasma membrane. Again, we do not know whether this is a functional entry mechanism or an atypical event.

UNCOATING

In order that at least the early viral genes may become available for transcription it is necessary that the virion be at least partially uncoated. With viruses that enter the cell by fusion of their envelope with either the plasma membrane or the membrane of a lysosomal vesicle, it can be said that uncoating coincides with entry—the nucleocapsid is discharged directly into the cytoplasm. In the case of viruses with tubular (helical) nucleocapsids, transcription is initiated off viral RNA still associated with nucleoprotein. Poxviruses are uncoated in two stages: first, to a core, from which half the genome is transcribed; then completely, following the synthesis of a virus-coded uncoating protein. One of the best-studied models is reovirus; the outer capsid is removed by proteases in a phagosome, leaving a "subviral particle" (SVP) from which mRNA is transcribed throughout the multiplication cycle. Indeed, mRNA can even be transcribed from such SVPs *in vitro* (Plate 3-3).

A different mechanism of uncoating is seen with the picornaviruses. The process of attachment of virion to cell leads to a conformational change in the capsid, resulting in the loss of capsid proteins VP4 then VP2, and rendering the particle susceptible to proteases and RNase. It is tempting to postulate that, for icosahedral viruses, the attachment step itself triggers the process of uncoating.

For certain viruses that replicate in the nucleus there is evidence that

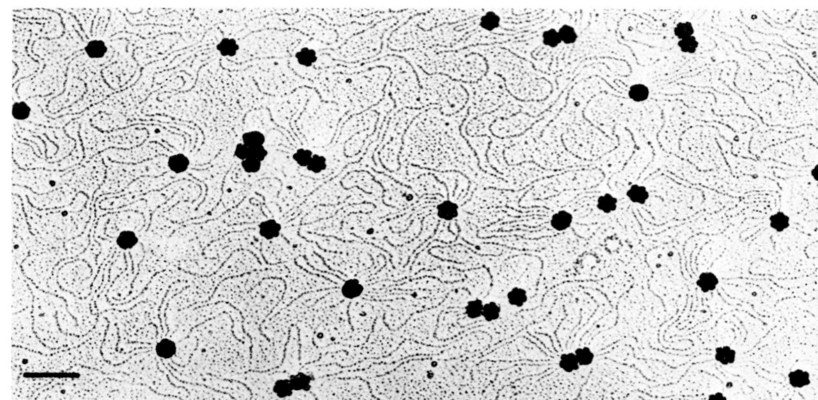

PLATE 3-3. *Reovirus messenger RNA (bar = 200 nm). Reovirus "reaction cores" that have synthesized mRNA for 8 minutes at 37°C were prepared for electron microscopy by the Kleinschmidt technique, stained with uranyl acetate and shadowed at a low angle with platinum–palladium. The results of polyacrylamide gel electrophoretic analysis of similar mRNA molecules are illustrated in Fig. 3-6. [From N. M. Bartlett, S. C. Gillies, S. Bullivant, and A. R. Bellamy,* J. Virol. **14,** *315 (1974), courtesy Dr. A. R. Bellamy.]*

the later stages of uncoating occur there, rather than in the cytoplasm. We know almost nothing about how this happens, nor indeed how virions, subviral particles, or viral macromolecules are transported from cytoplasm to nucleus and back again.

TRANSCRIPTION

As summarized under "Strategy of Expression of the Viral Genome," the vRNA of ssRNA viruses of positive polarity binds directly to ribosomes and is translated in full or part without the need for any prior transcriptional step (Fig. 3-4). All other classes of viral genome must be transcribed to mRNA. In the case of certain DNA viruses that replicate in the nucleus the cellular DNA-dependent RNA polymerase II is employed for this purpose. All other viruses require unique and specific types of transcriptase which are not only virus coded but must be carried into the cell as an integral component of the infecting virion in order that transcription can proceed at all. Cytoplasmic dsDNA viruses carry a DNA-dependent RNA polymerase, whereas dsRNA viruses have dsRNA-dependent RNA polymerase, while negative-stranded RNA viruses require a ssRNA-dependent RNA polymerase.

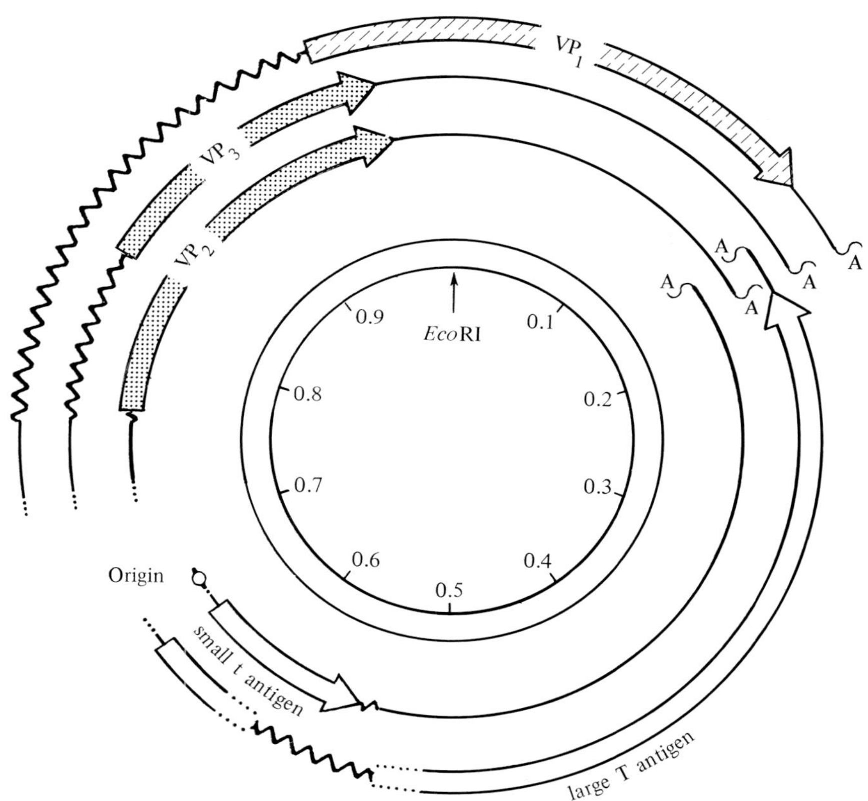

FIG. 3-5. *Genetic map of the SV40 viral DNA molecule indicating the transcription strategy. The circular double-stranded DNA molecule is oriented with the* EcoRI *restriction endonuclease cleavage site at zero and the origin of DNA replication at 0.66. The direction of transcription of the early genes (counterclockwise on one DNA strand) and of the late genes (clockwise on the other strand) is indicated by the arrows. The thin lines indicate regions of the transcribed mRNA that are not translated into protein, while the zigzag thin lines indicate regions of the mRNA that are spliced out. The 3′ terminal poly(A) tail of each mRNA is labeled A. It will be seen that the genes for the early proteins, small-t and large-T overlap, as do those for the late proteins VP1, VP2, and VP3. Also note that large-T is coded by two noncontiguous regions of DNA. The amino acid sequence of VP3 corresponds with the C-terminal half of VP2. However, VP1 shares no part of its amino acid sequence with VP2 or VP3 because its mRNA is translated in a different reading frame. [Reproduced in slightly modified form from Fiers* et al. *(1978). Reproduced by permission from* Nature (London) **273,** *113. Copyright © 1978 Macmillan Journals Limited.]*

Strategy of Transcription in DNA Viruses

Transcription of the DNA viral genome is programmed such that not all genes are read simultaneously or throughout the multiplication cycle. Particular sections of the chromosome are transcribed sequentially, the socalled *early* genes being read first, and the *late* genes later in the cycle. Indeed, in the case of the adenoviruses and the herpesviruses there are more than two rounds of transcription, requiring quite complex regulatory mechanisms. Enzymes required for replication of viral nucleic acid are generally encoded by early genes, and structural proteins required for assembly of new virions are encoded by late genes.

In 1978 Fiers and his colleagues presented us with the first complete description of the genome of an animal virus (Fig. 3-5). Analysis of the 5224 nucleotide-pair sequence of the circular dsDNA molecule of the papovavirus SV40 and its transcription program revealed some remarkable facts. First, the early genes and the late genes are transcribed, by the host cell's RNA polymerase II, in opposite directions, from different strands of the DNA. Second, certain genes overlap, so that their products share part of their amino acid sequences in common. Third, some regions of the viral DNA are read in two overlapping reading frames, so that two completely different amino acid sequences are obtained. Fourth, at least 15% of the viral DNA consists of *intervening sequences (introns)*, which are transcribed but never translated into protein because they are excised from the primary transcript; indeed, up to three distinct proteins can result from mRNAs derived from a common primary transcript by different splicing protocols.

Regulation of Transcription

Sophisticated studies with adenoviruses have elucidated the nature of the mechanisms that regulate the expression of viral genomes. These operate principally, but not exclusively, at the level of transcription. Because of the complications arising from posttranscriptional cleavage of mRNA and posttranslational cleavage of precursor proteins in eukaryotic cells, it is no longer adequate to talk of a "gene" and its "gene product." More appropriate perhaps is to think in terms of the *transcription unit*, i.e., that region of the genome beginning with the transcription initiation site, extending right through to the transcription termination site, and including all introns and exons in between; "simple" transcription units may be defined as those encoding only a single protein, whereas "complex" transcription units code for more than one. There are many adenovirus transcription units. At different stages of the viral multiplication cycle, "preearly," "early," "intermediate," and "late," the various transcription units are transcribed in a given temporal sequence. A product of the early region *E1A* induces the other early regions including *E1B*, but, following viral DNA replication, there is a 50-fold increase in the rate of transcription from the major late promoter (MLP) relative to early promoters such as *E1B*, and a decrease in *E1A* mRNA levels. A second control operates at the point of termination rather than initiation of transcription. Transcripts controlled by the P16 promoter terminate at a particular point early in infection but read through this termination site later in infection to produce a range of longer transcripts with different polyadenylation sites.

Posttranscriptional Processing of mRNA

Primary RNA transcripts from eukaryotic DNA are subject to a series of posttranscriptional alterations in the nucleus, known as *processing*, prior to export to the cytoplasm as mature mRNA. First, a *cap*, consisting of 7-methylguanosine (m^7Gppp) is added to the 5′ terminus of the primary transcript; the cap is thought to facilitate the formation of a stable initiation complex on the ribosome. Second, a sequence of 50–200 adenylate residues is added to the 3′ end; this *poly(A) tail* appears to stabilize mRNA against degradation in the cytoplasm. Third, a methyl group is added at the 6 position to about 1% of the adenylate residues throughout the mRNA (*methylation*). Fourth, particular noncoding *intervening sequences (introns)*, which begin with GU, end with AG, and contain stop codons, are removed from the primary transcript, a procedure known as *splicing;* the precise mechanism is not known but may well involve excision of the introns by endonucleases, followed by ligation of the exposed ends of the exons by ligases.

Splicing. The first three of these posttranscriptional events are seen with RNA transcripts from both DNA and RNA mammalian viruses. However, splicing seems to be restricted to those DNA viruses that multiply in the nucleus, plus the retroviruses, which are characterized by the transcription of mRNA from an integrated cDNA template, and the orthomyxoviruses, which have a special involvement with the nucleus (see below). Splicing is another important mechanism for regulating gene expression in nuclear DNA viruses. A given RNA transcript can have two or more splicing sites and be spliced in several different ways to produce a variety of mRNA species coding for distinct proteins; both the preferred poly(A) site and the splicing pattern may change in a systematic fashion as infection proceeds.

Yet another level of regulation is the rate of degradation of mRNA. Not only do different mRNA species display different half-lives but the half-life of a given mRNA species may change as the viral multiplication cycle progresses.

Strategy of Transcription in RNA Viruses

Minus-Stranded RNA Viruses. Primary transcription from the minus-stranded RNA viruses requires that the viral RNA is still within the helical nucleocapsid, in intimate association with the nucleoprotein as well as the transcriptase. Particular sequences of 10–20 nucleotides, located at or near the ends of each RNA molecule, may serve as recognition signals for the RNA polymerases.

The rhabdoviruses and paramyxoviruses are closely similar in their transcription strategies, as well as in the consensus sequences at the 3′ and 5′ termini of their vRNA, suggestive of a common ancestry. The polycistronic negative-stranded parental RNA is transcribed in two distinct ways: the *replication mode* and the *transcription mode.* Transcription in the replication mode produces full-length + strands (*antigenomes*) which are used as templates for the synthesis of new viral (−)RNA. Transcription in the transcription mode produces five *subgenomic* + strands, each polyadenylated and capped, and corresponding to a *monocistronic mRNA.* It is still not certain what dictates whether the transcriptase reads right through from the 3′ to 5′ end of the (−)RNA template, ignoring all internal termination signals, etc. along the way (replication mode). Nor is it known precisely how the monocistronic mRNAs are generated (transcription mode). There is some evidence that the transcriptase may have a certain probability of "falling off" its template as it proceeds from cistron to cistron; the five mRNAs are made in decreasing amounts, reading from the 3′ end of the parental RNA. This *modulation* is sometimes called the *polarity effect.*

Mention should be made of a remarkable requirement for the initiation of synthesis of mRNA from the negative-stranded segmented RNA viruses of influenza (orthomyxoviruses). A virion-associated endonuclease amputates a short segment from the capped 5′ terminus of eukaryotic (cellular) mRNA; the viral transcriptase then utilizes this small fragment as a primer to initiate the transcription of viral mRNA. Bunyavirus mRNAs also carry nonviral sequences at their 5′ termini, which may similarly turn out to represent cellular mRNA primers.

Ambisense RNA Viruses. The other families with segmented minus-stranded RNA, namely the bunyaviruses and arenaviruses, display a different type of strategy again. In general, each discrete vRNA segment, unlike those of the orthomyxoviruses, codes for more than one protein. Furthermore, as mentioned earlier, the S segment, at least, of arenaviruses and phleboviruses is ambisense. Thus, the replication strategy of ambisense RNA viruses, like the polarity of their genomes, is mixed. It is described in Chapters 24 and 25.

Plus-Stranded RNA Viruses. Transcription from the plus-stranded RNA of the picornaviruses yields only genome-length cRNA which serves only as a template for replication of more vRNA; the only messenger is the full-length plus strand, from which a single polyprotein is translated, then progressively cleaved.

Togaviruses of the alphavirus genus also contain a single polycistronic RNA molecule of + sense, but the transcription strategy differs from that of the picornaviruses. Only about two-thirds of the vRNA (the 5′ end) is translated; the resulting polyprotein is then cleaved into four nonstructural proteins, two of which form the RNA polymerase. This enzyme then transcribes a full-length − strand, from which two types of + (messenger) strand are copied: full-length (vRNA) and one-third length. The latter corresponds with the 3′ end of vRNA and is translated into a polyprotein precursor of three or four structural proteins. The caliciviruses have not been extensively studied but also produce both genome-length and subgenomic mRNA species.

Flaviviruses, on the other hand, have recently been accorded the status of a separate family because, unlike the togaviruses, they employ neither subgenomic RNA nor polyproteins in their replication. There is evidence that individual proteins are translated directly from full-length vRNA. This would require initiation of translation at multiple internal sites along the polycistronic mRNA (vRNA), a phenomenon not previously described with mammalian viruses (see Chapter 20).

Coronaviruses employ a unique mechanism of transcription. A *nested set* of overlapping subgenomic mRNAs is transcribed from genome-length complementary (−)RNA and only the unique (nonoverlapping) sequence in each mRNA is translated (see Chapter 23 for details).

Double-Stranded RNA Viruses. The double-stranded RNA genome of reoviruses is segmented, each molecule comprising a unique gene. Monocistronic mRNAs are transcribed from the partly uncoated subviral particle (Plate 3-3); this mRNA complexes with a protein and is itself transcribed, once only, to produce a double-stranded RNA molecule, which serves as the template for further transcription of mRNA.

Regulation. Transcription from RNA viral genomes is generally not as rigorously regulated as with DNA viruses. In particular, the temporal separation into early genes transcribed before the replication of viral nucleic acid, and late genes transcribed thereafter is not nearly so clear

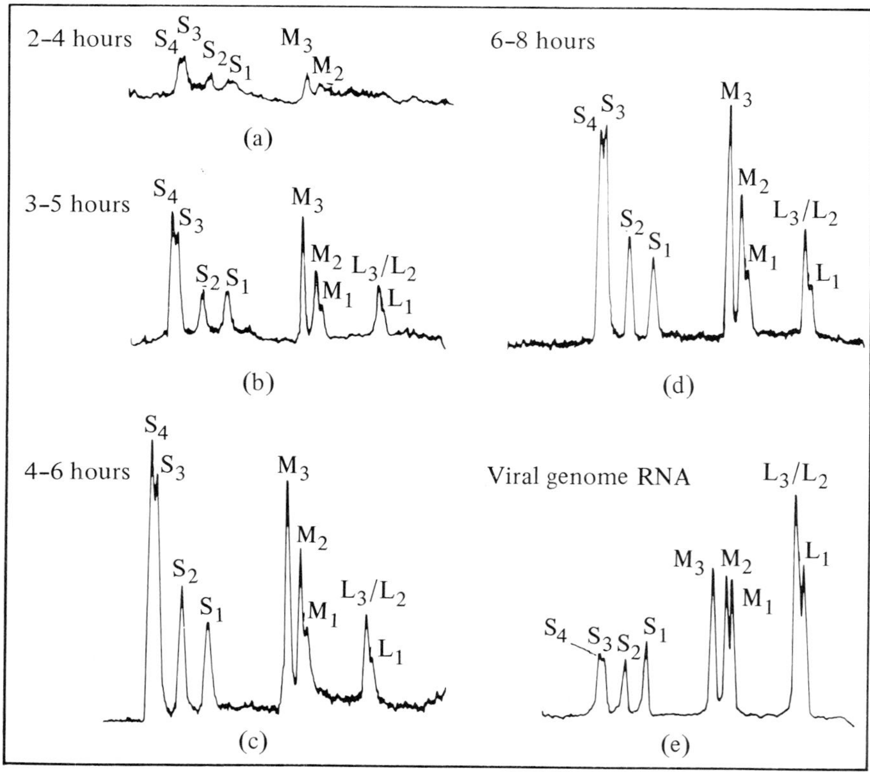

FIG. 3-6. *Transcription of mRNA from double-stranded viral RNA of reovirus by the viral transcriptase. Tracings from polyacrylamide gel autoradiograms showing relative rates of formation of reovirus mRNA species during the infection cycle (a–d), as determined by hybridizing labeled RNA from the cytoplasm of infected cells to an excess of genome RNA; compared with the tracing of genome RNA derived from virions (e). [From H. J. Zweerink and W. K. Joklik,* Virology **41,** *501 (1970).]*

(Fig. 3-6). Nevertheless, a subtle form of control can modulate the relative abundance of mRNAs for different proteins. For instance, in the case of the nonsegmented negative-strand RNA rhabdoviruses and paramyxoviruses, where the whole genome is transcribed simultaneously into monocistronic mRNA species coding for the five structural proteins, the "polarity" of the linear transcription by the viral transcriptase, described above, results in favored synthesis of mRNA for the proteins coded by the 3′ end of vRNA.

TRANSLATION

Capped and polyadenylated monocistronic viral mRNAs, following splicing when necessary, bind to ribosomes in the cell cytoplasm and are translated into protein in essentially the same fashion as eukaryotic mRNA, presumably utilizing cellular tRNAs and initiation factors. The sequence of events has been closely studied for reovirus. Each monocistronic mRNA molecule binds via its capped 5′ terminus to the 40 S ribosomal subunit, which then moves along the mRNA molecule until stopped at the first initiator codon (AUG) flanked by the favored sequence $^{\mathrm{A}}_{\mathrm{G}}\mathrm{XX}\underline{\mathrm{AUG}}\mathrm{G}$. The 60 S ribosomal subunit now binds also, together with methionyl transfer RNA and various initiation factors. Translation then proceeds. Despite the fact that mRNA is transcribed from all the monocistronic dsRNA reovirus genes at the same rate, there are pronounced differences in the amounts of each protein made, indicating the existence of some mechanism of regulation at the level of translation.

Early Proteins

As mentioned earlier, the proteins translated from the transcripts of early viral genes of DNA viruses include enzymes and other proteins required for the replication of viral nucleic acid, as well as proteins that regulate host cell RNA and protein synthesis. The function of most early viral proteins however is still unknown. Several have important regulatory functions, in controlling the transcription of the viral genome.

Late Proteins

The late viral proteins are translated from late mRNA, most of which is transcribed only following the replication of the viral nucleic acid. Most of the late proteins are structural proteins which are intended for incorporation into new virions (but they are often made in considerable excess). Some of these double as regulatory proteins that modulate the transcription—and, perhaps in some instances, the translation—of cellular or early viral genes.

Regulation

As we have seen, the temporal order and amount of synthesis of particular proteins encoded by the genome of DNA viruses are regulated mainly at the level of transcription. With RNA viruses also there is a tendency for nonstructural proteins to be made early and structural proteins later, but the control is generally not rigorous and occurs at the level of translation. For instance, in the case of caliciviruses, coronaviruses, and togaviruses, only the 5′ end of the + stranded vRNA, which codes for the nonstructural proteins, is translated immediately after infection. Hence the RNA polymerase is made promptly and the production of complementary (−)RNA can commence. This in turn serves as the template for transcription of subgenomic (+)mRNA corresponding to the 3′ end of vRNA, from which is translated the structural proteins required in abundance later in infection.

Posttranslational Cleavage of Polyproteins

In the special case of the plus-stranded picornaviruses, the polycistronic vRNA is translated directly into a single *polyprotein* which carries proteinase activity. This virus-coded protease cleaves the polypeptide in a defined sequence into smaller proteins. The first cleavage steps are carried out while the polyprotein is still nascent on the polyribosome. Some of the intermediates exist only ephemerally; others have important functions but are subsequently cleaved to produce smaller proteins with other functions.

Posttranslational cleavage is a less prominent but important feature of the production of particular proteins of several other RNA virus families. In the case of the togaviruses and caliciviruses, polyproteins corresponding to only part, but a large part, of the genome are translated from polycistronic mRNA, then cleaved. With several other families, cleavage of particular proteins late in the multiplication cycle is an essential prerequisite for the maturation of an infectious virion.

Migration of Proteins

Newly synthesized viral proteins must migrate to the various sites in the cell where they are needed, e.g., back into the nucleus in the case of those viruses that replicate there. Whether this occurs solely by random diffusion or by some sort of active or guided transport, perhaps involving the cytoskeleton, is unknown. In the special case of glycoproteins, the polypeptide backbone is translated on membrane-bound ribosomes, i.e., in rough endoplasmic reticulum; various co- and posttranslational modifications, including acylation, proteolytic cleavage, and addition and subtraction of sugars, occur sequentially as the protein moves in vesicles to the Golgi apparatus then to the plasma membrane (see below).

REPLICATION OF VIRAL NUCLEIC ACID

DNA Replication

The papovavirus genome, with its associated cellular histones, morphologically resembles chromatin. Its replication also closely mimics the replication of eukaryotic chromosomal DNA and utilizes host cell enzymes, probably including DNA polymerase α. An early viral protein, the large-T antigen, binds to three adjacent sites in the "control" region of the viral genome, initiating replication of the DNA. Thereafter, no other virus-coded protein appears to be required for viral DNA synthesis. Replication of this circular dsDNA commences from a unique palindromic origin on the viral chromosome and proceeds simultaneously in both directions at the same rate (Fig. 3-7). At each growing fork, both continuous and discontinuous DNA synthesis occur (on leading and lagging strands, respectively). As in the replication of mammalian DNA ,the discontinuous synthesis of the lagging strand involves repeated synthesis of short oligoribonucleotide primers, which in turn initiate short nascent strands of DNA; these *Okazaki fragments* are then covalently joined to form the growing nascent strand.

The replication of adenovirus DNA is somewhat simpler, but quite different. Adenovirus DNA is linear, the 5′ ends of the two strands being mirror images of one another (terminally repeated inverted sequences) and being covalently linked to a protein molecule. Uniquely, the primer for adenoviral DNA synthesis is not another nucleic acid but a precursor to this virus-coded protein, adenovirus preterminal protein (pTP). DNA replication proceeds from both ends, continuously but asynchronously, in a 5′ to 3′ direction, using a virus-coded DNA polymerase, and does not require the synthesis of Okazaki fragments.

RNA Replication

Replication of RNA is, as far as we know, a phenomenon unique to viruses. Transcription of RNA from an RNA template requires a type of enzyme not found in mammalian cells, namely an RNA-dependent RNA polymerase. Such enzymes must therefore be virus coded. It is not yet certain whether the polymerase required to transcribe (+)RNA from a (−)RNA template is necessarily different from that needed to make (−)RNA from (+)RNA. Both are essential because the replication of viral RNA (vRNA) requires first the synthesis of cRNA, which then serves as a template for making more vRNA. Where the vRNA is of negative polarity the cRNA is of positive polarity and the RNA polymerase in-

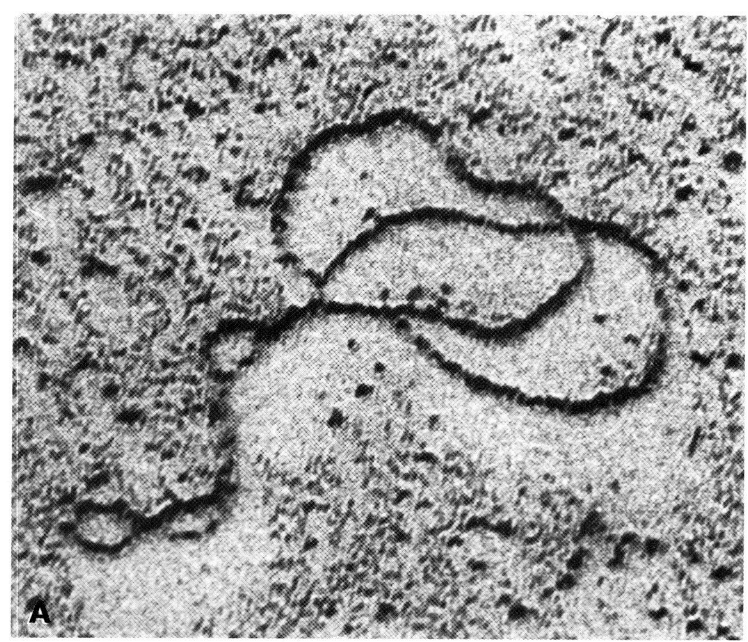

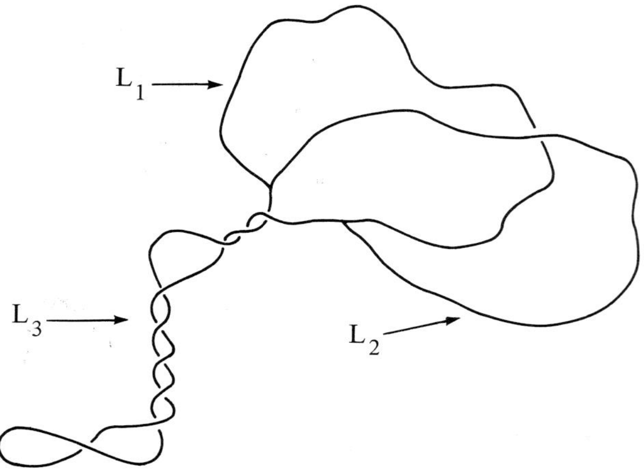

FIG. 3-7. *Replication of circular viral DNA. (A) Electron micrograph of a replicating molecule of SV40 DNA. Magnification: 1.5 × 10^5. In an interpretative drawing of the molecule (B) the two replicating branches are designated L_1 and L_2. The superhelical unreplicated section is designated L_3. [From E. D. Sebring* et al., J. Virol. **8**, *478 (1971).]*

volved is the same transcriptase as is responsible for transcription of mRNA. Note however that, whereas the primary transcript from such (−)vRNA is subsequently cleaved (in most cases) to produce mRNA, some must remain uncleaved to serve as a full-length template for vRNA synthesis.

In the case of the plus-stranded RNA viruses the cRNA is minus-stranded; the polymerase responsible for making vRNA from the cRNA template is sometimes called the *replicase,* although this laboratory jargon can cause confusion. Several replicase molecules can operate on a single cRNA template to transcribe a number of molecules of vRNA simultaneously. The resulting structure, known as the *replicative intermediate* (RI) is therefore partially double-stranded, with single-stranded tails (Fig. 3-8). Initiation of replication of picornavirus and calicivirus RNA, like that of adenovirus DNA, requires a protein, rather than a ribonucleoside triphosphate, as primer. This small protein, VPg, is covalently attached to the 5′ terminus of nascent (+) and (−)RNA strands, as well as vRNA, but not mRNA.

Little is known about what determines whether a given picornavirus plus-stranded RNA molecule will be directed (1) to a *replication complex,* bound to smooth endoplasmic reticulum, where it serves as a template for transcription by RNA-dependent RNA polymerase into minus-stranded cRNA; or (2) to a polyribosome, free in the cytoplasm or on rough endoplasmic reticulum, where it serves as mRNA for translation into *polyprotein;* or (3) to a *procapsid,* with which it associates to form a virion.

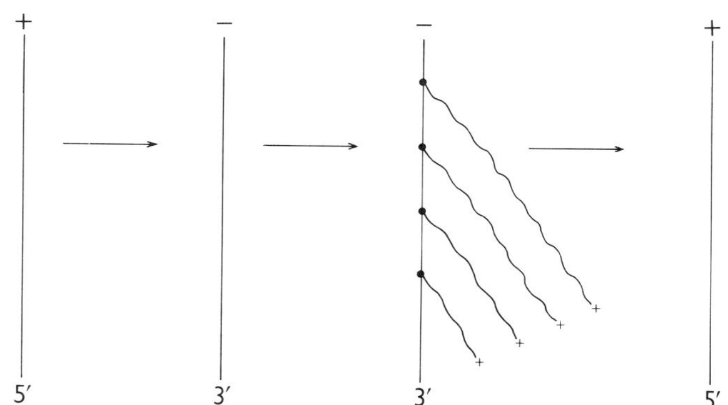

FIG. 3-8. *Mechanism of replication of single-stranded RNA. The "replicative intermediate" consists of several plus strands being transcribed simultaneously from one negative strand by separate molecules of RNA-dependent RNA polymerase.*

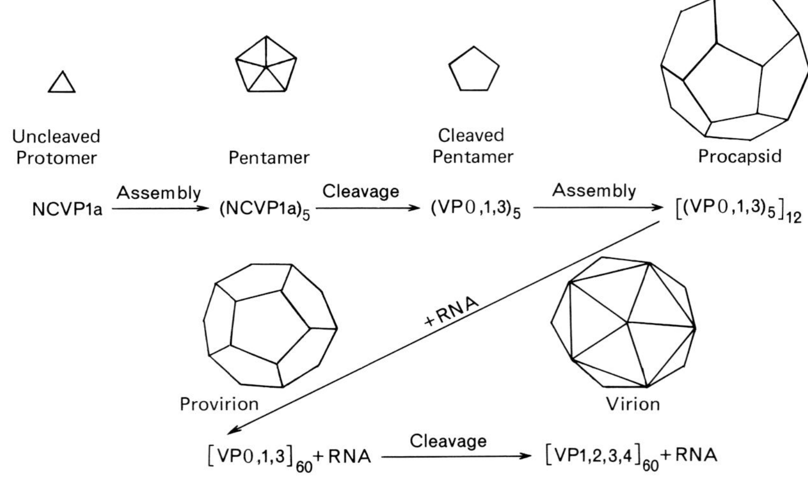

FIG. 3-9. *Assembly of picornavirus.*

ASSEMBLY (MORPHOGENESIS)

Assembly of Icosahedral Viruses

Structural proteins of simple icosahedral viruses can aggregate spontaneously to form capsomers, which in turn may assemble into empty procapsids. Somehow, the viral nucleic acid now enters this structure. Completion of the virion often involves proteolytic cleavage of one or more species of capsid protein.

One of the best studied examples is the picornavirus (Fig. 3-9). The precursor protein NCVP1a aggregates to form pentamers, within which each of the five NCVP1a molecules is then cleaved into VP0, VP1, and VP3. Twelve such pentamers now aggregate into a *procapsid* which associates with a viral RNA molecule to form a *provirion*. A final proteolytic event cleaves each of the VP0 molecules into VP2 and VP4, so that the mature virion is a dodecahedron with 60 capsomers, each of which is made up of one molecule each of VP1, 2, 3, and 4.

The mechanism of entry of viral nucleic acid into a preconstructed empty viral capsid has been elucidated for adenovirus. One end of the viral DNA molecule is characterized by a nucleotide sequence (*packing sequence*) which enables the DNA to enter the procapsid together with basic core proteins. Some of the virion proteins are then cleaved to make the mature particle.

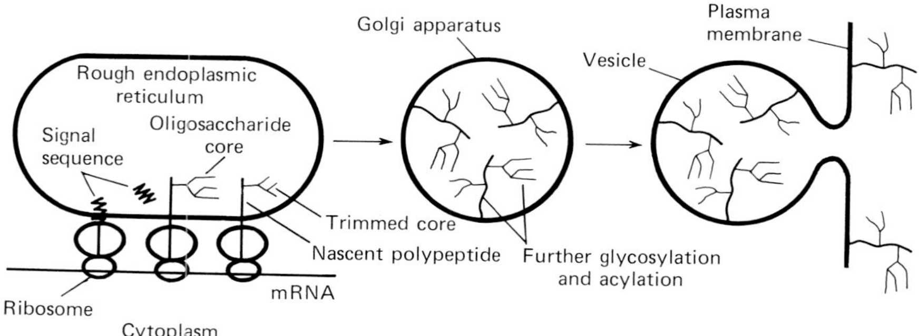

FIG. 3-10. *Glycosylation of viral protein.*

Morphogenesis of Enveloped Viruses

All mammalian viruses with helical nucleocapsids, as well as some with icosahedral nucleocapsids, acquire an envelope by *budding* through cellular membrane, usually plasma membrane. Since such envelopes always contain viral glycoproteins we will begin by discussing the mechanism of glycosylation of these proteins.

Glycosylation of Envelope Proteins. Much of our understanding of the glycosylation of viral proteins comes from studies with the rhabdovirus VSV and the togavirus SFV (Fig. 3-10). The fact that essentially comparable results have been obtained with ortho- and paramyxoviruses suggests that the essential steps are much the same for all enveloped viruses, hence again a composite picture can be presented here. Once more, we shall see that viruses exploit existing cellular pathways—in this case, those normally used for the synthesis and export of secretory and membrane glycoproteins.

The amino-terminal end of viral envelope proteins initially contains a sequence of 15–30 hydrophobic amino acids, known as the *signal sequence,* which characterizes the protein as one destined for insertion into membrane and/or export from the cell. The hydrophobicity of the signal sequence facilitates binding of the growing polypeptide chain to a receptor site on the cytoplasmic side of the rough endoplasmic reticulum (RER) and its passage through the lipid bilayer to the luminal side. A signal peptidase then removes the signal sequence. Oligosaccharides are added to the nascent polypeptide in the lumen of the RER by what now seems to be the standard mechanism for attaching sugars to asparagine, namely *en bloc* transfer of a mannose-rich core of preformed oligosaccharides from a lipid-linked intermediate, an oligosaccharide pyrophosphoryldolichol. Glucose residues are then removed by glycosidases, a process known as "trimming" of the core. At this stage the viral glycoprotein is transported from the RER to the Golgi apparatus, probably inside a coated vesicle. Here the core carbohydrate is further modified by the removal of several mannose residues, and the addition of further *N*-acetylglucosamine, plus galactose, then the terminal sugars, sialic acid and fucose. The completed oligosaccharide side chains are a mixture of the simple (high mannose) and the complex types. Also while in the Golgi apparatus the glycoprotein may become *acylated,* i.e., fatty acids, such as methyl palmitate, are covalently attached to the hydrophobic membrane-attachment end of the molecule. Another coated vesicle now transports the glycoprotein to the plasma membrane, in the case of viruses that bud from that site.

Cleavage of Envelope Proteins. At about this time a cellular protease usually cleaves the envelope protein into two polypeptide chains, which nevertheless remain covalently linked together by disulfide bonds. This final cleavage step is analogous to the maturation of many prohormones and blood proproteins. While it is not essential for the release of new virions (indeed, is not effected by certain types of host cell), it is absolutely essential for the production of infectious virions—certainly in the case of the orthomyxoviruses, which require cleavage of the hemagglutinin and the paramyxoviruses, which require cleavage of both the HN (hemagglutinin/neuraminidase) and the F (fusion) glycoprotein. Following fusion of the coated vesicle with the plasma membrane, the hydrophilic N-terminal end of the amphipathic glycoprotein now finds itself projecting from the external surface of the membrane, while the hydrophobic domain near the C-terminal end remains anchored in the lipid bilayer.

Transport of Glycoproteins. One fascinating aspect of the transport of glycoproteins around the cell in coated vesicles is the question of what determines the destination of particular glycoproteins. Different viruses bud from different domains in the plasma membrane of polarized epithelial cells, e.g., some from the apical and others from the basolateral surface, while yet others bud from intracytoplasmic smooth ER or from nuclear membrane. Presumably some feature of the molecule serves as a "postal address" ("zip code").

Budding. Budding is a form of exocytosis (Fig. 3-11). The process begins with the insertion of the newly completed viral glycoprotein into membrane. Because proteins are free to move laterally in the "sea of

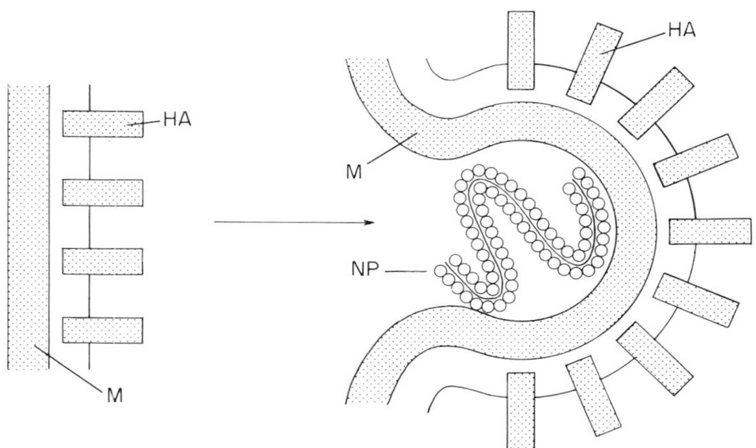

FIG. 3-11. *Budding of virus from plasma membrane. HA, Hemagglutinin; M, matrix protein; NP, nucleoprotein.*

lipid" that constitutes the plasma membrane, cellular proteins are displaced from the patch of membrane into which viral glycoproteins have been inserted. It is not known for certain whether there is any selection of particular lipids for incorporation into the viral envelope, but, to all intents and purposes, the mix of phospholipids, glycolipids, and cholesterol is essentially the same as that characterizing the membrane of that particular type of host cell. The monomeric, cleaved viral glycoprotein molecules associate into oligomers, to form the typical rod-shaped *peplomer* or *spike* with a prominent hydrophilic domain projecting from the external side of the membrane, a hydrophobic domain spanning the membrane near the C-terminal end, and a short hydrophilic sequence right at the C-terminus which projects slightly into the cytoplasm. In the special case of the icosahedral togaviruses (Plate 3-4A), each protein molecule of the viral nucleocapsid binds directly to the C-terminal end of a glycoprotein oligomer. Multivalent attachment of numerous spikes, each to an underlying molecule on the surface of the icosahedron, moulds the membrane around the nucleocapsid, forcing it to bulge progressively outward until finally the nucleocapsid is completely enclosed in a tightly fitting envelope and the new virion buds off. The more typical enveloped virus has a helical nucleocapsid, which does not bind directly to viral glycoprotein but to a *matrix (M) protein* which has first attached to the cytoplasmic side of the membrane beneath patches of

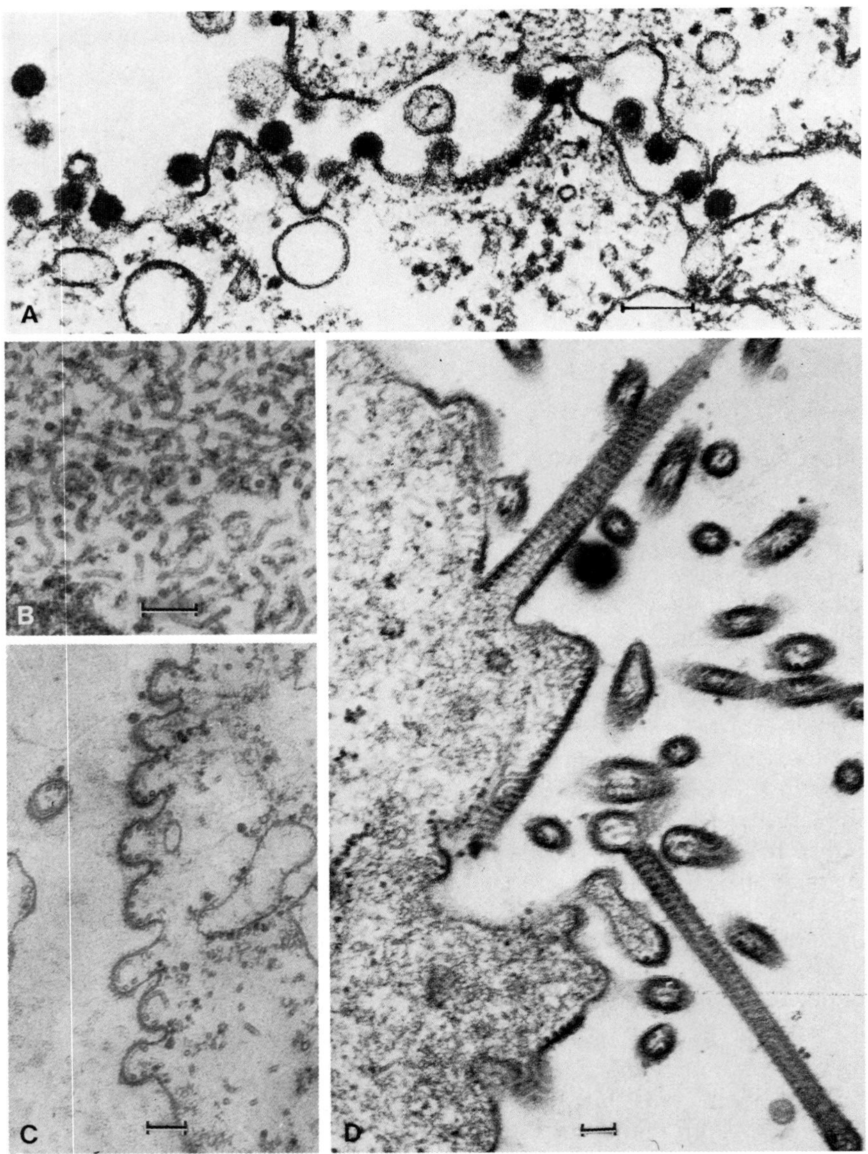

PLATE 3-4. *Viruses budding from the plasma membrane. (A) Togavirus. (B) Accumulation of paramyxovirus SV5 nucleocapsids. (C) and (D) Budding of SV5 from the plasma membrane, with some filamentous forms (bars = 100 nm). [A, courtesy Dr. A. Helenius; B, C, and D, from R. W. Compans* et al., Virology **30,** *411 (1966), courtesy Dr. P. W. Choppin.]*

aggregated viral glycoprotein (Fig. 3-11; Plate 3-4B–D). The putative role of the cytoskeleton in the process of budding is still to be established.

Not all enveloped viruses bud from plasma membrane. Coronaviruses and bunyaviruses bud from rough endoplasmic reticulum and the Golgi apparatus. Uniquely, the envelope of the icosahedral DNA herpesviruses is acquired by budding through the inner nuclear membrane; the virus then is thought to pass directly from the space between the two nuclear membranes to the exterior of the cell via cytoplasmic channels. The envelope of the complex DNA poxviruses is not acquired by a typical budding process, but by *de novo* synthesis of an inner "membrane" followed by acquisition of a true envelope during passage through the Golgi apparatus to the plasma membrane.

INHIBITION OF SYNTHESIS OF CELLULAR MACROMOLECULES

Whereas some budding viruses are noncytocidal, nonbudding viruses generally kill the cell; indeed they presumably require disintegration of the cell in order to escape from it. Capsid proteins of icosahedral viruses are often responsible for inhibition of host protein synthesis. However, in general, the mechanism(s) of what is often known as *cell shutdown* are not clearly understood. In those cases where the inhibition of cellular protein synthesis develops gradually and only late in the viral multiplication cycle, it is possibly due simply to competition for ribosomes by the large excess of viral mRNA. Even when viral mRNA is not in excess, it may have a selective advantage in binding to ribosomes and initiating translation. Late shutdown of cellular protein synthesis by adenovirus is at least partially due to inhibition of the transport of cellular mRNA out of the nucleus, while certain herpesviruses actually bring about selective degradation of cellular mRNA. Furthermore, the gross distortion of the topological arrangement of DNA seen in the nucleus of cells late in the multiplication cycle of many viruses might well explain the decrease in transcription of cellular mRNA demonstrable at that time. Inhibition of cellular DNA synthesis is doubtless also a secondary consequence of inhibition of protein synthesis.

The role of virus-coded proteins in cell shutdown seems to differ from virus to virus. Capsid proteins of some viruses, e.g., adenoviruses, are toxic for cells and may be responsible for the cytocidal effects seen late in infection (or early, following infection at high multiplicity). Yet other viruses seem to code for proteins still to be identified, which selectively inhibit the translation of eukaryotic mRNA.

EPILOGUE

This attempt to produce a single integrated picture of viral multiplication has, of necessity, glossed over significant differences between individual viruses. It is impossible in just a few pages to do justice to the fascinating replication strategies of every one of the 18 virus families. Each is unique and each boasts its own vast literature. Readers interested in delving more deeply into the biochemistry of the multiplication of particular viruses are referred first to the relevant chapters in Part II of this volume, and then to the books and reviews listed under Further Reading.

FURTHER READING

General

Alberts, B., Bray, D., Lewis, J., Raff, M., Roberts, K., and Watson, J. D. (1983). "Molecular Biology of the Cell." Garland, New York and London.

Baltimore, D. (1971). Expression of animal virus genomes. *Bacteriol. Rev.* **35,** 235.

Bishop, D. H. L., and Compans, R. W., eds. (1984). "Nonsegmented Negative Strand Viruses: Paramyxoviruses and Rhabdoviruses." Academic Press, New York.

Compans, R. W., and Bishop, D. H. L., eds. (1984). "Segmented Negative Strand Viruses: Arenaviruses, Bunyaviruses, and Orthomyxoviruses." Academic Press, New York.

Darnell, J. E. (1982). Variety in the level of gene control in eukaryotic cells. *Nature (London)* **297,** 365.

Joklik, W. K. (1981). Structure and function of the reovirus genome. *Microbiol. Rev.* **45,** 483.

Klenk, H.-D., and Rott, R. (1980). Cotranslational and posttranslational processing of viral glycoproteins. *Curr. Top. Microbiol. Immunol.* **90,** 19.

Kozak, M. (1983). Comparison of initiation of protein synthesis in procaryotes, eucaryotes and organelles. *Microbiol. Rev.* **47,** 1.

Lonberg-Holm, K., and Philipson, L. (1981). "Animal Virus Receptors. Series B: Receptors and Recognition." Chapman & Hall, London.

McGeogh, D. J. (1981). Structural analysis of animal virus genomes. *J. Gen. Virol.* **55,** 1.

Maniatis, T., Fritsch, E. F., and Sambrook, J. (1982). "Molecular Cloning: A Laboratory Manual." Cold Spring Harbor Lab., Cold Spring Harbor, New York.

Shatkin, A. J., ed. (1981). "Initiation Signals in Viral Gene Expression." Springer-Verlag, Berlin and New York.

Simons, K., Garoff, H., and Helenius, A. (1982). How an animal virus gets into and out of its host cell. *Sci. Am.* **246,** 46.

Strauss, E. G., and Strauss, J. H. (1983). Replication strategies of the single-stranded RNA viruses of eukaryotes. *Curr. Top. Microbiol. Immunol.* **105,** 1.

Individual Families of Viruses

Berns, K. I., ed. (1983). "The Parvoviruses." Plenum, New York.

Bishop, D. H. L. (1979). "Rhabdoviruses," Vols. 1–4. CRC Press, West Palm Beach, Florida.

Bishop, D. H. L. (1985). Replication of arenaviruses and bunyaviruses. *In* "Virology" (B. N. Fields *et al.*, eds.), p. 1083. Raven Press, New York.

Bishop, D. H. L., and Shope, R. E. (1979). Bunyaviridae. *In* "Comprehensive Virology" (H. Fraenkel-Conrat and R. R. Wagner, eds.), Vol. 14, p. 1. Plenum, New York.

Choppin, P. W., and Compans, R. W. (1975). Reproduction of paramyxoviruses. *In* "Comprehensive Virology" (H. Fraenkel-Conrat and R. R. Wagner, eds.), Vol. 4, p. 95. Plenum, New York.

Dales, S., and Pogo, B. G. T. (1981). "Biology of Poxviruses," Virol. Monogr., Vol. 18. Springer-Verlag, Berlin and New York.

Garoff, H., Kondor-Koch, C., and Riedel, H. (1982). Structure and assembly of alphaviruses. *Curr. Top. Microbiol. Immunol.* **99,** 1.

Ginsberg, H. S., ed. (1984). "The Adenoviruses." Plenum, New York.

Joklik, W. K., ed. (1983). "The Reoviridae." Plenum, New York.

Lamb, R. A., and Choppin, P. W. (1983). The gene structure and replication of influenza virus. *Annu. Rev. Biochem.* **52,** 467.

Marion, P. L., and Robinson, W. S. (1983). Hepadna viruses: Hepatitis B and related viruses. *Curr. Top. Microbiol. Immunol.* **105,** 99.

Perez-Bercoff, R., ed. (1979). "The Molecular Biology of Picornaviruses." Plenum, New York.

Rawls, W. E., and Leung, W.-C. (1979). Arenaviruses. *In* "Comprehensive Virology" (H. Fraenkel-Conrat and R. R. Wagner, eds.), Vol. 14, p. 157. Plenum, New York.

Roizman, B., ed. (1982–1984). "The Herpesviruses," Vols. 1–4. Plenum, New York.

Schaffer, F. L. (1979). Caliciviruses. *In* "Comprehensive Virology" (H. Fraenkel-Conrat and R. R. Wagner, eds.), Vol. 14, p. 249. Plenum, New York.

Sturman, L. S., and Holmes, K. V. (1983). The molecular biology of coronaviruses. *Adv. Virus. Res.* **28,** 35.

Tooze, J. (1980). "Molecular Biology of Tumor Viruses: DNA Tumor Viruses." Cold Spring Harbor Lab., Cold Spring Harbor, New York.

Weiss, R. A., Teich, N., Varmus, H. E., and Coffin, J., eds. (1982). "RNA Tumor Viruses." Cold Spring Harbor Lab., Cold Spring Harbor, New York.

CHAPTER 4

Viral Genetics and Evolution

Introduction 91
Mutation 92
Abortive Infections and Defective Viruses 98
Genetic Interactions between Viruses 98
Interactions between Viral Gene Products 102
Genetic Engineering 104
The Genetic Basis of Virulence 108
Genetic Variation in Nature 112
Further Reading 118

INTRODUCTION

Animal virology developed largely from the need to control viral diseases in man and his domesticated animals. Since direct experimentation with the natural hosts is often expensive or impossible, much effort has been devoted to two ends: (1) adaptation of viruses of medical and veterinary importance to small laboratory animals and cell culture, and (2) attenuation of viruses by serial passage in an unnatural host to obtain live-virus vaccines. These processes of adaptation and attenuation operate via spontaneous mutation or, more rarely, recombination or reassortment, followed by selection. As is usual in science, useful practical results were obtained empirically long before the genetic or molecular mechanisms involved in these phenomena were understood. This is well illustrated by the development of an effective attenuated rabies vaccine by Pasteur a hundred years ago. Today, however, it is important to understand what these processes are and how they occur.

Following the spectacular advances in molecular genetics that emerged from the study of bacterial viruses, and later, tumor-producing viruses, efforts have been made to understand the genetic properties and processes of other animal viruses. This work has involved the selection of appropriate mutants, the construction of genetic maps by recombination and complementation tests, and study of the functions of the products of the genes in which mutations occur.

In the 1970s several discoveries were made which have revolutionized the study of genetics, including viral genetics.The exploitation of these discoveries resulted in the development of recombinant DNA technology, or what is popularly called genetic engineering. Genetic engineering has already greatly expanded our understanding of the structure and function of viral genomes; it promises great practical advances in the diagnosis, prevention, and treatment of viral infections.

MUTATION

The most important and universal changes in the nucleic acid sequences of viral genomes are due to mutations. In every viral infection of a human or animal host or cell culture, one or a small number of virus particles multiplies several million times, so that mutations occur during the course of every infection. Whether such mutations emerge in the genotype of virus perpetuated in nature and thereby are of practical importance depends upon two factors: (1) whether the resultant change in the polypeptide affords the mutant virus some selective advantage, or whether it is neutral or disadvantageous, and (2) whether the mutation happens to occur early or late in the course of the infection.

In the laboratory, reasonable genetic constancy of viral stocks (e.g., those used for making viral vaccines or retained as reference strains) is achieved by (1) isolating a clone, i.e., a population of virus particles originating from a single particle, usually by growth from a single plaque in cell culture, followed by replaquing, then, (2) growing seed stock from the pure clone under carefully controlled conditions, and (3) as far as practicable avoiding serial passage of the virus. If passage is unavoidable, a single plaque is again selected to grow up a new seed stock. Sometimes, however, mutations are so common that there is a significant chance that the selected clone will be a mutant virus rather the parental type.

Mutation Rates

Rates of mutation involving base substitutions (point mutations) are probably the same in DNA viruses (and in retroviruses, which replicate through a DNA intermediate) as they are in DNAs of prokaryotic and eukaryotic cells, since multiplication usually occurs in the nuclei of such cells and is subject to the same "proofreading" exonuclease error correction as operates in cells. Such errors are estimated to occur at a rate of 10^{-8} to 10^{-9} per incorporated nucleotide.

Point mutations in the third nucleotide of a triplet often do not result in an altered amino acid, because of coding redundancy, and certain other mutations are lethal, because they produce a nonsense or stop codon. A mutation that is neutral or deleterious in one host may provide a positive selective advantage in a different host.

RNA is not the repository for genetic information in eukaryotic cells, hence there is no a priori reason why they should contain a "proofreading" mechanism for RNA, and so far no such exonucleases have been discovered. Evidence from studies of several different RNA viruses suggests that mutation rates of viral RNA are very much higher than with DNA, perhaps as high as 10^{-6} per nucleotide duplication, or 10^{-2} per molecular duplication in a 10-kb RNA virus. Of course, most of these base substitutions would be deleterious and the genomes containing them would be lost. Even so, there is growing evidence that mutational changes can be incorporated in the genome of RNA viruses at a very rapid rate. For example, an outbreak of poliomyelitis in 1978–1979 was traced from the Netherlands to Canada and then to the United States; oligonucleotide mapping of the RNAs obtained from successive isolates of the virus showed that over a period of 13 months of epidemic transmission about 100 base changes had become fixed in the poliovirus genome.

Classification of Mutations

Mutations can be classified according to the type of change in the nucleic acid, which may be nucleotide substitutions, small deletions or large deletions (Table 4-1). The physiological effects of mutations depend not only on the kind and location of the mutation but also on the activity of other genes. The phenotypic expression of a mutation in one gene may be reversed not only by a back-mutation in the substituted nucleotide but, alternatively, by a suppressor mutation occurring elsewhere in the same, or even in a different gene. For example, temperature-sensitive mutants of influenza virus developed as potential avirulent live vaccines reverted to virulence by virtue of a quite independent suppressor mutation in an apparently unrelated gene, which by some unknown mechanism negated the effect of the original mutation.

Mutations can also be classified by their phenotypic expression—hence temperature-sensitive, cold-adapted, host-range, plaque-size,

TABLE 4-1
Classification of Mutations

TYPE	CHANGE IN NUCLEIC ACID	EFFECT ON POLYPEPTIDE SEQUENCE	EFFECT ON POLYPEPTIDE FUNCTION
Silent mutation	Nucleotide substitution	None (e.g., redundancy in third base of triplet code)	None
Missense mutation	Nucleotide substitution	Amino acid substitution	Variable (depends upon amino acid and its location in polypeptide)
Nonsense mutation	Nucleotide substitution (extragenic suppression may occur)	Premature termination	Usually lost
Frame-shift mutation	Small deletion (intragenic suppression common)	Premature termination	Usually lost
Deletion mutation	Large deletion (do not revert)	Premature termination	Lost

etc., mutants have been described. Each of these kinds of mutants has been used for the analysis of viral functions, temperature-sensitive mutants being particularly useful (see below). Cold-adapted and temperature-sensitive mutants have been extensively used in attempts to produce attenuated live-virus vaccines (see Chapter 10). Mutations affecting antigenic sites (epitopes) on virion surface proteins may be strongly selected for when viruses multiply in the presence of antibody, and are of importance both in persistently infected animals (see Chapter 7) and epidemiologically, as in human influenza (see below). Space does not permit a description of all types of mutant, but two that are of particular importance in laboratory studies need to be described: temperature-sensitive conditional lethal mutants and defective interfering particles.

Conditional lethal mutants are produced by mutations which so affect a virus that it cannot grow under certain conditions, determined by the experimenter, but can replicate under "normal" conditions. Their importance is that a single selective test can be used to obtain and analyze mutants in which the lesion may be present in any one of several different genes. The conditional lethal mutants most commonly used by animal virologists are those showing temperature sensitivity of viral development, the so-called *temperature-sensitive mutants,* in which the selective condition used is a high temperature of incubation of the infected cells. Substitution of a nucleotide in the nucleic acid leads to an amino acid substitution in the resulting polypeptide, which weakens the internal bonding of the protein such that its structure, hence its function, though effectively normal at the permissive temperature, is radically changed by a rise in temperature of a few degrees.

Defective interfering (DI) mutants have been demonstrated in all families of RNA viruses and in some DNA viruses also. They occur when viruses are passed at high multiplicity, and the ratio of DI to infectious particles increases dramatically on serial passage of the progeny. Their common properties are that they are defective (i.e., they cannot multiply alone, but can in the presence of a *helper virus*), and they decrease the yield of wild-type virus (*interference*) but are themselves enriched in cells coinfected with DI and wild-type virus.

All DI particles that have been studied are primarily deletion mutants. In the case of influenza virus and other viruses with segmented genomes (e.g., reovirus), the defective virions lack one or more of the largest viral genes and contain instead small segments consisting of an incomplete portion of the missing gene(s). Similarly, in the case of viruses with a nonsegmented genome, the DI particles are found to contain RNA which is greatly abbreviated in length—as little as one-third of the original sequence in the *T (truncated) particles* of vesicular stomatitis virus (a rhabdovirus). Morphologically, DI particles resemble the homologous parent virions, sometimes being smaller, but having a comparable envelope or capsid. Sequencing studies have revealed a great diversity of structural rearrangements, in addition to simple deletions, in the RNA of DI particles.

The tendency of DI particles to increase preferentially with serial passage in cultured cells is explained as follows. The shortened RNA molecules that characterize DI particles (1) require less time to be replicated, (2) are less often diverted to serve as templates for transcription of mRNA, and (3) have enhanced affinity for the viral replicase, giving them a competitive advantage over their full-length counterparts from infectious virus. Hence, it can be seen why DI particles interfere with the replication of the parental virus with progressively greater efficiency on passage.

Mutagenesis

Mutations occur naturally because of chance errors in replication, the occurrence of which is probably influenced by natural background ionizing radiation. Their frequency can be enhanced by treatment of virions or isolated viral nucleic acid with physical agents like UV or X irradiation or with chemicals such as nitrous acid, hydroxylamine, ethylmethyl sulfonate, or nitrosoguanidine. Base analogs, such as 5-fluorouracil or 5-bromodeoxyuridine, are mutagenic only when virus is grown in their presence because to exert their mutagenic effect they need to be incorporated into the viral nucleic acid. *Chemical mutagenesis* of viruses is an important tool for laboratory studies in viral genetics, since it greatly increases the mutation rate and makes it possible to select mutants that would otherwise occur very rarely.

Directed mutagenesis, which enables the experimenter to introduce mutations at a selected site in a DNA molecule (a DNA genome or copy DNA transcribed from an RNA genome), is effected in the following way. A nick made with the appropriate restriction endonuclease is extended with exonuclease to expose a short single-stranded region, which is mutagenized using a single-strand-specific mutagen, and the gap is then repaired with DNA polymerase. This technique opens up new vistas in virology, for example (1) the function of individual genes and the proteins for which they code, or of particular regions of those molecules, can be explored; (2) mutations can be introduced into particular genes, e.g., those concerned with viral virulence, to produce avirulent mutants suitable for use as live vaccines.

Pleiotropism

Mutants selected for a certain phenotypic alteration are often found to be changed in other properties, reflecting the relevance of a single protein to several properties of the virus. For example, a single change in the hemagglutinin of influenza virus can affect properties such as hemagglutination, susceptibility to glycoprotein inhibitors, antigenicity, and ability to grow in particular hosts.

Pleiotropism is useful in certain applications of viral genetics, e.g., the selection of an attenuated variant for a vaccine. Initial selection can be based on *in vitro* tests, reserving the more expensive animal testing for final characterization of the selected strain. For instance, an attenuated poliovirus mutant could be selected on the basis of its ability to grow better at low than at high temperatures before being subjected to tests for lack of neurovirulence in primates.

Adaptation to New Hosts

Classically, the essential prerequisite for the study of an animal virus and the disease it causes was growth of the virus in a laboratory host. Some viruses produce typical signs the first time that the virus is inoculated, e.g., coxsackieviruses in infant mice. In other cases minimal signs of infection are observed initially, but after serial passage, sometimes prolonged, a lethal infection is regularly produced, as, for example, in the "adaptation" of poliovirus and dengue virus to rodents. Likewise, newly isolated viruses at first often fail to grow in certain kinds of cultured cells, but can be adapted by serial passage. Most modern virological research is performed with strains of virus adapted to grow rapidly to high yield and to produce plaques or cytopathic changes in continuous cell lines. A frequent byproduct of such adaptation to a new experimental host is the coincident attenuation of the virus for its original host.

Although growth in cell culture is a great advantage, it is not essential for the study of viral nucleic acid and proteins, as long as reasonable amounts of highly purified virions can be obtained. For example, the complete sequences of the DNAs of hepatitis B and human papilloma viruses have been determined although these viruses have not yet been grown in cultured cells. Once the viral DNA is available, recombinant DNA technology makes it possible to produce virtually unlimited quantities of any required nucleic acid sequence or any polypeptide for chemical and biological analysis.

ABORTIVE INFECTIONS AND DEFECTIVE VIRUSES

Viral infection is not always *productive,* i.e., does not always lead to the synthesis of infectious progeny; some or even all viral components may be synthesized but not assembled properly. This is called *abortive infection.* It may be due to the fact that certain types of cells are *nonpermissive,* i.e., lack some enzyme or other requirement essential for the multiplication of a particular virus, whether it be the wild-type or a host-dependent conditional lethal mutant.

Abortive infections due to genetically *defective viruses* are of considerable theoretical interest. For example, temperature-sensitive mutants are defective at high temperatures, but can be *rescued,* i.e., helped to yield infectious progeny, by coinfection of the cell with a helper virus, which is usually, but not necessarily, a related virus. Defective interfering particles have already been described.

Adeno-associated viruses (see Chapter 18), which can be recovered from the human throat, appear to be absolutely defective, in that they fail to multiply unless an adenovirus or herpesvirus is also multiplying in the same cells. In nature, they have been recovered only from persons concurrently infected with an adenovirus.

The extreme state of defectiveness is seen in cells transformed by certain DNA viruses (see Chapter 8). Here part or all of the viral genome is integrated with the cellular genome and replicates with it. The genetic information in part of the viral genome may be expressed, for virus-specific proteins are often synthesized, but ordinarily no infectious virus is produced.

GENETIC INTERACTIONS BETWEEN VIRUSES

When two different virions simultaneously infect the same cell, a variety of types of genetic interaction may occur between the newly synthesized nucleic acid molecules: recombination, reassortment, reactivation (when one of the virions has been inactivated), and marker rescue.

Recombination

Recombination involves the exchange of nucleic acid sequences between different but usually closely related viruses. The term is restricted to the rearrangement of sequences within a single nucleic acid molecule (*intramolecular recombination*) (Fig. 4-1A; Table 4-2). It occurs with all dsDNA viruses, presumably because of strand switching by the viral

TABLE 4-2

Nucleic Acid Interactions: Genetic Recombination and Reactivation[a]

PHENOMENON	PARENT 1	PARENT 2	PROGENY	COMMENT
Intramolecular recombination	A**B**C	AB**C**	ABC, A**BC**	With mutants of DNA viruses and *Picornaviridae*
	ABC	AST	ABT, ASC	With different strains of DNA viruses
	ABCD	XYZ	ABCZ	Loss of adenovirus genes (D), addition of SV40 genes (Z)
	ABC	123	12ABC3	Oncogenic viruses; integration of viral genes (ABC) into genome of cell (123)
	ABC	*onc*	A*onc*BC	Transduction of cellular oncogene by retroviruses
Gene reassortment	A/**B**/C	A/B/**C**	A/**B**/**C**	RNA viruses with segmented genomes
	A/B/C	A/S/T	A/B/T, A/S/C	
Cross-reactivation				
Between UV-inactivated virus and virus of a different but related strain	A̸B̸C	AST	ASC	Rescue of gene C from inactivated parent, by intramolecular recombination or gene reassortment
	A̸/B̸/C	A/S/T	A/S/C	
Multiplicity reactivation				
Between virions of same virus inactivated in different genes	ABC̸	AB̸C	ABC	Recombination or reassortment of genes B and C to yield viable virus
	A/B/C̸	A/B̸/C	A/B/C	

[a] A, etc., active viral genes; 1, etc., active cellular genes; **B**, etc. (boldface) mutant genes; A̸, etc. (slashed) inactivated genes; *onc*, cellular oncogene; ABC, continuous linear genome; A/B/C, segmented genome.

DNA polymerase. Among RNA viruses it has been demonstrated only in the *Picornaviridae* (with poliovirus and foot-and-mouth disease virus).

In rare cases, recombination occurs between unrelated viruses; the best example is between SV40 (a papovavirus) and adenoviruses. Both SV40 and adenovirus DNAs recombine readily with cellular DNA, so that it is perhaps not surprising to find that when rhesus monkey cells which harbor a persistent SV40 infection are superinfected with an adenovirus, not only does complementation occur (see below), the SV40 acting as a helper in an otherwise abortive adenovirus infection, but recombination occurs between SV40 and adenovirus DNAs to yield hybrid DNA within adenovirus capsids (Table 4-2).

Recombination is an essential part of the replication cycle of most retroviruses. Although the genome of these viruses is positive-strand RNA, replication does not occur until this is transcribed into DNA by the virion-associated reverse transcriptase and the resultant copy DNA is integrated into the cell's DNA. Furthermore, retroviruses may pick up *cellular oncogenes* by recombination; such oncogenes, with minor alteration, then become *viral oncogenes,* which confer the property of oncogenicity to the retrovirus (see Chapter 8). Recombination of viral and cellular DNAs is an essential part of the transformation process in cells transformed by adenoviruses, hepadnaviruses, and polyomaviruses (but not by papillomaviruses or EB virus, a herpesvirus) (see Chapter 8).

Reassortment

What was once called high frequency recombination, but is best described as *reassortment* (Fig. 4-1B, Table 4-2), occurs with viruses that have segmented genomes, whether these consist of 2 segments (*Arenaviridae*), 3 (*Bunyaviridae*), 8 (*Influenzavirus* A and B), or 10–11 (*Reoviridae*). In cells mixedly infected with two related viruses, there is an exchange of homologous segments with the production of different stable reassortants. Reassortment has been shown to occur in nature, and is an important source of genetic variability among viruses with segmented genomes.

Reactivation

The term *multiplicity reactivation* is applied to the production of infective virus by a cell infected with two or more virions of the same strain, each of which has suffered a lethal mutation in a different gene, e.g., after exposure to UV irradiation (Table 4-2). The phenomenon has been demonstrated with viruses of several families: for example, poxviruses, orthomyxoviruses, and reoviruses. Multiplicity reactivation could theoretically lead to the production of living virus if humans were to be

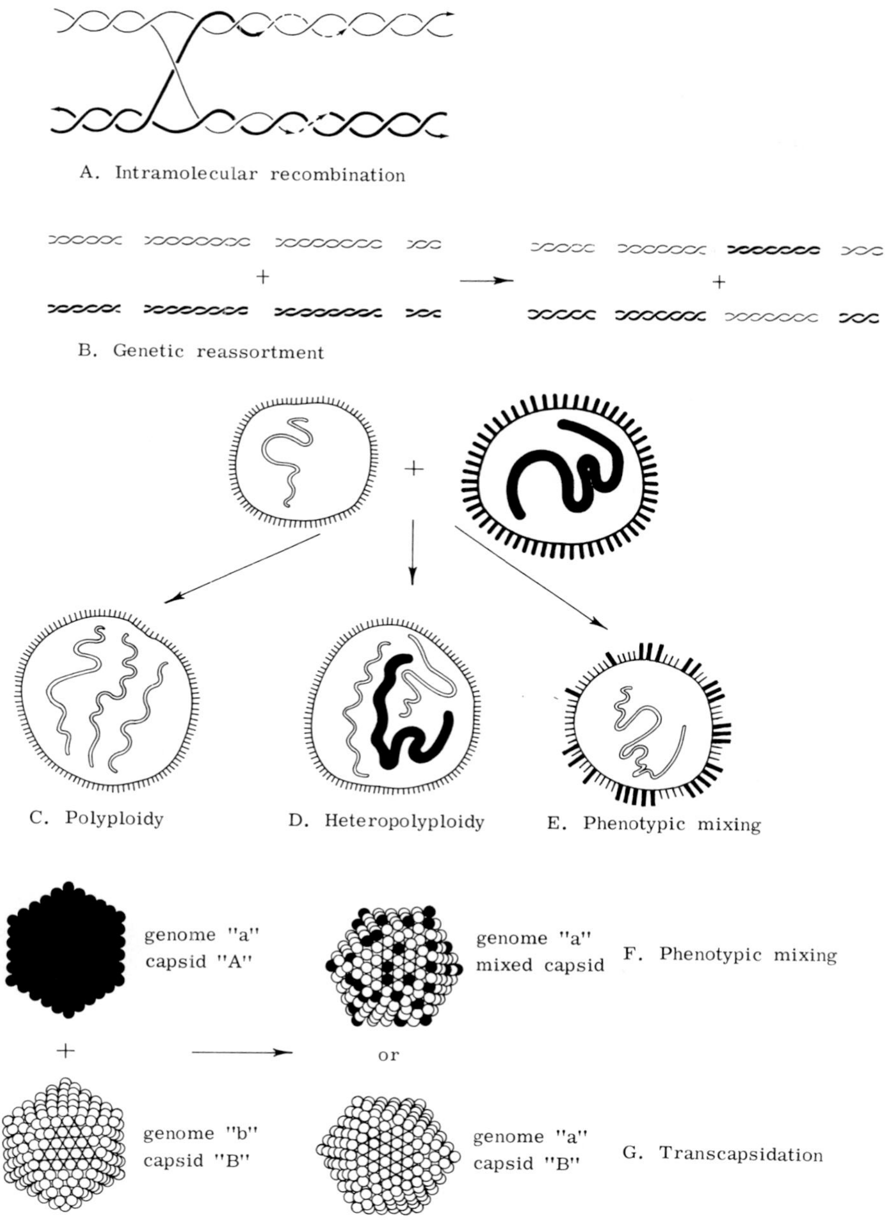

A. Intramolecular recombination

B. Genetic reassortment

C. Polyploidy

D. Heteropolyploidy

E. Phenotypic mixing

genome "a"
capsid "A"

genome "a"
mixed capsid

F. Phenotypic mixing

or

genome "b"
capsid "B"

genome "a"
capsid "B"

G. Transcapsidation

FIG. 4-1. *Genetic recombination, polyploidy, phenotypic mixing, and transcapsidation. (A) Intramolecular recombination. (B) Reassortment of genome fragments, as in* Reoviridae *and* Orthomyxoviridae. *(C) Polyploidy, as seen in unmixed infections with* Paramyxoviridae. *(D) Heteropolyploidy, as may occur in mixed infections with* Paramyxoviridae *and other enveloped RNA viruses. (E–G) Phenotypic mixing: (E) with enveloped viruses; (F) viruses with icosahedral capsids; (G) extreme case of transcapsidation or genomic masking.*

inoculated with UV-irradiated vaccines; accordingly this method of inactivation is not used for vaccine production.

Cross-reactivation or *marker rescue* are terms used to describe genetic recombination between an active virus and an inactivated virus of a related but distinguishable genotype. By the same processes of exchange of nucleic acid described above (intramolecular recombination or genetic reassortment), some infectious virions are produced which contain active genes from both viruses (Table 4-2).

A special kind of marker rescue is now extensively used for correlating functional and physical maps of viral DNA. Called *fragment rescue,* it involves the introduction into a susceptible cell by *transfection* of a fragment of DNA containing a specified mutation, produced by recombinant DNA techniques (see below), together with the parental virus or its complete genome. Alternatively, ss mutant viral DNA molecules (e.g., of SV40) can be annealed with fragments of wild-type ss viral DNA and cells transfected under restrictive conditions. If the wild-type fragment corresponds to the site of the mutation in the intact mutant strand, DNA replication and segregation of daughter molecules lead to plaque production.

INTERACTIONS BETWEEN VIRAL GENE PRODUCTS

Complementation

Complementation is the term used to describe all cases in which interaction between viral gene products in multiply or mixedly infected cells results in an increased yield of one or both viruses (Table 4-3). Unlike genetic recombination, complementation does not involve the exchange of viral nucleic acid, but reflects the fact that one virus provides a gene product that the other cannot make, thus enabling the latter to multiply in the mixedly infected cell. The gene as a unit of function can be defined by complementation tests that determine whether two mutants, which may exhibit similar phenotypes, are defective in the same or different functions. In experimental virology, complementation between related viral mutants is useful as a preliminary step to genetic mapping, to sort out mutants into groups. Temperature-sensitive mutants, for example, can usually be complemented by other mutants of the same virus that are not temperature sensitive, and different temperature-sensitive mutants can be allocated to functional groups by testing which pairs of *ts* mutants can complement one another.

Complementation can occur between unrelated viruses, e.g., between an adenovirus or herpesvirus and adeno-associated virus, or between SV40 and an adenovirus in monkey cells.

TABLE 4-3

Gene Product (Protein) Interactions: Complementation and Phenotypic Mixing[a]

PHENOMENON	PARENT 1	PARENT 2	PROGENY	COMMENT
Complementation				
(a) Between conditional lethal mutants of the same virus (under restrictive conditions)	A **B** C ↓ ↓ ↓ a **b** c	A B **C** ↓ ↓ ↓ a b **c**	A**B**C, AB**C**, (ABC)	Reciprocal; both mutants rescued; sometimes recombination also
(b) Between defective virus and unrelated helper virus	A B̸ C ↓ ↓ ↓ a c	B Y Z ↓ ↓ ↓ b y z	AB̸C and BYZ	Defective virus is rescued by gene product "b" of helper BYZ
Phenotypic mixing				
(a) Enveloped viruses	$\frac{\underline{\text{ABC}}}{\text{a}}$	$\frac{\underline{\text{XYZ}}}{\text{x}}$	$\frac{\underline{\text{ABC}}}{\text{ax}}$, $\frac{\underline{\text{XYZ}}}{\text{ax}}$	Mixed peplomers in envelopes, genomes unaltered; also parental phenotypes
(b) Nonenveloped viruses	$\frac{\underline{\text{ABC}}}{\text{a}}$	$\frac{\underline{\text{XYZ}}}{\text{x}}$	$\frac{\underline{\text{ABC}}}{\text{ax}}$, $\frac{\underline{\text{XYZ}}}{\text{ax}}$	Mixed capsomers in capsids
	$\frac{\underline{\text{ABC}}}{\text{a}}$	$\frac{\underline{\text{XYZ}}}{\text{x}}$	$\frac{\underline{\text{ABC}}}{\text{x}}$, $\frac{\underline{\text{XYZ}}}{\text{a}}$	Genome of one parent with capsid of the other (transcapsidation). Not always reciprocal

[a] A, etc., active viral genes; **B,** etc., mutant genes; B̸, defective gene B; a, etc., product of gene A, etc.; **b,** etc., product of mutant gene **B**; $\underline{\text{ABC}}$, genome; ax, proteins in envelope (or capsid).

Phenotypic Mixing

Following mixed infection by two viruses which share certain common features, some of the progeny may acquire phenotypic characteristics from both parents, although their genotype remains unchanged. For example, when cells are infected with antigenically different strains of influenza virus, or with influenza virus and a paramyxovirus, the envelopes of some of the progeny particles contain viral antigens characteristic of both of the parental viruses. However, each virion contains the nucleic acid of only one of the parents, so that on passage it will produce only virions resembling that parent (Fig. 4-1, Table 4-3).

With nonenveloped viruses phenotypic mixing can take the form of what has been called *transcapsidation* (Fig. 4-1G), in which there is very extensive or even complete exchange of capsids. For example, poliovirus nucleic acid may be enclosed within a coxsackievirus capsid, or adenovirus type 7 genome may be enclosed within an adenovirus type 2 capsid.

Polyploidy

With the exception of the retroviruses, which are diploid, all viruses of vertebrates are haploid, i.e., they contain only a single copy of each gene. However, among viruses that mature by budding from cellular membranes it is commonly found that several nucleocapsids (and thus genomes) are enclosed within a single envelope (Fig. 4-1C). If cells are coinfected with recognizably different strains of such viruses (for example, most paramyxoviruses), many of the progeny particles are heteropolyploid (Fig. 4-1D) and they may also have phenotypically mixed envelope antigens.

GENETIC ENGINEERING

Restriction Endonucleases

During the 1960s, it was shown that a phenomenon that had been called "restriction" in bacteriophages, whereby certain phages failed to multiply in particular species of bacteria, was due to the rapid degradation of the injected viral DNA by specific bacterial endonucleases, which were therefore called *restriction endonucleases*. Subsequently several hundred of these enzymes have been identified and purified from various bacteria. Each recognizes a distinct short sequence of nucleotides, generally of either four or six nucleotide pairs. Depending upon the frequency with which such sequences occur in any particular DNA molecule, different endonucleases can be used to cleave the molecules into fragments of various sizes, which can be separated from each other by gel electrophoresis. Restriction endonucleases have been used to compare species of orthopoxviruses and strains of herpesviruses. With herpesviruses, the resulting maps can be used to follow chains of transmission, since they are characteristic of strains of virus that are otherwise indistinguishable. However, the greatest importance of restriction endonucleases lies in the construction of recombinant DNA molecules.

Recombinant DNA

The identification of the recognition sequences and cleavage sites of the restriction endonucleases, and the development of knowledge of other enzymes involved in DNA synthesis (ligases, transferases), opened up the possibility of deliberately introducing specific foreign DNA sequences into other DNA molecules. When these recombinant molecules multiply, there is a corresponding amplification of the foreign DNA; the process is called *molecular cloning*. When the multiplying recombinant molecules are placed in a situation where their genetic information can be expressed, the polypeptide specified by the foreign DNA is produced. These results are usually achieved by incorporating the foreign DNA into a bacteriophage or a plasmid (Fig. 4-2), which is called the *vector*, and infecting bacteria with this vector. Vectors that multiply in bacteria, yeasts, animal cells, and even in intact animals have now been developed, e.g., retroviruses are being contemplated as vectors for the introduction of human genes into children with congenital biochemical defects. The development of recombinant DNA methodology was facilitated by great improvements in the techniques of sequencing DNA and the perfection of methods of making copy DNA (cDNA), from either viral RNA or mRNA (Fig. 4-3).

Uses of Genetic Engineering

Genetic engineering made possible studies of animal virus genomes that could not be contemplated before it became feasible to produce, at will, large quantities of selected fragments of viral nucleic acid.

1. Complete sequencing of the genome of many DNA viruses and of critical parts of the genomes of very large DNA viruses.
2. Complete sequencing of cDNA corresponding to the genome of many RNA viruses.
3. Recognition of the number and sequence of portions of viral DNA that are integrated into the DNA of transformed cells and cancers.

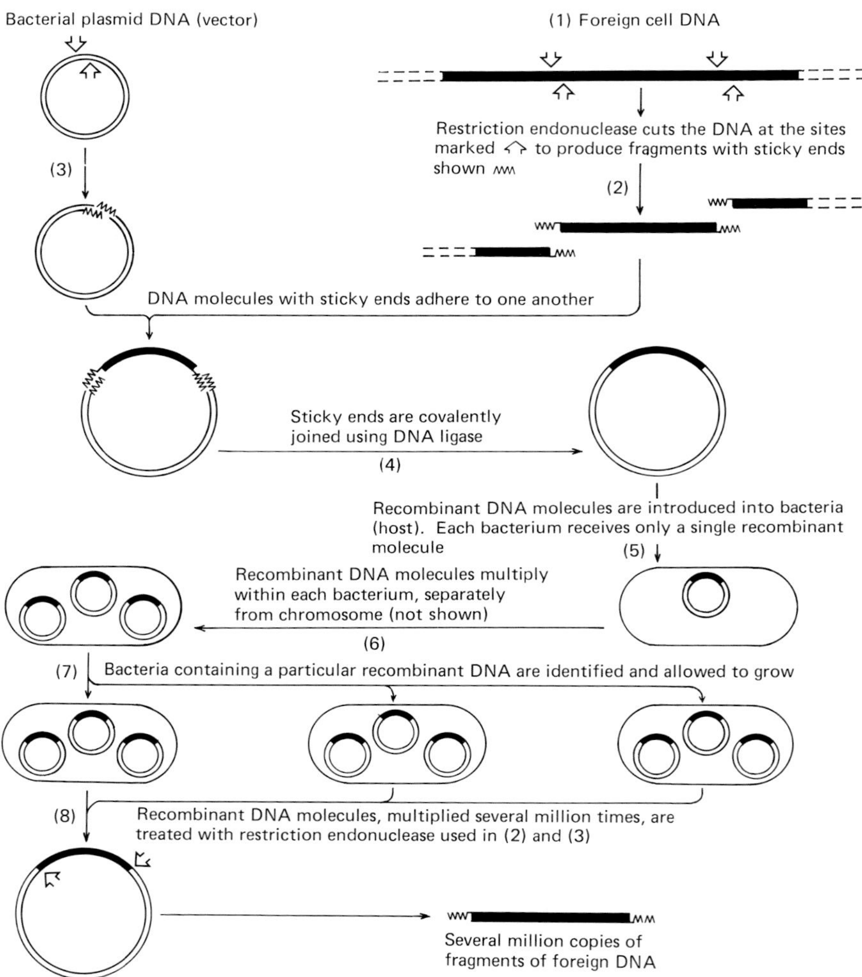

FIG. 4-2. *The steps in obtaining recombinant DNA. In parallel, foreign DNA from the donor cell (1) is cut into fragments by a selected restriction endonuclease (2) and the circular DNA molecule of the vector (plasmid or bacteriophage λ) is cut with the same endonuclease (3). The foreign DNA is then incorporated into the vector DNA molecule, which is thus circularized again, by a ligase (4). The vector is then introduced into the host bacterium by transformation (for plasmids) or infection (for bacteriophage λ) (5). Multiplication of the vector produces many copies of foreign DNA (6). Bacteria containing the desired foreign DNA are identified and allowed to grow (7). Vector DNA is isolated from the bacteria and the foreign DNA excised (8) by use of the same restriction endonuclease as was employed in steps (2) and (3). In this way a specified gene may be multiplied several millionfold.*

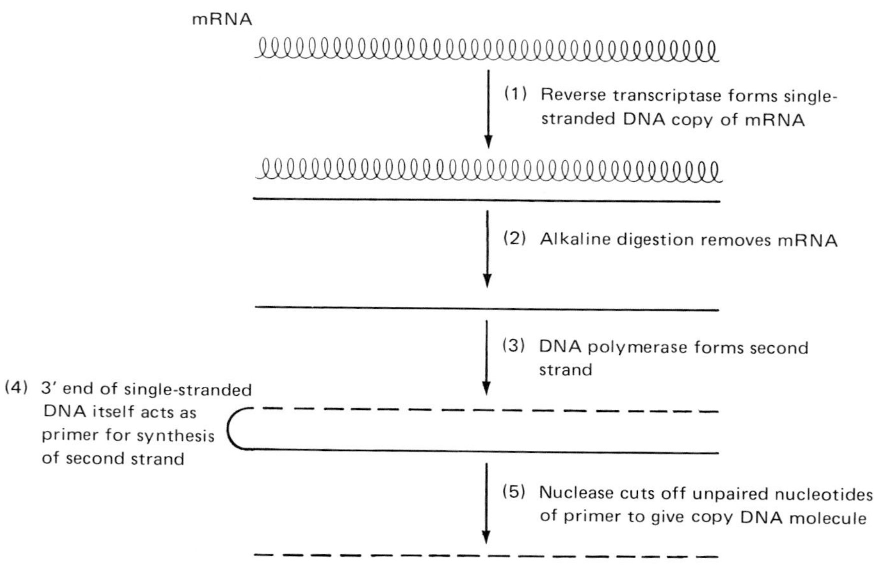

FIG. 4-3. *Preparation of copy DNA. The appropriate viral RNA or mRNA is exposed to reverse transcriptase to produce a ssDNA copy (1). The RNA is removed by alkaline digestion (2). The ssDNA is treated with DNA polymerase to form a second single strand of DNA (3), the 3′ end of the first DNA strand acting as a primer (4). A nuclease cuts off the unpaired nucleotides of the "primer" to give a "blunt-ended" copy DNA molecule (5), which is treated with terminal transferase to provide the molecule with "sticky ends."*

4. Marker rescue by transfection, as a method of genetic mapping.
5. Production of proteins coded by specific genes, in bacteria, yeasts, and animal cells or by cell-free translation.

Practical outcomes of genetic engineering of animal viruses include the development of nucleic acid probes for diagnosis by nucleic acid hybridization (see Chapter 12) and novel methods for the production of vaccines (Chapter 10) or interferons (Chapter 11).

Methodology for Obtaining a Viral Protein by Molecular Cloning

The desired gene, or a cDNAcopy of an RNA gene, is recombined with a suitable vector, such as a plasmid or phage. It is not essential to know the identity of the nucleotide sequence required; one can simply clone a random population of nucleic acid fragments, then screen for the gene product. This is commonly done by radioimmunoassay or autoradiography on the transformed bacterial clones, using a labeled monoclonal antibody. The particular recombinant plasmid found to carry the desired viral gene can then be manipulated further to optimize yield. Moreover, it may not even be necessary to produce the protein in its native form; commonly, the viral gene is inserted into the vector adjacent to all or part of the gene for a bacterial protein that is made in large amounts, causing the production of a "fused polypeptide" containing part of the bacterial protein and the viral protein in question. Its yield may be enhanced by a number of technical procedures such as the use of multicopy vectors and high efficiency promoters. However, if the protein is made in too high a concentration it may be toxic for the cell. Moreover, some polypeptides are unstable, and bacteria cannot glycosylate proteins, so that such products may be vulnerable to proteases. Ways of circumventing these and other technical problems are rapidly being developed.

More recently, cloning has been accomplished in eukaryotic cells, both yeasts and animal cells. The viral gene may first be incorporated, for convenience, into a bacterial plasmid, then into a "divalent" vector able to replicate autonomously in eukaryotes as well as prokaryotes. The small circular viral DNA molecules of bovine papilloma virus and SV40 are now widely used as cloning vectors in mammalian cells. A useful method for augmenting the expression of a cloned gene is cotransformation of an enzyme-deficient mutant cell line with the gene for the missing enzyme. For instance, if a viral gene is recombined into a suitable vector alongside the gene for dihydrofolate reductase (DHFR), then a $DHFR^-$ Chinese hamster ovary cell line transfected with this vector will, in the presence of methotrexate, produce a yield of the corresponding viral protein amounting to 1% of the cells' total output.

THE GENETIC BASIS OF VIRULENCE

Unraveling the genetic basis of virulence at the level of the individual human patient has long been one of the major goals of medical virology, and one of the most difficult to achieve, since many factors are involved (see Chapters 5, 6, and 7). With the recent advances in molecular genetics it has been possible to attack the problem in more meaningful ways.

The most detailed studies of this kind are those carried out with retroviruses and oncogenic DNA viruses to determine the genetic basis of cellular transformation (see Chapter 8). Experiments with ordinary viral infections are less advanced, but a start has been made. Viruses with segmented genomes have provided the best experimental material, since each segment of the genome of influenza virus and reovirus, for example, is in general equivalent to one gene, and genetic reassortment takes place with high frequency. In addition, a major advance has been achieved with poliovirus, using the vast experience of the live virus vaccine program, and a few suggestive results have been obtained with the large and complex virus of herpes simplex.

Influenza Virus

Experiments with influenza virus in mice have confirmed the view advanced by the pioneer in the field, Sir Macfarlane Burnet, that virulence is multigenic, i.e., that in general no one gene is responsible for virulence. Subsequent work established one absolute requirement for pathogenicity in the chicken and neurovirulence in the mouse, namely that only strains in which the hemagglutinin is cleaved in a broad spectrum of cells are pathogenic. However, when reassortants were made between the highly pathogenic fowl plague strain and avirulent strains of influenza virus, exchange of any of the eight RNA segments modified the pathogenicity. Further studies with a variety of reassortants showed that for each reassortant, an optimal combination of genes was selected which favored survival in nature and determined virulence.

Reovirus

Detailed analyses of virulence are being carried out with reassortant reoviruses in mice, based on the fact that the gene product of each of the 10 genes has been isolated and analysis of reassortant viruses has enabled the functions of many of them to be identified (Table 4-4). Three genes (S1, M2, and S4) encode the three polypeptides that are found on the outer capsid (Fig. 4-4). Each plays a role in determining virulence. Gene S1 specifies the hemagglutinin (protein σ1), which is located on the vertices of the icosahedron, and is responsible for cellular and tissue tropism. Gene M2 specifies polypeptide μ1C, which determines sensitivity of the virion to chymotrypsin and thus growth in the intestine. Gene S4 specifies polypeptide σ3, which inhibits protein and RNA synthesis in infected cells; mutations in this gene play a role in establishing persistent infection in cultured cells.

Poliovirus

The genome of poliovirus is a single molecule of ssRNA, whose entire nucleotide sequence has now been determined. Comparison of the nu-

TABLE 4-4

Correspondence between Reovirus Genes and Polypeptides and the Effects of Mutations of Genes for Capsid Polypeptides on Virulence[a]

GENE SEGMENT	POLYPEPTIDE	LOCATION IN VIRION	FUNCTION	RESULT OF ATTENUATING MUTATIONS
L1	λ3	Core		
L2	λ2	Core spike		
L3	λ1	Core		
M1	μ2	Core		
M2	μ1/μ1C	*Capsid*	A. Growth at mucosal surfaces	Decreased capacity to grow in intestine
			B. Immune induction	Loss of immune stimulation of suppressor T cells
			C. Growth in target tissue	Relative decrease in neurovirulence
M3	μNS	Nonstructural		
S1	σ1 (HA)	*Capsid*	Hemagglutinin; binds to receptors	
			A. Nerve cells	Loss of neurovirulence
			B. Lymphoid cells	Altered specificity in humoral and cytotoxic T cell responses
S2	σ2	Core		
S3	σNS	Nonstructural		
S4	σ3	*Capsid*	Inhibition of RNA and protein synthesis	Persistent infection Interference

[a] Based on B. N. Fields and M. I. Greene (1982). Reprinted by permission from *Nature* (*London*) **300,** 19.

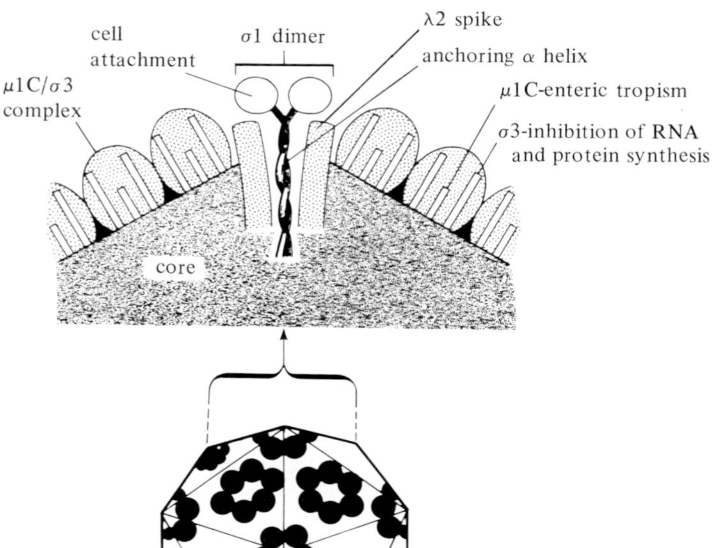

FIG. 4-4. *The location in the reovirus capsid of the polypeptides that play major roles in virulence (see Table 4-4). σ1 is located at the vertices of the icosahedron and consists of two components: a globular dimer at the surface, which is responsible for hemagglutination and cell attachment, and an α-helical region which anchors the hemagglutinin by interaction with the λ2 spike protein; μ1C and σ3 are associated with each other (in the ratio of one molecule of μ1C to two molecules of σ3) on the surfaces of the icosahedral capsid. [Modified from R. Basel-Duby, A. Jayasuriya, D. Chatterjee, N. Sonenberg, J. V. Maizel, Jr., and B. N. Fields (1985). Reprinted by permission from* Nature (London), **315,** *421. Copyright © 1985 Macmillan Journals Limited.]*

cleotide sequence of the poliovirus type 1 vaccine strain with that of the virulent parental strain (Mahoney) from which Sabin derived it showed that there were differences in 57 of the 7441 bases, scattered all over the genome. Twenty-one of these base substitutions resulted in amino acid changes, in several viral proteins. On the other hand, with poliovirus type 3 vaccine strain, which reverts to virulence with a frequency at least 10 times that of type 1 vaccine, there were only 10 nucleotide differences from the parental strain, of which only 3 resulted in amino acid changes, in 3 different proteins.

Herpes Simplex Virus

The herpesviruses have large genomes that do not readily lend themselves to sequencing, but since they cause a large number of important human diseases, considerable resources have been devoted to the analysis of the relationship between various restriction endonuclease fragments (and ultimately genes) and pathogenic behavior. A few suggestive results have emerged so far. The thymidine kinase (TK) gene does not affect the ability of herpes simplex virus to replicate in standard cell cultures, but TK-deficient mutants tend to be avirulent, multiply poorly in experimental animals, and do not as readily establish latent infections.

In another study, the pattern of ocular disease produced in rabbits by two strains of herpes simplex virus and various recombinants between them showed that the type of ocular disease (stromal or epithelial) was determined by genes located within a particular restriction endonuclease fragment.

GENETIC VARIATION IN NATURE

As well as providing the basis for the development of live virus vaccines, genetic variation by mutation, reassortment, and intramolecular recombination is the mechanism that, under the influence of natural selection, has produced the great variety characteristic of the families of viruses that infect man. There is growing evidence that reassortment is an important mechanism for producing sudden major genetic changes in nature, among the bunyaviruses, the orbiviruses, the rotaviruses, and the influenza A viruses. Mutation is also a potent mechanism for changing the properties of viruses, especially with the RNA viruses, which have a much higher base substitution rate than the DNA viruses. If the selective process is sufficiently rigorous, even DNA viruses can undergo relatively rapid changes in their behavior. We will conclude this chapter with a brief examination of how one DNA virus, myxoma, and one RNA virus, influenza A, have evolved over the past few decades.

Changes in Virus and Host in Myxomatosis

Observations made on myxomatosis in wild European rabbits in Australia and Great Britain illustrate the occurrence of evolutionary changes in a virus and its mammalian host, a process which may have occurred many times in both viruses and the animals that they infect.

Myxomatosis is a highly lethal disease of European rabbits (*Oryctolagus cuniculus*), which occurs naturally as a trivial disease of the tropical forest rabbit (*Sylvilagus brasiliensis*) of South America, in which it produces a localized fibroma. Infection is due to mechanical transfer, by biting arthropods, of the virus in the skin tumors.

The wild European rabbit was introduced into Australia in 1859 and rapidly spread over the southern part of the continent, where it became the major animal pest of the agricultural and pastoral industries. Myxoma virus from South America was successfully introduced into Australia in 1950 to help control the rabbit pest. The virus originally liberated in Australia produced case fatality rates of over 99%. This highly virulent virus spread readily from one animal to another during the summer, when the mosquito population was abundant. Farmers ran "inoculation campaigns" to introduce virulent myxoma virus into the wild rabbit populations every spring and summer. It might have been predicted that the disease would disappear after each summer, due to the absence of susceptible rabbits and the greatly lowered opportunity for transmission during the winter because of the scarcity of vectors. This must often have occurred in localized areas, but for the continent as a whole the outcome proved to be very different. The capacity to survive the winter conferred a great selective advantage on viral mutants able to cause a less lethal disease that enabled an infected rabbit to survive in an infectious condition for weeks instead of days. Within 3 years such mutants became the dominant strains throughout the country. Some inoculation campaigns with the virulent virus produced localized highly lethal outbreaks, but in general, the viruses that spread through the rabbit populations each year were the somewhat attenuated strains, which offer more prolonged opportunities for mosquito transmission. The better transmission of such strains even during summer epizootics was attested by field experiments; their advantages for overwintering are obvious. Thus the highly lethal introduced virus has now been replaced by a heterogeneous collection of strains of lower virulence.

Rabbits that recover from myxomatosis are immune to reinfection for the rest of their lives. However, since most wild rabbits have a life span of less than a year, herd immunity is not a very important feature in the epidemiology of the disease. This contrasts sharply with the importance of herd immunity in man (see Chapter 9).

Even in the initial outbreaks some rabbits recovered from the infection, suggesting that selection for genetically more resistant animals might operate rapidly. Sometimes recovery was due to environmental factors, e.g., the high ambient temperature that often occurs in Australia during the summer greatly reduces the severity of myxomatosis. In general, however, it was found that in areas of continued exposure to epizootics of myxomatosis the genetic resistance of the surviving rabbits steadily increased. The early appearance of the "attenuated" virus strains, which allowed 10% of genetically unselected rabbits to recover, was an important factor in allowing such genetically resistant rabbit populations to build up. In areas of annual exposure to myxomatosis, the genetic resistance of wild rabbits changed so that the case fatality rate after infection under laboratory conditions with a particular strain of virus fell from 90 to 25% within 7 years.

As might be expected, genetic resistance increased more rapidly in areas where epizootics occurred every year than in areas where several years elapsed between successive outbreaks of myxomatosis. Strains of myxoma virus recovered from the field in areas where the resistance of rabbits is high are now more virulent than those recovered from outbreaks in areas where the genetic resistance has increased more slowly. Natural selection is operating to maximize the occurrence of the kind of disease that promotes transmission.

Influenza A Virus: Antigenic Drift and Antigenic Shift

Human influenza virus was first isolated in 1933. Since that time influenza viruses have been recovered from all parts of the world and their antigenic constitution has been exhaustively studied. This has provided the opportunity for observing continuing evolutionary changes in the influenza viruses. The mortality from influenza is too low, and the generation time of man too long, for any significant observable change to have occurred in the genetic resistance of man.

Influenza A virus periodically causes epidemics in man and epizootics in horses, pigs, birds, and occasionally in other animals such as seals. Subtypes are classified according to the surface antigens, the hemagglutinin (HA, or H) and neuraminidase (NA, or N). All of the 13 HA subtypes have been found in birds, 3 of them also in man and 2 each in horses and pigs. The 9 NA subtypes show a similar distribution.

The outstanding feature of influenza A virus is its antigenic variability in the surface glycoproteins, HA and NA, which undergo two types of change, caused by different mechanisms and known as *antigenic drift* and *antigenic shift*. Antigenic drift occurs within a subtype and involves a series of minor changes (usually point mutations), producing variants ("strains") each antigenically slightly different from its predecessor. Antigenic shift involves the acquisition of a gene for a completely new hemagglutinin or neuraminidase. A new human subtype and a pandemic may result.

Antigenic Drift. After antigenic shift introduces a new pandemic strain, mutations subsequently accumulate in all its RNA segments. Mutations in the gene that codes for the HA include some which alter the antigenic sites on the HA molecule. When antiserum against the previously prevalent strain no longer neutralizes the variant, a new strain has emerged. Most of the significant changes in the HA are clus-

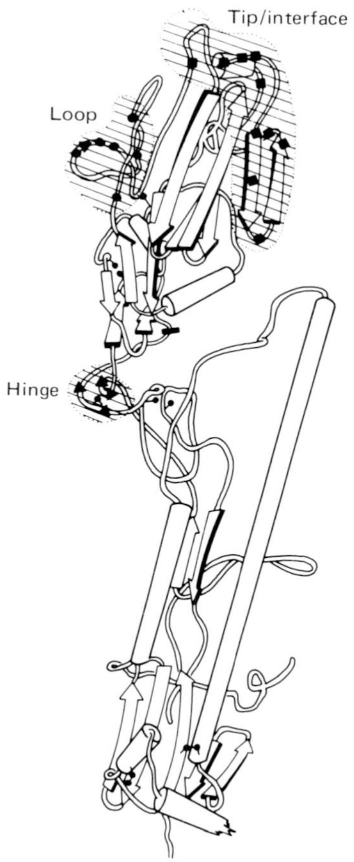

FIG. 4-5. *Three-dimensional model of the monomeric HA subunit of the trimeric hemagglutinin molecule of influenza virus. The dots indicate the positions of those amino acids that are substituted in novel strains of influenza virus arising within the H3 subtype either by natural antigenic drift or by* in vitro *selection in the presence of monoclonal antibodies. It will be noted that the epitopes tend to cluster into three or four major antigenic domains on the distal end ("head") of the molecule, designated "tip," "loop," "interface," and "hinge." [Modified from D. C. Wiley, I. A. Wilson, and J. J. Skehel, (1981). Reproduced with permission from* Nature (London) **289,** *373. Copyright © 1981 Macmillan Journals Limited.]*

tered into the four regions of the molecule which are thought to be the antigenic sites (Fig. 4-5). Substitution of a single amino acid in a critical antigenic site may totally abolish the capacity of the prevalent antibody to bind to that site. On the other hand, some regions of the HA are conserved in all human and avian strains that have been sequenced, presumably because they are essential for the maintenance of the structure and function of the molecule. The important feature of antigenic drift in human influenza viruses is that in immune populations the new strains have a selective advantage over their predecessors such that they tend to displace them. Although minor variants can cocirculate, the general rule is that the novel strain supplants previous strains of that subtype.

Antigenic Shift. Since influenza A virus was first isolated in 1933, sudden replacement of the prevalent human strain with a new subtype occurred in 1957 (H1N1 to H2N2) and 1968 (H2N2 to H3N2). When H2N2 ("Asian flu") appeared in China in 1957 it rapidly spread round the world, as did the "Hong Kong flu" (H3N2) after 1968, each displacing the then prevalent subtype. In 1977 the H1N1 subtype mysteriously reappeared and since then the two subtypes H3N2 and H1N1 have cocirculated (see Fig. 21-1).

The hemagglutinin molecules of different subtypes are, by definition, quite distinct serologically. Further information about their degree of relatedness, and thus possibly their evolutionary history, may be derived from the nucleotide sequences of the HA genes. Figure 4-6 is based on the sequence of some 350 nucleotides of the HA gene (from which the N-terminal amino acid sequence of the HA_1 polypeptide may be derived) for 32 type A influenza viruses, including representatives of each of the 13 known HA subtypes. The dendrogram indicates the percentage of amino acids that differs between each pair of strains. The preservation of cysteine residues and certain other amino acids in all sequences indicates that the 13 subtypes evolved from a common ancestor and share a common basic structure. However, there is only about 30% homology between the amino acid sequences of H3 and H2, whereas the homology between strains within each subtype is usually over 90%. In other words, the many strains that emerged by antigenic drift within the H2 subtype between 1957 and 1968 are closely related genetically, and the H3 strains which have become prevalent since 1968 are also closely related to one another, but there are major differences between the two groups. Clearly, a sharp discontinuity in the evolutionary pattern occurred with the emergence of the new subtype in 1968, as it had previously in 1957. What is the mechanism of this radical type of change, which we know as antigenic shift? The answer is derived by comparing the sequences of these HAs from human influenza subtypes with those obtained from birds and animals. Such studies revealed much closer homology between, say H3 (human) and a particular avian or equine influenza H3 than between H3 (human) and H2 (human). Since laboratory studies have clearly demonstrated the ease with which

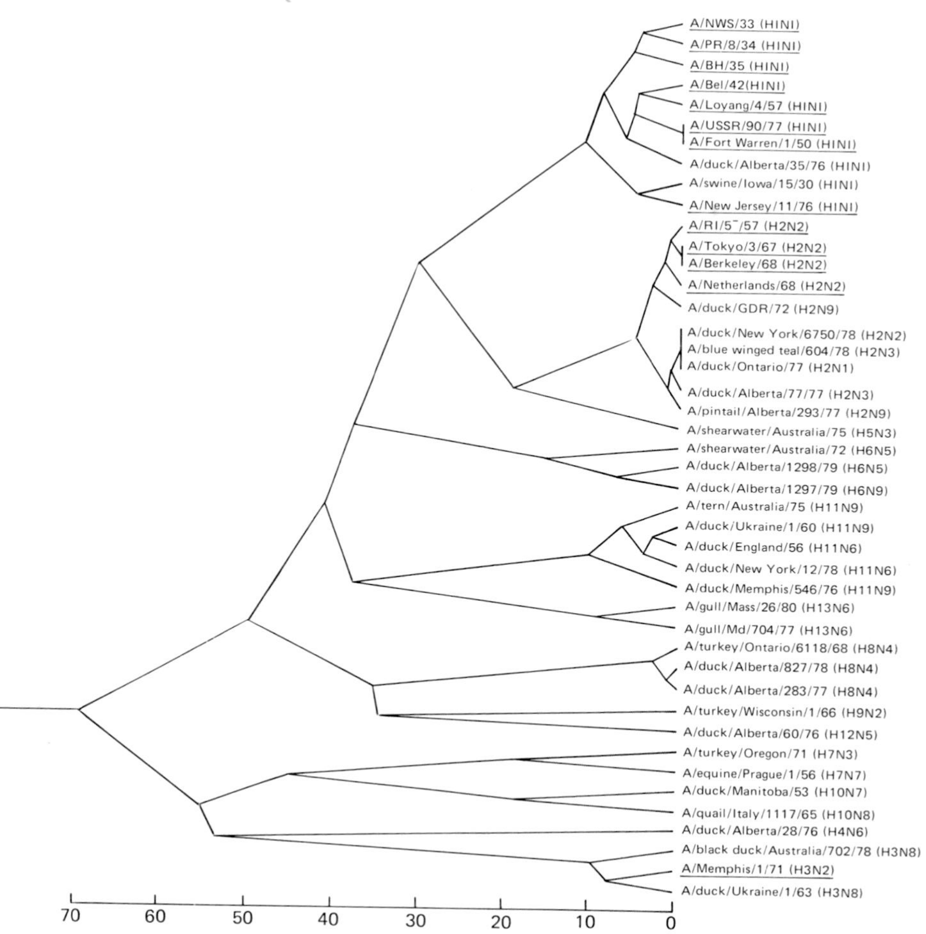

FIG. 4-6. *Dendrogram showing the relationships between the N-terminal amino acid sequences of* HA_1 *molecules, deduced from genomic sequences of the HA genes of viruses representing the 13 known HA subtypes of influenza A viruses. The positions of each bifurcation on the dendrogram indicate the mean percentage sequence differences of the strains connected through that point. Human isolates (H1N1, H2N2, and H3N2) are underlined. (Courtesy Dr. A. J. Gibbs.)*

genetic reassortment can occur in animals or birds, as well as in embryonated eggs or cultured cells, it seems reasonable to suppose that new subtypes of human influenza virus are derived by such reassortment in nature.

One exception appears to be the shift back to H1N1 in 1977. The strain recovered in May 1977 resembled very closely indeed a virus that circulated in man in 1950. Subsequently this subtype has circulated world wide among persons born after H1N1 was replaced by H2N2 in 1957. Meanwhile H3N2 has continued to be common. The explanation of the reemergence of H1N1 20 years after its disappearance from the human population is still a mystery.

FURTHER READING

Books

Fields, B. N., Jaenisch, R., and Fox, C. F., eds. (1980). "Animal Virus Genetics," UCLA Symp. Mol. Cell. Biol., Vol. 13. Academic Press, New York.

Fraenkel-Conrat, H., and Wagner, R. R., eds. (1977). "Comprehensive Virology," Vol. 9. Plenum, New York.

Maniatis, T., Fritsch, E. F., and Sambrook, J. (1982). "Molecular Cloning: A Laboratory Manual." Cold Spring Harbor Lab., Cold Spring Harbor, New York.

Palese, P., and Kingsbury, D. W., eds. (1983). "Genetics of Influenza Viruses." Springer-Verlag, Berlin and New York.

Palese, P., and Roizman, B., eds. (1980). "Genetic Variation of Viruses." Ann. N.Y. Acad. Sci., Vol. 354. N.Y. Acad. Sci., New York.

Watson, J. D. (1976). "Molecular Biology of the Gene," 3rd ed. Benjamin, New York.

Reviews

Fenner, F. (1983). Biological control, as exemplified by smallpox eradication and myxomatosis. *Proc. R. Soc. London, Ser. B* **218,** 259.

Fields, B. N., and Greene, M. I. (1982). Genetic and molecular mechanisms of viral pathogenesis: Implications for prevention and treatment. *Nature (London)* **300,** 19.

Holland, J., Spindler, K., Horodyski, F., Graham, E., Nichol, S., and Van de Pol, S. (1982). Rapid evolution of RNA genomes. *Science* **215,** 1577.

Joklik, W. K. (1981). Structure and function of the reovirus genome. *Microbiol. Rev.* **45,** 483.

Nathans, D. (1979). Restriction endonucleases, simian virus 40 and the new genetics. *Science* **206,** 903.

Perrault, J. (1981). Origin and replication of defective interfering particles. *In* "Initiation Signals in Viral Gene Expression" (A. J. Shatkin, ed.), pp. 151–207. Springer-Verlag, Berlin and New York.

Williams, D. A., Lemischka, I. R., Nathan, D. G., and Mulligan, R. C. (1984). Introduction of new genetic material into pluripotent haematopoietic stem cells of the mouse. *Nature (London)* **310,** 476.

CHAPTER 5

Pathogenesis and Pathology of Viral Infections

Introduction 119
Effects of Viruses on Cells 120
Pathogenicity and Virulence 125
Mechanisms of Tissue Damage 127
Routes of Entry 128
Mechanisms of Spread in the Body 134
Virus Shedding 143
Further Reading 144

INTRODUCTION

At the molecular and cellular levels, viruses behave quite differently from bacteria and protozoa, but this distinction largely disappears when we consider viruses at the levels of the whole organism and the community. In this chapter and in Chapters 6–8 we consider the ways in which viruses produce disease, i.e., the pathogenesis of viral infections. This involves four types of interaction between viruses and their vertebrate hosts: the effects of viruses on cells, the way in which viruses spread through the body and cause disease, the immune response and its influence on viral infections, and the many nonimmunological factors which affect virus–host interactions. The first two subjects are discussed in this chapter; the other two are discussed in Chapter 6. Chapter 7 is concerned with persistent infections, and Chapter 8 deals with virus-induced cancers.

TABLE 5-1
Virus–Cell Interactions

TYPE OF INFECTION	EFFECT ON CELL	YIELD OF INFECTIOUS VIRUS
Cytocidal	Morphological changes in cells (CPE); inhibition of protein, RNA and DNA synthesis; changes in plasma membrane; cell death	+
Persistent		
Steady-state	No CPE; little metabolic disturbance; cells continue to grow and divide. May be associated with loss of the special functions of differentiated cells	+
Transformation	Alteration in cell morphology but cells survive and can be passaged indefinitely; may produce tumors in experimental animals	− (DNA oncogenic viruses) + (RNA oncogenic viruses, usually)

EFFECTS OF VIRUSES ON CELLS

Ultimately, the disturbances of bodily function that we observe as the signs and symptoms of viral disease result from the effects of viruses on cells. As Table 5-1 indicates, a variety of types of interaction between virus and cell can occur, the outcome in any particular case being determined by the genetic makeup of the cell and virus concerned.

Cytocidal Infection

The terminal event of cytocidal infections is cell death, and the histological appearance of the damage produced in cultured cells by particular cytocidal viruses, known as the cytopathic effect (CPE), is often sufficiently characteristic to be used as a diagnostic criterion (see Chapter 12). Gross biochemical changes occur in cells infected with cytocidal viruses. Early virus-coded proteins often shut down host protein synthesis, an event which is incompatible with survival of the cell. Large numbers of viral macromolecules accumulate late in the infectious cycle; some of these, particularly certain capsid proteins, may be directly toxic. Viral proteins or virions themselves sometimes congregate in large crystalline aggregates or inclusions that visibly distort the cell. Infected cells usually swell substantially and there are changes in membrane permeability.

Thus there are numerous changes in the virus-infected cell that, individually or cumulatively, are lethal. In a sense, only the first lethal change is relevant, even though it may not necessarily be the one that produces the first visible CPE or the final dissolution of the cell. Cell damage by cytocidal viruses is not necessarily accompanied by multiplication of the virus. It may occur even when late stages of the expression of the viral genome, essential for the production of infectious virus, are blocked, as for example by UV damage to the viral nucleic acid or by chemical inhibitors.

Shutdown of Cellular Protein Synthesis. Most cytocidal viruses code for early proteins which shut down the synthesis of cellular protein; cellular RNA and DNA synthesis are usually affected secondarily. The shutdown is particularly rapid and severe in infections of cultured cells by picornaviruses, some poxviruses, and herpesviruses, all of which are rapidly cytopathogenic. With viruses of several other families the shutdown is late and gradual, while with noncytocidal viruses such as arenaviruses and retroviruses there is, by definition, no shutdown and no cell death.

Cytopathic Effects of Viral Proteins. Although toxins are not a feature of viral infections, viral capsid proteins in high dosage are often toxic and may be the principal cause of CPE. This may follow the accumulation of viral proteins late in the multiplication cycle after infection at low multiplicity or it may be seen quite early, as a laboratory artifact resulting from the experimental use of very large inocula. The best studied example is adenovirus, which contains two capsid antigens capable of causing CPE. Purified adenovirus penton produces reversible cell clumping and detachment from the substrate, whereas the purified fiber depresses cellular synthesis of RNA, DNA, and protein and inhibits the multiplication of adenoviruses and other viruses in the treated cells. This generalized shutdown of all cellular and viral activity is not to be confused with the more specific shutdown of cellular (but not viral) RNA and protein synthesis by virus-coded early proteins which is an important feature of the multiplication cycle of many cytocidal viruses.

Inclusion Bodies. The most characteristic, but by no means invariable, morphological change in virus-infected cells is the *inclusion body*—an area of the nucleus or cytoplasm with altered staining behavior (Plate 5-1). Depending on the causative virus, such inclusions may be single or

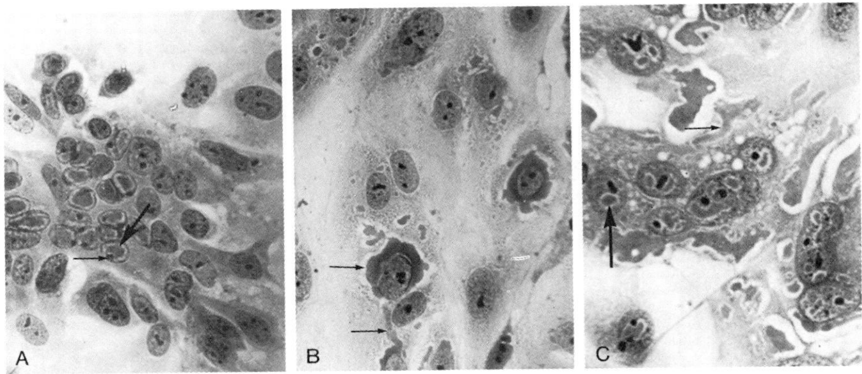

PLATE 5-1. *Types of viral inclusion body. H and E stain. ×200. (A) Intranuclear inclusions; cells form syncytium—herpesvirus. Small arrow, nucleolus; large arrow, inclusion body. Note also margination of nuclear chromatin. (B) Intracytoplasmic inclusions—reovirus. Arrows indicate inclusion bodies in perinuclear locations. (C) Intranuclear and intracytoplasmic inclusions; cells form syncytium—measles virus. Small arrow, intracytoplasmic inclusion body; large arrow, intranuclear inclusion body. (Courtesy Dr. I. Jack.)*

multiple, large or small, round or irregular in shape, intranuclear or intracytoplasmic, acidophilic or basophilic.

The most characteristic viral inclusion bodies are the intracytoplasmic inclusions found in cells infected with poxviruses, paramyxoviruses, reoviruses, and rabies virus (Negri bodies), and the intranuclear inclusion bodies produced by herpesviruses and adenoviruses (Fig. 5-1). Some viruses (cytomegalovirus, measles virus) produce both nuclear and cytoplasmic inclusion bodies. Most such inclusions have now been shown, by fluorescent-antibody staining and electron microscopy, to be sites of viral synthesis; some, such as the intranuclear inclusion bodies of herpesviruses, are late degenerative changes which confer altered staining characteristics on the cell. In a few instances (e.g., adenoviruses, reoviruses) the inclusions may represent crystalline aggregates of virions.

Although they are most commonly due to viruses, inclusion bodies are not diagnostic of viral infection, since similar histological appearances may be produced by chemicals or occasionally by bacteria.

Cell Fusion. Live or UV-irradiated paramyxoviruses can cause rapid fusion of cultured cells (or lysis of red cells) if applied at high multi-

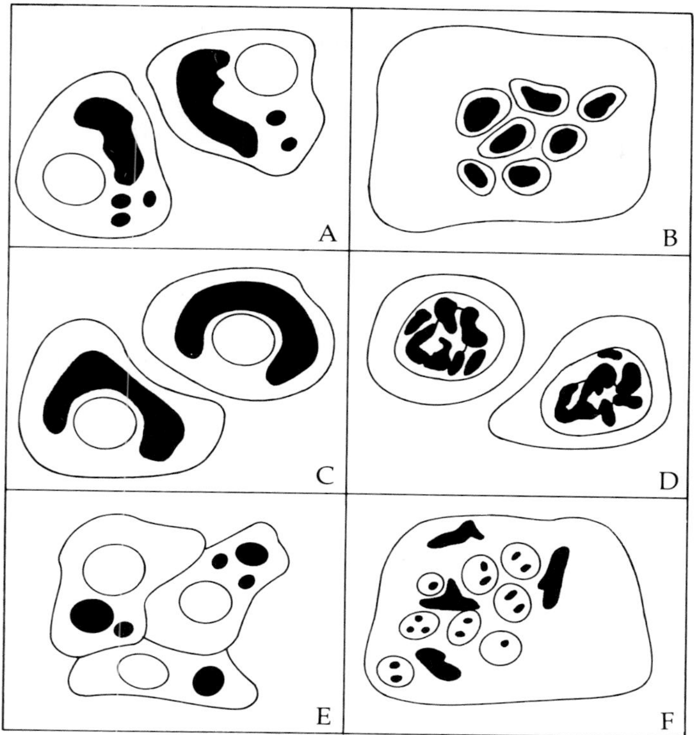

FIG. 5-1. *Inclusion bodies in virus-infected cells. (A) Vaccinia virus—intracytoplasmic acidophilic inclusion. (B) Herpes simplex virus—intranuclear acidophilic inclusion; cell fusion produces syncytium. (C) Reovirus—perinuclear intracytoplasmic acidophilic inclusion. (D) Adenovirus—intranuclear basophilic inclusion. (E) Rabies virus—intracytoplasmic acidophilic inclusions (Negri bodies). (F) Measles virus—intranuclear and intracytoplasmic acidophilic inclusions; cell fusion produces syncytium.*

plicity. The process is called *early polykaryocytosis.* Cell biologists have used this phenomenon to produce functional heterokaryons by fusing different types of cell; for example UV-inactivated Sendai virus was used to produce "hybridoma cells" by fusion of antibody-producing B lymphocytes with myeloma cells in the pioneering experiments that produced the first monoclonal antibodies.

Most of the viruses that rapidly fuse cells when artifically large doses are used also produce *late polykaryocytosis,* i.e., late in their multiplication cycle they cause changes in the cell membrane that result in the fusion of the infected cell with neighboring uninfected cells. These *syncytia*, i.e., giant cells with many nuclei, are a feature of the pathology of certain viral infections in man, as well as being a conspicuous result of infection of cell monolayers by paramyxoviruses and herpesviruses.

New Antigens on the Cell Surface. In most viral infections new antigens, specified by the viral genome, appear in the plasma membrane of infected cells. For example, the plasma membranes of cells infected with enveloped RNA viruses (e.g., influenza virus) incorporate viral hemagglutinin, so that red blood cells can be adsorbed to their surface (*hemadsorption*). Such virus-coded antigens on the surface of infected cells constitute a target for the body's specific immune mechanisms, both humoral and cellular, which destroy the cell and hasten recovery, while at the same time (in many cases) precipitating symptoms of disease (see Chapter 6).

Another class of new antigens found on the cell membrane are the *transplantation antigens* in cells that are transformed by viruses. Transformation induced by physical, chemical, or viral agents is always accompanied by the appearance of new antigens at the cell surface. With physical and chemical carcinogens these antigens are different in every instance; with the oncogenic viruses, however, the transplantation antigens are characteristic of the virus, being proteins specified by viral genes (see Chapter 8).

Nonspecific Histological Changes. In addition to changes due to the specific effects of viral multiplication, most virus-infected cells also show nonspecific lesions, very much like those induced by physical or chemical agents. The most common early and potentially reversible change is what histopathologists call "cloudy swelling," which is associated with changes in the permeability of the plasma membrane. Electron microscopic study of such cells reveals diffuse changes in the nucleus, plasma membrane, endoplasmic reticulum, mitochondria, and polyribosomes. Later the chromatin moves to the edge of the nucleus and becomes condensed (pyknosis). Much of the late CPE is the autolytic consequence of the leakage of lysosomal enzymes into the dying cell.

Steady-State Persistent Infection

In steady-state infection, the infected cells produce and release virus but cellular metabolism is little affected and the infected cells can grow and divide. This type of cell–virus interaction does not occur with DNA viruses, but is found with several RNA viruses: arenaviruses, retroviruses, and some paramyxoviruses, for example. In all these infections virions are released by budding from the plasma membrane of the cell. Although viral replication does not kill the cell and infected virus-yielding cells may grow and divide in culture for long periods, in the animal the antigenic changes in the cell membrane usually provoke an immunological response, which may destroy the cells (see Chapter 6). Further, in the intact animal such infected cells may lose the capacity to carry out their specialized functions as differentiated cells, e.g., the production of hormones or immunoglobulins.

The existence of steady-state infection is demonstrable by a variety of laboratory procedures, such as staining with labeled antibody to viral antigens, hemadsorption, or interference with the multiplication of a superinfecting virus.

Carrier Cultures. There is another way in which cultured cells can coexist with viruses. It could be regarded as a laboratory artifact that may be caused by a variety of factors. In essence, cytocidal viral multiplication proceeds in a few cells only, whereas most cells in the culture are not infected, either because they are not susceptible or because they are protected by interferon or other host factors.

Transformation

Most viruses that are capable of producing tumors in animals, either naturally or experimentally, can also *transform* cells *in vitro* (Plate 5-2). In such cells there is a persistent virus–cell interaction which is *nonproductive* (of infectious virus) with DNA viruses but usually *productive* with retroviruses (the only RNA viruses that transform cells). The characteristics of transformed cells are summarized in Table 5-2; their significance in relation to oncogenesis is discussed in Chapter 8.

In transformed cells the viral DNA (or with retroviruses, the cDNA) is usually, but not invariably, integrated in a covalent fashion into the host DNA. With DNA oncogenic viruses this occurs only if the cell is nonpermissive, i.e., viral replication does not go on to completion. For the retroviruses, integration of cDNA is part of the life cycle; viral replication occurs concomitantly with transformation.

PATHOGENICITY AND VIRULENCE

It is relevant here to consider two much-abused words, *pathogenicity* and *virulence*, as they apply to viruses. Neither term can be used without reference to the host, for they apply to the interaction between host and

TABLE 5-2
Characteristics of Cells Transformed by Viruses

1. Capacity to produce malignant neoplasms when inoculated into isologous or severely immunosuppressed animals
2. Greater growth potential *in vitro*
   (a) Formation of three-dimensional colonies of randomly oriented cells in monolayer culture, usually due to loss of contact inhibition
   (b) Capacity to divide indefinitely in serial culture
   (c) Higher efficiency of cloning
   (d) Capacity to grow in suspension or in semisolid agar (anchorage independence)
   (e) Reduced serum requirement for growth
3. Altered cell morphology
4. Altered cell metabolism
   (a) Increased rate of sugar transport
   (b) Greater production of organic acids
   (c) Greater production of acid mucopolysaccharides on the plasma membrane
   (d) Changes in the glycolipid and glycoprotein composition of the plasma membrane and enhanced agglutinability by lectins
   (e) Increased depolymerization of cytoskeletal proteins
   (f) Induction of fetal antigens
   (g) Release of proteases and protease activators
5. Chromosomal abnormalities
6. Novel (virus-specified) antigens and DNA
   (a) New surface antigens (transplantation antigens)
   (b) New intracellular antigens (e.g., T antigens)
   (c) Some viral DNA sequences present, integrated with cellular DNA or as episomes
   (d) Sometimes the transforming virus can be rescued by helper virus or fusion with a susceptible cell

parasite. A virus is *pathogenic* for a particular host if it can infect that host and produce signs and symptoms of disease in it. But infection is not synonymous with disease; many infections, even with virulent viruses, may be *subclinical (inapparent)*. A given strain of virus is said to be more *virulent* than another if it regularly produces more severe disease in a host in which both strains are pathogenic. Live vaccines, for example, are usually attenuated strains, i.e., they are less virulent than the wild-type virus. The word "virulent" should not be used to describe the killing of cultured cells by viruses; the correct word is *cytocidal*. Cytocidal viruses are not always virulent in intact animals (e.g., echoviruses, which are highly cytopathogenic *in vitro*, are generally harmless *in vivo*); conversely, noncytocidal viruses such as rubella, rabies, or the leukemia viruses may cause severe disease.

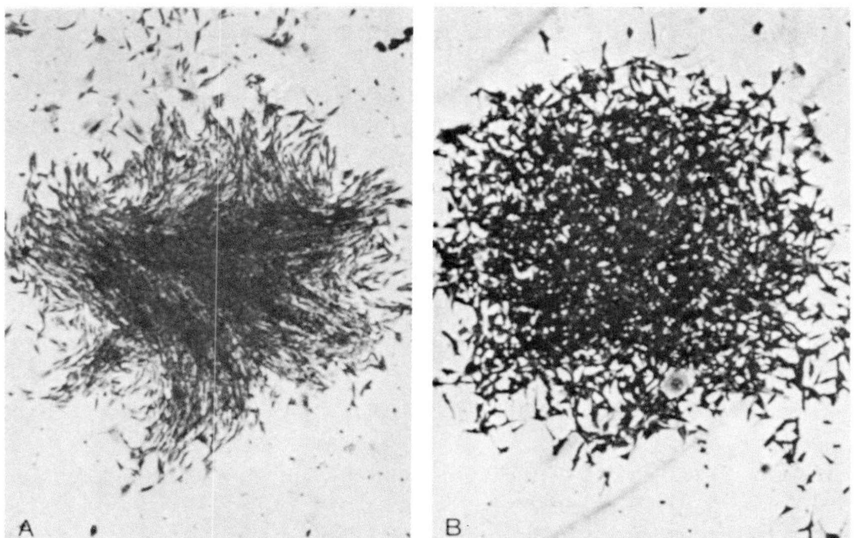

PLATE 5-2. *Transformation of cultured BHK21 cells by polyoma virus. (A) Normal colony, illustrating the regular parallel arrangement of elongated fibroblastic cells. (B) Transformed colony, illustrating the criss-cross random orientation of more rounded cells. ×25. [From W. House and M. G. P. Stoker,* J. Cell Sci. **1,** *169 (1966).]*

MECHANISMS OF TISSUE DAMAGE

In spite of the present sophisticated state of our understanding of molecular virology, we know very little about how viruses cause disease. As implied above, the symptoms and signs of disease cannot be explained simply by destruction of cells by virus. Of course, destruction of a substantial proportion of the cells of a vital organ such as the heart, lung, liver, or brain gives symptoms and signs referable to that organ (e.g., tachycardia, respiratory distress, jaundice, convulsions) and will obviously have serious consequences. Furthermore, a degree of cellular or tissue edema that is tolerable in most tissues may have serious consequences if it occurs in organs such as the brain or lung. Damage to the endothelial cells of small blood vessels can also have wide-ranging effects if it leads to anoxia. On the other hand, less vital tissues with great powers of regeneration, such as intestinal epithelium, can recover rapidly from extensive damage; in this particular example viral multiplication generally occurs without even producing diarrhea. Symptoms and signs can nevertheless occur without cellular destruction; physiological changes may result from sublethal impairment of "luxury" functions of specialized cells, e.g., enzyme or hormone production. The explanation of the more general symptoms that are a feature of many viral infections, e.g., "toxemia," fever, malaise, anorexia, lassitude, still eludes us. Certainly fever is induced quite readily by interleukin-1, by interferons, and by certain viral capsid proteins, and this temperature rise may be beneficial to the infected host, e.g., by enhancing immune responses or slowing viral replication. Another major mechanism of tissue damage, namely immunopathology, will be considered in Chapter 6.

ROUTES OF ENTRY

In order to infect man, a virus must first attach to and infect cells of one of the body surfaces: skin, alimentary tract, respiratory tract, urogenital tract, or conjunctiva (Fig. 5-2). Parenteral injection, either by

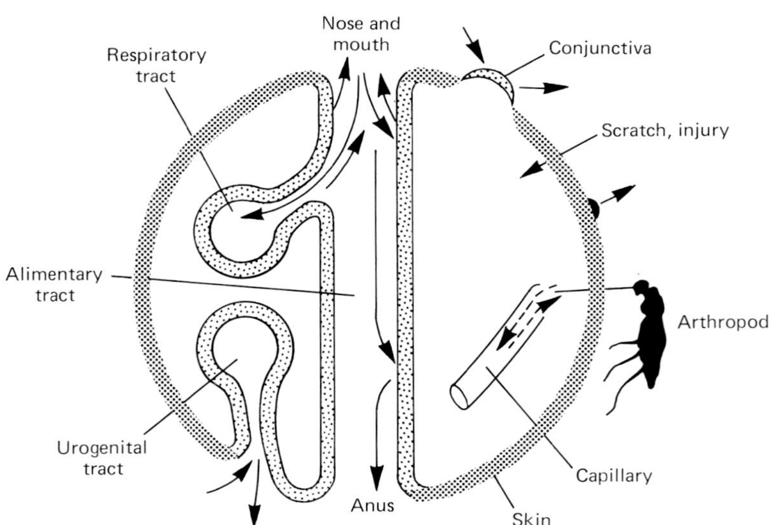

FIG. 5-2. *The surfaces of the body in relation to the entry and shedding of viruses. (Modified from C. A. Mims and D. O. White, "Viral Pathogenesis and Immunology." Blackwell Scientific Publications, Oxford, 1984.)*

needle or by the bite of an arthropod or vertebrate, bypasses the body surfaces. Localized pathological changes may be produced when viruses breach a body surface, although invasion may also occur without the development of a local lesion. After the surface has been breached the infection may remain localized, or it may spread through the organism via the lymphatics, blood vessels, or nerves.

Infection via the Skin, Conjunctiva, and Urogenital Tract

The largest organ in the body, the skin, provides a tough and impermeable barrier to the entry of viruses. However, a few viruses multiply in the skin to produce local lesions after entry through minute abrasions or by artificial puncture (Table 5-3). These include several different poxviruses: cowpox, orf, tanapox, and molluscum contagiosum, and the

TABLE 5-3
Viruses of Man That Initiate Infection via the Skin, Oral Mucosa, Genital Tract, or Eye

ROUTE	VIRUS	
Minor trauma	*Hepadnaviridae*	Hepatitis B
	Papillomavirus	All types
	Herpesviridae	Herpes simplex 1 and 2
	Poxviridae	Molluscum contagiosum, cowpox, orf, milkers' nodes, vaccinia
Arthropod bite		
Mechanical	*Poxviridae*	Tanapoxvirus
Biological	*Alphavirus*	All species
	Flaviviridae	All species
	Orbivirus	Colorado tick fever
	Bunyaviridae	La Crosse, sandfly fever, Rift Valley fever
Bite of vertebrate	*Rhabdoviridae*	Rabies
	Herpesviridae	B virus
Injection	*Hepadnaviridae*	Hepatitis B
	Herpesviridae	Cytomegalovirus, EB virus
	Filoviridae	Ebola
	Retroviridae	HTLV-III
	Unclassified	Hepatitis non-A, non-B
Genital tract	*Papillomavirus*	Genital types
	Herpesviridae	Herpes simplex 2
	Retroviridae	HTLV-III
Conjunctiva	*Adenoviridae*	h8 and others
	Picornaviridae	*Enterovirus* 70

several strains of papillomavirus that cause various kinds of warts. More commonly, however, viruses are introduced through the skin by the bite of an arthropod vector—generally a mosquito, tick, sandfly, or gnat; most of these *arboviruses* belong to the families *Togaviridae* (genus *Alphavirus*), *Flaviviridae, Bunyaviridae,* or *Reoviridae* (genus *Orbivirus*). More rarely, lethal infection is acquired via the bite of an animal—a carnivore or bat in the case of rabies, a monkey in the case of herpes B virus. Finally, the skin penetration may be *iatrogenic*—the result of human intervention—particularly with hepatitis viruses (B and non-A, non-B), which are transmissible by blood transfusion, injection, and other manipulations such as tattooing and acupuncture. Two major epidemics of African hemorrhagic fever, in Sudan and Zaire, were due to transmission of Ebola virus mainly by contaminated syringes and needles. Widespread infection of the skin in the generalized exanthemata (measles, chickenpox, rubella, human monkeypox, and several arboviral and enteroviral infections) is due to virus spread via the bloodstream which initiates infection of the skin "from below."

The conjunctiva, although much less resistant to viral invasion than the skin, is constantly cleansed by the flow of secretion (tears) and is wiped by the lids. Three viruses characteristically infect man by this route, producing conjunctivitis: *Adenovirus 8, Enterovirus 70,* and as a rare zoonosis, Newcastle disease virus.

The urinary tract is regularly flushed with urine and is not a significant route of entry. However, the female genital tract is not cleansed in this way. Herpes simplex type 2 and genital warts (due to papillomavirus) are two of the commonest sexually transmitted diseases of man, and AIDS (due to HTLV-III) is spreading alarmingly.

Infection via the Respiratory Tract

Although lined by cells that are directly susceptible to infection by many viruses, the respiratory tract is ordinarily protected by effective cleansing mechanisms (Fig. 5-3). A mucociliary blanket lines the nasal cavity and most of the lower respiratory tract, and inhaled foreign particles deposited on this surface are trapped in mucus and carried from the nasal cavity or lungs to the back of the throat, and then swallowed. Large particles are filtered off by the hairs lining the nostrils. Particles 10 μm in diameter or thereabouts are usually deposited on the nasal mucosa that overlies the turbinate bones, which project into the nasal cavity and act as baffle plates (Fig. 5-3B). Smaller particles (5 μm or less) are usually inhaled directly into the lungs and some may reach the alveoli, where virions may be destroyed by alveolar macrophages or,

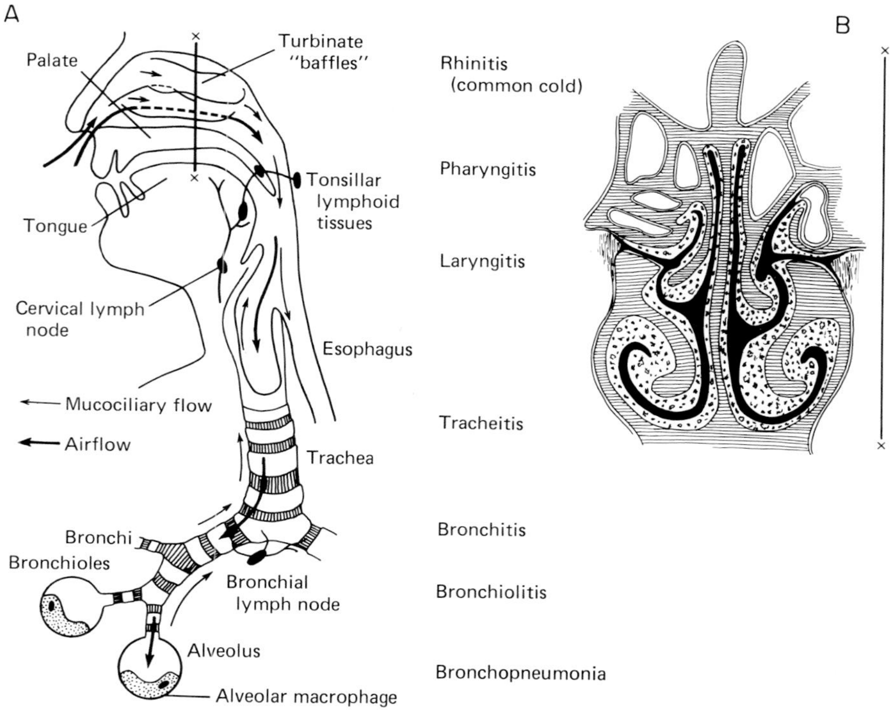

FIG. 5-3. *(A) Pathways of infection and mechanical protective mechanisms in the respiratory tract. On right, clinical syndromes produced by infection at various levels of respiratory tract. (B) Section of turbinates (×——×) magnified 2.5 times, showing the narrow and complicated pathway of inspired air, and thus the ease with which slight cellular swelling "blocks the nose." [(A) Reproduced from C. A. Mims and D. O. White, "Viral Pathogenesis and Immunology." Blackwell Scientific Publications, Oxford, 1984.]*

more rarely, productively infect such cells, or cells of the adjacent alveolar lining.

In spite of these protective mechanisms, more kinds of viruses initiate human infection via the respiratory tract than by any other route (Table 5-4). Most of these infections remain localized in the upper or lower respiratory tract, but some viruses that enter via the respiratory tract produce their principal effects following systemic spread (e.g., measles, rubella, chickenpox).

The sequence of events in respiratory tract infections is exemplified by influenza. Virus particles alighting on the film of mucus that covers the epithelium of the upper respiratory tract may undergo one of several

TABLE 5-4
Viruses of Man That Initiate Infection via the Respiratory Tract

With the production of local respiratory symptoms	
Orthomyxoviridae	Influenza
Paramyxoviridae	Parainfluenza, respiratory syncytial virus
Picornaviridae	Rhinoviruses, some enteroviruses
Coronaviridae	Most species
Adenoviridae	Most species
Herpesviridae	EB virus, herpes simplex, cytomegalovirus
Producing generalized disease, usually without initial respiratory symptoms	
Paramyxoviridae	Mumps, measles
Togaviridae	Rubella
Herpesviridae	Varicella
Picornaviridae	Some enteroviruses
Papovaviridae	BK and JC viruses
Bunyaviridae	Hantaan virus
Arenaviridae	Lymphocytic choriomeningitis
Poxviridae	Smallpox (extinct)

fates. If the individual has previously recovered from an infection with that strain of influenza virus, antibody (mainly IgA) in the mucous secretion may combine with the virus and neutralize it. Mucus also contains glycoprotein inhibitors that combine with virus and prevent it from attaching to the specific receptors on the host cell; eventually these inhibitors are destroyed by the viral neuraminidase. The successful infection of a few cells by influenza virus, and the passage of newly synthesized virions from these cells into the respiratory mucus and then into other cells may lead to progressive infection which is aided by the transudation of fluid that follows cellular injury and helps to disperse the virus. On the other hand, the inflammation also results in an increased diffusion of plasma constituents, including cytotoxic T cells, macrophages, NK cells, antibody, complement, and nonspecific inhibitors, which may inactivate the virus or destroy virus-infected cells and cut short the infection (see Chapter 6). Interferon may also play a role in limiting the spread of infection. However, influenza virus often overcomes these protective barriers and multiplication proceeds. Usually damage is confined to the epithelial cells of the upper respiratory tract, but in cases of pneumonia, fluorescent antibody staining shows that there are foci of infection in the epithelial cells of the bronchi, bronchioles, and alveoli and in alveolar macrophages, but not in the vascular endothelium. Damage to the respiratory epithelium by influenza virus lowers its resistance to secondary bacterial invaders, especially pneumococci and staphylococci.

Infection via the Alimentary Tract

Several viruses, such as herpes simplex and EB viruses, probably initiate infection via the mouth or oropharynx. The esophagus appears to be spared direct infection, probably because of its tough stratified squamous epithelium and the rapid passage of swallowed material over its surface. The intestinal tract is partially protected by mucus, but virus particles may readily lodge in sequestered places in the crypts and if there are specific receptors on the cells there, initiate infection. Virions may also be taken up by specialized cells that overlie Peyer's patches in the ileum.

The other protective mechanisms in the intestinal tract are the various compounds utilized for the digestion of food; from the stomach downward, acid, bile, and proteolytic enzymes. The kinds of virus that initiate infection of the intestinal tract (Table 5-5) are all acid and bile resistant: enteroviruses, rotaviruses, adenoviruses, and caliciviruses. Rotaviruses and caliciviruses are now recognized as major causes of diarrhea. The great majority of enterovirus and adenovirus infections are asymptomatic, but some of the enteroviruses (polioviruses, hepatitis A virus) are

TABLE 5-5
Viruses of Man That Initiate Infection via the Alimentary Tract

Via mouth or oropharynx	
Herpesviridae	Herpes simplex, EB virus, cytomegalovirus
Via intestinal tract	
Producing enteritis	
Reoviridae	Rotaviruses
Caliciviridae	Norwalk agent and related viruses
Adenoviridae	Some adenoviruses
Producing generalized disease, usually without local symptoms	
Picornaviridae	Many enteroviruses including polioviruses and hepatitis A virus
Usually symptomless	
Picornaviridae	Some enteroviruses
Adenoviridae	Some adenoviruses
Reoviridae	Reoviruses
Coronaviridae	Rarely

important causes of systemic infection even though they do not produce symptoms referable to the intestinal tract.

MECHANISMS OF SPREAD IN THE BODY

Viruses may remain localized to the body surface through which they entered: skin, genital tract, conjunctiva, intestine, or respiratory tract, or they may cause generalized infections, which are often associated with subsequent localization in particular organs.

Local Spread on Epithelial Surfaces

Poxviruses and papillomaviruses that enter via the skin usually proliferate locally, to produce a localized pustular lesion or a small tumor. Viruses that enter the body via the alimentary or respiratory tracts may spread rapidly over the epithelial surfaces, aided by the movement of mucus across them. After infections of the respiratory tract by rhinoviruses, coronaviruses, and influenza virus, or the intestinal tract by rotaviruses or caliciviruses, there is little or no invasion of the underlying tissues and the incubation period is quite short. Although these viruses usually enter the lymphatics and thus have the potential to spread, they do not appear to multiply in the deeper tissues, possibly because the appropriate receptors are restricted to the epithelial surfaces, or because viral development is inhibited at the higher temperature of the interior of the body.

Subepithelial Invasion and Lymphatic Spread

A network of lymphatics lies beneath all the surface epithelia. Virions that enter the lymphatic capillaries are carried to the local lymph nodes, which act as filters. As they enter, virions are exposed to the macrophages lining the marginal sinus and are usually engulfed. They may be inactivated and digested and their component antigens presented to the underlying lymphoid cells in such a way that the immune response is initiated (see Chapter 6). Some viruses, however, multiply in the macrophages (e.g., measles virus, some adenoviruses, and some herpesviruses); some virions may pass straight through the lymph node and eventually enter the blood stream. Since the macrophages and lymphocytes in the lymph node are constantly circulating in the lymph and blood vascular systems, virus may be passively transported around the body in these cells.

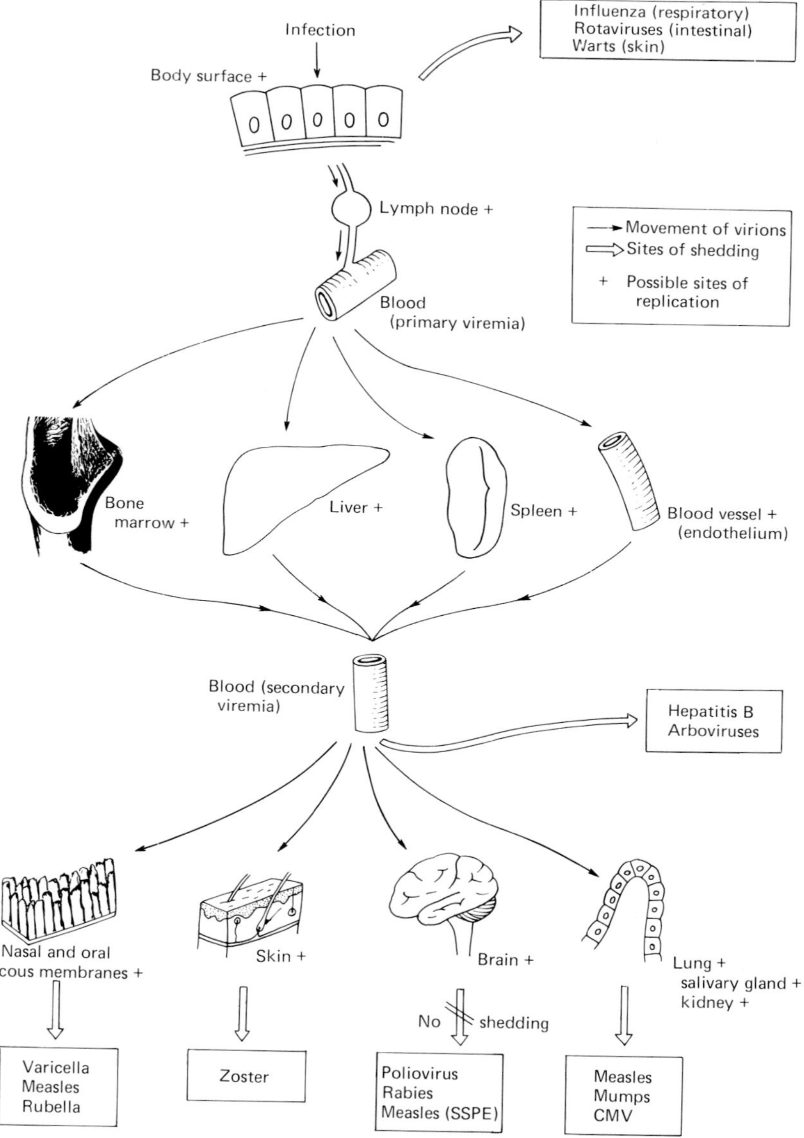

FIG. 5-4. *The spread of virions through the body in human viral infections. + Indicates possible sites of viral replication. Large arrows indicate sites of shedding of virus, with illustrative examples of diseases in which that route of excretion is important. Transfer from blood is by transfusion, etc., with hepatitis B, and by mosquito bite in certain arboviral infections. (Modified from C. A. Mims and D. O. White, "Viral Pathogenesis and Immunology." Blackwell Scientific Publications, Oxford, 1984.)*

Spread by the Bloodstream

In almost all systemic infections the blood serves as the main vehicle for the transport of virions (Fig. 5-4); the presence of virus in the blood is called *viremia*. In the blood, virions may be free in the plasma or may be associated with particular types of leukocytes, with platelets, or with erythrocytes. Leukocyte-associated viremia is a feature of several types of infection, including measles and smallpox, for example. Several viruses multiply in macrophages; others, e.g., EB virus, multiply in lymphocytes. Occasionally virions are adsorbed to erythrocytes, as in Rift Valley fever, Colorado tick fever, and lymphocytic choriomeningitis. All the flaviviruses and togaviruses, and the enteroviruses that cause viremia, circulate free in the plasma.

Whether virions circulate free in the plasma or are cell associated affects their passage from the circulation to extravascular sites and their removal from the circulation. Leukocytes can pass through the walls of small vessels by diapedesis, and infected leukocytes can thus initiate infection in various parts of the body. On the other hand, virus in the plasma may escape from circulation after uptake by macrophages in organs with sinusoids, or by localization in the endothelium of small capillaries and venules. The latter process is much slower than phagocytosis by macrophages, and neurotropic viruses (poliovirus and certain arboviruses) reach the CNS from the blood only if the viremia is adequate in level and duration.

Since virions circulating in the blood are continually removed by cells of the reticuloendothelial system, viremia can be maintained only if there is a continued release of virus into the blood from cells in contact with it, or if the clearance system is grossly impaired. The circulating leukocytes can themselves constitute a source of replicating virus; indeed blood leukocytes maintained in culture (from which polymorphonuclear leukocytes are rapidly lost) support limited replication of many viruses. However, viremia is maintained primarily by organs with extensive sinusoids, like the liver, spleen, and bone marrow, the endothelial cells of the blood vessels themselves, and the lymphoid tissues (via the thoracic duct). Cells of the voluntary muscles may be an important site of multiplication of some enteroviruses and togaviruses; virions are transferred to the blood via the lymph.

Invasion of the Skin

Rashes are more easily seen in human beings than in furred and feathered beasts, hence most of the descriptive data are derived from human infections. The individual lesions in generalized rashes are de-

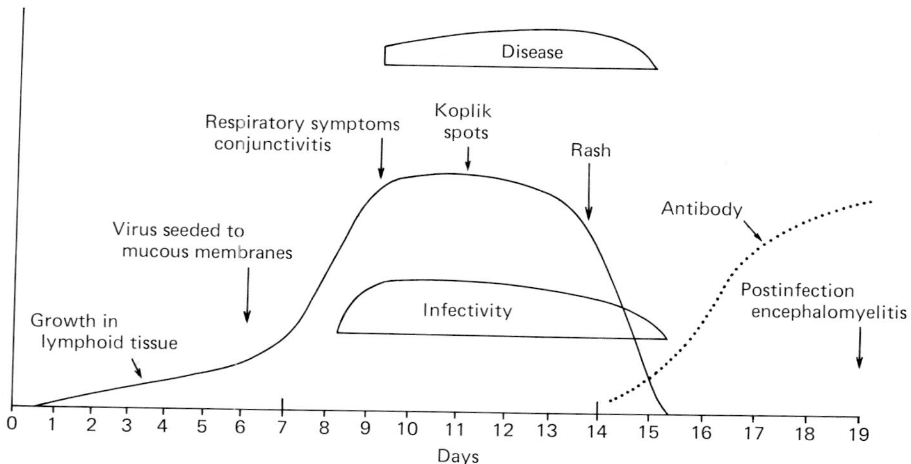

FIG. 5-5. *Diagram illustrating the pathogenesis of measles. Postinfectious encephalomyelitis occurs in about 1 in every 1000 cases of measles. (From C. A. Mims and D. O. White, "Viral Pathogenesis and Immunology." Blackwell Scientific Publications, Oxford, 1984.)*

scribed as macules, papules, vesicles, or pustules. A lasting local dilation of subpapillary dermal blood vessels produces a macule, which becomes a papule if there is also edema and an infiltration of cells into the area. Primary involvement of the epidermis usually results in vesiculation, ulceration, and scabbing, but prior to ulceration a vesicle may be converted to a pustule if there is a copious cellular exudate. Secondary changes in the epidermis may lead to desquamation. More severe involvement of the dermal vessels may lead to hemorrhagic and petechial rashes, although coagulation defects and thrombocytopenia may also be important in the genesis of such lesions.

Early work on the pathogenesis of the generalized exanthemata was carried out with mousepox and used as a basis for understanding the pathogenesis of smallpox. Now that this disease is extinct, it is more appropriate to describe the sequence of events in measles (Fig. 5-5), which is still a very common infection almost everywhere in the world except in the United States and Czechoslovakia.

Measles virus infects children via the respiratory tract, but no lesion occurs at the site of initial implantation. Virions enter the local lymphatics, either free or associated with macrophages, and are transported to the local lymph nodes. Here the virus multiplies without producing much cell damage and there is early spread to other lymph nodes and to the spleen. Infected mononuclear cells give rise to multinucleated giant cells, and T lymphocytes are susceptible to infection when they are mitotically active. About 6 days after the initial infection viremia occurs and virions are seeded to all the epithelial surfaces of the body: oropharynx, conjunctiva, skin, respiratory tract, bladder, and alimentary canal. Virus deposited in these sites from the local blood vessels produces foci which spread toward the epithelial surface. Because the epithelia of the conjunctiva and respiratory tract are only one or two cells deep, they undergo necrosis first, some 9–10 days after infection. There is an abrupt onset of illness with a cough, running nose, and inflamed conjunctivae. It seems likely that immune responses contribute to the respiratory damage, malaise, and fever that appear at this stage and get steadily worse until the rash appears. Mucosal foci in the mouth ulcerate on about the eleventh day, to produce the characteristic Koplik's spots. By the fourteenth day, just as circulating antibodies become detectable, the characteristic maculopapular rash appears and the fever falls. This skin rash, unlike that of smallpox which is due directly to viral multiplication in the epidermis, is due in large part to cell-mediated immune responses to viral antigens. The focal infection in measles rarely spreads far from the small blood vessels and the skin lesions do not ulcerate. Measles decreases the resistance of the respiratory epithelium to secondary bacterial infection, hence pneumonia, sinusitis, or otitis media may supervene. It also provides the classic example of the increased severity of a disease due to the effects of malnutrition (see Chapter 6).

Invasion of the Central Nervous System

Because of its critical physiological importance and its vulnerability to damage by any process that increases intracranial pressure, invasion of the central nervous system (CNS) by viruses is always a serious mater. Viruses can spread to the brain by two routes, via the bloodstream (Fig. 5-6) or via peripheral nerve fibers (Fig. 5-7). Access from the blood may, in turn, occur in a number of ways. Growth through the endothelium of small cerebral vessels has been clearly demonstrated in several systems, and there is suggestive evidence that virions may sometimes be passively transferred across the vascular endothelium. Perhaps the more common route, seen especially with viruses that cause meningitis rather than encephalitis, is by traversing the blood–CSF junction in the meninges or choroid plexus. This may open the way to more extensive spread throughout the brain and spinal cord.

The other important route of infection of the CNS is spread via the peripheral nerves, seen in rabies, varicella, and herpes simplex (Fig. 5-7). In the two herpesvirus infections, virions move at different times in each direction, from the body surface to the sensory ganglia or, in reac-

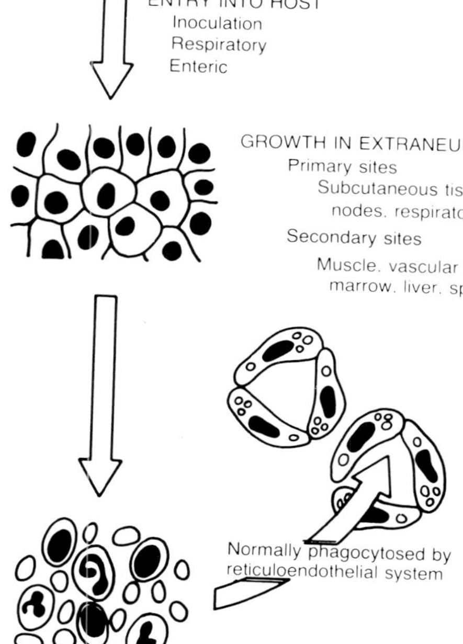

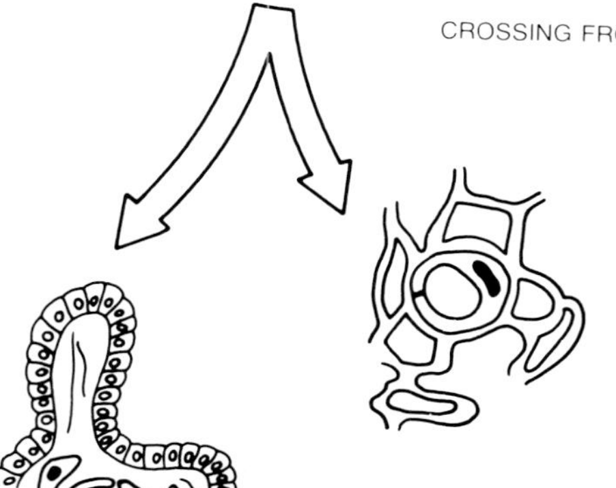

FIG. 5-6. *Steps in the hematogenous spread of virus into the central nervous system. (From R. T. Johnson, "Viral Infections of the Nervous System." Raven Press, New York, 1982.)*

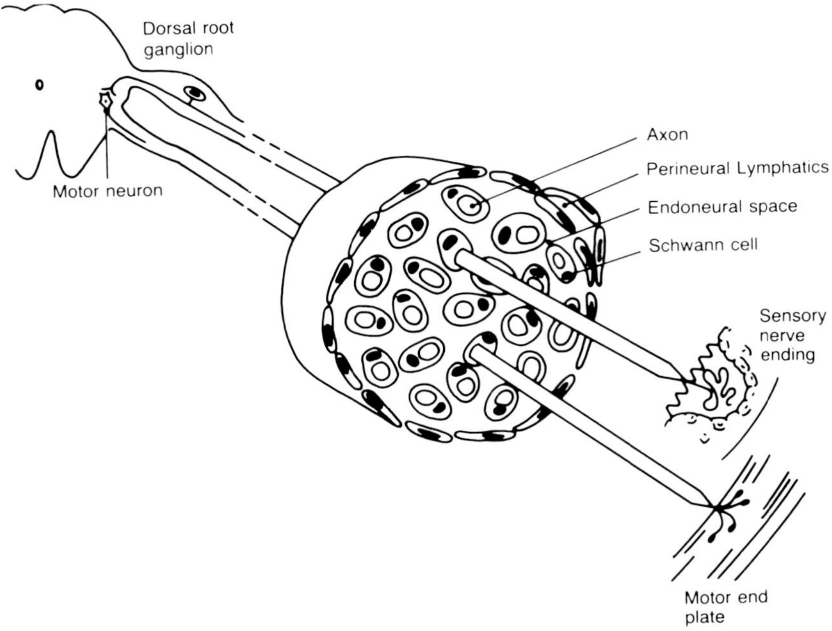

FIG. 5-7. *Neural pathways for CNS infection. Virions can be taken up at sensory or motor endings and moved within axons, endoneural space, or perineural lymphatics or by Schwann cell infection. If movement is via axons, viruses taken up at sensory endings will be delivered selectively to dorsal root ganglion cells and those at motor endings to motor neurons. (From R. T. Johnson, "Viral Infections of the Nervous System." Raven Press, New York, 1982.)*

tivation of herpes simplex or zoster, from the ganglia to the skin. Herpesvirions may travel in the axon or by sequential infection of Schwann cells. Rabies virus passes up the axon, especially from motor axon terminals at neuromuscular junctions, or after exposure to rabies aerosols (as with speleologists in some parts of the world) up the olfactory nerve.

Cytocidal infections of neuronal cells, whether due to poliovirus, flavivirus, togavirus, or herpesvirus, are characterized by the three hallmarks of encephalitis: cell necrosis, phagocytosis by glial cells (neuronophagia), and perivascular infiltration of inflammatory cells, which is an expression of cell-mediated immunity (see Chapter 6). The cause of symptoms in other CNS infections is more obscure. Rabies virus is noncytocidal in cultured cells; in infected animals it evokes none of the inflammatory reactions or cell necrosis found in the encephalitides, yet it is highly lethal for most species of animal. With certain other viruses, infection of neurons causes no symptoms; for example, the extensive CNS infection of mice congenitally infected with lymphocytic choriomeningitis virus, readily demonstrable by fluorescent antibody staining, has no recognizable deleterious effect. Still other pathological changes are produced by some of the viruses that cause slowly progressive diseases of the CNS (see Chapter 7). In scrapie of sheep, for example, there is slow neuronal degeneration and vacuolization; in visna (another chronic disease of sheep), changes in cell membranes lead to demyelination.

Postinfectious encephalitis is most commonly seen after measles (about 1 per 1000 cases), more rarely after rubella and varicella, and in about 1 per 80,000 primary vaccinations against smallpox. The pathological picture is predominantly demyelination without neuronal degeneration—changes unlike those produced by the direct action of viruses on the CNS. Allied with the usual failure to recover virus from the brain of fatal cases this has led to the view that postinfectious encephalitis is probably an autoimmune disease.

Invasion of Other Organs

Almost any organ in the body may be infected with one or another kind of virus. Among the more important are infection of the testis or accessory sexual organs, which may be associated with excretion of virus in the semen and thus the potential for transfer during sexual activities. In like manner, localization of circulating virus in the salivary glands, mammary glands, kidney tubules, and lungs leads to excretion in the saliva, milk, urine, and respiratory secretions. Viral infection of the pancreas has been suggested as one cause of early onset diabetes. Infection of muscle cells occurs with several different togaviruses and coxsackieviruses, while infection of synovial cells by rubella virus or Ross River alphavirus produces arthritis. Infection of the liver is of course extremely important because of the severity of the acute hepatitis that results, for example, in yellow fever and hepatitis A, B, and non-A, non-B, and the possibility with hepatitis B of the establishment of a chronic carrier state that may eventually result in hepatocellular carcinoma (see Chapters 8 and 13).

Infection of the Fetus

Most viral infections of the mother have no harmful effect on the fetus, but in some diseases virus crosses the placenta to reach the fetal circulation, probably after establishing foci of infection in the placenta. Severe cytopathogenic infections cause fetal death and abortion, a result

TABLE 5-6
Congenital Viral Infections

SYNDROME	DISEASE	HOST
Fetal death and abortion	Smallpox (extinct)	Man
	Bluetongue vaccine	Sheep
Severe neonatal disease	Cytomegalovirus	Man
	Rubella	Man
Congenital defects	Rubella	Man
	Varicella	Man
	Hog cholera vaccine	Pig
Inapparent, with lifelong carrier state	Lymphocytic choriomeningitis	Mouse
Inapparent, with integrated viral genome	Murine leukemia	Mouse
	Avian leukosis	Chicken

that was common in smallpox, for example. More important, paradoxically, are the effects of less lethal viruses such as rubella and cytomegalovirus (Table 5-6).

Since Gregg's initial observations in 1941, it has been recognized that there is an association between certain congenital abnormalities and maternal rubella contracted in the early months of pregnancy. A variety of abnormalities commonly occur, of which the most severe are deafness, blindness, and congenital heart and brain defects. These defects may not be recognized until after the birth of an apparently healthy baby, or they may be associated with severe neonatal disease—hepatosplenomegaly, purpura, and jaundice—to comprise the "congenital rubella syndrome." Little is known of the pathogenesis of congenital abnormalities in rubella. Immunological tolerance does not develop; children who have contracted rubella *in utero* display high titers of neutralizing antibodies throughout their lives, but there may be some diminution in cell-mediated immunity. Rubella virus is relatively noncytocidal; few inflammatory or necrotic changes are found in infected fetuses. The retarded growth may be due to slowing of cell division leading to the reduced numbers of cells observed in many of their organs. Clones of persistently infected cells might be unaffected by the maternal antibody that develops during the first weeks after the maternal infection, even though such antibody could limit fetal viremia.

Cytomegalic inclusion disease of the newborn results from infection acquired congenitally from mothers suffering an inapparent cytomegalovirus infection during pregnancy. The important clinical features in neonates include hepatosplenomegaly, thrombocytopenic purpura, hepatitis and jaundice, microcephaly, and mental retardation.

Apart from infection of the fetus via the placenta, germ-line transmission of the retroviruses, as integrated cDNA, occurs in many species of animals, but apparently not in man. More important in human medicine because of its epidemiological implications (see Chapter 9) is the transovarial transmission of arboviruses, notably bunyaviruses and some flaviviruses, in mosquitoes and ticks.

VIRUS SHEDDING

The last stage in pathogenesis is the shedding of infectious virions, which is necessary to maintain infection in populations of animals or human beings (see Chapter 9: Epidemiology). In some infections, such as rabies in man, shedding does not occur—the infection of man with rabies virus is a dead end infection. Shedding usually occurs via one of the body surfaces involved in the entry of viruses. With localized infections the same surfaces are involved in both entry and exit (see Fig. 5-2); in generalized infections a greater variety of modes of shedding is often found (see Fig. 5-4).

Skin

The skin is an important source of virus in diseases in which transmission is by direct contact via small abrasions in the skin, e.g., molluscum contagiosum, warts, and genital herpes. Several poxviruses may be spread from animals to man, and sometimes from man to animal, by contact with skin lesions, e.g, the viruses of cowpox, vaccinia, orf, and milker's nodes.

Although skin lesions are produced in several generalized diseases, only a few species of viruses are shed from these skin lesions in a way that leads to transmission. Virions are not excreted from the maculopapular skin lesions of measles, nor from the rashes associated with picornavirus, togavirus, or flavivirus infections. Smallpox and herpesvirus infections, on the other hand, produce a vesiculopustular rash in which virus is plentiful in the fluid of the lesions. Even here, however, virions in saliva are much more important than those in the skin lesions as far as transmission is concerned.

Respiratory Tract

The many different viruses that cause localized disease of the respiratory tract are shed in saliva or mucus and fluid expelled from the respiratory tract during coughing, sneezing, and talking.

Viruses are also shed from the respiratory tract in several systemic infections, e.g., measles, chickenpox, and rubella. The herpesviruses that cause infectious mononucleosis and cytomegalovirus disease are shed mainly in oropharyngeal secretions.

Enteric Tract

All viruses that infect the enteric tract are shed in the feces. Large numbers of virions of the viruses that infect the respiratory tract are swallowed, but the only ones found in an infectious state in the feces are adenoviruses and enteroviruses, since, at least in man, rhinoviruses and most enveloped respiratory viruses are inactivated by the acidic stomach contents or the bile salts encountered in the duodenum.

Urogenital Tract

A number of viruses, e.g., mumps and measles, are shed in urine, and viruria is often prolonged in cytomegalovirus and polyomavirus infections and congenital rubella. Viruria is lifelong in arenavirus infections of rodents.

Other Routes of Shedding

Milk. Several kinds of virus are excreted in milk, which may serve as a route of transmission of cytomegaloviruses and some of the tick-borne flaviviruses.

Semen. The most important virus excreted in semen is the retrovirus HTLV-III, the cause of the acquired immune deficiency syndrome.

Blood and Internal Organs. The blood is important as the source from which arthropods acquire arboviral infection and blood may also be the route of transfer of viruses to the ovum or fetus. Blood transfusion used to be an important source of hepatitis B and HTLV-III before proper screening methods were developed; it is still a major mode of infection with non-A, non-B hepatitis and less commonly other viruses, e.g., cytomegalovirus. Contaminated syringes play a similar role among drug addicts and, in western society, are an important source of hepatitis B, delta, and non-A, non-B.

FURTHER READING

BOOKS

Johnson, R. T. (1982). "Viral Infections of the Nervous System." Raven Press, New York.
Mims, C. A., and White, D. O. (1984). "Viral Pathogenesis and Immunology." Blackwell, Oxford.

Notkins, A. L., and Oldstone, M. B. A., eds. (1984) "Concepts in Viral Pathogenesis." Springer-Verlag, Berlin and New York.

REVIEWS

Bablanian, R. (1975). Structural and functional alterations in cultured cells infected with cytocidal viruses. *Prog. Med. Virol.* **19,** 40.

Fenner, F. (1948). The pathogenesis of the acute exanthems. *Lancet* **2,** 915.

Halstead, S. B. (1981). The pathogenesis of dengue. Molecular epidemiology in infectious disease. *Am. J. Epidemiol.* **114,** 632.

Mills, E. L. (1984). Viral infections predisposing to bacterial infections. *Annu. Rev. Med.* **35,** 469.

Mims, C. A. (1981). Vertical transmission of viruses. *Microbiol. Rev.* **45,** 267.

Shatkin, A. J. (1983). Molecular mechanisms of virus-mediated cytopathology. *Philos. Trans. R. Soc. London, Ser. B* **303,** 167.

Sweet, C., and Smith, H. (1980). Pathogenicity of influenza virus. *Microbiol. Rev.* **44,** 303.

Tyrrell, D. A. J. (1983). How do viruses invade mucous surfaces? *Philos. Trans. R. Soc. London, Ser. B* **303,** 75.

CHAPTER 6

Host Responses to Viral Infections

Basic Immunology 147
Antiviral Immunological Mechanisms 159
Recovery from Viral Infection 167
Immunity to Reinfection 170
Immunopathology 172
Determinants of Host Resistance 175
Interferons 182
Further Reading 190

BASIC IMMUNOLOGY

Classically, the immune response was considered to have two arms: (1) the antibody response, referred to as *humoral immunity*, and (2) a T lymphocyte-mediated response, known as *cell-mediated immunity (CMI)*. Today, the distinction is blurred by the realization that cells, including but not exclusively T lymphocytes, are vitally involved in the generation of both the "humoral" and the "cell-mediated" response. Indeed, a cardinal feature of the immune response is its dependence on cooperation, usually via soluble mediators, between numerous subsets of cells in an intricate and delicately controlled network. The cells involved in antiviral immune responses are shown diagrammatically in Fig. 6-1.

Lymphocytes are the prime movers in the immune response. Any given lymphocyte displays receptors with specificity for only a single *antigenic determinant (epitope)*. This is the basis of the most fundamental characteristic of the immune response to antigens, namely its remarkable specificity.

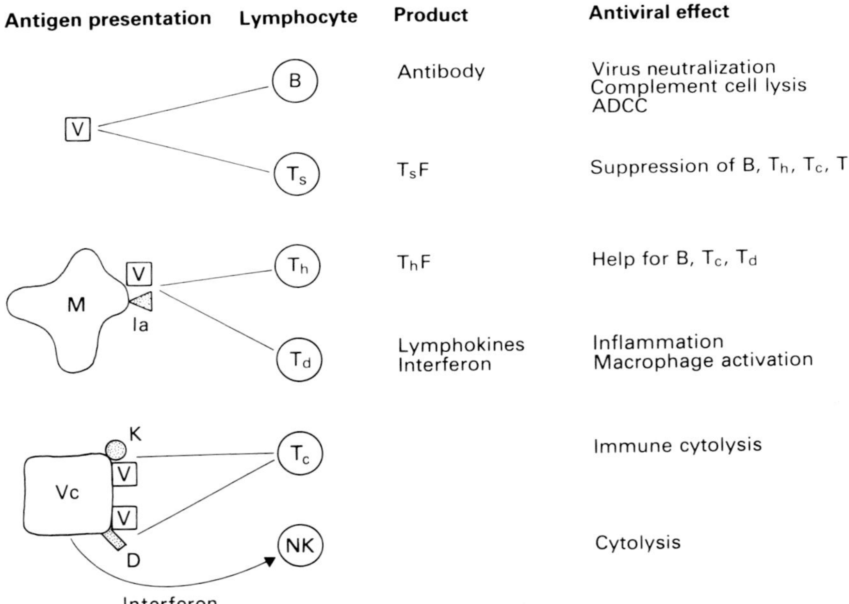

FIG. 6-1. *Cells involved in antiviral immune responses. V, Virus or viral antigen; M, antigen-presenting macrophage; Vc, virus-infected cell. (From C. A. Mims and D. O. White, "Viral Pathogenesis and Immunology." Blackwell Scientific Publications, Oxford, 1984.)*

T lymphocytes (*T cells*), so named because of their dependence on the thymus for their maturation from the pluripotent hemopoietic stem cell, are distinguishable from B lymphocytes (*B cells,* derived from the bursa in birds or its equivalent, the bone marrow, in mammals) not only by their somewhat different antigen receptors, but also by other surface markers and by completely different functions. Both T and B cells respond to the binding of antigen through their antigen receptors by differentiating into physiologically active lymphoblasts (*blasts*) and by dividing to form an enlarged clone of cells (*clonal expansion*), some of which are long-lived and are involved with immunological *memory.* But there the similarity ends. B blasts (plasma cells) secrete specific antibody, whereas T blasts secrete antigen-specific and nonspecific soluble factors known as *lymphokines* or *interleukins.* Furthermore, unlike B cells, most T cells cannot bind soluble antigen; they recognize a foreign antigen only when it is presented in association with "self" *MHC (major histocompatibility) antigen* on the surface of another cell.

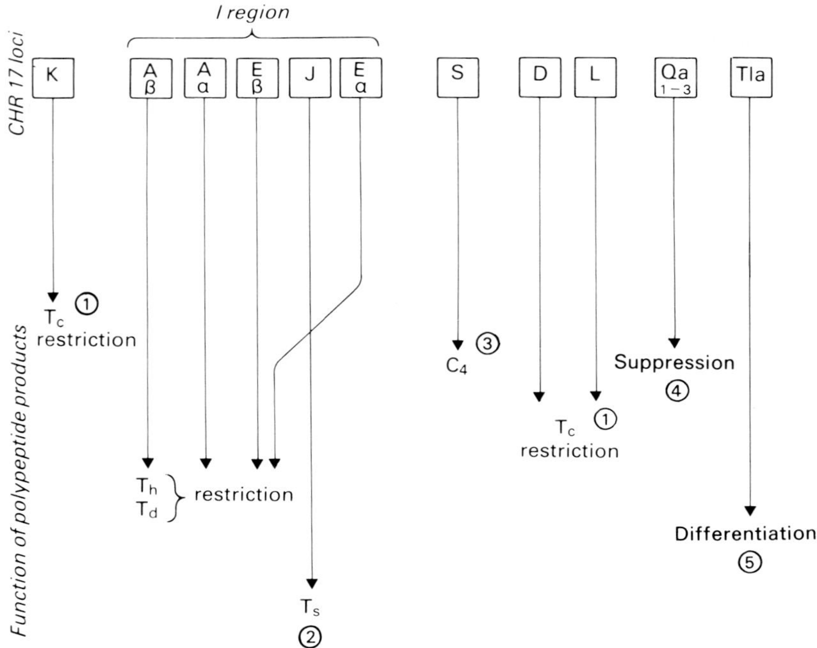

FIG. 6-2. *MHC (H2) loci in the mouse. DNA cloning methods show that there are more loci than illustrated here. There are three classes of loci, or genes: class I—K, D, L, Qa, Tla; class II—I region; class III—S. KDL antigens are displayed on most cells in the body. I Region (Ia) antigens are displayed on antigen-presenting cells, e.g., macrophages. (1) Also may be T_d cell restriction. (2) Suppressor T cells bear I-J. (3) Controls complement activity. (4) T cells with this antigen generate suppressor effects. (5) T cell differentiation antigen. (From C. A. Mims and D. O. White, "Viral Pathogenesis and Immunology." Blackwell Scientific Publications, Oxford, 1984.)*

Functionally, T lymphocytes are classified as *helper, suppressor, cytotoxic,* or *delayed-type hypersensitivity* T cells on the basis of these four distinct physiological activities. Recent evidence with T cell clones suggests that a single cell type can discharge more than one of these functions but that it may recognize viral antigen in association with only a particular class of self MHC antigen (see Fig. 6-2). *Helper T (T_h or T_H) cells,* otherwise known as *T inducer* or *amplifier* cells, recognize foreign (e.g., viral) antigen only in conjunction with self *MHC class II* antigen (D/DR in man; I-A or I-E in mouse), usually on the surface of a *macrophage, dendritic cell,* or *B cell.* Various subsets of T_h cells secrete antigen-specific and nonspecific *helper factors (ThF)* that bind to and activate B cells, T_d cells, and T_c cells, respectively. *Cytotoxic T (T_c) cells* on the other hand, generally recognize foreign antigen in association with self *MHC class I* antigen (HLA-A/B/C in man; H-2D, K, or L in the mouse); their function is to lyse virus-infected or cancer cells displaying foreign antigen on their surface. *Delayed-type hypersensitivity T (Tdth or T_d or T_D) cells* recognize antigen in association with either class I or class II antigen. T_dCells secrete a wide variety of lymphokines which attract and activate macrophages and other T cells, so greatly augmenting the immune response. *Suppressor T (T_s) cells* can probably bind viral antigen directly; they secrete antigen-specific, nonspecific, or idiotype-specific (antiidiotype) *suppressor factors (TsF)* which can regulate the immune response by suppressing the activity of B, T_d, T_h, or T_c cells after they have fulfilled their function and are no longer needed.

Several additional cell types are implicated in the immune response to viral infection. They include *NK (natural killer) cells, K (killer) cells,* and *polymorphs* as well as cells of the *macrophage/monocyte* series. Furthermore, there are important, though immunologically nonspecific, molecules other than the lymphokines mentioned so far. The *complement* system consists of a series of serum components that can be activated, particularly in the presence of antigen–antibody complexes, to "complement" the immune response. The *interferons* are a family of proteins with immunomodulatory as well as antiviral effects, the production of which is induced by viral infection of any cell, or by antigenic stimulation of lymphocytes. The role of each of these cells and molecules in the immune response will now be described in more detail.

T Lymphocytes

T lymphocytes are most easily distinguished from B lymphocytes and other cells by the presence of certain surface "markers," namely characteristic antigens recognizable by antibodies (Table 6-1). In man these are the so-called OKT antigens, and in mice the Thy 1, Lyt, and Qa antigens. Since far more is known about the immunology of the humble mouse, it will be the subject of much of what follows.

All murine T cells carry Thy 1. Most thymocytes are also Lyt $1^+2^+3^+$ (Ly 123 for short); the short-lived nonrecirculating "T_1" population retains this phenotype, whereas about 30–35% differentiate into Lyt $1^+2^-3^-$ (Ly 1), and 5–10% into Lyt $1^-2^+3^+$ (Ly 2,3) cells. More properly perhaps these populations should be designated Ly 1^{high} 2, 3^{low} and Ly-1^{low} 2, 3^{high}, as these Ly markers are not completely absent from any T cells that are screened by sufficiently sensitive methods. The Ly 1 population includes the T_h and most of the T_d cells, sometimes called

TABLE 6-1
Surface Markers of Murine Lymphocytes[a]

LYMPHOCYTE	Thy 1	Ly 1	Ly 2,3	I-A	I-J	Ig	FcR[b]	C3R[c]
T_h	+	+	−	+	−	−	−	−
T_s	+	−	+	+	+	−	−	−
T_c	+	−[d]	+	−	−	−	−	−
T_d	+	+ or −	− or +	+	−	−	−	−
B	−	−	−	+	−	+	+	+
NK	−(±)	−	−	−	−	−	±	−
Mφ	−	−	−	+	−	−	+	+

[a] From C. A. Mims and D. O. White, "Viral Pathogenesis and Immunology." Blackwell Scientific Publications, Oxford (1984).
[b] Fc receptor.
[c] C3 receptor.
[d] Some activated T_c claimed to be Ly $1^+2,3^+$.

"inducers" or "amplifiers," while the Ly 2,3 population contains mainly T_c ("effector" T cells), T_s, and some T_d. The Qa antigens, like Lyt, are exclusive to T cells. T_s cells generally bear I-J antigen. However, though extremely useful for the purposes of identifying, purifying, or removing particular subpopulations of T cells, surface markers are not entirely reliable indicators of the specialized function of the cell in question. Such markers change during ontogeny, differentiation, antigen activation, or physiological state. Moreover, individual cells may carry out more than one function; for example, cloned T_c cells may effect delayed hypersensitivity.

However, there is a solid body of evidence showing that the function of a T cell is closely correlated with the particular type of MHC antigen it expresses on its surface and, correspondingly, on the soluble antigen-specific factors it secretes, e.g., I-A and I-E gene products are expressed on activated helper T cells and their factors, but I-J on suppressor cells and their factors (Table 6-1). Moreover, distinct classes of MHC antigen on target cells direct particular classes of T cells to them (Figs. 6-1 and 6-2): in general H-2K/D/L are recognized (in association with viral antigen) by T_c cells, and I-A and I-E by T_h and T_d cells. In this sense the mode of presentation of antigen predetermines the nature of the immune response. Since class I MHC antigens are present on all cells, viral antigens on the surface of infected cells elicit a T_c response, which can in turn destroy those target cells. On the other hand, since class II MHC antigens are present on macrophages and related types of cells, presentation of viral antigens in this context turns on T_h cells, which are needed to set in train the antibody synthetic pathway, and T_d cells, which augment the immune response by attracting and activating macrophages. The role of T cells may be envisaged as that of cell surface surveillance, constantly browsing over the plasma membrane searching for "altered self."

T lymphocytes differentiate into specialized effectors following binding of specific antigen. Clearly, at least two or three steps are involved: antigen-specific recognition, then activation of the cell to differentiate and divide. As a result of cell division the response is amplified, and also memory cells are generated. Little is known about whether separate signals are required to trigger the induction of different physiological functions, or the development of immunological memory. The effector response is generally transient, e.g., T_c and T_d activities have usually peaked by 1 week after delivery of antigen and disappeared by 2–3 weeks. This may be attributable to the fact that immunological destruction of infected cells removes the antigenic stimulus. It is also possible that the key to immunoregulation may be suppressor T cells, which are first demonstrable at about the time the other cell types decline in number.

T lymphocyte populations seem to contain a larger proportion of cells responding to cross-reactive determinants on the eliciting antigen than do B cells. For example, whereas immunization with the purified hemagglutinin molecule of a given strain of influenza virus elicits antibodies that are predominantly strain specific, the T_h, T_d, and T_c cells are predominantly (though not exclusively) cross-reactive between strains and even between subtypes. This may facilitate recovery following subsequent infection with a different serotype. In so far as such cross-reactive T cells would not prevent the virus from successfully establishing infection and spreading to other hosts, they would at the same time protect the host against serious disease yet permit the long term survival of the parasite.

T cells act principally via the soluble factors (*lymphokines* or *interleukins*) which they secrete following antigen binding. Some of them are antigen-specific factors, others nonspecific (in their action, though specific in their induction). Among many T cell-derived nonspecific lymphokines that have been described are (1) *interferons* α and γ which are important immunomodulators as well as antiviral agents; (2) *interleukin-2 (IL-2),* a mitogen that promotes the clonal expansion of antigen-activated T lymphocytes; (3) the late-acting *T cell-replicating factor (TRF);* and (4) the *B cell growth factor (BCGF),* which amplifies the clone of B cells following its activation.

The *antigen-specific helper and suppressor factors* constitute the principal mediators of specific T–T cell and T–B cell interactions. Each factor is characterized by two distinct components: (1) a variable segment with specificity for antigen, and (2) an *I* region gene product (generally I-A or I-J). The variable region serves to focus the molecule onto cell-bound antigen; the constant region dictates the physiological effector function. Thus, most helper factors carry I-A or I-E, whereas most suppressor factors carry I-J. Some antigen-specific T cell factors are MHC restricted (see below); others are not.

One of the more fascinating discoveries of modern immunology has been that the unique amino acid sequence that comprises the antigen-combining site in the variable (V) region of an antibody molecule can itself be recognized as a foreign antigen. Hence we mount an *antiidiotypic* immune response against the unique *individual idiotype (IdI)* characterizing the antigen-combining site, as well as against the *cross-reactive idiotypes (IdX)* present on the surrounding framework of the variable region of the H chain. T lymphocytes bearing antiidiotypic receptors recognize the corresponding idiotype on a particular antibody molecule or on antibody-like (idiotypic) receptors on B cells or on other T cells; antiidiotypic helper factors are secreted and assist the corresponding clone of B cells to make antiidiotypic antibody, and so on. In theory, the whole process could be extrapolated indefinitely as in a hall of mirrors. Furthermore, antiidiotypic antibodies and T cells can either activate or suppress T cells with the corresponding idiotype, depending on other characteristics of the cells and factors involved. Indeed, some immunologists believe that idiotype–antiidiotype interactions contribute to an intricate web of positive and negative feedback controls which are needed to regulate all aspects of the immune response, and without which the immune response would run out of control because of the cascade effect of the numerous known amplifying signals. The mind boggles at the potential complexity of such a cybernetic network.

B Lymphocytes

Some of the multipotential hemopoietic stem cells originating from fetal liver and adult bone marrow differentiate via pre-B cells into immature B cells, which are characterized by specific antigen-binding receptors of the IgM isotype, and by receptors for complement (C3′) and for the Fc portion of immunoglobulin, and other cell surface components involved in triggering (Table 6-1). On encounter with antigen, the particular clones of B cells bearing receptors (antibody) complementary to any of the several epitopes (antigenic determinants) on that antigen bind it and, after receiving the appropriate antigen-specific and nonspecific signals from helper T cell factors (ThF), respond by division and differentiation into antibody-secreting plasma cells. B cells may also process antigen and present peptides, in association with class II MHC antigen, to T cells, which in turn respond by secreting helper factors. Each clone of plasma cells secretes antibody of only a single specificity and avidity, corresponding to the particular receptors it expresses. Early in the response, when large amounts of antigen are available, there is an opportunity for antigen-reactive B cells to be triggered even if their receptors fit the epitope with relatively poor *affinity* hence bind the antigen with low *avidity*. Later on, when only small amounts of antigen remain, B cells with receptors binding antigen with high affinity are selected, hence the avidity of the antibody secreted increases correspondingly.

Clonal expansion of B lymphocytes on exposure to antigen also generates a population of memory cells. On reexposure to the same antigen even years later, these memory cells respond promptly with renewed multiplication and the production of larger amounts of specific antibody, mainly of the IgG class, after a delay of only a day or two.

Antibody synthesis takes place principally in the spleen and lymph nodes, gut-associated lymphoid tissues (GALT), and bronchus-associated lymphoid tissues (BALT). The spleen and lymph nodes receive foreign antigens via the blood or lymphatics and synthesize antibodies mainly restricted to the IgM (early in the response) and IgG classes. On the other hand, the submucosal (interstitial) lymphoid concentrations (follicles) of the respiratory and alimentary tracts, such as the tonsils and Peyer's patches, receive antigens directly from the overlying epithelial cells,and make antibodies mainly of the IgA class. Plasma cells are more numerous in the lamina propria of the small intestine than anywhere else in the body and the great majority of these secrete only antibody of the IgA class, directly into the mucus which coats the gut. Indeed, IgA-specific B, T_h, and contrasuppressor lymphocytes preferentially home to and localize in submucosal lymphoid tissue such as GALT, BALT, and exocrine gland interstitia.

Antibody

The different classes (*isotypes*) of immunoglobulin, with some of their properties, are shown in Table 6-2. The major circulating class of antibody is immunoglobulin G (IgG). It continues to be synthesized for many years, and often for life, following systemic viral infection and is the principal mediator of protection against reinfection. The four subclasses of IgG in man differ in the "constant" region of their heavy chains and consequently in their biological properties such as placental

TABLE 6-2
Properties of Immunoglobulin Classes in Man[a]

PROPERTY	IgG	IgM	IgA[b]	IgE	IgD
Molecular weight	150,000	900,000	385,000 (170,000)	200,000	185,000
Heavy chain	γ	μ	α	ϵ	δ
Subclasses[c]	γ1,2,3,4	μ1,2	α1,2	—	—
Half-life (days)[d]	25	5	(6)[e]	2	3
Percentage of total Ig	80	6	(13)	0.002	0–1
Complement fixation	+	++	±	—	?
Transfer to offspring	Via placenta	Nil	Via milk	Nil	Nil
Functional significance	Major systemic Ig	Appears early Multivalent	Mucosal antibody	Allergenic responses	Present on B cells

[a] From C. A. Mims and D. O. White, "Viral Pathogenesis and Immunology." Blackwell Scientific Publications, Oxford (1984).

[b] Data for secretory IgA; serum IgA in parentheses.

[c] Differ in heavy chains and biological activity.

[d] Half-life generally shorter for smaller animals, e.g., 2.5 days (IgG) and 0.5 days (IgM) in the mouse.

[e] Strictly speaking, the half-life of secretory IgA on mucosal surfaces is measured in minutes rather than days, because it is soon carried away in mucus secretions.

passage, complement fixation, and binding to phagocytes. Much antiviral IgG is of the IgG_1 and IgG_3 subclasses.

IgM is a particularly avid class of antibody that is formed early in the immune response, before being replaced by IgG. Hence, specific antibodies of the IgM class are diagnostic of recent or chronic infection. IgM antibodies are also the first to be found in the fetus as it develops immunological competence in the second half of pregnancy. Hence the presence of antiviral IgM antibodies in a newborn infant suggests intrauterine infection, because the only maternal antibodies that can pass the placenta to reach the fetus are IgG.

Secretory IgA is the principal immunoglobulin on mucosal surfaces and in milk and colostrum. In man, most serum IgA is in the monomeric form and of the IgA_1 subclass, whereas most secretory IgA is polymeric and is about equally divided between IgA_1 and IgA_2. Whereas most secretory IgA is synthesized locally in submucosal lymphoid tissue, a significant minority is made elsewhere and selectively transported across intestinal, mammary, or salivary gland epithelial cells into intestinal mucus, milk, and saliva, respectively (acquiring, in the process, part of the epithelial cell's IgA receptor, sometimes known as the "secretory piece"). IgA antibodies are important in resistance to infection of the mucosal surfaces of the body, particularly the respiratory, intestinal, and urogenital tracts, and IgA antibody responses are much more effectively elicited by oral or respiratory delivery than by systemic administration of antigen, a matter of importance in the design of certain vaccines (see Chapter 10).

Complement

The classical complement pathway consists of a complicated system of at least 20 distinct proteins and 9 numbered components present in normal serum, which functions by mediating and amplifying immune and inflammatory reactions. The first component (C1) is activated after combining with an antigen–antibody complex, which may be free or on the surface of an infected cell. As a result of the combination of antigen with the Fab binding site of an antibody molecule, the Fc moiety of the Ig molecule is slightly altered to expose an attachment site for the first complement component, C1q. The activated enzyme C1qrs, a protease, cleaves the next component in the complement pathway, and so on. The progressive increase in the number of molecules of successive components produces a cascade reaction (Fig. 6-3). A single molecule of activated C1 generates thousands of molecules of the later components and the final response is thus greatly amplified. The later components have various biological activities, including inflammation and cell destruction, so that an immunologically specific reaction at the molecular level can lead to a relatively gross response in the tissues.

When C3b is generated it becomes bound to the antigen–antibody complex, and the whole complex can now attach to C3b receptors present on macrophages and polymorphs. Phospholipase is formed from the last components (C8 and C9), and if the complex is on the surface of a cell, the phospholipase causes membrane damage and cell lysis.

The system just described is the *classical complement pathway*. There is also an *alternate* (or *alternative*) *complement pathway,* which can be triggered independently of antigen–antibody reactions.

Macrophages

Cells of the macrophage series play a central role in the immune response to viruses. Here, we will confine our attention to the role of macrophages in the normal immune response. In considering the so-called "extrinsic" antiviral activities of macrophages, as well as their intrinsic capacity to digest virions, one must appreciate that the virucidal, cytotoxic, cytostatic, and monokine-secreting potential of this

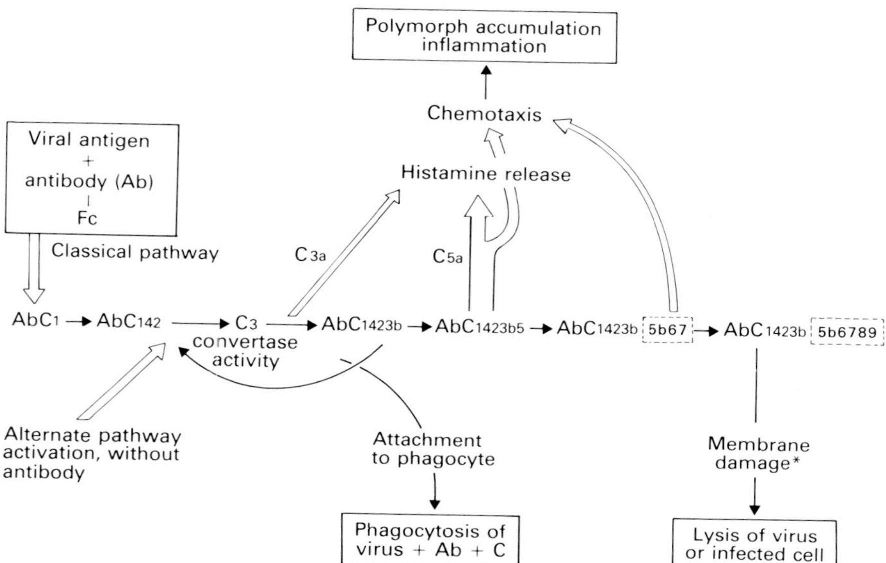

FIG. 6-3. *Diagram to show complement activation sequence and antiviral actions. Unfortunately the components were numbered before this sequence of action was elucidated. Histamine release is also termed "anaphylatoxin activity." *Even if the virus is not lysed, its infectivity is reduced when complement components on its surface interfere with adsorption to cells. Complement thus enhances the action of neutralizing antibody. (From C. A. Mims and D. O. White, "Viral Pathogenesis and Immunology." Blackwell Scientific Publications, Oxford, 1984.)*

diverse class of cells depends very much on the genetics and age of the host, the lineage and state of differentiation of the macrophages, and their state of *activation*. Macrophages attracted to the site of infection and activated by lymphokines secreted by T_d lymphocytes acquire a range of new properties. These include increased powers of phagocytosis, perhaps enhanced efficiency as effector cells, and a tendency toward reduced intrinsic susceptibility to viral multiplication.

Once regarded as mere "accessory cells" of only peripheral interest to immunologists preoccupied with the central problem of immunological specificity, macrophages are now recognized as essential to the determination of that specificity, for only antigen presented in close association with MHC class II (I-A or I-E) antigen on the surface of a macrophage is recognized by T_d and T_h cells, hence only in that way does the immune response commence. It is still not absolutely certain which particular cells of the macrophage/monocyte series are most efficient or most crucial in antigen presentation. *Fixed tissue macrophages* are certainly involved, but evidence is accumulating that *Langerhans cells* in the skin and *dendritic cells* in T cell areas of lymphoid tissues may be just as important. Unlike typical macrophages, these dendritic cells are nonphagocytic and lack Fc receptors; in other respects, including morphology, they resemble macrophages and are Ia^+. Indeed, the central requirement is that the presenting cell must express the Ir gene product; Ia^- macrophages are not involved.

Nor is anything known about the way in which macrophages *process* antigen before they *present* it on their surface. Some degree of proteolytic digestion seems to occur and perhaps some selectivity (by Ia association?) in the display of particular peptides (*determinant selection*).

Macrophages secrete over 50 soluble products (*monokines*). One of the best known is *interleukin-1 (IL-1),* which is secreted by both Ia^+ and Ia^- macrophages and serves as an essential signal for T cell maturation in response to antigen; this vital hormone now appears to be required for the initiation and amplification of all T cell-dependent immune responses and the inflammatory response. Another monokine is *interferon* (see below).

Natural Killer Cells

Natural killer (NK) cells are immunologically nonspecific leukocytes with the capacity to kill tumor cells, virus-infected cells, and, to a lesser extent, some normal cells. Most "endogenous" (nonactivated) NK cells are small to medium lymphocytes which, upon activation, undergo blastogenesis to large, granular lymphocytes, comprising about 5% of the total splenic or peripheral blood leukocytes. Their precise lineage is still a mystery. It is not even known whether NK cells differentiate from T lymphocyte or from myelomonocytic precursors. They are certainly not typical T lymphocytes as they are plentiful in athymic mice, but they have been reported to express T lymphocyte markers in trace amounts and to differentiate from Ly 2 T cells. They are nonphagocytic and only somewhat adherent to surfaces, but they can (weakly) express Fc receptors for IgG and may at times be capable of acting as K cells via ADCC (see below).

The spontaneous, relatively nonspecific cytotoxicity of NK cells is rapidly and dramatically augmented by interferon inducers, especially viruses. Virus-infected cells and cancer cells are in general, but not exclusively, more susceptible to NK-mediated cytolysis than are normal cells. The basis of this partial selectivity is unknown. NK cells do not need to recognize MHC antigens, nor to be armed with antibody, and they demonstrate almost no immunological specificity as we generally understand it.

ANTIVIRAL IMMUNOLOGICAL MECHANISMS

T Lymphocytes

Cytotoxic T Cells. Cytotoxic T (T_c) cells kill cells displaying "foreign" antigen on their surface, including cancer cells and virus-infected cells. While it has been long recognized that virus-coded glycoproteins are incorporated into the plasma membrane of cells infected with viruses that subsequently acquire an envelope by budding, it has only recently become clear that viral antigens are also expressed on the surface of cells infected with certain nonbudding viruses such as adenoviruses, papovaviruses, and reoviruses, rendering these cells also susceptible to T_c-mediated cytolysis. Because at least some viral protein appears in the membrane before any virions have been produced, lysis of the cell at this stage, or shortly thereafter, brings viral replication to a halt—nipping infection in the bud, as it were. The phenomenon is readily demonstrable in cultured cells, using the release of ^{51}Cr as an assay for cytolysis.

In 1974 Zinkernagel and Doherty reported a most unexpected discovery that has had wide-ranging implications for immunology. They observed that virus-infected cells could be lysed only by T_c derived from an animal histocompatible with the target cell, specifically in the K/D region of the H-2 complex in the mouse, since extended to include the L region (Figs. 6-1 and 6-2). The observation has been widely confirmed in other animals, including man, where the *MHC restriction* is to the HLA-A, -B, -C antigen complex. Sophisticated studies with inbred congenic, recombinant, and mutant strains as well as chimeric and neonatally tolerant mice have since confirmed that the effector cell recognizes neither the viral antigen nor self MHC antigen alone, but both in association. A given clone of T_c cells recognizes viral antigen in conjunction with H-2K or -D, but not both. Some viral proteins can associate with K or with D, while others can associate with only one of them. The phenomenon can be mimicked by incorporating purified proteins into liposomes. Mixtures of liposomes containing only viral glycoprotein with others containing only H-2 antigen cannot induce T_c cells; the two antigens must be present in close apposition in the same liposome. Moreover, uninfected histocompatible cells can be rendered susceptible to virus-induced T_c cells by fusion with liposomes containing viral antigen. Recently, T_c cells recognizing viral antigen in association with MHC class II antigen have also been described.

A fundamental question still to be resolved, however, is whether the T cell carries two distinct receptors (one recognizing the MHC and the other separately recognizing the viral antigen) or only one (recognizing both epitopes situated close together, or alternatively recognizing a composite epitope or "neoantigenic determinant" spanning the putative viral–MHC complex). These are sometimes called the *dual recognition* and *altered self* hypotheses. The contribution of virologists in this area has had a seminal influence on the immunologists' pursuit of their holy grail—the T cell receptor. Recent descriptions of putative receptors isolated from cloned antigen-specific T cells may quickly lead to definitive chemical characterization of the antigen-binding paratope of the T cell receptor and the epitope(s) to which it attaches on the surface of the presenting cell.

In most virus infections specific T_c cell activity is demonstrable after 3–4 days, reaches a peak after about 1 week, and declines to a very low level by 2 weeks; thereafter, memory T_c cells persist essentially for life, in the mouse at least. The cytotoxic T cell response is therefore seen to be an early defence mechanism, developing well ahead of the antibody response. MHC restriction ensures that the T_c lymphocyte addresses the infected cell, which it can kill, rather than the virion, which it cannot.

Delayed(-Type) Hypersensitivity. DTH (or DH) is a T cell-mediated inflammatory reaction characterized by infiltration of mononuclear cells. The T_d cells respond to antigen by proliferating and releasing lymphokines which attract blood monocytes to the site and induce them to proliferate and differentiate into activated macrophages. In many experimental viral infections there is no difficulty in demonstrating DTH by the classical methods: (1) swelling of the footpad or ear of a mouse 1–2 days after local intradermal inoculation of antigen; (2) inhibition of migration of macrophages *in vitro* as a result of secretion of MIF by lymphocytes; or (3) proliferation of lymphocytes in response to antigen as monitored by [^{3}H]thymidine incorporation *in vitro*. It is not at all certain that all three techniques measure the same phenomenon.

The classical T_d (Tdth) cell is Lyt $1^+2^-3^-$, and IA (or I-E) restricted, but there is a distinct T_d population which is Lyt $1^-2^+3^+$, and K/D restricted. It had not been appreciated that a T_h or T_c cell could produce a typical DTH response until cloned T_h and T_c cell lines were demonstrated to display both properties. One must now ask whether there is in fact any such thing as a distinct T_d cell.

The fact that inoculation of mice with ultraviolet-inactivated influenza virus gives a much greater DTH response than does live virus suggests that the DTH response is suppressed in normal infection. This suppression may be brought about by a particular subset of T_s cells.

Helper T Cells. Most viral antigens, like the majority of nonviral antigens, are *T dependent*, i.e., they require help from T lymphocytes to induce a satisfactory antibody response. Helper T cells, probably of different subsets, also augment T_c and T_d responses to viral infections. The observed breadth of reactivity of influenza virus-primed T_h cells in infections with heterologous strains of the virus might play a role in the phenomenon of original antigenic sin (see below).

Suppressor T Cells. Antigen-specific T_s cells have not been easy to demonstrate in virus-infected animals so far. In general, T_s responses are preferentially turned on by inoculation of virus orally, or of high doses of UV-inactivated virus systemically. *In vitro,* certain viruses, perhaps acting as mitogens, cause primed or unprimed T_s lymphocytes to proliferate; such cells can have a generalized immunosuppressive effect.

Antibody

Specific antibodies can exert antiviral effects via several distinct mechanisms, some involving interaction with the virion itself, others with the virus-infected cell.

Neutralization of Virus. While specific antibody of any class can bind to any accessible epitope on a surface protein of a virion, only those antibodies which bind with reasonably high avidity to particular epitopes on a particular protein in the outer capsid or envelope are capable of neutralizing the infectivity of the virion. Antibodies directed against irrelevant or inaccessible epitopes of surface proteins, or against internal proteins of the virion, or virus-coded nonstructural proteins (e.g., virus-coded enzymes) can sometimes indirectly exert immunopathological effects but generally play no role in elimination of the infection. In fact, certain nonneutralizing antibodies not only form damaging circulating "immune complexes" but may actually impede access of neutralizing antibody and enhance the infectivity of the virion for some cells, e.g., macrophages (see below).

1. *Classical neutralization* results when antibody binds to the virion and thereby prevents infection of a susceptible cell. The mechanism of neutralization is not fully understood. The particular protein involved is not necessarily the most plentiful on the viral surface; for example, in the case of reovirus, it represents less than 2% of the total mass in the outer capsid. The key protein is usually the one by which the virion is known to adsorb to receptors on the host cell. But neutralization is not simply a matter of coating the virion with antibody, nor indeed of blocking attachment to the host cell. Except in the presence of such high concentrations of antibody that most or all accessible antigenic sites on the surface of the virion are saturated, neutralized virions generally attach satisfactorily to susceptible cells. The block occurs at some point following

adsorption and entry. One hypothesis is that, whereas the virion is normally uncoated intracellularly in a controlled way that preserves its infectivity, a virion–antibody complex tends to be destroyed by cellular enzymes, presumably in the phagosome. It has been demonstrated that antibody to a surface protein of poliovirus can distort the conformation of the capsid *in vitro* such that enzymes can penetrate and destroy the infectivity of the virion.

Our understanding of the antigenicity and immunogenicity of viral proteins has been revolutionized within the space of the last few years by the concerted application of the techniques of modern molecular biology and immunochemistry to an intensive study of the hemagglutinin (HA) of influenza virus. Laver and Webster opened up the field by characterizing the HA molecules from a considerable number of influenza viruses that have arisen sequentially in nature by antigenic drift and shift (see Chapter 4). First, peptide maps, and subsequently complete cDNA or protein sequences revealed that the minor antigenic changes characteristic of drift can be attributed to the gradual accumulation of point mutations in the HA gene, expressing themselves as single amino acid substitutions in the primary sequence of the protein. Then Gerhard and colleagues made a major contribution to immunochemistry with the application of monoclonal antibodies to the antigenic characterization of viral proteins. They analyzed the "reactivity patterns" (in hemagglutination inhibition, radioimmunoassay, or enzyme immunoassay) of a set of anti-HA monoclonal antibodies, to provide an indication of the diversity of distinguishable antigenic determinants (epitopes) on the HA molecule of different strains of influenza virus. When monoclonal antibodies were used to select viral mutants *in vitro*, the mutants were generally found to contain only a single amino acid substitution, this being sufficient to prevent the monoclone in question from binding. Other monoclones, recognizing topographically distinct antigenic sites, could still bind and neutralize the mutant, as could polyclonal antisera against the parent strain. Then came the publication by Wiley and colleagues of the first three-dimensional model of a viral protein—again influenza HA (Fig. 4-5). The positions of the particular amino acids known to vary in naturally occurring or monoclone-selected strains were found to be located predominantly on certain prominent regions on the surface of the exposed "head" of the molecule. Most of the epitopes occur in a large antigenic area extending from the "tip" and "interface" of the molecule to a conspicuous "loop" a little further down. In so far as all these sites are in the immediate vicinity of the "cleft" (just left of the "tip" in Fig. 4-5) which is thought to represent the receptor-binding site, it seems plausible that antibodies binding to any of these epitopes may neutralize the infectivity of the virion (and inhibit hemagglutination) by steric hindrance of adsorption of the virion to cells. The other major antigenic site, the "hinge," is closer to the site of cleavage of the HA monomer into its two component polypeptides HA_1 and HA_2, which is considered to be vitally involved in the fusion of the infecting virus with lysosomal membrane at pH 5; it is possible that neutralizing antibodies binding to this region of the molecule may block the entry and/or uncoating of the virion.

Antibody to the neuraminidase (NA) of influenza virus, though non-neutralizing in the traditional sense, can inhibit release of progeny particles from the plasma membrane of infected cells and significantly slow viral spread. Similarly, antibody to the fusion (F) protein of paramyxoviruses, though unable to prevent initiation of infection, does block the direct cell-to-cell spread of newly formed virions once infection has been established.

2. *Enhancement of neutralization by "enhancing" antibody* is a more recently described phenomenon. Better known perhaps as *rheumatoid factor, enhancing antibody* is directed against novel epitopes on the immunoglobulin moiety of the virion–antibody complex which are revealed as a consequence of antigen binding. Such antibody, usually of the IgM class, is probably made late in most viral infections, particularly in chronic infections, where antigen–antibody complexes are plentiful. Enhancing antibody presumably strengthens, and consequently perhaps broadens, the specificity of antigen–antibody interactions, and also, no doubt, antiidiotype–antibody complexes.

3. *Enhancement of neutralization by complement* is a well-documented phenomenon that itself can occur in a number of different ways which will be discussed below (see Complement).

4. *Opsonization and agglutination* of virions by antibody accelerate clearance by phagocytic cells.

Immune Cytolysis of Infected Cells. Antibodies may also be involved in the destruction or inhibition of virus-infected host cells by a number of potentially important mechanisms.

1. *Antibody–complement-mediated cytotoxicity,* readily demonstrable *in vitro* even at very low concentrations of antibody, is mediated by any antibody that recognizes antigen expressed on the surface of an infected cell and that is capable of activating the complement sequence. Such antigens may include not only the envelope proteins of budding viruses but other structural or nonstructural proteins which may be expressed on the plasma membrane, and even "neoantigenic determinants" formed by association between viral and MHC proteins. Because virus-

coded proteins are expressed on the cell surface before new virions have been produced, *immune cytolysis* of infected cells may play a major role in elimination of the infection. The alternate complement pathway appears to be particularly important in this phenomenon, which is discussed in more detail below (see Complement).

2. *Antibody-dependent cell-mediated cytotoxicity (ADCC)* is mediated, in the presence of antibody, by certain types of normal leukocyte. Unlike cytotoxic T lymphocytes, these effector cells are not immunologically specific; they bind only to target cells to which antibody of the IgG class is specifically attached. At least three types of normal leukocyte can effect ADCC in this way because they bear receptors for the Fc end of IgG molecules (FcR). Some are non-B–non-T lymphocytes known as K (killer) cells, while polymorphs and macrophages also mediate the phenomenon very efficiently.

3. *Inhibition of viral multiplication by antibody* to viral antigens expressed on the surface of infected cells can also occur, in the absence of leukocytes or complement, and without involving lysis of the cell. Neither the mechanism nor the *in vivo* significance of the phenomenon is known.

4. *Modulation of viral antigen on cell surface.* Antibody can also by itself *modulate* viral antigens expressed on the cell surface. Antibody combines with antigen and the complex is redistributed (*patching* and *capping*), then either shed from the cell (*exocytosis*) as an immune complex or internalized (*endocytosis*). This does not totally prevent viral multiplication, but may interfere with assembly. When the fusion protein of measles virus is removed from the cell surface in this way, cell-to-cell fusion and spread may be impeded. The phenomenon may be important in virus persistence (Chapter 7).

Complement

Activation of complement by the *classical pathway*, dependent on specific antigen–antibody interaction, has antiviral effects that can readily be demonstrated *in vitro.*

1. *Activation of complement following attachment of antibody to virions* may result in the following:

a. *Damage to or lysis of the envelope of enveloped virions (virolysis)* resulting in leakage or destruction of viral nucleic acid.
b. *Enhancement of neutralization* of nonenveloped or enveloped viruses by antibody, probably because complement components provide further coverage of viral proteins required for attachment to susceptible cells. In some viral infections the first antibodies that are

formed, which are of low avidity as well as low titer, depend on complement for neutralizing activity.

c. *Opsonization* of the virus particle following which it is more readily ingested and disposed of by phagocytes. In the case of macrophages that are susceptible to infection, uptake via C3b receptors perhaps makes infection of the cells less likely.

2. *Activation of complement following attachment of antibody to infected cell* may result in lysis of that cell (see above). Sometimes this is not sufficient for cell lysis unless the alternative pathway is also activated.

3. *Activation of complement following interaction of antibody with viral antigens in tissues* leads to inflammation and the accumulation of leukocytes, with possible antiviral effects (see below).

Potentiation of the effect of antibody by complement would be more likely to be important early in the course of infection, when antibody is of lower avidity and present in smaller quantities. Some of the above effects, however, can also be mediated following the activation of complement by the *alternative pathway,* independent of antibody. Several viruses, or virus-infected cells, directly activate the alternative complement pathway. This is an important phenomenon because it can occur immediately after viral invasion, before immune responses have been generated.

Macrophages

The antiviral activities of macrophages differ with the genetic background and age of the host, the particular subpopulation of cells involved, their state of activation, and the capacity of the virus to multiply in them.

Viral Growth in Macrophages. Some viruses infect macrophages. Indeed, as we shall see in the next chapter, several viruses notorious for their ability to establish lifelong chronic infections grow preferentially or exclusively in macrophages. Certain other viruses undergo an abortive cycle of replication in macrophages, i.e., the infection is nonproductive, but the macrophage dies nevertheless. There are some classical genetic studies demonstrating that the intrinsic susceptibility of macrophages to a particular virus as tested *in vitro* can determine the susceptibility of that strain of mouse to that virus. In another study the virulence of strains of ectromelia (mousepox) viruses correlated with their ability to grow in liver macrophages. Moreover, young mice, which are susceptible to herpes simplex virus, can be protected by transfusion of macrophages from older mice, which are resistant. Finally, macrophages normally susceptible to destruction by a particular virus may become refractory following activation. While it is probably exceptional to find that susceptibility or resistance of an animal can be determined wholly by the genetics, age, or state of activation of its macrophages, it is nevertheless apparent that the susceptibility or resistance of macrophages to infection is an important consideration in determining the outcome of the encounter.

Phagocytosis ± Opsonization. The uptake of virions by macrophages is increased in the presence of antiviral antibody. Generally such opsonization will also ensure that the virus is neutralized, hence will fail to grow in the macrophage even if it had that capacity originally. There is, however, one fascinating exception. In the presence of a subneutralizing concentration of antibody or an excess of nonneutralizing antibody, togaviruses (and certain others) are actually taken up more efficiently by macrophages (via the Fc receptors on the macrophage, to which the virus–antibody complex binds) and the virus grows intracellularly to high titer. Such *enhancing antibody* may underlie the pathogenesis of dengue hemorrhagic fever (see Chapter 20).

Interferon Production. Macrophages rank among the most productive sources of interferon in the body. The subject will be discussed further below.

Antigen Presentation. As with noninfectious immunogens, viral proteins are presented to T cells in association with Ia antigens on the surface of Ia^+ cells of the macrophage series. We know virtually nothing of the nature of the *processing* that proceeds this *presentation,* nor of the relative efficiency of presentation by macrophages that destroy the virus versus those that are infected, abortively or productively.

ADCC. Macrophage-mediated ADCC, a mechanism whereby these nonspecific effectors, once "*armed*" with antibody, can specifically kill virus-infected targets, has been described above.

Direct Cytotoxicity. There are indications that adherent, macrophage-like cells can at times preferentially destroy virus-infected cells in the absence of specific antibody. Macrophages bind directly to the glycoproteins of certain viruses on the surface of infected cells.

Natural Killer Cells

Virologists have been prominent in the study of NK cells since their discovery that NK cell activity is greatly enhanced by virus infection. Indeed, NK cytotoxic activity peaks within 1–2 days—earlier than any of the responses to infection other than interferon synthesis. It is significant then that interferon is the principal inducer of NK cell activation (although the glycoproteins of some viruses can activate NK cells directly). While it is tempting to surmise that NK cells represent the first line of defense against viral infections, persuasive *in vivo* evidence to support such a role is not currently available.

RECOVERY FROM VIRAL INFECTION

Cell-mediated immunity (CMI), antibody, complement, phagocytes, and interferon are all doubtless involved in the response to all viral infections and are together responsible for recovery. These components, illustrated diagrammatically in Fig. 6-4, all normally operate together, and to some extent any attempt to assess their relative importance is an artificial exercise. Indeed, it has proved remarkably difficult to assign a determinative role *in vivo* to any single antiviral immunological mechanism in recovery from any single viral disease. The fact that a particular cell, substance, or phenomenon is unequivocally demonstrated to produce antiviral effects in culture does not prove its importance *in vivo*. The reductionist approach simply allows us to analyze the phenomenon in isolation using relatively defined reagents. Having alerted ourselves in

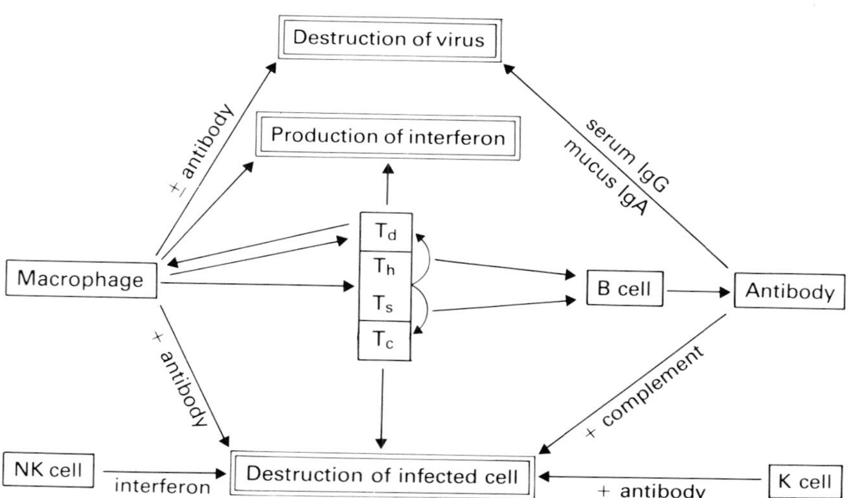

FIG. 6-4. *Immune response to virus infection. (From C. A. Mims and D. O. White, "Viral Pathogenesis and Immunology." Blackwell Scientific Publications, Oxford, 1984.)*

this way to the potential importance of the mechanism in question we must return to the living animal for proof of its biological reality. Yet even this is not sufficient. The fact that one can save the life of an animal by adoptive transfer of primed cytotoxic T cells is dramatic indeed, but does not establish the *relative* importance of this cell in recovery under natural circumstances. For instance, the dosage, timing, or state of activation of the artifically transferred cells may be quite out of proportion to the natural counterpart. Or the transferred cell may have a second, more crucial effect or function *in vivo,* or secrete an unknown product. Indeed, the cell may be capable of curing infection yet be of negligible importance in natural infection compared with another mechanism not currently under test. In spite of these important reservations, however, convincing data are accumulating, particularly following the development of techniques for the cloning of T lymphocytes.

The experimental approach least subject to laboratory artifacts is simple clinical observation of the occurrence and progress of viral infections in children or animals suffering from congenital immunodeficiencies. There are many such diseases but few of them present as a "pure" T or B cell deficit. Athymic ("nude") mice, which have almost no T cells, have been widely used to evaluate T cell responses, but T-dependent antibody responses are also affected, and the increased NK activity of these mice must be taken into account. Similarly, attempts have been made to remove selectively one arm of the immune response, but it is often difficult to eliminate one component without affecting the others. For instance, cyclophosphamide and antithymocyte serum have been used to eliminate antibody synthesis and CMI responses, respectively, and have provided useful information, but these treatments lack precision, and such experiments often leave ambiguities. An alternative strategy is to ablate completely all immune potential, then add back one or more of the separate components. Adult mice are thymectomized to eliminate the source of T cells, X irradiated to eliminate lymphoid cells, and then reconstituted with bone marrow cells to restore hematopoiesis. The separate components of the immune response can now be restored one at a time to these animals and the effect on resistance to virus infections studied.

On the whole, the data obtained from these various experimental approaches indicate a key role for T lymphocytes, or CMI in recovery from systemic viral infections. Lymphocytes and macrophages normally predominate in the cellular infiltration of virus-infected tissues; in contrast with bacterial infections, polymorphs are not at all plentiful. Patients with severe T cell deficiencies due to thymic aplasia, lymphoreticular neoplasms (e.g., Hodgkin's disease), or posttransplant immunosuppression show increased susceptibility to cytomegalovirus, herpes simplex, varicella-zoster, measles, and certain other viral infections. When infants with thymic aplasia become infected with measles virus there is no rash but an uncontrolled and progressive growth of virus in the respiratory tract, leading to fatal giant cell pneumonia. Evidently, in the normal child, the CMI response controls infection in the lung and also plays a vital role in the development of the characteristic skin rash. Similarly, infants with thymic aplasia, when vaccinated against smallpox with live vaccinia virus, fail to control this normally trivial localized lesion, which continues to enlarge, eventually destroying an extensive area of tissue and often proving fatal.

In mice, T cell depletion by neonatal thymectomy or antilymphocyte serum treatment has been shown to increase susceptibility to ectromelia, herpes simplex, cytomegalovirus, coxsackievirus, and other infections. Athymic ("nude") mice which are congenitally deficient in T cells are also more susceptible to many viral infections. Blanden first demonstrated greatly increased susceptibility of mice to ectromelia virus when T cells were depleted by treatment with antilymphocyte serum. The T cell-depleted mice failed to show the usual inflammatory mononuclear cell infiltration in the liver following infection, developed extensive liver necrosis, and died, in spite of the production of antiviral antibodies and interferon. Virus titers in liver and spleen of infected animals could, however, be greatly reduced by transfer of immune T cells taken from previously infected donors; this process was K/D restricted, implicating the cytotoxic class of T cells.

Strong support for this conclusion was later forthcoming from experiments showing that mice could be protected against intranasal challenge with influenza virus by intravenous transfer of a particular subpopulation of T cells from mice infected with influenza virus a week earlier; Lyt $1^-2^+3^+$ T cells displaying K/D-restricted cytotoxic activity against influenza virus-infected target cells *in vitro* were protective. In contrast, Lyt $1^+2^-3^-$ influenza-immune T cells with no cytotoxic activity but with I-restricted DTH activity actually accelerated the demise of infected mice. While not denying a role for DTH in the early phase of at least some types of viral infection, the findings do indicate that inflammation, leading as it does to consolidation in organs such as the lung, can be detrimental.

More recently, it was demonstrated that the lives of influenza virus-infected mice could be saved by transfer of cloned K/D-restricted, Lyt $1^-2^+3^+$ influenza HA-specific T lymphocytes which displayed cytotoxic activity. This strongly suggests a key role for T_c cells, although it must be added that the clone also produced interferon and displayed DTH activity. A careful comparison between the biological properties and the protective capacities of a large number of distinct T cell clones should soon settle this important issue.

The role of NK cells in recovery from viral infection is not yet certain. "Beige" mice, which have little or no NK activity, or normal mice depleted of NK cells by treatment with NK-specific antiserum, show increased replication of some viruses but not others. Athymic mice have plenty of NK cells but die of systemic viral infections.

Viruses producing systemic disease characterized by a "plasma" viremia, e.g., yellow fever or poliovirus, appear to be controlled principally by circulating antibody. Children with severe hypogammaglobulinemia recover normally from measles or varicella, or from smallpox vaccination, but are about 10,000 times more likely than normal individuals to develop paralytic disease after vaccination with live poliovirus vaccine. They have normal CMI and interferon responses, normal phagocytic cells and complement, but cannot produce antibody, which is essential if virus spread to the central nervous system via the bloodstream is to be prevented.

While there is reasonably good evidence that antibody plays a key role in recovery from picornavirus, togavirus, and papovavirus infections, it does not necessarily follow that the antibody is acting solely by neutralizing virions. Indeed, it has been shown that certain nonneutralizing monoclonal antibodies can save the lives of mice, presumably by ADCC or antibody–complement-mediated lysis of infected cells, or by opsonization of virions for macrophages.

IMMUNITY TO REINFECTION

Whereas a plethora of interacting phenomena may contribute to recovery from viral infection, the mechanism of acquired immunity to reinfection with the same agent is probably much simpler. The first line of defense is antibody. If this succeeds in neutralizing the challenge, that is the end of the matter. Only if the antibody defenses are breached does infection become established; under these circumstances the several mechanisms that contribute to recovery are called into play again, the principal differences on this second occasion being that (1) the dose of infecting virus has been reduced by antibody, and (2) preprimed memory T and B lymphocytes generate a more rapid (*anamnestic*) response.

There is abundant evidence of the efficacy of antibody in preventing infection. Passive immunization with antibody parenterally protects man very satisfactorily against challenge with measles, polio, rabies, or hepatitis viruses (see Chapter 10). Furthermore, passive immunity is routinely transmitted from mother to child via maternal antibodies which cross the placenta and protect the newborn infant against most of the infections that the mother experienced. There is also transfer of secretory IgA antibodies via the milk; transfer in colostrum (early milk) is a particularly important route in cows, sheep, horses, and other animals in which there is no placental transfer. This maternal "umbrella" of antibodies lasts for about 6 months in man, hence the infant encounters many viral infections while still partially protected. Under these circumstances the virus multiplies, but only to a limited extent, stimulating an immune response without causing significant disease. The infant thus acquires active immunity while partially protected by maternal immunity.

Passive immunity acquired from the mother lasts only a few months because of the limited life of immunoglobulin *in vivo*. In contrast, the antibody acquired by active infection with systemic viruses continues to be synthesized for the life of the individual, and protects against reinfection for many years. For instance, individuals who suffered an attack of yellow fever during an epidemic in Virginia in 1855 were found to have circulating antibodies 75 years later despite the total absence of yellow fever from the area ever since. Similarly, evidence from isolated Eskimo communities in Alaska showed that antibody to poliovirus persisted for 40 years in the absence of any possible reexposure. Furthermore, survivors of shipwrecks on certain remote Atlantic islands were found to have retained immunity to measles despite no possible contact with the virus for decades. This lasting immunity is attributable to serum IgG, which must be encountered during the viremic stage of reinfection with these viruses.

As a general rule the secretory IgA antibody response is short-lived compared to the serum IgG response. Accordingly, resistance to reinfection with respiratory viruses tends to be limited to a few years at most. Repeated attacks of common colds or influenza reflect infection with antigenically distinct strains of virus, but even with particular serotypes of parainfluenza virus or with respiratory syncytial virus, reinfections are common.

The immune response to the first infection with a virus can have a dominating influence on subsequent immune responses to antigenically related viruses. The second virus often induces a response that is directed mainly against the antigens of the original virus strain. Thus, the immune response of a human to sequential infections with different strains of influenza type A is dominated by antibody to the antigens characterizing the particular strain of influenza type A which that individual first experienced. This phenomenon is called, somewhat irreverently, *original antigenic sin* and is seen not only in influenza but also in enterovirus, paramyxovirus, and togavirus infections. The immunological basis of original antigenic sin remains unexplained. It has important implications for interpretation of seroepidemiological studies, for immunopathology, and particularly for vaccination strategy against diseases such as influenza.

IMMUNOPATHOLOGY

The immune response to viral infection may frequently contribute to the pathology of the disease itself. The characteristic rash of measles, which is absent in children with a congenital absence of T cells, is a good example, discussed earlier. Inflammation with accompanying cellular infiltration is of course a regular feature of viral infection. Common signs such as erythema, edema, and enlargement of lymph nodes may be said to have an immunological basis in this sense. But there are viral diseases in which the cardinal manifestations are caused by the body's immune response. In the extreme case, the disease may be prevented by suppressing the immune response.

The prototype is the disease that results following intracerebral inoculation of adult mice with lymphocytic choriomeningitis (LCM) virus. This fatal meningitis can be completely prevented by immunosuppression with X irradiation, antilymphocyte serum, cytotoxic drugs, or neonatal thymectomy. The disease is entirely due to the mouse's own vigorous CMI reaction. An immune response that in most extraneural sites would be beneficial proves to be lethal when it occurs in the brain. Two things must be added, however. First, the evidence does not tell us whether it is cytotoxic T cells or DTH T cells that inflict the damage. Second, intracerebral inoculation of adult mice constitutes a highly artificial experimental situation; under normal circumstances the virus is acquired by infant mice, via a more natural route, and though the virus multiplies extensively in the body there is no immunopathology and no disease. Nevertheless, this classical model demonstrates more dramatically than any other that the immune response is a two-edged sword—too much of a good thing can be bad, especially in vital organs in confined spaces such as the brain or the lung.

Mention has already been made of evidence that the DTH response to influenza contributes to the pathology of influenza virus infection in the mouse lung model. Intravenous transfer of Ly 1^+ T cells with DTH activity greatly increased the amount of cellular infiltration and edema in the lung and accelerated the death of the animals. Preincubation of these

TABLE 6-3

Possible Mechanisms of Immunopathology in Viral Disease

DTH → inflammation, mononuclear accumulation, tissue damage	
Immune complexes → deposition in glomeruli, arterioles, joints	
Immune cytolysis → lysis of infected cells by	cytotoxic T cells antibody + complement ADCC
IgE on mast cells → anaphylaxis, bronchospasm	

T_d cells with influenza-specific $T_{s(dth)}$ cells suppressed the DTH response and reduced the severity of the disease. This leads to the hypothesis, for which there is currently no evidence, that in natural infections of man the DTH response, which is needed to recruit macrophages to the site early in the infection, is subsequently turned off by T_s cells before it does too much damage.

Theoretically, immune cytolysis of virus-infected cells by any of the mechanisms described in this chapter might have pathological consequences (Table 6-3). Cytotoxic T cells are often incriminated, but antibody–complement-mediated cytolysis, or ADCC mediated by K cells, polymorphs, or macrophages are just as capable of destroying infected cells. However, it has perhaps been too readily assumed that immune cytolysis of infected cells would necessarily increase the extent of cell destruction. After all, any cell that is susceptible to immune cytolysis is destined to be destroyed as a result of the viral infection itself just a few hours later, except in the special case of infections by noncytocidal viruses such as arenaviruses (e.g., LCM) and retroviruses (e.g., murine leukemia).

One particularly well documented immunopathological mechanism is the deposition of viral antigen–antibody complexes in kidneys and arterioles to produce *immune complex disease.* Circulating immune complexes are doubtless present transiently during most acute viral infections. When deposited in the walls of small blood vessels in the skin and joints they may be responsible for the prodromal rashes seen in exanthematous viral infections. If the vascular changes are more marked they give rise to the condition called erythema nodosum, characterized by deposits of antigen, antibody, and complement in vessel walls and tender red nodules in the skin. When small arteries are severely affected, as in some patients with hepatitis B, this causes periarteritis nodosa.

There are several chonic viral infections of animals in which immune complex disease is much more severe. For example, in mink infected with Aleutian disease virus, very large amounts of antibody of low neutralizing capacity are formed. This antibody fails to eliminate the virus but binds to it, and to soluble viral antigens, forming immune complexes which are deposited over many months in arteries and glomeruli, causing arteritis and glomerulonephritis. Severe glomerulonephritis is also a characteristic feature of chronic infections with a number of other viruses, including equine infectious anemia and LCM. The subject will be discussed further in the next chapter.

In addition to their local effects, antigen–antibody complexes may also generate systemic reactions, the most serious of which is *disseminated intravascular coagulation,* seen for example in some of the hemorrhagic fevers. The coagulation cascade is activated, leading to histamine release, increased vascular permeability, and fibrin deposition in kidneys, lungs, adrenals, and pituitary, causing multiple thromboses with infarcts. Depletion of platelets, prothrombin, and fibrinogen gives rise to scattered hemorrhages.

Quite possibly, still other types of hypersensitivity occur as either a normal or an atypical feature of many viral infections. For example, the observation that respiratory syncytial virus (RSV) infections are regularly more severe in very young infants with high levels of maternal antibody suggests that the disease (bronchiolitis/pneumonitis) may have an immunological basis. There are several possible mechanisms, including a type of IgE-mediated local anaphylaxis.

Autoimmunity

If virus-induced immune responses were directed against normal host components, the stage would be set for a different type of immunopathological event, namely *autoimmunity.* In an acute infection this would merely lead to increased but ephemeral tissue damage, but if the infection became chronic or if the autoimmune phenomena continued over a longer period more extensive damage would be possible. There is little direct evidence that autoimmunity plays a major part in viral dis-

TABLE 6-4
Theoretical Mechanisms of Induction of Autoimmune Disease by Viruses

Polyclonal activation by virus of B cells, some of which make anticellular antibody
Viral destruction of the T_s cells that normally suppress production of antibodies to self antigens
Antibody to the idiotype on antiviral antibody reacts with virus receptors on cells
Fortuitous sharing of antigens by virus and host
Release, exposure, or altered presentation of self antigens on virus-infected cells

eases, but much speculation about its possible role in degenerative diseases of the central nervous system of unknown etiology such as multiple sclerosis. Research is just beginning in this area and the whole topic is perhaps too esoteric to explore in this short text. Table 6-4, though somewhat Delphic, lists possible ways in which autoimmune disease could theoretically be induced by viral infection.

DETERMINANTS OF HOST RESISTANCE

Susceptibility to infection or disease, or its reciprocal, *resistance*, can be simply measured in experimental animals by determining the dose of virus required to cause infection, disease, or death in 50% of the test group. The lethal dose is expressed in LD_{50}. Less commonly measured is susceptibility to infection itself, regardless of disease, as indicated for instance by seroconversion, and expressed in ID_{50}. Different strains of inbred mice may differ many thousandfold in their susceptibility/resistance to a given virus. The outcome of an infection can be viewed as the product of host susceptibility × viral virulence (see Chapter 5). A highly virulent strain of virus is less lethal for a highly resistant strain of animal than for a susceptible one; conversely, a relatively avirulent strain of virus may be lethal for an unusually susceptible host. The severity of an infection depends upon the interplay between the virulence of the virus and the resistance of the host. One can regard an infection as a race between the ability of the virus to multiply, spread, and make its exit from the body, and the ability of the host to curtail and control these events.

The variability in the response of individual human beings to infection with a given virus can be regularly observed during epidemics; one person may die of paralytic poliomyelitis, another merely develop a temporary stiff neck, and a third a completely subclinical infection the only evidence of which is a sharp rise in antibody and lifelong immunity to reinfection. Of course, the dose of virus taken in by each individual will be influential but there is far more to it than that. An unusual opportunity to observe the results of a viral infection when a standard dose is administered to large numbers of healthy young adults occurred in 1942 when more than 45,000 United States military personnel received a standard dose of yellow fever vaccine which was inadvertently contaminated with hepatitis B virus. Only 914 developed clinical hepatitis and there were only 33 with severe disease. Of course, a random population of humans is in no way analogous to a genetically inbred strain of mice, and it is impossible to decide whether the observed variations in response should be ascribed to genetic differences in susceptibility or to physiological differences—no doubt both would have played a part.

Genetic differences in susceptibility are most obvious when different species are compared. Common viral infections tend to be less pathogenic in the natural host species than in previously unexposed species. For instance, yellow fever, which originated in Africa and did not arrive in the Americas until the seventeenth century, is more pathogenic for South American primates than for the original host, the African primate. In Africa, rinderpest virus causes a severe disease in Indian buffaloes or in European cattle but a relatively mild infection in native African (Zebu) cattle. Rarely, it has proved possible, using inbred strains of mice, to identify a single host gene that seems to hold the key to susceptibility or resistance. More generally, no doubt, multiple genes are involved.

Physiological factors, such as age, sex, and hormones in the host often play a role in influencing the outcome of an infection. Individuals, even from a given inbred strain of mouse, show marked differences in susceptibility. Furthermore, if a single animal could be repeatedly tested it would doubtless suffer a different disease on different occasions, due to physiological changes in susceptibility.

Genetic Determinants

There is a good deal of circumstantial evidence that human racial groups differ in their genetic susceptibility to certain viral infections, such as yellow fever, to which West Africans appear to be more resistant than Europeans, and measles, which seems to be particularly severe in inhabitants of the Pacific Islands and Africa. Accurate human genetic data on resistance to infection are almost unobtainable, because genetic, physiological, and environmental differences are generally confounded.

In mice, however, it has been possible to study the genetics of resistance to viral infection in some detail. Two situations have been found in which resistance is associated with the response of *macrophages.* (1) Susceptibility to certain flaviviruses is under single gene control, with resistance dominant; in this system the yield of virus from cultures of splenic macrophages from the susceptible strain of mouse is 100–1000 times greater than in cultures from resistant strains. (2) Susceptibility to mouse hepatitis virus (a coronavirus), which is also associated with capacity of macrophages to support multiplication of the virus, is under single gene control with susceptibility dominant.

In a few instances it has been shown that the susceptibility of animals, or particular organs, is a direct consequence of the presence of the *cellular receptors* for that virus. The best genetic evidence relates to the susceptibility of different strains of chickens to Rous sarcoma virus. Here susceptibility is attributable to a single gene that codes for the cellular receptor; susceptibility is dominant. Poliovirus provides an example of the importance of cellular receptors at the species level. It ordinarily infects only primates; mice and other nonprimates are insusceptible because their cells lack the receptors for this virus. However, poliovirus nucleic acid, either naked or enclosed within the capsid of a mouse-pathogenic coxsackievirus, will undergo a single cycle of multiplication in mouse brain. Since the progeny particles have the capsids of normal poliovirions, they are unable to initiate a second cycle.

Immunological responsiveness to particular antigens differs greatly from one strain of mouse to another, being under the control of specific *Ir (immune response) genes*. There are many of these genes, most of them situated in the MHC gene complex [H-2 in the mouse (see Fig. 6-2) and HLA in man]. Individuals with a genetically determined poor immune response to a given viral antigen, especially the critical surface antigen, would presumably have difficulty in controlling infection with that particular virus. In the mouse at least, absence of a specific response is generally recessive. Susceptibility to cytomegalovirus, leukemia virus, and lymphocytic choriomeningitis virus infections has been shown to be H-2 associated, but it must be said that most of the genetic determinants of virus susceptibility map outside the *H-2* region. In man, certain HLA genes have been shown or claimed to be associated with increased or decreased immune responses to particular viruses, but this is an area that requires very careful controlled studies on large numbers of individuals; further work is needed.

Immunosuppression

Viral infections are often more frequent and/or more severe in individuals whose immune system in general has been weakened as a result of some congenital or acquired immunodeficiency. There are several *congenital immunodeficiencies* that render children extremely vulnerable to lethal infection with otherwise harmless viruses. Earlier in this chapter we contrasted thymic aplasia, a T cell deficiency which renders infants unusually susceptible to infection with herpes simplex, varicella, and measles viruses, with hypogammaglobulinemia, a B cell deficiency which predisposes infants to paralysis following immunization with avirulent live poliovaccine.

Acquired immunodeficiency also predisposes to infection. For example, children with leukemia may die following exposure to measles or varicella, and respiratory syncytial virus infections may be more prolonged.

When the immune system is deliberately suppressed by cytotoxic drugs or irradiation as treatment for lymphoreticular tumors, or for organ transplantation, latent infections with cytomegalovirus, varicella-zoster virus, herpes simplex virus, adenovirus, or polyomavirus are commonly *reactivated*. This important subject will be discussed further in Chapters 7, 14, and 16.

Virus-Induced Immunodepression. It has long been known that when tuberculin-positive individuals suffer from measles they temporarily become tuberculin-negative, and studies of Eskimos in Greenland have shown that measles exacerbates preexisting tuberculosis. *In vitro* studies revealed that cultured human T lymphocytes will support the growth of measles virus, and such infected cells fail to respond to tuberculin, like lymphocytes taken from individuals rendered temporarily tuberculin-negative by measles. Since then, direct tests for immune function have shown that infections with many different viruses depress humoral and/or cell-mediated immune responses in man and experimental animals. The reduced response is to unrelated antigens, and is thus distinct from the immunologically specific effect seen in tolerance.

The mechanism of virus-induced immunodepression is generally not fully understood, but may result from the multiplication of virus in lymphocytes and/or macrophages, or from induction of suppressor T cells. Many viruses are capable of replication in macrophages (see Chapter 7, and Table 7-4), and several viruses have now been shown to grow in T cells that have been stimulated by mitogens (concanavalin A or phytohemagglutinin) to divide *in vitro* (Table 7-6), while EB virus grows nonproductively in, and transforms, cultured B cells. The immune responses are depressed, but not always abolished; individual lymphocytes infected with murine leukemia virus, for instance, can at the same time produce antibody to sheep red cells. Furthermore, several viruses are nonspecific mitogens and may turn on suppressor cells which suppress immune responses to unrelated antigens.

The most dramatic and clear-cut example of virus-induced immunosuppression is that of AIDS (acquired immune deficiency syndrome), which is caused by HTLV-III (human T cell lymphotropic virus, type III). AIDS is characterized by a profound depletion of helper T cells, with normal or increased numbers of suppressor T cells. The victims generally die within 2 years, from unusual infections such as *Pneumocystis* pneumonia, or tumors, notably Kaposi's sarcoma or malignant lymphoma. Unlike other human retroviruses, HTLV-III multiplies productively in OKT4^{+} T cells, destroying them. The virus and the disease are described in Chapter 28.

While AIDS represents a striking instance of a viral infection suppressing the immune system to the point where superinfection with otherwise harmless bacteria, fungi, or protozoa may be lethal, there are many more common examples of viral infections predisposing to *secondary bacterial infection* and some of these are not necessarily due to immunosuppression. For instance, respiratory viruses regularly increase the susceptibility of the respiratory tract to invasion by bacteria that are normal commensals in the nose or throat. Influenza or measles virus destroys ciliated epithelium, allowing *Streptococcus pneumoniae* or *Haemophilus influenzae* to invade the lung and cause secondary bacterial pneumonia (see Chapter 29). Similarly, rhinovirus, measles virus, or respiratory syncytial virus damaging the mucosa in the nasopharynx and sinuses predisposes to bacterial superinfection which commonly leads to purulent nasal discharge, sinusitis, and/or otitis media, especially in children.

Enhanced susceptibility of mice to secondary bacterial infection following influenza or parainfluenza virus infection has been ascribed to virus-induced defects in macrophage and/or polymorph function. The pathogenicity of murine hepatitis virus is enhanced by concurrent infection of the mice with a normally harmless blood parasite, *Eperythrozoon coccoides,* probably due to the effect of the latter on macrophages. (It must be added, however, that certain other studies in mice indicate that concurrent infections with certain other parasites may actually enhance resistance to several viruses, by activation of macrophages.) Veterinary virologists are particularly conscious of the potentiating effects on viral diseases of coinfection with parasites. Animals are almost universally infected with protozoa and/or worms. A high *parasitic load* generally lowers the resistance of the animal. This may also be a major factor in lowering the resistance of children in the developing countries of the tropics.

Malnutrition

Malnutrition can interfere with any of the mechanisms that act as barriers to the multiplication or progress of viruses through the body. It has been repeatedly demonstrated that severe nutritional deficiencies will interfere with the generation of antibody and CMI responses, with the activity of phagocytes, and with the integrity of skin and mucous membranes.Often, however, nutritional deficiencies are complex, and the identification of the important food factor is difficult. This is reflected in the use of inclusive terms such as "protein–calorie malnutrition." Also at times it is impossible to disentangle the nutritional effects from socioeconomic effects such as poor housing, crowding, inadequate hygiene, and microbial contamination of the environment.

Children with protein deficiency, the extreme form being represented by the clinical condition called kwashiorkor, are uniquely susceptible to measles. Measles is up to several hundred times as lethal in certain of the developing countries of tropical Africa and South America as it is in the countries of northern Europe and North America, the case fatality rate approaching 10%, and in severe famines even 50%. Yet, severely malnourished children in Asia do not appear to develop severe measles virus infections. This paradox, as well as recent immunological studies suggesting that antibody production to measles virus and certain parameters of cell-mediated immunity are not abnormal in malnourished children as was previously supposed, take us back to square one in our understanding of this important subject.

Breast milk is an important determinant of resistance of infants to viral infections. Milk not only provides a uniquely balanced diet but also confers immunologically specific resistance to viral infections. This is largely mediated by the transfer of maternal antibodies. In human colostrum 97% of the protein is secretory IgA and in human milk about 1 g of IgA is transferred to the infant each day. As indicated earlier, milk or colostrum is especially important in those animal species where there is no transfer of antibodies via the placenta. When the antibodies are secretory IgA, they are not absorbed from the gut but can protect the infant against intestinal infections. In ruminants colostrum also contains IgG antibodies, which are absorbed from the intestine for a short period after birth, protecting the infant against a wider variety of viral infections.

Just as malnutrition can exacerbate viral infections, so viral infections can exacerbate malnutrition. Children in developing countries can thus be trapped in a vicious cycle. For example, a study of children in Guatemalan villages showed that measles and other respiratory and intestinal infections in infancy are responsible for periodic interruptions in weight gain. Body weights at 2 years of age were sometimes little more than half those of American children and there was often almost no weight gain during the second and third years of life.

Age

The high susceptibility of newborn animals to many viral infections has been of considerable importance in laboratory studies of viruses. Thus the human coxsackieviruses were discovered by the use of suckling mice, which are also sensitive hosts for the recovery of arboviruses. The oncogenic papovavirus, polyoma, induces tumors only when inoculated into newborn rodents; older animals are infectible but develop no apparent disease. In laboratory animals the first few weeks of life are a period of very rapid physiological change. During this time mice pass from a stage of immunological nonreactivity (to many antigens) to immunological maturity. This change profoundly affects their reaction to viruses such as lymphocytic choriomeningitis, which induces a tolerant state when inoculated into newborn mice.

The human species is reasonably mature immunologically at the time of birth, but still very susceptible to infection with viruses against which the mother has no antibody. The umbrella of maternal antibody is what protects infants against infection for the first few months of life. If this is missing, the newborn, especially the premature, are particularly vulnerable to infections with viruses such as herpes simplex, varicella, or coxsackie B. Older infants and children respond well to viral infections. Although most viral infections are acquired in childhood (see Chapter 9) they tend to be subclinical. Adults, in contrast, tend to suffer more severely. For example, varicella (chickenpox), a trivial disease in children, often causes a severe pneumonia in adults. EB virus produces inapparent infection in childhood but infectious mononucleosis (glandular fever) in adolescents and adults. Similarly mumps may be complicated by orchitis in adolescents. Hepatitis A and polioviruses generally infect children subclinically, but adults have a higher probability of developing serious disease (for discussion, see Chapter 19). Adults exposed to such viruses for the first time in a "virgin soil" epidemic (see Chapter 9), or by travel to a developing country are particularly at risk. At the far end of life, the aged become very susceptible to respiratory viruses, especially influenza. This is probably attributable to loss of elastic tissue around alveoli, weaker respiratory muscles, and a poorer cough reflex, predisposing to secondary bacterial pneumonia. The susceptibility of the elderly to reactivation of latent viruses such as herpes zoster may have more to do with waning cell-mediated immunity.

Hormones

There are few striking differences in the susceptibility of males and females to viral infections (except in the obvious instances of viruses with a predilection for tissues such as testis, ovaries, or mammary glands). *Pregnancy,* however, significantly increases the likelihood of severe disease following infection with hepatitis viruses or polioviruses. Furthermore, certain latent viral infections, e.g., with the human papovaviruses, cytomegalovirus, and possibly HSV-2, display a greater tendency to reactivate during pregnancy.

Corticosteroids exacerbate many viral infections. Adult mice, normally resistant to coxsackievirus B1, die if also inoculated with cortisone. In man, infections with herpesviruses such as varicella are often exacerbated in patients given corticosteroids; indeed, incautious use of corticosteroids in the treatment of HSV keratoconjunctivitis can lead to blindness. The precise mechanism of these adverse effects is not understood. Corticosteroids reduce inflammatory and immune responses, and depress interferon synthesis, but it is also clear that adequate production of these hormones is vital for the maintenance of normal resistance to infection.

Fever

Fever so regularly accompanies viral infections that it is natural to suppose it has some antiviral action. The principal mediator of the febrile response appears to be the macrophage product, interleukin-1 (IL-1), previously known as endogenous pyrogen; IL-1 has been shown to enhance T cell proliferation and antibody production at 39°C *in vitro*. Interferons are also pyrogenic when present in sufficiently high concentration; their antiviral and immunomodulatory functions will be discussed below.

Lwoff suggested many years ago that fever constitutes a natural defense against viruses, and that virulent strains of virus have evolved with the ability to multiply in the host at temperatures achieved during fever (indeed, latent infections with one virulent virus, herpes simplex, are actually reactivated by fever, producing "fever blisters"—see Chapter 7). Temperature-sensitive (*ts*) viral mutants might therefore be expected to be less virulent, and this correlation has now been observed with *ts* mutants of many viruses, some of which are being put forward as potential vaccine strains (see Chapter 10).

INTERFERONS

In 1957 Isaacs and Lindenmann described a protein, "interferon," which was secreted by virus-infected cells and could protect uninfected cells against infection with the same or unrelated viruses. Interferon was postulated, and subsequently proven, to be the principal mediator of the well-known phenomenon of *interference*. We now know that there are in fact more than a dozen different human interferons, falling into three antigenically and chemically distinct types, known as α, β, and γ (Table 6-5). In a spectacular demonstration of the power of modern molecular biology the last 5 years have seen the discovery and cloning by recombinant DNA technology of separate genes for at least a dozen subtypes of

TABLE 6-5
Physicochemical Properties of Human Interferons[a]

PROPERTY	INTERFERON		
	α	β	γ
Previous designations	Le-IFN Type I	F-IFN Type I	Immune IFN Type II
Subtypes	>12	1	1
MW[b]			
Major subtypes	16,000–23,000	23,000	20(–25),000
Cloned[c]	19,000	19,000	16,000
Glycosylation	No[d]	Yes	Yes
pH 2 stability	Stable[d]	Stable	Labile
Induction	Viruses	Viruses	Mitogens
Principal source	Epithelium, leukocytes	Fibroblast	Lymphocyte
Introns in gene	No	No	Yes
Homology with Hu-IFN-α	80–95%	30–50%	<10%

[a] From D. O. White, "Antiviral Chemotherapy, Interferons and Vaccines." Karger, Basel (1984).

[b] Molecular weight of monomeric form. Interferons often occur as polymers.

[c] Nonglycosylated form, as produced in bacteria by recDNA technology.

[d] Most subtypes, but not all.

human IFN-α (Hu-IFN-α), one IFN-β, and one IFN-γ. Most of these human genes have now been cloned in bacteria, yeasts, and/or mammalian cells and the resulting protein products purified and sequenced (for detailed description, see Chapter 11). IFN-β and -γ, but generally not -α, are glycosylated (in mammalian cells). Most of the human interferons have molecular weights in the range 16,000–23,000, though some are larger. IFN-α and -β have the unusual property of being resistant to inactivation at pH 2, although some atypical acid-labile IFN-αs have recently been described.

Interferon α and β are not made constitutively in significant amounts but are induced by any virus, multiplying in virtually any type of cell, in any species of vertebrate that has been tested, whether mammal, bird, amphibian, reptile, or fish. IFN-γ is atypical in that it is made only by lymphocytes and only following antigen-specific or nonspecific (e.g., mitogen) stimulation; it can be regarded as a lymphokine with immunoregulatory functions. Some interferons (especially β and γ) display a degree of host species specificity; e.g., rabbit interferons are ineffective in man. There is no viral specificity, in the sense that IFN-α, -β, or -γ induced by a paramyxovirus is fully effective against a togavirus; but certain cloned IFN subtypes may be much more effective against some viruses than against others.

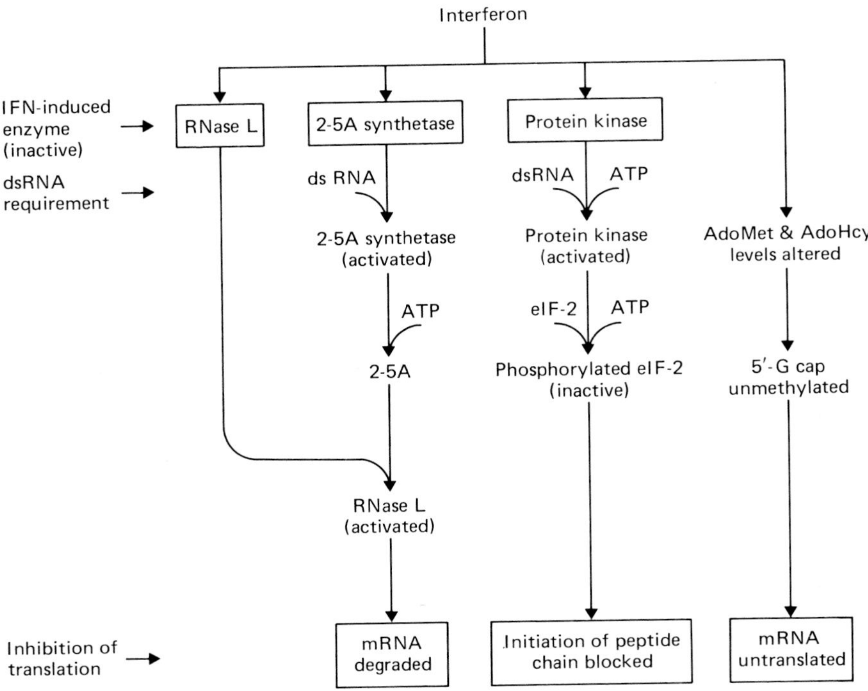

FIG. 6-5. *Antiviral action of interferon: postulated pathways. (From D. O. White, "Antiviral Chemotherapy, Interferons and Vaccines." Monographs in Virology* **16,** *1–112. Karger, Basel, 1984.)*

Antiviral Action of Interferons

The mechanism of antiviral action of IFN-α and -β has recently been elucidated, in part at least (Fig. 6-5). Interferon binds to a specific receptor on the cell surface; there appears to be a common receptor for IFN-α and -β and another for IFN-γ. Such binding triggers a cascade of biochemical events. No fewer than three new enzymes are induced: $(2'\text{-}5')(A)_n$ synthetase (often abbreviated to 2-5A synthetase), RNase L (otherwise known as RNase F, or endoribonuclease), and a 73K protein kinase. Both the 2-5A synthetase and the protein kinase require for their activation double-stranded RNA, which is therefore assumed to be an intermediate or by-product in the replication of DNA as well as RNA viruses. Once activated, 2-5A synthetase catalyzes the synthesis from ATP of an unusual family of short-lived oligonucleotides known as $(2'\text{-}5')\text{pppA(PA)}_n$, or 2-5A for short. In turn, 2-5A activates the latent endonuclease, RNase L, to destroy messenger RNA, hence to inhibit protein synthesis in interferon-treated virus-infected cells. The 73K protein kinase, following activation by dsRNA, also inhibits protein synthesis, but in a completely different way, namely by inactivating (by phosphorylation) the peptide chain initiation factor, eIF-2.

Numerous other biochemical changes have been described following interferon treatment of cells, including, for example, inhibition of methylation of the 5′ terminal guanosine "cap" of viral mRNA. At the time of writing, it is not possible to assign any of these biochemical changes a critical role in the antiviral action of interferon. Indeed, recent data, while not disputing the reality of the several biochemical consequences of exposure to interferon, do throw into question the relevance of either the 2-5A synthetase/RNase L or the protein kinase pathway to the induction of the antiviral state. Moreover, there is evidence that the critical mediator(s) of IFN action are not dependent on dsRNA and have yet to be discovered.

Interferons as Hormones

Interferon was discovered as an antiviral agent, defined accordingly, and generally regarded as such by virologists for the next two decades. Yet, ever since about 1970, one voice in particular, that of Gresser, has been crying in the wilderness that interferon exerts a wide range of other effects on cells and that its antiviral effect may not necessarily be its principal function in nature. It is now widely acknowledged that the interferons probably represent a family of hormones which bind to specific receptors on the cell surface and initiate a cascade of biochemical events leading, perhaps under defined circumstances, not only to inhibition of viral replication, but also to inhibition of cell division, changes in the plasma membrane, modulation of the immune system, and several additional phenomena (Table 6-6).

Inhibition of Cell Division. Interferons can inhibit cell division. Interferons α, β, and γ, and the various cloned subtypes, differ in their relative efficacy as antiproliferative agents and may potentiate one another. Normal cells appear to be just as susceptible as are their malignant counterparts *in vitro*. *In vivo* examples of the growth-inhibitory effects of interferon are also on record. For instance, interferon blocks the regeneration of liver in partially hepatectomized mice, and also the multiplication of bone marrow cells in irradiated mice. The interferon levels attained during LCM virus replication in newborn mice are such as to inflict progressive organ damage, e.g., glomerulonephritis, and runting; it should be added, however, that such effects have not been

TABLE 6-6
Some Biological Effects of Interferons

Inhibition of multiplication of viruses	
Inhibition of cell division ?	
Immunomodulation	
All cells	Increased expression of MHC antigens and F_c receptors
B cell	Antibody production decreased or increased
T cell	Proliferation suppressed, lymphokine release enhanced
T_d cell	Delayed hypersensitivity decreased or increased
T_c cell	Cytotoxicity increased
NK cell	Maturation, recycling, and cytoxicity increased
Macrophage	Activated

seen in other animal species, and their significance is being questioned. The anticancer effects of interferons, clearly demonstrable *in vivo* both in animals and in man, appear to be principally attributable to another phenomenon, namely their immunomodulatory effects, particularly the activation of natural killer cells, macrophages, and cytotoxic T cells (see below).

Changes in the Plasma Membrane. Interferon induces a variety of changes referable to the plasma membrane. These include increases in the expression of class I and II MHC antigens, Fc receptors, and other membrane components. The organization of microfilaments and fibronectin is disturbed. Effects on the plasma membrane may also explain the inhibition of budding of murine leukemia virus from interferon-treated cells.

Immunomodulation

The effects of interferons on the immune system are manifold and sometimes apparently contradictory. "Null" lymphocytes, NK cells, and macrophages are themselves major producers of interferons following viral infection. In addition, interferons α and γ are secreted by both T and B lymphocytes following binding of lectins or specific antigens. It is not yet clear whether certain T cell subsets are more efficient interferon producers than others.

Whether secreted by lymphocytes or by other types of cell, interferons may inhibit the replication of lymphocytes as well as other cell types. This is the explanation of the well-known antimitogenic effect of interferon; interferon secreted by lectin-treated lymphocytes suppresses the proliferation of the lectin-activated population as a whole. Furthermore, interferons in general, and IFN-γ especially, seem to inhibit antibody synthesis when administered before the immunogen, but to stimulate it if added some days later. Similarly, interferon was found to inhibit the development of delayed hypersensitivity if administered before the immunogen, but to enhance it if given a few hours later. The mechanisms involved have yet to be fully worked out. In natural viral infections of course interferon synthesis follows that of viral antigen, hence these interferons would provide a beneficial stimulus to both the humoral and the cell-mediated response.

Interferons have been shown to enhance the cytotoxicity of T_c lymphocytes and macrophages; indeed, IFN-γ appears to be the lymphokine previously known as macrophage-activating factor (MAF). Furthermore, interferon appears to be the principal factor controlling the life cycle of natural killer (NK) cells—it triggers their maturation from pre-NK cells, enhances their recycling capacity, and augments their lytic potential. Since NK cells also secrete interferon on contacting a virus-infected target cell, there exists a positive feedback mechanism accelerating the elimination of virus from the body.

Overall, it appears that interferons may stimulate or inhibit various arms of the immune response, depending on the timing and dosage. One can think of interferons as lymphokines, induced principally by viral infection, and playing a key role in the regulation of the immunological response to that virus. Interferons secreted by virus-infected target cells, as well as by lymphocytes, macrophages, and NK cells, activate T_c lymphocytes, macrophages, and NK cells in the immediate vicinity to develop their cytotoxic potential, while also enhancing the expression of both class I and class II MHC antigens on the surface of the cells with which such leukocytes interact. In turn, these effector cells not only destroy the target but produce more interferon following contact with viral antigen on its surface. The consequential cascade greatly amplifies the lytic arm of the immune response. Yet it is apparent that other important arms of the immune response are depressed by interferon, perhaps reflecting the propensity of interferon to inhibit the replication of lymphocytes. A full analysis of this complicated issue must await careful *in vitro* studies of the various actions of purified preparations of each cloned subtype of interferon on cloned populations of well-characterized lymphocyte subtypes. Armed with that knowledge, it should then be feasible to go back to the whole animal to document the biological relevance of each of interferon's diverse effects.

Not all of the immunomodulatory, growth-inhibitory, and membrane effects listed in Table 6-6 have been demonstrated to occur with each of the cloned subtypes. Careful examination of the relative potency of each subtype in each of these and other previously untested activities now constitutes an important priority, because it may well emerge that particular subtypes are relatively specialized in their natural role as antiviral agent, hormone, or lymphokine, and that these differences could be exploited in antiviral and anticancer therapy.

In Vivo Significance of Interferon

Genetic Control of Interferon Production and Susceptibility. The structural genes for most of the known human interferons (at least 10 IFN-α and IFN-β) are clustered contiguously on chromosome 9. A smaller number of others have been assigned to chromosomes 2 and 5, and the gene for IFN-γ is on chromosome 12. Most of the genes for interferons contain no introns, though that for Hu-IFN-γ does. A genetic locus on human chromosome 21 is essential for the expression of antiviral activity by IFN-α, -β, and -γ. Some evidence suggests that this gene may code for the IFN receptor.

The regulation of expression of the genes for interferons in mammalian cells is quite complex. It is a common observation of scientists assembling cDNA libraries for cloning purposes that there is differential expression of different IFN genes in different cell lines *in vitro*. Furthermore, various inbred strains of mouse differ in the amount of circulating interferon they produce following inoculation with different viruses; single but distinct autosomal loci control interferon levels. Additional genes may affect the interaction of interferons with cells of the immune system. A further level of complexity is suggested by the finding that a gene (known as *Mx*) confers resistance to influenza virus in a certain strain of mouse by enhancing the sensitivity of influenza virus (but not of other viruses) to the antiviral action of interferon.

Production and Distribution *in Vivo*. It is difficult to determine which cell types, or even which tissues and organs, are responsible for most of the interferon production *in vivo*, but, extrapolating from the findings with cultured cells, one can probably safely assume that most cells in the body are capable of producing interferon in response to viral infection. Certainly, interferon can be found in the mucus bathing epithelial surfaces such as the respiratory tract, while "fibroblast" (β) interferon is produced by most or all cells of mesenchymal derivation. Lymphocytes, especially T cells and "null"lymphocytes (presumably NK cells and K cells), as well as macrophages, produce large amounts of IFN-α and -γ, and probably comprise the principal source of circulating interferon in viral infections characterized by a viremic stage.

Role in Recovery from Viral Infection. There are data supporting a central role for interferon in the recovery of experimental animals and man following at least some viral infections. The most telling evidence that interferon can indeed be instrumental in deciding the fate of the animal following natural viral infection has been provided by Gresser and his colleagues, who have shown that mice infected with any of several nonlethal viruses, or with sublethal doses of more virulent viruses, die if antiinterferon serum is administered. Further support is given by the genetic evidence discussed above. In general however, we know very little about the relative importance of the various interferons in recovery from viral infections, or indeed in immunopathology, which could be important in diseases with an immunopathological component. A mouse mutant lacking the structural genes for some or all of the interferons, or for the regulation of their expression, would make a useful experimental model. But, teleologically, one could argue that such an animal may not survive, if interferon does indeed play a crucial role in defense against viral infection or in the regulation of growth or differentiation.

The recovery of any given individual will be influenced by the numerous genes regulating not only the amount of each species of interferon produced following infection, but also the antiviral efficacy of each interferon against that particular virus. Nothing is known of the extent of variation between individuals, or even between species, in the expression of these several genes, let alone of the causes or effects of such variation. Should prospective studies (or even post hoc tests during a particular illness) reveal a genetic or acquired deficiency in the production, regulation or action of some or all interferons in a particular individual, it may be feasible in the future to direct interferon therapy toward those most likely to benefit.

As might be expected, infants with certain congenital immunodeficiencies or infections, and patients otherwise immunosuppressed, e.g., for organ transplantation or as a consequence of malignancies of the lymphoid system, display diminished production of interferon α, and especially γ, when their lymphocytes are tested, e.g., by lectin stimulation *in vitro*. There are also some reports of diminished capacity to produce interferon by individuals who suffer from repeated or unusually severe infections, e.g., children with recurrent respiratory infections or patients with acute fulminating viral hepatitis.

While it is widely supposed that interferon constitutes the first line of defense in the process of recovery from viral infections, it would be naive to believe that it is the only, or even the most important factor. If this were so, one might expect that a systemic infection with any virus, or indeed, immunization with a live vaccine, might protect an individual, for a period at least, against challenge with an unrelated virus. While some data have been presented on this point, the phenomenon cannot be generally demonstrated. The evidence is somewhat stronger that infection of the upper respiratory tract with one virus will provide temporary and strictly local protection against others. Perhaps this distinction provides the clue that the direct antiviral effect of interferon is limited in both time and space. Its main antiviral role may be to protect cells in the immediate vicinity of the initial focus of infection.

FURTHER READING

Books

Mims, C. A. ,and White, D. O. (1984). "Viral Pathogenesis and Immunology." Blackwell, Oxford.

Möller, G., ed. (1981). "MHC Restriction of Anti-Viral Immunity," Immunol. Rev., Vol. 58. Munksguard, Copenhagen.

Nahmias, A. J., and O'Reilly, R. J. (1982). "Comprehensive Immunology," Vol. 9, Part 2. Plenum, New York.

Notkins, A. L., and Oldstone, M. B. A., eds. (1984). "Concepts in Viral Pathogenesis." Springer-Verlag, Berlin and New York.

van Regenmortel, M. H. V., and Neurath, A. R., eds. (1985). "Immunochemistry of Viruses: The Basis for Serodiagnosis and Vaccines." Elsevier, Amsterdam.

Vilcek, J., and de Maeyer, E. (1984). " Interferon and the Immune system." Elsevier, Amsterdam.

White, D. O. (1984). "Antiviral Chemotherapy, Interferons and Vaccines," Monogr. Virol. Vol. 16, pp. 1–112. Karger, Basel.

Reviews

Ada, G. L., Leung, K. N., and Ertl, H. (1981). An analysis of effector T cell generation and function in mice exposed to influenza A or Sendai viruses. *Immunol. Rev.* **58,** 5.

Blanden, R. V. (1974). T cell response to viral and bacterial infection. *Transplant. Rev.* **19,** 56.

Brinton, M. A., and Nathanson, N. (1981). Genetic determinants of virus susceptibility: Epidemiologic implications of murine models. *Epidemiol. Rev.* **3,** 115.

Burns, W. H., and Allison, A. C. (1975). Virus infections and the immune responses they elicit. *In* "The Antigens" (M. Sela, ed.), Vol. 3, pp. 480–574. Academic Press, New York.

Chandra, R. K. (1979). Nutritional deficiency and susceptibility to infection. *Bull. W. H. O.* **57,** 167.

Cooper, N. R. (1979). Humoral immunity to viruses. *In* "Comprehensive Virology" (H. Fraenkel-Conrat and R. R. Wagner, eds.), Vol. 15, p. 123. Plenum, New York.

Dimmock, N. J. (1984). Mechanisms of neutralization of animal viruses. *J. Gen. Virol.* **65,** 1015.

Doherty, P. C., Blanden, R. V., and Zinkernagel, R. M. (1976). Specificity of virus-immune effector T cells for H-2K or H-2D compatible interactions: implication for H-antigen diversity. *Transplant. Rev.* **29,** 89.

Hirsch, M. S., and Proffitt, M. R. (1975). Auto-immunity in viral infections. *In* "Viral Immunology and Immunopathology" (A. L. Notkins, ed.), pp. 419–434. Academic Press, New York.

Mandel, B. (1979). Interaction of viruses with neutralizing antibodies. *In* "Comprehensive Virology" (H. Fraenkel-Conrat and R. R. Wagner, eds.), Vol. 15, p. 37. Plenum, New York.

Oldstone, M. B. A. (1975). Virus neutralization and virus-induced immune complex disease. *Prog. Med. Virol.* **19,** 84.

Oldstone, M. B. A., Fujinami, R. S., and Lampert, P. W. (1980). Membrane and cytoplasmic changes in virus-infected cells induced by interactions of antiviral antibody with surface viral antigens. *Prog. Med. Virol.* **26,** 45.

Sissons, J. G., and Oldstone, M. B. A. (1980). Antibody-mediated destruction of virus infected cells. *Adv. Immunol.* **29,** 209.

Webster, R. G., Laver, W. G., Air, G. M., and Schild, G. C. (1982). Molecular mechanisms of variation in influenza viruses. *Nature (London)* **296,** 115.

Welsh, R. M. (1978). Mouse natural killer cells: Induction, specificity and function. *J. Immunol.* **121,** 1631.

CHAPTER 7

Persistent Infections

Introduction 193
Persistent Infections of Cultured Cells 194
Persistent Infections in Animals 197
Infections with Late Complications 198
Latent Infections 199
Chronic Infections 203
Slow Infections 207
Pathogenesis of Persistent Infections 211
Further Reading 215

INTRODUCTION

The thinking of clinicians and virologists about viral diseases has long been dominated by acute febrile diseases such as smallpox, measles, poliomyelitis, and influenza. In such diseases signs and symptoms appear after an incubation period that ranges from 1–2 days to 2–3 weeks, and the host usually eliminates the infecting agent within 2 or 3 weeks of the onset. Virus can ordinarily be isolated from the blood or secretions only in the short period just before and just after the appearance of symptoms.

In contrast to these acute infectious diseases, however, are some in which virus persists for months or years, i.e., *persistent viral infections*. Persistent infections are associated with a great variety of pathogenetic mechanisms and clinical manifestations, and it is difficult to classify them satisfactorily. For convenience, we shall subdivide them into four categories, recognizing that there is some overlap:

1. *Late complications of acute infections:* virus persists, usually in the brain, and lethal disease occurs years later (see Table 7-1).
2. *Latent infections:* persistent infections in which the virus is not demonstrable or is shed only intermittently; disease is absent except for episodes of reactivation (see Table 7-2).
3. *Chronic infections:* persistent infections in which virus is always demonstrable and often shed; disease is usually absent, or has an immunopathological basis (see Table 7-3).
4. *Slow infections:* persistent infections with a long incubation period followed by slowly progressive disease that is usually lethal (see Table 7-4).

The key distinctions between these four groups of persistent infections are illustrated diagrammatically in Fig. 7-1. In slow infections, the concentration of virus in the body builds up gradually over a prolonged period until disease finally becomes manifest. Chronic infections, on the other hand, can be regarded as acute infections (clinical or subclinical) following which the host fails to reject the virus; sometimes disease supervenes late in life as a result of an immunopathological or neoplastic complication. The distinction between chronic infections and latent infections, e.g., herpesviruses between recrudescences of endogenous disease, may be a fundamental one concerned with the state of the virus between attacks, or it may merely be a matter of the relative ease of demonstration of infectious virus. The distinction between them becomes blurred in the case of certain herpesviruses, such as cytomegaloviruses, where small amounts of virus are secreted from time to time. Late complications of infections have some of the characteristics of slow, chronic, and latent infections; virus persists in the brain following an earlier infection and years later a lethal brain disease develops.

Persistent infections are important for four reasons:

1. They may be reactivated and cause acute episodes of disease.
2. They may be associated with immunopathological diseases.
3. They may lead to neoplasia.
4. They are of epidemiological importance, since by persistent or recurrent shedding they often enable the virus to persist in populations (see Chapter 9).

PERSISTENT INFECTIONS OF CULTURED CELLS

Persistent infections are more complex and more difficult to study than acute infections, so it is useful to analyze some aspects of their pathogenesis by studies with cultured cells. Experimentally, persistent

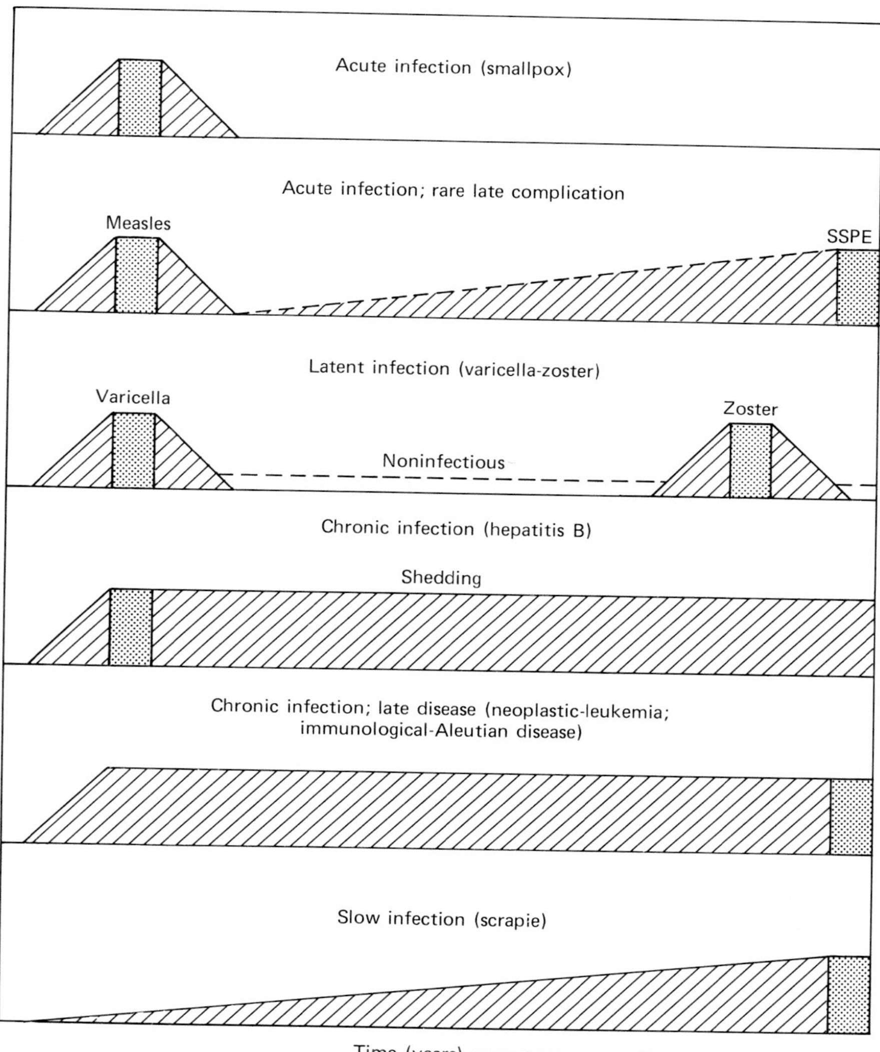

FIG. 7-1. *Diagram depicting acute infection and various kinds of persistent infection, as illustrated by the diseases indicated in brackets. Solid line, demonstrable infectious virus; dashed line, virus not readily demonstrable; vertical box, disease episode.*

infections in cell cultures can be categorized as steady-state infections, carrier cultures, or infections involving integrated viral genomes, by obtaining answers to the following questions:

1. Are all cells infected or are only a minority involved?
2. Do infected cells continuously produce virions, or not?
3. Do infected cells divide, or do they die?
4. Can the culture be "cured" of infection by cloning cells in the presence of antiviral antibody ?
5. Can the balance be altered toward cell destruction by washing inhibitory substances from the medium?

Steady-State Infections

Steady-state infections are not uncommonly found in cells infected with RNA viruses that mature by budding from the plasma membrane. In such cells large amounts of infectious virus may be continuously released from cells whose metabolism and multiplication are scarcely affected. Such latently infected cells can be superinfected with other viruses without noticeably affecting the growth of either virus. Alternatively, the yield of either the superinfecting virus or the noncytocidal virus may be enhanced (complementation) or reduced (interference) (see Chapter 4).

A typical example is the paramyxovirus SV5, a common contaminant of primary cultures of rhesus monkey kidney cells, in which it multiplies to high titer with little cytopathic effect. The infected cells survive and multiply, and produce large amounts of infective virus for many days. Cellular DNA, RNA, and protein synthesis are hardly affected by the virus; viral RNA synthesis amounts to less than 1% of the cellular RNA synthesis. Electron microscopic observations reveal little cellular damage despite the fact that large numbers of virions are maturing at the plasma membrane at all times (Plate 3-3). Infection with SV5 does not interfere with the growth of other viruses.

Similar steady-state infections have been described with most paramyxoviruses and with several rhabdoviruses, arenaviruses, and retroviruses, all of which are relatively noncytocidal viruses that do not shut down the metabolism of the cell and are released by budding from the plasma membrane. Most of the situations described in the literature share the following features: (1) most or all cells in the culture are infected, (2) virus is released continuously but at a slow rate from most or all cells, and (3) the cultures cannot be "cured" by treatment with antibody.

Carrier Cultures

Persistent viral infections not uncommonly arise in cells maintained by serial culture in the laboratory, after they have been deliberately infected with any of a variety of viruses. These carrier cultures differ from the steady-state infections that we have just described in that virus-free cells can always be recovered by cloning from a carrier culture, whereas this is impossible in steady-state infections. This enduring state of peaceful coexistence of virus and cells usually results from the presence in the culture of virus-inhibitory substances, which limit the spread of virus from cell to cell. The serum in the medium may contain antibody or other antiviral substances, or infected cells may be synthesizing enough interferon to keep viral multiplication and thus cell destruction in check. Sometimes the situation is complicated by the presence of some cells that are genetically resistant to infection. Often more than one mechanism of cell protection operates and sometimes the system undergoes an evolving series of virus–cell relationships.

Cells Carrying Integrated Viral Genomes

Cells transformed by viruses are persistently but usually nonproductively infected. Every cell in such a culture contains a partial or complete copy of the viral genome (or with retroviruses a cDNA copy), usually integrated into the cellular DNA but occasionally as an episomal element (see Chapter 8).

PERSISTENT INFECTIONS IN ANIMALS

Viruses may also produce persistent infections in the intact animal. As in cultured cells, such infections may be accompanied by the production of infectious virus or may be characterized by integration of viral nucleic acid into the cellular DNA. There are a few viruses (e.g., herpes simplex virus) that produce cytocidal infections in most cultured cell systems but persistent as well as cytocidal infections *in vivo*.

As outlined in the Introduction, we have grouped persistent infections into four categories which, for convenience, can be called late complication, latent, chronic, and slow infections. With the help of appropriate tables (Tables 7-1 to 7-5) we shall first describe illustrative examples of infections thus classified, and then consider the pathogenetic mechanisms that allow such viruses to persist in the infected host.

INFECTIONS WITH LATE COMPLICATIONS

The paradigm of this type of persistent infection is subacute sclerosing panencephalitis (SSPE), which occurs between 1 and 10 years after recovery from measles. It is rare (about one case per million) and can be diagnosed before death by the discovery of a high level of antibody to measles virus, but not to the M protein, in the cerebrospinal fluid. As measles disappears following widespread immunization of children in the United States, SSPE has also declined; hence it does not appear to be produced after infection with vaccine.

At the time of death, certain nerve cells are full of measles virus nucleocapsids, but complete virions are not produced because the cells fail to synthesize the viral matrix (M) protein. Nevertheless, the complete viral genome must be present, as measles virus can be produced by cocultivation of brain cells with permissive cells *in vitro*. Measles virus glycoproteins are synthesized but may be stripped from infected cells or endocytosed by them following binding of antibody to the cell surface.

Progressive multifocal leukoencephalopathy (PML) has a somewhat similar pathogenesis, being a very rare late complication of an almost universal but subclinical and unrecognized infection of children with human polyomavirus. Infectious virions are often excreted in the urine long after the original infection, and symptomless urinary excretion may be reactivated by immunosuppression or pregnancy. Following immunosuppression, e.g., for kidney transplantation or as a result of lymphoreticular malignancy, a large-scale infection of oligodendrocytes in the brain may occur, leading to progressive demyelination and finally death. The nuclei of oligodendrocytes are full of virus particles; infectious polyomavirus can sometimes be demonstrated by cocultivation with fetal brain cells.

TABLE 7-1
Acute Human Infections with Late Neurological Complications

VIRUS	DISEASE	ANTIBODIES	SYMPTOMS
Measles	Subacute sclerosing panencephalitis	++, in serum and CSF	Acute measles, recovery, then slowly progressive encephalitis years later
Human polyomavirus	Progressive multifocal leukoencephalopathy	+	Initial subclinical infection, then progressive encephalopathy, following immunosuppression, years later

LATENT INFECTIONS

We use this expression for a small but important group of herpesvirus infections, of which the best understood in human subjects are herpes simplex and varicella-zoster (Table 7-2). Following the acute primary infection, infectious virus seems to disappear, yet years later acute disease recurs, often more than once, and infectious virus is then readily demonstrable once more. There is ample evidence that the recurrent episodes of acute disease result from exacerbation of endogenous infections that have lain dormant for weeks, months, or years.

Herpes Simplex

First infections with herpes simplex virus type 1 may be subclinical or may be manifest as an acute stomatitis in early childhood, or tonsillitis or pharyngitis in adolescents. Typically, at intervals of months or years after recovery from the primary infection, blisters from which the virus can be recovered appear, usually around the lips and nose (known variously as "herpes simplex," "herpes labialis," "herpes facialis," "fever blisters," or "cold sores"). A variety of stimuli can trigger this recurrent viral activity, e.g., exposure to ultraviolet light, fever, menstruation, nerve injury, or emotional disturbances. Virus can be recovered from the saliva of about 2.5% of asymptomatic adults at any given time, suggesting low-grade chronic infection, with periodic release of infectious virus. There is no relationship between antibody levels and the frequency of either virus shedding or recurrent attacks.

Herpes simplex type 2 causes comparable initial and recurrent infections of the male and female genitalia; changing sexual practices have made it a relatively common and troublesome disease in western society, often with a dozen or more recurrences a year.

In both types of herpes simplex the genome of the virus, but evidently not the virion itself, persists in dorsal root ganglion cells. Presence of the viral DNA can be recognized in such cells by the use of nucleic acid probes, but infectious virus can be recovered only following cocultivation *in vitro*. Periodically, reactivation occurs in the ganglion cells, and infectious virions then travel down the peripheral nerves and infect cells in the skin. The mechanisms that initiate periodic reactivation and peripheral spread are not understood.

Herpes Zoster

This disease, familiarly known as "shingles," is characterized by a rash which is usually limited to an area of skin (or mucous membrane)

TABLE 7-2

Latent Infections: Little or No Infectious Virus Demonstrable between Acute Episodes of Disease

		SITES OF INFECTION			
VIRUS	DISEASE	BETWEEN DISEASE EPISODES	DURING ACUTE ATTACK	VIRUS SHEDDING	ANTIBODIES
Herpes simplex	1. Primary oral or genital herpes	As DNA, in cerebral or dorsal root ganglion neurons	1. Epithelial cells	Sporadically in saliva or genital secretions between attacks; plentiful in recurrent herpes vesicles	+
	2. Recurrent herpes simplex or genitalis		2. Axons, then cells of corresponding dermatome		+
Varicella-zoster	1. Generalized primary infection: varicella	As DNA, in cerebral or dorsal root ganglion neurons	1. Generalized with rash	1. Varicella: from throat and skin lesions; no shedding thereafter	+
	2. Later skin eruption: herpes zoster		2. Axons, then cells of corresponding dermatome	2. Zoster: from skin lesions. Contact may contract varicella	+
Cytomegalovirus	Usually subclinical	As DNA, in salivary gland cells and leukocytes	Various epithelial cells	Sporadically throughout life in saliva and urine	+
EB virus	1. Glandular fever	As DNA, in B cells	Lymphoid tissue and epithelial cells	Saliva in acute phase	+
	2. Burkitt's lymphoma, nasopharyngeal carcinoma				+

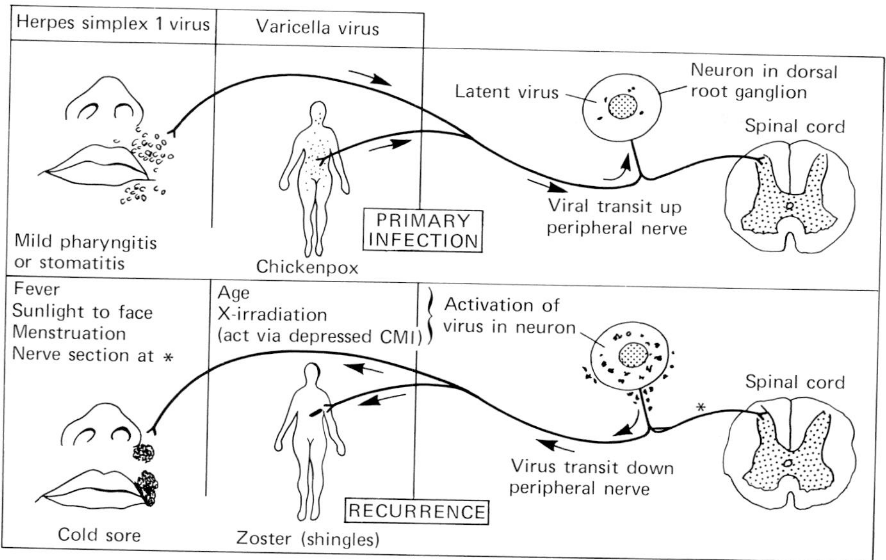

FIG. 7-2. *Mechanisms of latency in herpes simplex and zoster infections. Primary infection in childhood or adolescence; latent virus in cerebral or spinal ganglia; later activation causes recurrent herpes simplex or zoster. (From C. A. Mims and D. O. White, "Viral Pathogenesis and Immunology." Blackwell Scientific Publications, Oxford, 1984.)*

served by a single sensory ganglion. It occurs predominantly in older people, and only in persons who have already had varicella in childhood. Epidemiological evidence, supported by virological and serological studies, shows that zoster and varicella are two clinical manifestations of the activity of a single virus. Zoster represents a reactivation of virus that has remained latent since an attack of varicella, perhaps many years ealier, viral DNA persisting in neuronal cells of one or more of the dorsal root or cranial nerve ganglia. Reactivation appears to be associated with the waning cell-mediated immunity associated with aging, in lymphoreticular neoplasms or immunosuppressed patients, or after X irradiation of the spine, which was once performed for the treatment of arthritis.

Figure 7-2 illustrates the likely pathogenesis of zoster and herpes simplex. All data are consistent with the view that during a primary attack of herpes simplex or varicella, virions move to the ganglia along the sensory nerves, probably by spread in Schwann cells of the nerve sheath. It persists in the neurons as viral DNA, some of which may be transcribed but little or none of which is translated except when reactivation and the production of infectious virus occurs. In recurrent herpes simplex and in zoster virions move down the sensory nerves again until they reach the skin, where further proliferation occurs and produces vesicles.

Cytomegalovirus and Epstein–Barr (EB) Virus

Two other herpesviruses that cause human infections, cytomegalovirus and EB virus, are characterized by initial prolonged excretion of infectious virus followed by a state of latency. They are thus chronic infections which also exhibit latency (Table 7-2).

Generalized cytomegalovirus infection is occasionally seen in hospitals, and may result from either of two alternative sources of infection. In individuals with lymphoreticular disease, or undergoing prolonged immunosuppression, e.g., for kidney or bone marrow transplantation, a generalized cytomegalovirus infection may result from reactivation of an endogenous latent infection or as an exogenous primary infection resulting from the transmission of cytomegalovirus during the transfusion of large volumes of fresh blood. Both situations reflect the widespread occurrence of healthy carriers of cytomegalovirus in the general population. Indeed, surveys show that at any time tests are made about 10% of children under 5 years of age are excreting cytomegalovirus in their urine. When primary infection with cytomegalovirus occurs during pregnancy, infection of the fetus may occur, causing mental retardation, hearing loss, and visual defects. Infection of the fetus due to reactivation of a latent infection during pregnancy or at the time of birth is also common but does not generally have such serious repercussions. The site of latency is not known with certainty, but lymphocytes and macrophages are the cells suspected.

The EB virus, the causative agent of infectious mononucleosis, is another ubiquitous herpesvirus carried asymptomatically by large numbers of clinically healthy people. Infectious virus is excreted in saliva for some months following infection, then persists in B cells, mostly in episomal form, but partly as integrated viral DNA. Infection in childhood is usually symptomless, but in adolescents primary infection causes glandular fever. Polyclonal activation of B cells in the acute infection results in the production of a miscellany of antibodies. EB virus may be asymptomatically reactivated in immunosuppressed patients, but of greater importance is its probable involvement in two kinds of malignancy: Burkitt's lymphoma and nasopharyngeal carcinoma (see Chapter 8).

CHRONIC INFECTIONS

A persistent infection is said to be chronic rather than latent when infectious virus continues to be produced for very long periods, perhaps for life. Culture of cells from various animal tissues, especially monkey kidney cells, once used so widely for purposes of diagnosis and vaccine production, has revealed many viruses that may cause such infections (Table 7-3). Often organs that do not yield infectious virus if ground up and inoculated directly into appropriate cultures or experimental animals do eventually produce infectious virus in explant cultures, or by cocultivation.

In many chronic infections the common features are the absence of disease but persistent or recurrent presence of infectious virus, which may be shed, or transferred by blood transfusion, so that infection of susceptible contacts can occur in the absence of apparent disease in the transmitters.

Hepatitis B and Non-A, Non-B

The most important chronic viral infection of man, which affects some hundreds of millions of persons, especially in Asia and Africa, is hepatitis B, due to human hepadnavirus. During acute infections, the virus replicates in the liver and circulates in the plasma, usually in association with a great excess of smaller particles composed of viral surface antigen (HBsAg). In most individuals HBsAg and virions are cleared from the circulation, but in 5–10%, including over 95% of those infected in infancy, a persistent infection is established, which can extend for many years, often for life. The carrier state is characterized by the persistence of HBsAg and infectious particles in the bloodstream, hence the danger of such persons as blood donors. The liver does not necessarily suffer serious damage, but some carriers go on to develop chronic hepatitis and cirrhosis. Furthermore, hepatitis B virus is almost certainly an important cause of primary hepatocellular carcinoma. Integrated sequences of hepadnavirus DNA are found in the tumor cells. It seems likely that a cocarcinogen that is more common in Africa and Asia than in Europe or America is also involved (see Chapter 8).

With the decline in incidence of posttransfusion hepatitis B following the introduction of radioimmunoassay for screening donated blood, non-A, non-B hepatitis has emerged as the commonest form of post-transfusion hepatitis in western countries. Intravenous drug users are also especially vulnerable. Though clinically milder than hepatitis B, non-A, non-B infection more frequently progresses to chronic hepatitis,

TABLE 7-3

Chronic Infections: Virus Demonstrable for Long Periods; Sometimes No Disease

VIRUS	HOST	SITES OF INFECTION	VIRUS SHEDDING	ANTIBODIES	DISEASE
Various viruses	Monkeys	Kidneys	Variable	+	Generally none recognized; viruses found in explanted cells
Hepatitis B[a]	Man	Liver	Serum	+	Transmitted by transfer of serum; may cause liver cancer
Hepatitis non-A, non-B	Man	Liver	Serum	+	Chronic hepatitis common
Rubella	Man, after congenital infection	Widespread	Urine Respiratory	+	Rubella syndrome; teratogenic effects

[a] May also occur as latent infection, viral DNA being either episomal or integrated, in liver cells.

and antigen–antibody complexes circulate in the blood for a considerable time.

Lymphocytic Choriomeningitis (LCM)

LCM is the classical chronic virus infection, contemplation of which was important for the formulation by Burnet of the concept of "immunological tolerance," for which he received the Nobel Prize in 1960. LCM virus is transmitted congenitally to every mouse in an infected colony. The mice are born normal and appear normal for most of their lives, although careful study of the physiological activity of some endocrine glands shows that their specialized functions may be impaired. The mice have persistent viremia and viruria; almost every cell in the mouse is infected and remains so throughout the life of the animal. Circulating free antibody cannot be detected, but immunological tolerance is not complete; virions circulate in the bloodstream as virion–IgG–complement complexes, which are infectious. However, there is no cell-mediated immune response to the virus. Late in life inbred mice (but not wild house mice) may exhibit "late disease" due to the deposition of viral antigen–antibody complexes in the glomeruli. Other members of the family *Arenaviridae* cause similar chronic infections of wild rodents in Africa and South America and produce severe disease, with hemorrhagic symptoms, when man is accidentally infected (Lassa fever, Bolivian and Argentinian hemorrhagic fevers).

Other Chronic Infections of Animals

A few other viruses that produce chronic infections with late immunopathological or neoplastic disease in various animals are included in Table 7-4. Although no comparable diseases are known with certainty to occur in man, there is suggestive evidence that some chronic or "autoimmune" human diseases may turn out to have a similar etiology.

Lactic dehydrogenase virus infection of mice is unusual in that viral titers during acute infection reach 10^{11} infectious particles per milliliter of plasma, without causing ill effects. Subsequently the titer falls to about 10^5 infectious particles per milliliter and persists at this level for life. Antibodies are produced rather late, but fail to neutralize viral infectivity and do not cause significant immunopathology. The infection is confined to macrophages and affects some of their functions such as the ability to remove endogenous lactate dehydrogenase from the plasma—hence the name.

Aleutian disease of mink is caused by a parvovirus which is highly pathogenic for a strain of mink that is homozygous for a recessive gene

TABLE 7-4

Chronic Infections: Virus Always Demonstrable; Late Immunopathological or Neoplastic Disease

VIRUS	HOST	SITES OF INFECTION	NONNEUTRALIZING ANTIBODIES	DISEASE
Lymphocytic choriomeningitis (arenavirus)	Mouse	Widespread	+	Glomerulonephritis
Lactic dehydrogenase virus (togavirus)	Mouse	Macrophages	+	Immune complexes in glomeruli, but no disease
Aleutian disease (parvovirus)	Mink	Macrophages	++	Arteritis, glomerulonephritis, hyperglobulinemia
Equine infectious anemia (retrovirus)	Horse	Macrophages	+	Anemia, vasculitis, glomerulonephritis
Murine leukemia[a] (retrovirus)	Mouse	Widespread	+	Immune complexes in glomeruli; leukemia occasionally
Avian leukosis[a] (retrovirus)	Chicken	Widespread	+	Leukemia occasionally

[a] Also occur as latent infections with integrated provirus; transmitted congenitally and not causing disease.

conferring the Aleutian coat color. On experimental infection of other strains of mink, virus multiplies in the macrophages and then a persistent infection is established, with excretion of virus in saliva, urine, and feces persisting for years. Very high levels of IgG are produced, which are nonneutralizing, and immune complexes deposited in the glomeruli and blood vessels produce glomerulonephritis and arteritis.

Chronic Infections with Late Neoplasia

Infections of some species of animals (e.g., chickens, mice) with RNA tumor viruses (genus *Oncovirus*, family *Retroviridae*) are unique in that they occur both in a horizontally transmitted mode and as DNA provirus integrated into the genome of every cell of all individuals of the species. The situation with avian lymphoid leukemia viruses in chickens is represented in Fig. 7-3. Natural transmission can occur by three pathways:

1. Horizontal contact transmission causes a transient viremia, production of neutralizing antibodies, but generally no disease.
2. Congenital transmission of infectious virions may occur and produce chronic infection with persistent viremia and a high incidence of lymphoid leukemia. Such animals may pass the virus congenitally from one generation to the next and act as reservoirs for horizontal infection.
3. Unique to the retroviruses is a second mode of vertical transmission, namely integrated provirus through the germ line—endogenous chicken retrovirus. Such infection rarely produces leukemia.

Mammalian oncoviruses also show three methods of transmission, the classical case of vertical transmission being Bittner's "milk factor," or mammary tumor virus, which in some strains of mice is inherited as a DNA provirus. The circumstances under which oncoviruses produce malignancy are described in Chapter 8.

SLOW INFECTIONS

Originally used to describe slowly progressive viral diseases found in sheep in Iceland, the term *slow infections* is now used to categorize infections that have very long incubation periods, then cause a slowly progressive, lethal disease. They are chronic infections in that virus can be recovered from infected animals during the incubation period and also after symptoms have appeared. Two different kinds of agent are included in this group (Table 7-5): retroviruses of the *Lentivirinae* sub-

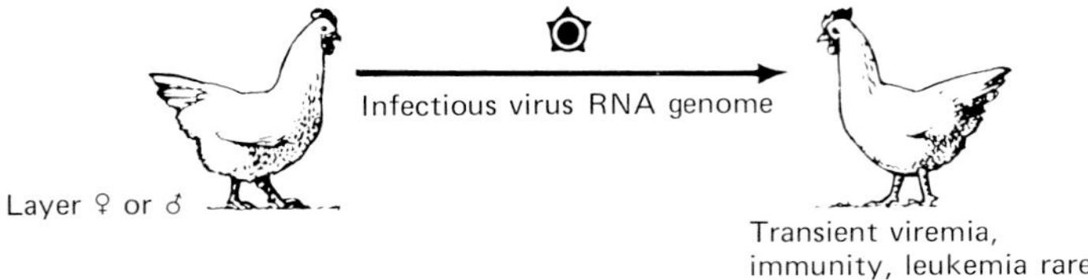

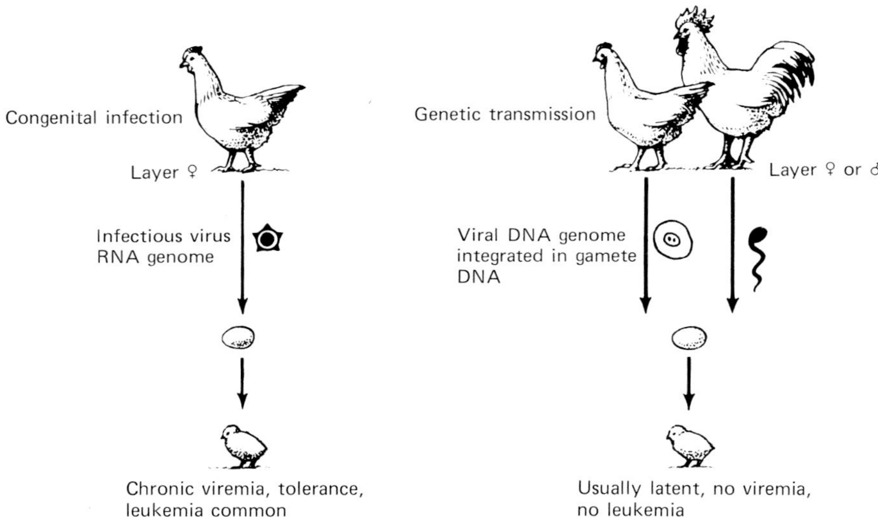

FIG. 7-3. *Horizontal and vertical transmission of avian oncoviruses. [From R. A. Weiss,* In *"Virus Persistence," 33rd Symposium Soc. Gen. Microbiol. (B. W. J. Mahy, A. C. Minson, and G. K. Darby, eds.), p. 267. Cambridge University Press, Cambridge, 1982.]*

family, and the unclassified agents that cause the subacute spongiform encephalopathies.

Lentiviruses

The nononcogenic retroviruses of the *Lentivirinae* subfamily cause chronic demyelinating disease, chronic pneumonitis or chronic arthritis in sheep. They multiply productively in lymphocytes and may also exist as integrated DNA provirus. A feature of visna, and of another per-

TABLE 7-5

Slow Infections: Long Incubation Period, Slowly Progressive Fatal Disease

GROUP	VIRUS	HOST	SITES OF INFECTION	ANTIBODIES	DISEASE
Lentiviruses	Visna	Sheep	Brain	+	Slowly progressive, recurrent viremia, immunopathological component
	Maedi		Lung		
	Progressive pneumonia		Lung		
	HTLV-III/LAV	Man	Helper T cells	±	AIDS
Subacute spongiform encephalopathy	Scrapie	Sheep	CNS and lymphoid tissue	–	Slowly progressive encephalopathy
	Mink encephalopathy	Mink			
	Kuru	Man			
	Creutzfeldt–Jakob	Man			

sistent lentivirus infection, equine infectious anemia, is the occurrence of antigenic drift in surface proteins during the progress of infection in a single animal. HTLV-III/LAV, the etiological agent of AIDS, is also a lentivirus. It multiplies in helper T lymphocytes and is demonstrable in the blood for the remainder of the individual's life. Within 1–10 years about 10% of victims develop AIDS and die from the immunosuppressive effects of the virus (see Chapter 28).

Subacute Spongiform Viral Encephalopathies

It has been suggested that one of the many synonyms of Creutzfeldt–Jakob disease, namely subacute spongiform encephalopathy, should be used as a generic name for four diseases that have strikingly similar clinocopathological features and causative agents, namely, scrapie of sheep and goats, mink encephalopathy, and kuru and Creutzfeldt–Jakob disease in man. The basic neurocytological lesion is a progressive vacuolation in the dendritic and axonal processes of the neurons, and, to a lesser extent, in astrocytes and oligodendrocytes, an extensive astroglial hypertrophy and proliferation, and finally a spongiform change in the gray matter.

Scrapie. Scrapie is a natural infection of sheep in which transmission occurs with difficulty by contact, but more commonly vertically from ewe to lamb. The incubation period is very long, up to 3 years, and once symptoms have appeared the disease progresses slowly but inevitably to paralysis and death. Research was limited until it was discovered that mice could be infected, with incubation periods of a few months. Unusual features are the apparent absence of an immune response and the lack of effect of either interferon or measures that depress the immune response. The causative agent has not been positively identified as a virus; tiny filamentous protein rods visualized by electron microscopy apparently lack nucleic acid (see Chapter 28). To the extent that they had been tested, these unusual biological and physicochemical properties appear to be shared by the agents of the other three subacute spongiform encephalopathies.

Kuru and Creutzfeldt–Jakob Disease. Kuru is a disease confined to a group of 15,000 highland New Guineans; it is thought to have been spread by ritualistic cannibalism and is now disappearing as that habit has disappeared. Its significance lies in the fact that it was the first human degenerative disease of the central nervous system to be shown to have a "viral" etiology, a finding whose importance has now been amplified by the demonstration that the rare but cosmopolitan presenile dementia of the Creutzfeldt–Jakob type can also be transmitted to chimpanzees. Insofar as they have been studied, the causative agents and the histopathology of these two human diseases closely resemble those of scrapie. They are discussed at greater length in Chapter 28.

PATHOGENESIS OF PERSISTENT INFECTIONS

It is clear from the foregoing account that a wide variety of different conditions is included under the umbrella of persistent infections. However it is worthwhile enquiring whether there are any common mechanisms whereby the viruses that cause these infections bypass the host defenses that ensure the elimination of virus in the acute viral infections. Several mechanisms appear to be involved, but in the current state of knowledge we can only speculate about why some of these factors play a dominant role in certain persistent infections but not in others. They include factors related to the virus on the one hand, and to the host defenses on the other.

Unique Properties of the Virus

Nonimmunogenic Agents. The uncharacterized agents that cause the subacute spongiform encephalopathies seem to be relatively or completely nonimmunogenic, and not to induce interferon, nor to be susceptible to its action. If this turns out to be true there may be no mechanism whereby the host can control the multiplication and pathological effects of these agents.

Integrated Genomes. Retroviruses whose cDNA is integrated are maintained indefinitely as part of the genome of the host, and are of significance as viruses only when induced to replicate. Several DNA viruses (herpesviruses, papovaviruses, adenoviruses) also persist as viral DNA, either integrated or as episomes, and in this form are secure from host rejection. Under certain conditions they may be reactivated and may then cause disease.

Antigenic Variation. Visna and equine infectious anemia are retroviruses which may persist as DNA proviruses, but also show a mechanism of resistance to host defenses when in the expressed (viral) state. During persistent infections, a succession of antigenic variants develops within a single animal, by antigenic drift in the viral envelope glycoprotein, thus enabling each successive variant to evade the immune re-

TABLE 7-6
Ineffective Immune Responses in Persistent Viral Infections[a]

PHENOMENON	MECHANISM	EXAMPLES[b]
No antibody	Nonimmunogenic agent	Scrapie, kuru, Creutzfeldt–Jakob disease agents
Nonneutralizing antibody	Small amounts, or low affinity, or reacting with irrelevant epitopes	LCM virus, Aleutian disease virus
Blocking antibody	Nonneutralizing antibody blocks neutralizing antibody and immune T cells	LCM virus
Enhancing antibody	Antibody attached to virus enhances infection of macrophages	Cytomegalovirus Lactate dehydrogenase virus
Disturbance of lymphocyte functions	Infection of lymphocytes	EB virus Cytomegalovirus HTLV-III
Disturbance of macrophage functions	Infection of macrophages	Lactate dehydrogenase virus
Antigen-specific suppression	Induction of suppressor T cells or clonal deletion of T cells	LCM virus Herpes simplex virus
Avoidance of immune lysis	1. Little or no antigen on cell membrane	Herpes simplex in ganglion cell EB virus in B cell
	2. Loss of viral antigen by stripping/endocytosis	Measles virus in neurons
	3. Antigen on inaccessible membrane	Cytomegalovirus, rabies virus
	4. Fc receptor induced on infected cell; IgG binds nonspecifically	Herpes simplex virus Cytomegalovirus
Antigen saturation	Excess viral antigens combine with antibodies and immune cells	Hepatitis B virus
Antigenic variation	Antigenic drift within host	Visna virus Equine infectious anemia virus

[a] Some of these mechanisms are also operative in nonpersistent infections.
[b] Speculative only; in most of these instances an association exists but no cause-and-effect relationship has been demonstrated between the immunological phenomenon and the persistent infection listed.

TABLE 7-7
Reactivation of Persistent Viral Infections of Man[a]

CIRCUMSTANCE	VIRUS	FEATURES	CLINICAL SEVERITY
Old age	Varicella-zoster	Rash	+
Pregnancy	Polyomavirus JC, BK	Viruria	–
	Cytomegalovirus	Replication in cervix	–
	Herpes simplex 2	Replication in cervix	(Danger to newborn)
Immunosuppression with cytotoxic drugs	Herpes simplex	Vesicles, can be severe	+
	Varicella-zoster	Vesicular rash	+
	Cytomegalovirus	Fever, hepatitis, pneumonitis	+, –
	EB virus	Increased shedding of virus from throat	–
	Polyomavirus BK, JC	Very common, viruria	–
	Hepatitis B	Viremia	–
Lymphoreticular tumors	Varicella-zoster	Rash, common	+
	Polyomavirus JC	Progressive multifocal leukoencephalopathy	++

[a] From C. A. Mims and D. O. White "Viral Pathogenesis and Immunology." Blackwell Scientific Publications, Oxford (1984).

sponse. In equine infectious anemia, clinical signs occur in cycles, each cycle being initiated by a new antigenic variant of the virus.

Avoidance of Host Defense Mechanisms

The most important ways whereby viruses persist relate to avoidance of the defence mechanisms of the host, of which the most important is the immune response. This is achieved in a variety of ways (Table 7-6), some of which are also seen in nonpersistent viral infections. Persistent viral infections often become clinically manifest (reactivated) when another disease, form of treatment, or physiological condition interferes in some way with an immune response which was formerly operating sufficiently effectively to prevent symptoms (Table 7-7).

Growth in Protected Sites. During their latent phase, herpes simplex and varicella viruses avoid immune elimination by remaining within cells of the nervous system, as DNA in the ganglion cells during the intervals between disease episodes, and within the axons prior to acute recurrent episodes of disease. Likewise, other herpesviruses such as human cytomegalovirus and EB virus bypass immune elimination, in this instance by persisting in lymphocytes. Other viruses grow in cells on epithelial surfaces, e.g., kidney tubules, salivary gland, or mammary gland, and are persistently shed in the appropriate secretions. Most such viruses are not acutely cytopathogenic, and perhaps because they are released on the luminal borders of cells they do not provoke an immunological inflammatory reaction, hence the cells are not destroyed by T lymphocytes or macrophages. Secretory IgA, which does have access to the infected cells, does not cause complement activation and therefore fails to induce complement-mediated cytolysis or an inflammatory response.

Growth in Macrophages. As indicated in Table 7-4, in many chronic infections the virus appears to grow mainly in lymphoid tissue, especially in macrophages. This may have two effects relevant to persistence: (1) impairment of the antibody response and (2) impairment of the phagocytic and cytotoxic potential of the reticuloendothelial system.

Nonneutralizing Antibodies. Viruses that cause persistent plasma-associated viremia not only multiply in lymphoid tissue and macrophages but also characteristically induce production of nonneutralizing antibodies. These antibodies combine with viral antigens and virions in the serum to form immune complexes that may produce "immune complex disease" and block immune cytolysis of virus-infected target cells by T cells or complement-fixing antibodies.

Tolerance. Many persistent infections are associated with a very weak antibody response, especially in congenitally infected animals. Immunological tolerance is rarely complete, but there is a severe degree of specific hyporeactivity in conditions such as congenital LCM and congenital retrovirus infections. Tolerance to a viral antigen may be genetically determined, indeed immune responses to all viral antigens are presumably under the control of *Ir* genes.

Defective Cell-Mediated Immunity. Persistent infections could be caused by partial suppression of the host's CMI response, as a result of any one or combination of several factors: immunodepression by the causative virus, immunological tolerance, the presence of "blocking" antibodies or virus-antibody complexes, failure of immune lymphocytes to reach target cells, inadequate expression of viral antigens on the surface of the target cell, or stripping/endocytosis of surface antigens. These factors are probably important in persistent infections such as visna, SSPE, PML, and in those caused by herpesviruses. Finally, we may again note that many persistent viruses multiply extensively in macrophages and lymphocytes, and thus affect several parameters of the immune response. Indeed, depressed cell-mediated immunity may diminish the rate of destruction of infected cells and thereby prolong the release of viral antigens. The resulting protracted immunogenic stimulus would explain the very high antibody levels found in SSPE and Aleutian disease of mink, for example, which may produce a cascade effect by blocking the already inadequate T cell defenses.

FURTHER READING

Books

Epstein, M. A., and Achong, B. G., eds. (1979). "The Epstein-Barr Virus." Springer-Verlag, Berlin and New York.

Ho, M. (1982). "Cytomegaloviruses: Biology and Infection." Plenum, New York.

Johnson, R. T. (1982). "Viral Infections of the Nervous System." Raven Press, New York.

Mahy, B. W. J., Minson, A. C., and Darby, G. K., eds. (1982). "Virus Persistence," Symp. Soc. Gen. Microbiol., Vol. 33. Cambridge Univ. Press, London and New York.

Mims, C. A., and White, D. O. (1984). "Viral Pathogenesis and Immunology." Blackwell, Oxford.

Mims, C. A., Cuzner, M. L., and Kelly, R. E., eds. (1984). "Viruses and Demyelinating Diseases." Academic Press, London.

Notkins, A. L., and Oldstone, M. B. A., eds. (1984). "Concepts in Viral Pathogenesis." Springer-Verlag, Berlin and New York.

Reviews

Buchmeier, M. J., Welsh, R. M., Dukto, F. J., and Oldstone, M. B. A. (1980). The virology and immunobiology of lymphocytic choriomeningitis virus infection. *Adv. Immunol.* **30,** 275.

Gajdusek, D. C. (1977). Unconventional viruses and the origin and disappearance of kuru. *Science* **197,** 943.

Ganem, D. (1982). Persistent infection of humans with hepatitis B virus: Mechanisms and consequences. *Rev. Infect. Dis.* **4,** 1026.

Hill, T. J. (1985). Herpes simplex latency. *In* "The Herpesviruses" (B. Roizman, ed.), Vol. 3, p. 175. Plenum, New York.

Klein, R. J. (1982). The pathogenesis of acute, latent and recurrent herpes simplex infections. *Arch. Virol.* **72,** 143.

Lehmann-Grube, F., Peralta, L. M., Bruns, M., and Lohler, J. (1983). Persistent infection of mice with the lymphocytic choriomeningitis virus. *In* "Comprehensive Virology" (H. Fraenkel-Conrat and R. R. Wagner, eds.), Vol. 18, p. 43. Plenum, New York.

Merz, P. A., Rohwer, R. G., Kascsak, R., Wisniewski, H. M., Somerville, R. A., Gibbs, C. J., and Gajdusek, D. C. (1984). Infection-specific particle from the unconventional slow virus diseases. *Science* **225,** 437.

Norkin, L. C. (1982). Papovaviral persistent infections. *Microbiol Rev.* **46,** 384.

Oldstone, M. B. A., Fujinami, R. S., and Lampert, P. W. (1980). Membrane and cytoplasmic changes in virus-infected cells induced by interaction of antiviral antibody with surface viral antigen. *Prog. Med. Virol.* **26,** 45.

Stroop, W. G., and Baringer, J. R. (1982). Persistent, slow and latent viral infections. *Prog. Med. Virol.* **28,** 1.

CHAPTER 8

Oncogenic Viruses

Introduction 217
Retroviruses 218
Oncogenes 227
Herpesviruses 234
Papovaviruses 238
Hepadnaviruses 241
Role of Viruses in Human Cancer 243
Further Reading 245

INTRODUCTION

The "cause" of cancer remains a mystery in spite of the efforts of thousands of scientists and copious support from funding agencies over the past two decades. Clear evidence that particular types of cancer can result from exposure to carcinogens such as mutagenic chemicals and irradiation led to the hypothesis that cancer is due to somatic mutation. Such a mutation is assumed to have a certain low probability of occurring spontaneously in a relevant gene in any cell of the body, and a greater probability in the presence of an environmental mutagen. Recent data suggest that a second type of genetic change may also lead to cancer, namely transposition of genes, such that their expression, or that of their new neighbors, is deregulated. Now, the discovery that certain viruses, particularly retroviruses, long known as causes of cancer in animals, transduce cellular oncogenes from one host to another has brought a new sense of unity and purpose to cancer research.

Viruses have been known to cause certain cancers ever since Ellerman and Bang, then Rous demonstrated the transmissibility of avian leukosis

TABLE 8-1
Viruses Associated with Cancer in Man

VIRUS		
FAMILY	SPECIES	TUMOR
Retroviridae	Human T cell leukemia virus	Adult T cell leukemia/lymphoma
Herpesviridae	EB virus	Nasopharyngeal carcinoma
		Burkitt's lymphoma
	Herpes simplex virus type 2?	Cervical carcinoma
Papovaviridae	Human papillomaviruses	Epidermodysplasia verruciformis
		Cervical carcinoma
Hepadnaviridae	Hepatitis B virus	Hepatocellular carcinoma

and sarcoma, respectively, shortly after the turn of the century. Though viral theories of cancer have come and gone from time to time since then, there has been a steady stream of discoveries clearly incriminating viruses in a wide variety of malignant tumors of numerous species of animals, birds, amphibia, and reptiles.Many retroviruses and herpesviruses cause cancer under natural conditions, while several papovaviruses and adenoviruses are oncogenic when inoculated into baby rodents. More recently, strong circumstantial evidence for an etiological role for viruses in a small number of human cancers has been obtained. Again, it is noteworthy that retroviruses, herpesviruses, and papovaviruses (plus hepadnaviruses) are the only families involved (Table 8-1). It is doubtless significant that these are all DNA viruses, except for the retroviruses, the reverse transcriptase of which transcribes a DNA copy of the RNA viral genome for integration into host cell DNA.

RETROVIRUSES

Retroviruses are the major cause of leukemias, lymphomas, and sometimes sarcomas in many species of animals and birds, including chickens, mice, cats, cattle, and gibbon apes. In most of these cases the virus spreads from animal to animal via normal horizontal routes. Some of these *exogenous* retroviruses are oncogenic because they carry a particular gene, the viral *oncogene* (v-*onc*). The *endogenous* retroviruses, on the other hand, exist as a *provirus* (viral DNA) integrated into the genome of every cell of the host's body and are transmitted vertically to the next generation of that animal species via germ line DNA. Such endogenous retroviruses do not usually cause disease, indeed are rarely expressed, but may be activated (induced to replicate) under certain conditions, e.g. in some strains of inbred mice.

During the past couple of decades retroviruses have been subjected to intensive scrutiny by some of the world's most able virologists. As a result, a great deal is now known, particularly about the avian and murine retroviruses, which have been the principal models for study. Without question this is the most difficult area in the whole of virology.

Properties of Retroviruses

The virion has a unique three-layered structure. The genome–nucleoprotein complex, incorporating the enzyme, reverse transcriptase, is thought to have helical symmetry, and is enclosed within an icosahedral capsid, which in turn is surrounded by a host-derived envelope from which project typical virus-coded glycoprotein peplomers (Plate 8-1). Various morphological forms of the virion observed in sections of infected cells by electron microscopy in earlier days reflect differences in morphogenesis of particular retroviruses. The fine distinctions need not concern us here. Suffice to say that they gave rise to a nomenclature (A-, B-, C-, and D-type particles) which is still with us today, so some writers refer to the typical retrovirus as a *C-type particle.*

The genome (Table 8-2) is also unique among mammalian viruses in that it is diploid. Two identical molecules of single-stranded RNA of positive polarity are linked together by hydrogen bonds at their 5′ termini (Fig. 8-1). Another unusual feature is that of terminal redundancy—an identical sequence of about 20 nucleotides occurs at both the 3′ and the 5′ end of each haploid unit. Uniquely also, a molecule of cellular tRNA ($tRNA_{Trp}$ in the case of avian retroviruses; $tRNA_{Pro}$ in mammalian retroviruses) is hydrogen-bonded to a particular site near the 5′ end of the genome; this tRNA serves as a primer for the initiation of DNA synthesis. The 5′ terminus of each haploid unit is capped, the 3′ terminus is polyadenylated.

The genome of the typical nondefective leukemia virus contains two

TABLE 8-2
Properties of Retroviridae

Spherical virion, 80–110 nm
Helical ribonucleoprotein within icosahedral capsid within envelope
ssRNA genome, linear, positive polarity, diploid (inverted dimer), total MW 4–6 million, often defective, may carry oncogene
Reverse transcriptase transcribes cDNA → ds circular DNA → integration into cellular chromosomes
Provirus may never be expressed or may be inducible
Virions bud from plasma membrane

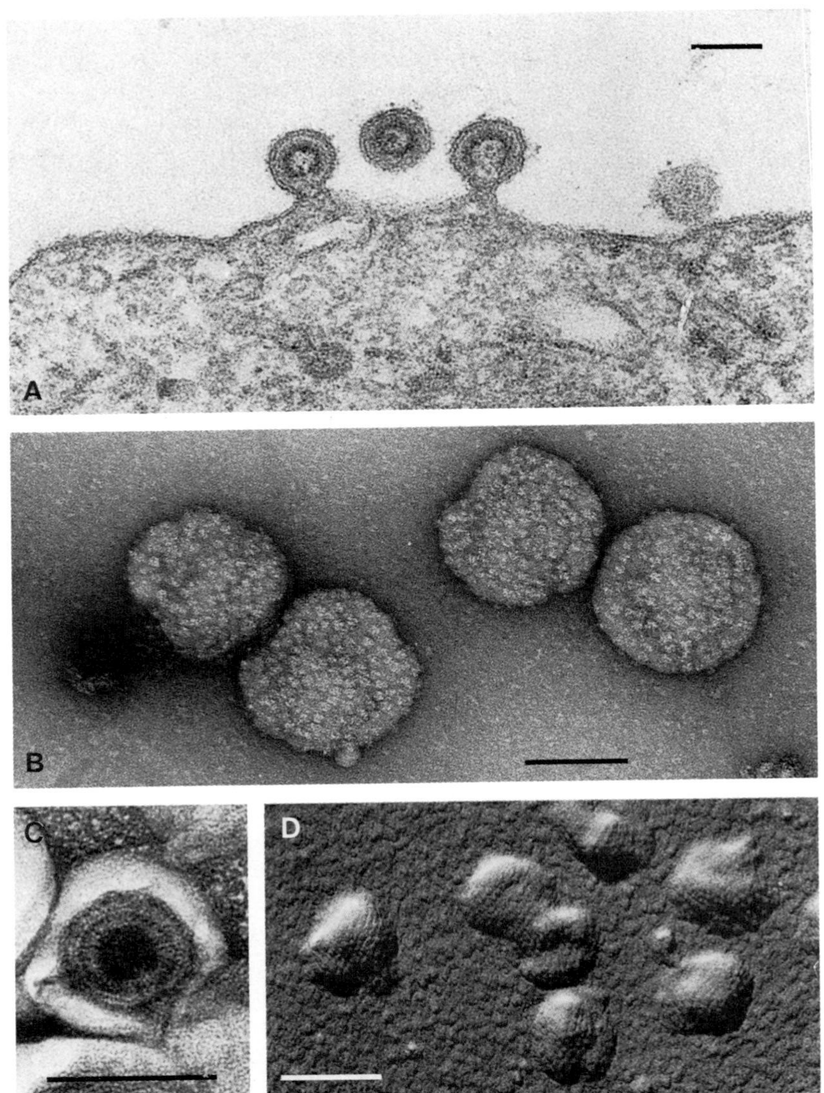

PLATE 8-1. Retroviridae, *subfamily* Oncovirinae, *murine leukemia virus. (A) Budding of virions from a cultured mouse embryo cell. (B) Virions negatively stained with uranyl acetate, showing peplomers on the surface. (C) Virion somewhat damaged and penetrated by uranyl acetate, so that the concentric arrangement of core, shell, and nucleoid becomes visible. (D) Cores isolated by ether treatment of virions, freeze-dried, and shadowed. The hexagonal arrangement of the subunits of the shell around the core is recognizable (bars = 100 nm). (Courtesy Drs. H. Frank and W. Schafer.)*

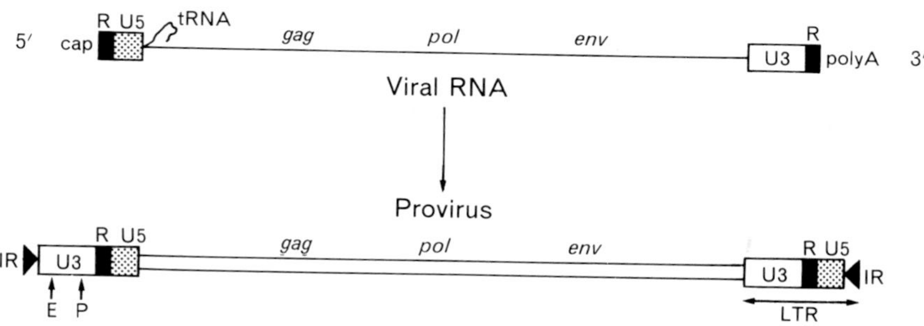

FIG. 8-1. *Structure of retrovirus genome and integrated provirus. The upper panel depicts one of the two identical ssRNA molecules that comprise the diploid retrovirus genome. The major coding regions,* gag, pol, *and* env, *encode the viral structural proteins. In the case of defective, rapidly transforming retroviruses, a portion of this coding sequence is replaced by an* onc *gene. The 5′ terminus is capped and the 3′ terminus polyadenylated. A short sequence, R, is repeated at both ends of the molecule, while unique sequences, U5 and U3, are located near the 5′ and 3′ ends, immediately proximal to R. A cellular tRNA molecule is attached in the vicinity of U5. The lower panel depicts the provirus as integrated into cellular DNA. The genome now consists of dsDNA. It is flanked at each end by a complex structure known as the long terminal repeat (LTR). Each LTR comprises U3, R, and U5, plus short inverted repeat sequences (IR) at the distal end. U3 contains the promoter (P) and enhancer (E), as well as several other sites with important functions.*

copies of each of three genes, known as *gag* (encoding the core proteins), *pol* (encoding the unique viral polymerase, reverse transcriptase), and *env* (encoding the two envelope glycoproteins). A fourth gene, *onc* (the oncogene responsible for oncogenesis) occurs only in certain exogenous retroviruses, the *acute(ly) transforming* retroviruses, which are capable of transforming cells *in vitro* and inducing cancer *in vivo* with high efficiency. Because the oncogene has usually been incorporated into the viral RNA in place of one or more normal viral genes (Fig. 8-1) the genome of most of these highly oncogenic retroviruses is *defective,* hence is dependent on a nondefective *helper* retrovirus for its replication.

The family *Retroviridae* is divided into three subfamilies: *Oncovirinae, Spumavirinae,* and *Lentivirinae,* only the first having oncogenic potential. Further classification of the *Oncovirinae* is extremely complicated (Table 8-3). There are tentatively three genera corresponding to what are otherwise known as *B-, C-, and D-type oncoviruses.* Of most importance are the type C oncoviruses, which in turn comprise three subgenera: the mammalian, avian, and reptilian C-type oncoviruses. *Interspecies* antigens are shared by members of the same subgenus but distinguish, say, avian from mammalian oncoviruses. *Group-specific* antigenic determinants on

TABLE 8-3
Classification of Retroviridae

Subfamilies: *Oncovirinae, Spumavirinae, Lentivirinae*
Genera of *Oncovirinae*
- B-type oncoviruses
- C-type oncoviruses
- D-type oncoviruses

Subgenera of C-type oncoviruses
- Mammalian
- Avian
- Reptilian

Species of mammalian C-type oncoviruses
- (1) Group-specific, then type-specific antigens define species
- (2) Host of origin, host range—ecotropic, xenotropic, amphotropic
- (3) Endogenous or exogenous
- (4) Pathogenicity, oncogenicity, tumor type
- (5) Defective or nondefective genome

the major internal *gag* protein are shared by oncoviruses from the same host species. *Type-specific* (neutralizing) determinants on the envelope glycoproteins define individual species of oncoviruses. Increasingly, direct sequencing of viral proteins and of the viral genome will be invoked to refine the taxonomy of the *Retroviridae*.

The numerous species comprising each of the oncovirus subgenera can be subdivided further in several different ways:

Endogenous or Exogenous. As discussed in Chapter 7, a complete DNA copy of the genome (*provirus*) of one, or often several species of retrovirus is transmitted in the germ line DNA from mother to baby (vertical transmission, Fig. 7-3) and is thus perpetuated in the DNA of every cell of every individual of most vertebrate species. The proviral genome is normally totally silent in feral animals (although the *env* gene is expressed in some strains of chicken, and virus is synthesized throughout life in certain inbred strains of mouse). However, the provirus can be activated (induced) by various procedures such as irradiation, exposure to mutagenic or carcinogenic chemicals, and hormonal or immunological stimuli. Under these circumstances virions are synthesized. Such retroviruses are said to be *endogenous* (Table 8-4).

In contrast, other species of retrovirus behave as more typical infectious agents, spreading horizontally to contacts. Such retroviruses are said to be *exogenous*. Many exogenous retroviruses are recombinants that have arisen in the laboratory, or perhaps following fortuitous coinfection of an animal; they do not occur in provirus form in nature.

TABLE 8-4
Comparison of Endogenous and Exogenous Retroviruses

PROPERTY	ENDOGENOUS	EXOGENOUS	
		SLOW TRANSFORMING	ACUTE TRANSFORMING
Transmission	Vertical (germ line)	Horizontal	Horizontal
Expression	Usually not, but inducible	Yes	Yes
Genome	Complete	Complete	Defective
Replication	Productive	Productive	Requires helper
Oncogene	No	No	Yes
Transformation	No	No	Yes
Oncogenicity	Nil, or rare leukemia	Weakly oncogenic leukemia	Highly oncogenic sarcoma, leukemia, carcinoma

Host Range. Even this is a complicated question. The host species from which a particular retrovirus was originally recovered may not necessarily be its native host. Some endogenous retroviruses are *ecotropic* (i.e., multiply only in the host species from which they originate), while others are *amphotropic* (can multiply in native and certain foreign hosts), and yet others, amazing as it may seem, are *xenotropic* (cannot multiply in the host species in whose genome they are carried as provirus but can in certain other, not necessarily related, species).

Pathogenicity (Oncogenicity). Most endogenous retroviruses never produce disease of any kind, cannot transform cultured cells, and contain no oncogene in their genome (see below). Most exogenous retroviruses, on the other hand, are oncogenic; some characteristically induce leukemia or lymphoma, others sarcoma, yet others carcinoma—usually displaying a predilection for a particular type of target cell (Table 8-5). Exogenous retroviruses can be further subdivided into the highly oncogenic (acute transforming) and the weakly oncogenic. The highly oncogenic retroviruses are the most rapidly acting carcinogens known. For example, some of the sarcomagenic viruses cause death in as short a time as 2 weeks in certain host species, and rapidly transform cultured cells *in vitro*. These properties are attributable to an oncogene which they carry in their genome. The weakly oncogenic (sometimes called slow transforming) retroviruses, on the other hand, contain no oncogene, but can induce B cell, T cell, or, more rarely, myeloid leukemia with low efficiency and after a much longer incubation period. For example, the avian leukemia/leukosis viruses, which are ubiquitous in many flocks of chickens, produce lifelong viremia which usually causes no disease but may lead later in life to a wide variety of diseases involv-

TABLE 8-5
Some Typical Retroviruses

HOST	VIRUS	TRANSFORMATION OF CULTURED CELLS	TUMOR *IN VIVO*
Chicken	Avian leukosis virus	No	B cell lymphoma
	Rous sarcoma virus	Yes	Sarcoma
	Avian erythroblastosis virus	Yes	Erythroleukemia; sarcoma
	Avian myeloblastosis virus	Yes	Myeloblastic leukemia
Mouse	Gross murine leukemia virus	No	Leukemia
	Moloney murine leukemia virus	No	Leukemia
	Harvey murine sarcoma virus	Yes	Sarcoma
	Bittner mammary tumor virus	No	Mammary carcinoma
Cat	Feline leukemia virus	No	Leukemia
Cattle	Bovine leukemia virus	No	Leukemia
Monkey	Baboon endogenous virus	No	Nil
	Squirrel monkey retrovirus	No	Nil
	Woolly monkey sarcoma virus	Yes	Sarcoma
	Gibbon ape leukemia virus	No	Leukemia
	Mason–Pfizer monkey virus	No	Nil
Man	Human T cell lymphotropic virus-I	No	Adult T cell leukemia
	HTLV-II	No	?
	HTLV-III	No; cytocidal	Nil; AIDS

ing the hemopoietic system: predominantly lymphomatosis, but sometimes erythroblastosis, myeloblastosis, and osteopetrosis, or rarely solid tumors (sarcomas, carcinomas, and endotheliomas).

Defective or Nondefective. As mentioned earlier the genome of the highly oncogenic exogenous retroviruses (including nearly all sarcoma viruses) has two important characteristics: (1) it carries an extra gene, the oncogene (*v-onc*), which confers oncogenicity; and (2) it has deletions in one or more of the three essential genes, hence is defective for replication (though capable of transformation, see Chapter 5 and Table 5-2). Such defective retroviruses can multiply only in the presence of a nondefective helper retrovirus which supplies the protein product(s) of the missing gene(s). Ironically, the only known sarcoma virus that is not defective is the Rous sarcoma virus, upon which many of the classical studies were made.

Replication of Retroviruses

The replication of retroviruses differs in a number of critical respects from that of all other families of viruses: (1) the ssRNA viral genome must be transcribed into a dsDNA copy; (2) this viral DNA must then become integrated into cellular DNA; (3) viral DNA may simply remain integrated and be transmitted to that cell's progeny, or may be transcribed into RNA leading to the production of virions; and (4) even such productive infection is noncytocidal—the cell survives and divides.

The key steps in the integration and replication of the retroviral genome are depicted in Fig. 8-2. First, the reverse transcriptase, using the tRNA as a primer, makes a ssDNA copy of haploid viral RNA. The parental RNA molecule is then removed from the DNA:RNA hybrid by RNase H, which is a second enzymatic activity carried by the reverse transcriptase molecule. By a complex process not yet fully elucidated, the newly made DNA minus strand is converted into dsDNA, which is colinear with vRNA except that each end is flanked by an additional identical sequence some hundreds of nucleotides long, known as the *long terminal repeat (LTR).* The LTR comprises sequences derived from both ends of the vRNA: U3 from the 3′ end, U5 from the 5′ end, and a short sequence, R, which is present at both ends of genomic RNA (Fig. 8-1). This *linear duplex DNA* now migrates from the cytoplasm to the nucleus where it becomes circularized. Several copies of this *closed duplex DNA* are then integrated into host chromosomal DNA at random or quasi-random sites. During integration, short direct repeats of cellular DNA are synthesized and joined to each end of the provirus (Fig. 8-2). As we shall see later, it is probably no coincidence that this structure is very similar to that of transposable elements (transposons) encountered in lower eukaryotic and prokaryotic organisms.

The *provirus* is a self-sufficient transcriptional unit, because the U3 sequence within the LTR contains a *promoter* which directs initiation of transcription, and an *enhancer* which enhances transcription. The enhancer may confer tissue specificity on the virus and its oncogene. Host RNA polymerase II transcribes RNA from the proviral DNA. Full-length, capped, and polyadenylated RNA (vRNA) is destined for incorporation into new virions. Some, however, is spliced to form subgenomic mRNA that becomes translated into proteins, which are then cleaved, phosphorylated, and glycosylated as appropriate to form the final protein products of *gag, pol, env,* ± *onc.* Following assembly in the cytoplasm and budding from the plasma membrane (Plate 8-1), the final steps in morphogenesis of the virion (maturation, involving further proteolytic cleavage and "condensation of the nucleoid") actually occur after release from the cell.

Genetic recombination is extremely common among retroviruses. Recombinants may acquire sequences from other retroviruses or from the cell itself. Their accurate identification requires nucleotide sequencing of the entire genome. Furthermore, coinfection with two retroviruses can

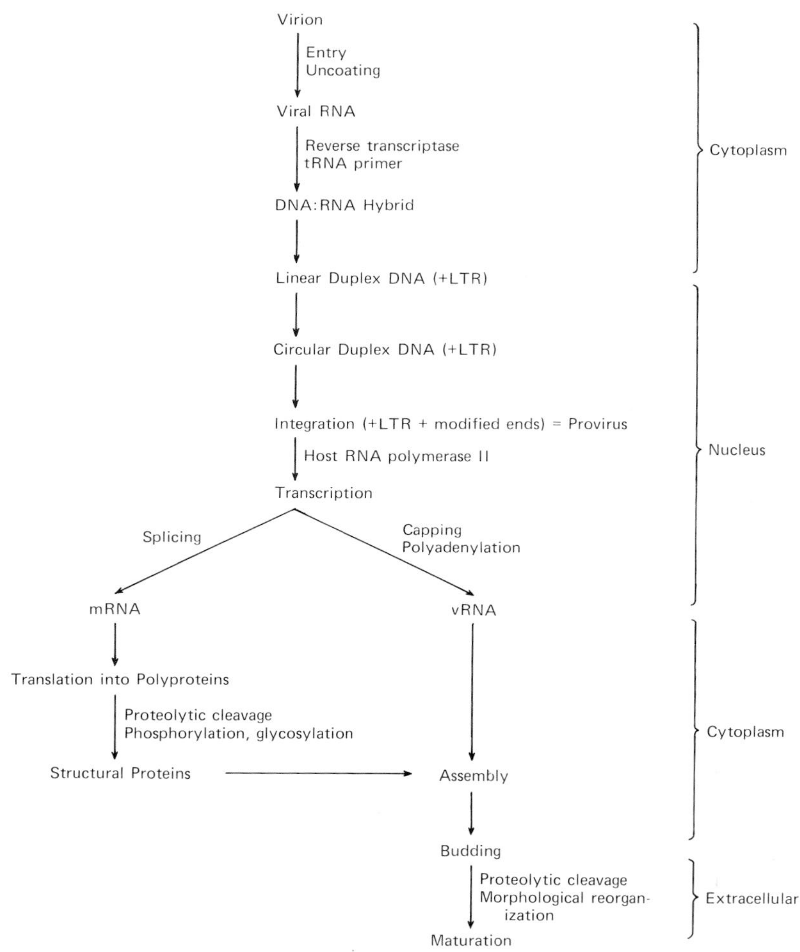

FIG. 8-2. *Replication of retroviruses.*

lead to *phenotypic mixing,* the phenomenon whereby the genome of one parent in a mixed infection can become enclosed in an envelope with surface glycoproteins coded by the genome of the other parent (see Chapter 4). Such virus particles are exceedingly difficult to characterize, as they are found by nucleotide sequencing to have the genome of one species of virus, but by neutralization to have the antigenic characteristics of another. The host range of the virus may be extended because the novel envelope glycoprotein(s) enable the virus to attach to receptors on cells of a different animal species (see Chapter 4).

Human T Cell Lymphotropic Viruses

From the viewpoint of human medicine one of the most exciting discoveries of tumor virology was that of the human T cell lymphotropic virus (HTLV-I) in 1980. A typical retrovirus, HTLV-I now appears to be the cause of a proportion of T cell leukemia in adults.

This highly malignant disease of young adults is characterized by hepatosplenomegaly, lymphoadenopathy, and, sometimes, skin lesions consisting of T cell infiltrates. Hypercalcemia is also a feature. The proliferating T blasts are OKT-4^{+} and have a characteristic lobulated nucleus.

The world distribution of the disease has yet to be determined, but it seems to be particularly prevalent in southern Japan and in the Caribbean, while a variant occurs in Africa. Infection is usually subclinical, as indicated by the fact that over 10% of the normal population in some parts of Japan and up to 50% of the close relatives of patients have antibody. Worldwide, less than 1% of people have antibody.

HTLV-I is unrelated antigenically or by nucleic acid hybridization to retroviruses of other animals, except for somewhat similar viruses in Old World monkeys and in cattle. The virus transforms human T cells *in vitro* to immortal lines with many of the characteristics of leukemic cells. Certain nonlymphoid cell lines also support the growth of the virus; the infection is noncytocidal but productive. Since HTLV-I cDNA is found only in transformed T cells and not in normal human cells, the virus is assumed to be an exogenous, not an endogenous, human retrovirus.

Recent research suggests a possible mechanism of tumorigenesis by HTLV-I in the absence of any known oncogene. An additional gene, X, encodes a protein which is believed to exert positive feedback, not only enhancing transcription of proviral DNA by binding to the LTRs, but also activating transcription of cellular genes by binding to tissue-specific enhancer sequences. Thus this phenomenon, known as *trans-acting transcriptional regulation,* may result in excessive production of a certain cellular protein which may lead to malignancy.

Two further types of HTLV have now been described. HTLV-II was isolated from a T cell variant of hairy-cell leukemia, but has not been causally linked to any disease. HTLV-III, recently hailed as the cause of acquired immunodeficiency syndrome (AIDS), is discussed in Chapter 28.

ONCOGENES

The highly oncogenic retroviruses owe their oncogenicity to a particular viral oncogene (or sometimes to two different oncogenes). Deletion of this single gene removes the oncogenicity of the virus. Furthermore, transfection of cultured cells with a DNA copy of this oncogene linked to the viral LTR which contains the promoter and enhancer sequences can transform these cells to the malignant state. For example, the so-called *src* gene of the famous Rous sarcoma virus confers oncogenicity on that virus, and a DNA copy of this oncogene in association with the appropriate LTR transforms chicken fibroblasts *in vitro* to malignant sarcoma cells. Moreover, cells transformed by a sarcoma virus containing a *ts* mutation in its *src* gene can be alternately transformed at the permissive temperature, returned to the normal phenotype at the nonpermissive temperature, then transformed again as the temperature is lowered once more. These findings demonstrate that the protein product of an oncogene is required for both the initiation and the maintenance of transformation.

Oncogenes are not necessary for viral replication. Indeed, they appear to serve no useful function as far as the virus is concerned, but in most cases render the virus defective. This observation provides a hint as to their real role and origin.

Perhaps the most startling revelation in the history of oncology was the discovery that the genome of every cell in every normal animal contains oncogenes very closely analogous to these viral oncogenes. In 1969 Huebner and Todaro put forward their "oncogene hypothesis", which postulated that retroviral oncogenes, present but unexpressed in all normal cells, may be activated by carcinogens, leading to cancer. While it now appears that the oncogene hypothesis in its original form was somewhat wide of the mark, it was of seminal importance in generating massive financial support for cancer research in the United States during the 1970s and in focusing the efforts of a whole generation of tumor virologists on the crucial significance of oncogenes.

As it has turned out, *cellular oncogenes* (c-*onc*), sometimes called *proto-oncogenes,* are not of viral origin—on the contrary, *viral oncogenes* (v-*onc*) are of cellular origin. Cellular oncogenes accidentally become incorporated by recombination into the genome of a retrovirus—a phenomenon known as *transduction.* Normal cells contain at least 20 distinct cellular oncogenes. Unlike viral oncogenes, (1) they are located at constant positions on the chromosomes of every cell of every individual of the species, (2) they contain introns, (3) they segregate as classical Mendelian loci, (4) they have been conserved with high fidelity through evolution (e.g., *src* genes from man and *Drosophila* display 95% homology), and (5) they are normally transcribed at low levels only, but their expression may increase substantially in particular tissues and/or at particular stages of differentiation. This strongly suggests that cellular oncogenes are essential genes and that they may play key roles in the normal regulation of cell growth, division, and differentiation.

Over 20 different viral oncogenes have now been described in various oncogenic retroviruses. They resemble, but are not identical with the corresponding cellular oncogenes, the principal differences resulting from (1) deletion of all introns and cellular transcriptional signals, (2) substitution of a small number of nucleotides (point mutations), and/or (3) deletion and/or rearrangement of some portion(s) of the coding sequence. The protein encoded by the viral oncogene often has a similar function and intracellular location to that encoded by the corresponding cellular oncogene from which the v-*onc* is presumed to have been derived.

Another remarkable episode in the oncogene saga was the discovery in 1982 that the DNA of malignant cells from chemically induced experimental animal cancers or naturally occurring human cancers contains a certain gene which (1) differs only very slightly from a corresponding protooncogene in nonmalignant cells, (2)displays considerable homology with a v-*onc* gene from a known oncogenic retrovirus, and (3) can itself transform cultured cells to the malignant state. The DNA of human bladder carcinoma cells was shown to differ from that of normal bladder epithelial cells by a single nucleotide substitution in a single gene. Transfection of NIH3T3 mouse fibroblasts in culture with this mutant gene transformed them into sarcoma cells. The protooncogene involved (now known as c-*Ha-ras*) was discovered to show a remarkable degree of homology with the v-*Ha-ras* oncogene found in the Harvey murine sarcoma virus. Since then, many of the transforming oncogenes detected by the transfection-focus assay in naturally occurring human cancers and chemically-induced animal cancers have been found to belong to the *ras* family. They fall into three subclasses: H-*ras* (homologous but not identical to the oncogene carried by Harvey murine sarcoma virus), K-*ras* (homologous to that of Kirsten murine sarcoma virus) and N-*ras* (for which no retroviral homologue has yet been described). They all encode similar proteins containing 189 amino acids. Point mutations leading to substitution of a single amino acid in either of two critical domains confer transforming ability upon the molecule.

These findings indicate that cellular oncogenes can become oncogenic by mutation occurring either *in situ* or as a consequence of recombination with the genome of a retrovirus which is subsequently reinserted into cellular DNA. There is evidence to suggest that cancer may also result from unregulated expression of normal protooncogenes. Possible mechanisms whereby these events could occur are the subject of intensive research today, for, as J. M. Bishop has surmised, it is not unlikely

that oncogenes may represent "the final common pathway to tumorigenesis . . . . An enemy has been found; it is part of us."

Function of Proteins Encoded by Oncogenes

The 20 or more viral oncogenes and their cellular counterparts can be classified into about half a dozen groups according to the nature of the protein for which they code (Table 8-6). The products of one class are found where they might be expected, namely in the nucleus; the *myc* and *myb* proteins have been shown to bind to DNA. By contrast, the proteins of the *ras* family are found in plasma membrane, bind to GTP and display GTPase activity; they are thought to be activated by GTP-binding and perhaps to serve as coupling factors, relaying signals to other parts of the cell. The *sis* gene-product is a growth factor related to PDGF (platelet-derived growth factor). On the other hand, the *erb-B* product, which is located in the plasma membrane, resembles not a growth factor but the receptor for a growth factor! Specifically, it resembles that part of the receptor for epidermal growth factor (EGF) which carries tyrosine-specific protein kinase (tyrosine kinase) enzyme activity. This is particularly interesting because approximately half of all known oncogenes, including *src*, the first to be described, encode proteins that either carry tyrosine kinase activity or display sequence homology with that enzyme. The common thread linking these apparently disparate observations can be perceived by considering the structure of receptors for growth factors such as EGF or PDGF.

The EGF receptor spans the plasma membrane. The external domain of the molecule constitutes the binding site for EGF. The intra-

TABLE 8-6
Some Typical Oncogenes

	PRODUCT	
ONCOGENE	FUNCTION	LOCATION
src	Tyrosine kinase	Plasma membrane[a]
erb-B	Growth-factor receptor[b]	Plasma membrane[a]
sis	Growth factor[c]	Cytoplasm
ras	GTP binding (GTPase)	Plasma membrane
myc	DNA binding	Nucleus

[a] Usually; sometimes other cytoplasmic membranes.
[b] Homologous with the tyrosine kinase domain of the receptor for epidermal growth factor (EGF).
[c] Related to the platelet-derived growth factor (PDGF).

cytoplasmic moiety caries tyrosine kinase activity. However, the enzyme is activated only by a conformational change that follows binding of EGF to the receptor. Then, various proteins in the vicinity (as well as the receptor itself) may be phosphorylated (at tyrosine residues, rather than at the conventional serine or threonine). Thus the apparent paradox that oncogenes encode both growth factors and their receptors is resolved. Intracellular proteins will be phosphorylated at an enhanced rate and at inappropriate times by either (1) autocrine stimulation by binding of a constitutively produced growth factor to its natural receptor, or (2) uncontrolled expression of protein kinase activity by a truncated or otherwise altered receptor which lacks its growth-factor-binding-site and is permanently in the activated state.

Presumably, such protein kinases are able to phosphorylate key proteins in their immediate vicinity and perhaps also some that are situated in, or subsequently migrate to distant parts of the cell. For instance, phosphorylation of elements of the cytoskeleton situated just beneath the plasma membrane might lead to a major perturbation of cellular structure and function. It has also been shown that the *src*-coded tyrosine kinase can phosphorylate a lipid in the plasma membrane, thereby accelerating its breakdown to diacylglycerol, which in turn activates a serine-specific protein kinase. Whatever the precise mechanism, it is postulated that phosphorylation of protein somehow culminates in the transmission of a message to the nucleus. There is evidence that binding of a mitogenic growth factor to its receptor induces the expression of nuclear oncogenes.

Postulated Mechanisms of Tumorigenesis by Oncogenes

Though still highly speculative, there is already evidence to support a number of alternative routes to carcinogenesis involving oncogenes. It is perhaps most logical to classify them into those mediated by viral oncogenes and those mediated by cellular oncogenes—although one must bear in mind that all v-*onc* genes were presumably derived originally by accident from c-*onc* genes during recombination between proviral and cellular DNA.

Transduction of Oncogene by Retrovirus. This is the best documented of all the mechanisms to be discussed. A cellular oncogene captured originally by recombination becomes an integral part of a retroviral genome and is now known as a viral oncogene. During the process of recombination which accompanies every successful integration of provirus, opportunities abound for point mutations, deletions, and various forms of rearrangement to occur. Therefore the v-*onc* gene generally differs significantly from its c-*onc* progenitor and may code for a somewhat different protein with a somewhat different function or enzyme specificity. Moreover, when later reinserted into the genome of another cell, the v-*onc* is expressed at an exceptionally high rate because its transcription is regulated by the powerful promoter and enhanced by the enhancer within the LTR. Thus, a viral oncogene may lead to cancer either because its protein product differs significantly from that encoded by the corresponding cellular oncogene, or because it is expressed at an excessively high rate.

Activation of Cellular Oncogene. There is now a great deal of evidence suggesting that cellular oncogenes as well as viral oncogenes can be responsible for malignant transformation. It is not difficult to imagine cancer arising from *over-expression of a normal c-onc,* or *inappropriate expression,* at the wrong place or time, e.g., in the wrong cell lineage or at the wrong stage of differentiation.

Insertional Mutagenesis. Integration of a provirus may greatly amplify the expression of a cellular oncogene in the immediate vicinity of the insertion site, if the expression of the c-*onc* now falls under the control of the newly integrated proviral promoter or enhancer. This is thought to be the likely mechanism of tumorigenesis by those retroviruses which themselves lack an oncogene. On those rare occasions when the weakly oncogenic avian leukoviruses do cause malignancy, the viral genome is generally found to have been integrated at a particular location, just upstream from a known cellular oncogene. Integration of such an avian leukosis provirus has been shown to increase the synthesis of the c-*myc* oncogene product (a nuclear protein) 30- to 100-fold. Indeed, only the LTR itself need be integrated, and furthermore, c-*myc* may be induced in cells in which this gene is never normally expressed. In a more direct demonstration of the postulated mechanism, certain normal cellular oncogenes, e.g., c-*ras*, have been shown to be capable of transforming cells *in vitro* if attached to the strong promoter in the LTR from a retroviral provirus.

The inference being drawn from these experiments is that a quantitative, rather than a qualitative difference in the protooncogene product can itself cause cancer. In view of the crucial implications of this hypothesis for future cancer research it is vital to ascertain whether absolutely normal (rather than premalignant) cells can be transformed to the fully malignant state *in vitro* and *in vivo* by such mechanisms (see below).

Translocation. Enhancement of protooncogene expression following the chromosomal translocations that are regularly observed in certain tumors may also be attributable to the gene coming under the control of an inappropriate promoter and/or enhancer. For instance, the 8:14 chromosomal translocation that characterizes Burkitt's lymphoma brings the c-*myc* gene into juxtaposition with immunoglobulin genes, thus perhaps under the control of the transcriptionally active sequence(s) governing immunoglobulin synthesis. It is possible that the EB virus is responsible for this particular translocation. Retroviruses could do the same.

Gene Amplification. Gene amplification, particularly of oncogenes, is a feature of many cancers of man and animals, e.g., a 30-fold increase in the number of c-*ras* gene copies is found in one human cancer cell line, while the *myc* gene is amplified in several human tumors. The increase in gene copy number leads to a corresponding increase in the amount of the oncogene's product. Again, this "dosage effect" may conceivably suffice to produce cancer.

Mutation. In a cellular oncogene, e.g., *c-ras,* mutation may alter the function of the protein for which it codes. Such mutation can occur either *in situ* as a result of chemical or physical mutagenesis, or in the course of recombination with retroviral DNA—which takes us full circle.

Cooperation between Oncogenes: Multistep Carcinogenesis. As is well known to classical oncologists, nonvirally induced cancer does not generally arise as the result of a single event, but by up to half a dozen or more steps leading to progressively greater loss of regulation. Significantly, the genome of some retroviruses (e.g., avian erythroblastosis virus) carries two different oncogenes (and that of polyoma virus, three), while two or more distinct oncogenes are activated in certain human tumors (e.g., Burkitt's lymphoma). Moreover, the NIH3T3 cells, which were so readily transformed by the mutated c-*ras* oncogene in the transfection-focus assay mentioned above, were already an immortalized cell line; transformation of primary cultures of normal fibroblasts could not be demonstrated. However, it has recently been shown that transfection of primary cultures of normal fibroblasts with two separate oncogenes of different complementation groups (see Chapter 4) quite regularly induces malignancy. One oncogene encoding a plasma membrane protein plus one encoding a nuclear protein are generally required. Cotransfection of normal rat embryo fibroblasts with the mutant *ras* gene plus the polyoma virus large T gene, or with *ras* plus the *E1A* early gene of oncogenic adenoviruses, or with *ras* plus *myc,* converted them into tumor cells. It should be noted that, whereas *ras* and *myc* are typical oncogenes, originally of cellular origin, the other two are assumed to be true viral genes. However, *E1A* has now been shown to have partial sequence homology with v-*myc* and may therefore have been derived from c-*myc* at some point much earlier in evolution. Recently, it has been demonstrated that a chemical carcinogen can be substituted for the viral gene in such cotransfection tests; following immortalization of cells *in vitro* by treatment with the carcinogen, a cloned oncogene was able to convert this continuous cell line to a tumor cell line. Such analogies resurrect earlier unifying theories of cancer causation that viewed viruses as analogous to other mutagenic carcinogens, both being capable of initiating a chain of two or more events leading eventually to the full expression of malignancy. If viruses or oncogenes are to be considered as cocarcinogens in a cooperative sequence of genetic events culminating in cancer, it will be important to determine whether their role is that of initiator or promoter, or whether they are interchangeable. The most plausible hypothesis may be that (1) a limited number of cellular oncogenes represent common genetic targets for carcinogenic chemicals, radiation, and tumor viruses, and (2) the full expression of malignancy requires the mutation or enhanced expression of more than one class of oncogene.

Other Genes Involved in Carcinogenesis. Lest all this appears too simple, oncologists are now beginning to speak of *anti-oncogenes* (or *tumor suppressors*)! They note that in several human tumors, e.g., retinoblastoma and neuroblastoma, both alleles of a particular gene are lost. Suppressor genes can evidently negate the transforming capacity of oncogenes. Additional genes no doubt modulate the invasiveness, metastasis, and immune rejection of neoplasms—though in regard to the latter, one would not expect foreign antigens on the surface of cancer cells arising other than via viral infection.

HERPESVIRUSES

The *Herpesviridae* (see Chapter 16) are the major family of DNA viruses to be incriminated in cancer. Typical herpesviruses have been shown unequivocally to be the etiological agents of lymphomas or carcinomas in hosts ranging from amphibia, through birds, to primates (Table 8-7). The human herpesvirus, EBV is regularly associated with a B cell lymphoma and a carcinoma of the throat in man.

Herpesvirus ateles and *Herpesvirus saimiri* cause inapparent infections in their natural simian hosts, but when transmitted to other species of monkey induce a fatal T cell lymphoma/lymphatic leukemia. The lymphomas induced by these simian viruses and Marek's disease virus (see below) may represent useful models for human Burkitt lymphoma. Lucké's renal adenocarcinoma of frogs, on the other hand, presents an interesting parallel to human nasopharyngeal carcinoma, in which an epithelial cell rather than a lymphocyte is the target.

TABLE 8-7
Oncogenic Herpesviruses

VIRUS	NATURAL HOST	NATURAL TUMOR	EXPERIMENTAL TUMOR[a]
Epstein–Barr virus	Man	Nasopharyngeal carcinoma, Burkitt's lymphoma	Monkeys
Herpesvirus papio	Baboon	Lymphoma	Primates
Herpesvirus ateles } *Herpesvirus saimiri* }	Monkey	Nil?	Lymphomas, leukemias in other primate species
Marek's disease virus	Chicken	Lymphoma	Chicken
Lucké frog virus	Frog	Renal adenocarcinoma	Tadpole
Herpesvirus sylvilagus	Rabbit	Lymphoma	Rabbit

[a] Induced by inoculation of virus.

Marek's disease of chickens is a classic example of a malignant condition unequivocally caused by a virus and completely preventable by immunization. Marek's disease virus (MDV) infects T lymphocytes, causing them to proliferate to produce a lethal generalized lymphomatosis. It also infects cells of the feather follicles, from which virus is shed to infect other chickens. The disease was a common and major problem for the poultry industry but is preventable by vaccination with a live attenuated variant of MDV, or, more usually today, by live unattenuated turkey herpesvirus, which is harmless in chickens but cross-protects. This striking instance of prevention of cancer by a vaccine serves as the light on the hill to guide the way and lift the hopes of those still groping in the dark for an answer to human cancer.

Epstein–Barr Virus (EBV)

Burkitt's lymphoma (BL) is a highly malignant B cell lymphoma of children, rare in most parts of the world but discovered by Burkitt to be common in East Africa (Plate 8-2). Epstein and colleagues subsequently isolated from cultured tumor cells a herpesvirus, now known as the Epstein–Barr virus (EBV), the properties and clinical significance of which are discussed in Chapter 16. Comprehensive studies over the past 20 years have confirmed that the EBV DNA genome is detectable in multiple copies in each cell of most African Burkitt lymphomas. Though sometimes integrated into host chromosomes, the EBV DNA is generally in the form of closed circles of the complete viral DNA molecule, found free in the cytoplasm as autonomously replicating *episomes*. The cells express EBNA (EB nuclear antigen) detectable by immunofluorescence, but do not produce virus until induced to do so by cultivation *in*

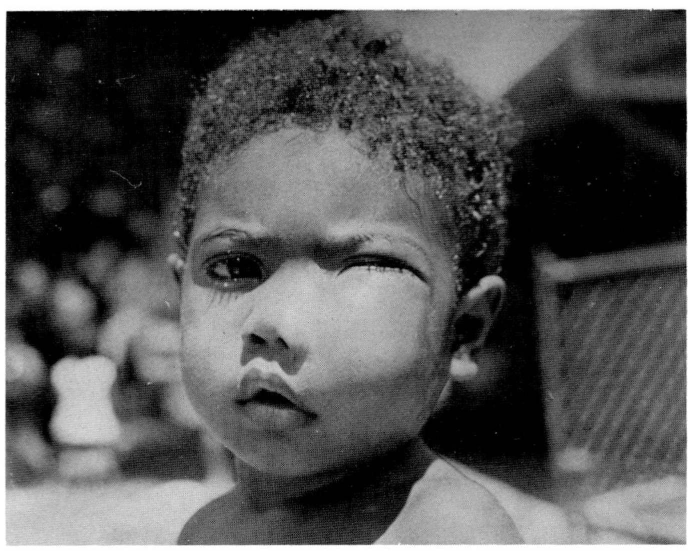

PLATE 8-2. *Burkitt lymphoma. Characteristic facial tumor in a child from New Guinea. (Courtesy Dr. J. Biddulph.)*

vitro. It must be added, however, that only a minority of histologically and karyologically comparable Burkitt lymphomas from other parts of the world contain EBV DNA or EBNA.

The malignant cells also contain a characteristic chromosomal aberration, namely an 8:14 translocation. Recent work on oncogenes, described above, has given this karyological abnormality a genetic interpretation which may explain the mechanism of carcinogenesis by EBV. The human c-*myc* oncogene, located on the distal segment of chromosome 8, is transposed to one of three chromosomes that contain genes for immunoglobulin—usually chromosome 14, sometimes 2 or 22. This leads to enhanced expression of the translocated c-*myc* gene (see Oncogenes, above).

EBV induces malignant lymphomas when injected into certain species of primate. The virus also transforms cultured B lymphocytes from man or other primates, conferring on them the capacity to propagate indefinitely *in vitro* (but not necessarily *in vivo*). The EBV genome persists in these lymphoblastoid cell lines, predominantly in episomal form, and certain antigens are expressed, notably EBNA and LYDMA, but most cell lines do not produce virus. The ability of EBV to immortalize lymphocytes has been widely exploited by immunologists, and by pharmaceutical companies for the manufacture of interferons (see Chapter 11).

Nasopharyngeal carcinoma (NPC), while uncommon generally, is the commonest cancer in certain parts of southern China and in Cantonese populations who have settled in other parts of South-East Asia. This undifferentiated type of NPC (but not the well-differentiated type more commonly seen in other parts of the world) contains EBV DNA, in episomal form, even more regularly than does Burkitt's lymphoma; the correlation is virtually 100%, regardless of country of origin. These epithelial cells also synthesize EBNA antigen, yield EBV on transplantation into athymic mice, and transform B lymphoctyes on cocultivation *in vitro*. Interestingly, EBV fails to transform, or replicate in, cultured epithelial cells, which seem to lack receptors for the virus; EBV antigens are made if EBV or EBV DNA is introduced into epithelial cells artificially.

The regular association between EBV and these two human cancers strongly suggests a cause-and-effect relationship. But it does not prove it. First, EBV is not associated with most of the non-African Burkitt's lymphoma. Second, it is a ubiquitous virus, infecting most children subclinically, hence it is quite possible that the association is fortuitous. Third, there is a delay of a few years between initial infection with EBV and the development of BL, and of up to 40 years with NPC. However, a long-term prospective seroepidemiological survey in Uganda has now firmly established that severe EBV infection acquired very early in life predisposes to Burkitt's lymphoma. In particular, a high antibody titer against the EBV capsid antigen VCA, arising early in childhood, indicates a high risk of subsequently contracting Burkitt's lymphoma. The age of primary infection with EBV does not seem to be so critical in NPC, but reactivation of a latent EBV infection in the nasopharynx, with a consequent rise in anti-VCA IgA, frequently precedes the development of the tumor. The striking racial and geographical distribution of BL and NPC suggests a genetic predisposition or a second environmental cofactor. For example, the prevalence of BL in malarious areas of East Africa has led to the proposition that concomitant malarial infection may be sufficiently immunosuppressive to lead to a failure to control the proliferation of EBV-transformed cells. However, this remains an unproven hypothesis—there are many other potential cofactors. While the data now available strongly indicates a causal connection between EBV and these two tumors, final verification may have to await the development of a satisfactory vaccine (see Chapter 16). A prospective study demonstrating that an EBV vaccine protects against either cancer would provide proof and prophylaxis simultaneously.

Herpes Simplex Virus (HSV-2)

A few years ago there was considerable enthusiasm for the hypothesis that HSV-2 caused cancer of the cervix. A disproportionately high percentage of patients with carcinoma of the cervix have clear evidence of prior infection with HSV-2, but several other factors are at least equally closely associated with this tumor—notably, commencement of sexual intercourse at an early age, sexual promiscuity and multiple partners. In the parlance of the epidemiologists, HSV-2 is a *risk factor* in cervical carcinoma. While it may yet turn out to be a contributing cause the spotlight has recently shifted from HSV-2 to genital papillomaviruses.

PAPOVAVIRUSES

The second family of DNA viruses to be incriminated in cancer is the *Papovaviridae* (see Chapter 14). Paradoxically, of the two genera, the one most intensively studied by tumor virologists for 20 years, *Polyomavirus*, is now thought to have little or nothing to do with cancer, while the other, *Papillomavirus*, largely neglected until recently, is today the center of attention since infection with warts viruses has been shown to predispose to carcinoma in several animals, including man.

Transformation by Nonhuman Polyomaviruses and Human Adenoviruses

During the heyday of experimental viral oncology—the 1960s and 1970s—two papovaviruses of the *Polyomavirus* genus, polyoma virus of mice and simian virus 40 (SV40), as well as certain human adenoviruses (types 12, 18, and 31), became the most popular models for biochemical investigation of the mechanism of viral carcinogenesis. They attracted attention because they were found to induce malignant tumors on inoculation into baby (but not adult) hamsters and other rodents. Furthermore, although they multiply productively in and destroy cultured cells of their native host species, they can transform cultured cells of certain other species. Though not necessarily malignant, these transformed cells display many of the properties of malignant cells (Table 5-2; Plate 5-2). As they can be readily cloned and grow to high density *in vitro* they represent ideal experimental material for detailed biochemical analysis. A remarkable amount is now known about the expression of the integrated viral genome in cells transformed by these viruses.

The polyomavirus- or adenovirus-transformed cell does not produce virus. Viral DNA is integrated at random sites on the cell's chromosomes. Most of the integrated viral genomes are complete in the case of the polyomaviruses, but defective in the case of the adenoviruses. Yet, only certain early viral genes are transcribed, albeit at unusually high frequency. Their products, demonstrable by immunofluorescence, are known as *T (tumor) antigens.* A great deal is now known about the role of these proteins in transformation. For example, the *"middle T" antigen* of polyoma virus (like the product of the *E1B* gene of adenoviruses or the v-*ras* of retroviruses) seems to bring about the change in cell morphology and enables the cells to grow in suspension in semisolid agar as well as on a solid substrate (*anchorage independence*), whereas the *"large T" antigen* (like the *E1A* gene of adenoviruses or the v-*myc* of retroviruses) is responsible for the reduction in dependence of the cells on serum and enhances their life span in culture (*immortalization*).

Virus can be "rescued" from polyomavirus-transformed cells, i.e., induced to multiply, by any of a number of manipulations, including irradiation, treatment with certain mutagenic chemicals, or cocultivation or fusion with certain types of permissive cell. This cannot be achieved with adenovirus-transformed cells; the integrated adenoviral DNA contains substantial deletions.

It should be stressed that integration of viral DNA does not usually lead to transformation. As discussed in Chapter 7, persistent infections with herpesviruses and parvoviruses are characterized by integration of the viral genome without any indication of cellular transformation. Many or most episodes of integration of papovavirus or adenovirus DNA probably have no important biological consequence at all. Transformation by these viruses is a rare event, requiring that the viral transforming gene happens to be integrated in the location and configuration needed for its expression. Even then, many transformed cells revert (*abortive transformation*). Furthermore, those cells displaying the overt characters of transformation (immortalization, anchorage independence, altered morphology and surface properties, growth to high saturation density, etc.—see Table 5-2) are not necessarily malignant. This needs to be demonstrated independently by transfer to athymic mice or preferably to syngeneic animals. As was discussed above (Oncogenes) recent evidence suggests that certain ostensibly normal cell lines commonly used for transformation assays, such as 3T3 cells, are in fact "premalignant"; transformation of truly normal cells to the fully malignant state may require the cooperation of more than a single oncogene, e.g., polyoma large T, adenovirus *E1A,* or retrovirus v-*myc,* together with at least one other.

In view of the elegance of the molecular biology that has given us such insight into the parameters of transformation by polyomaviruses and adenoviruses, it seems almost a pity that neither now appears to have anything to do with human cancer! Even the so-called "highly oncogenic" types (12, 18, and 31) of human adenoviruses now seem to be oncogenic only for newborn rodents; painstaking searches for evidence of association with cancer in man have produced no evidence to suggest any involvement whatever. Therefore we shall not afford ourselves the luxury of devoting any more space to the subject in a text on medical virology.

Papillomaviruses

Papillomaviruses, as their name implies, produce papillomas, or warts, in the skin or mucous membranes. These benign neoplasms are harmless hyperplastic outgrowths which generally regress spontaneously. Occasionally, however, they may progress to malignancy. There is evidence to suggest that the viral DNA may set the scene for the transition to the malignant state, but a cofactor may also be required.

One of the most instructive models is the bovine papillomavirus (BPV), of which there are six types. In the hot Australian sun, benign viral papillomas on the eye or on hairless or nonpigmented patches of skin often become malignant; the cofactor here may be ultraviolet irradiation. In the Scottish Highlands, multiple benign papillomas are common but only cattle consuming bracken fern go on to develop carcinoma of the alimentary tract or bladder. Virions are plentiful in the papillomas but absent from the carcinomas. However, *in situ* hybridization with a labeled BPV-1 DNA probe reveals that the cancer cells do contain the viral genome, not integrated but free, in the form of a closed circular molecule of DNA. The fact that viral DNA is also found in distant metastatic lesions tends to rule out the possibility that this viral DNA represents contamination from nearby papillomas. The fact that it is all episomal indicates that integration of viral DNA is not required for the induction of malignancy. Bovine papillomaviruses, as well as those from man and other animals, will transform bovine or murine cells *in vitro* and induce fibromas in rodents. Examination of these transformed cells also reveals no virions but episomal viral DNA in essentially every cell. Moreover, the viral DNA, which is fully infectious in its own right, will also induce tumors in rodents and transform cultured rodent cells. Indeed, transfection with a fragment of BPV-1 DNA representing only 69% of the genome satisfactorily transforms cultured cells. Only a small part of the papillomavirus genome is transcribed in bovine (or rabbit) carcinoma cells. Nevertheless, viral DNA may be essential for maintenance of the transformed state, as treatment with interferon has been reported to "cure" most cells transformed by BPV-1; revertants (untransformed cells) have lost their viral DNA.

A remarkable situation has recently been reported in the multimammate mouse, *Mastomys natalensis*. It is claimed that *in situ* hybridization reveals the presence of the genome of a novel papillomavirus in fetal cells, then in increasing copy number as the animals age. At about 18 months the adult rodent develops papillomas which frequently progress to carcinoma.

Man is host to a couple of dozen papillomaviruses responsible for various benign papillomas, some of which occasionally undergo malignant transformation (see Chapter 14). The most striking and convincing instance is that of *epidermodysplasia verruciformis*. This is an uncommon condition characterized by marked suppression of cell-mediated immunity and multiple skin warts which sometimes progress to squamous carcinoma. There are two particularly interesting aspects of the pathogenesis of this cancer. First, only certain types of human papilloma virus seem to be involved, as indicated by the type of episomal viral DNA found by *in situ* hybridization in metastatic cancer cells. Second, despite the widespread distribution of benign warts, cancer develops only on those areas of skin exposed to sunlight.

Genital warts (*condyloma acuminata*) are caused by HPV-6, 11, 16, and 18. Many patients with cancer of the cervix have a history of condylomas or colposcopic evidence of concurrent flat condylomas. The genome of HPV-16 or 18 is found in cervical intraepithelial neoplasia (CIN) with abnormal mitotic figures, and in invasive carcinoma. It has therefore been postulated that HPV-induced benign hyperplasia predisposes to CIN and then carcinoma. While the evidence for an etiological role for papillomaviruses in cancer of the cervix is stronger than that for herpes simplex virus, it is still only circumstantial.

HEPADNAVIRUSES

Some members of the newly defined *Hepadnaviridae* family (see Chapter 13) are also strongly associated with naturally occurring cancers in their native hosts. This is not surprising in view of certain features of their genome, replication, and life history. Like the papovaviruses, hepadnaviruses are characterized by a small circular DNA genome which commonly integrates into cellular DNA and persists indefinitely as a chronic, often asymptomatic infection.

An excellent model is the hepadnavirus associated with hepatocellular carcinoma in the woodchuck, *Marmota monax*. Most wild woodchucks are chronic carriers of the virus and a significant minority of them develop hepatoma. Viral DNA is found to be integrated into the chromosomes of the malignant liver cells. The parallel with hepatitis B virus and hepatocellular carcinoma of man is quite striking.

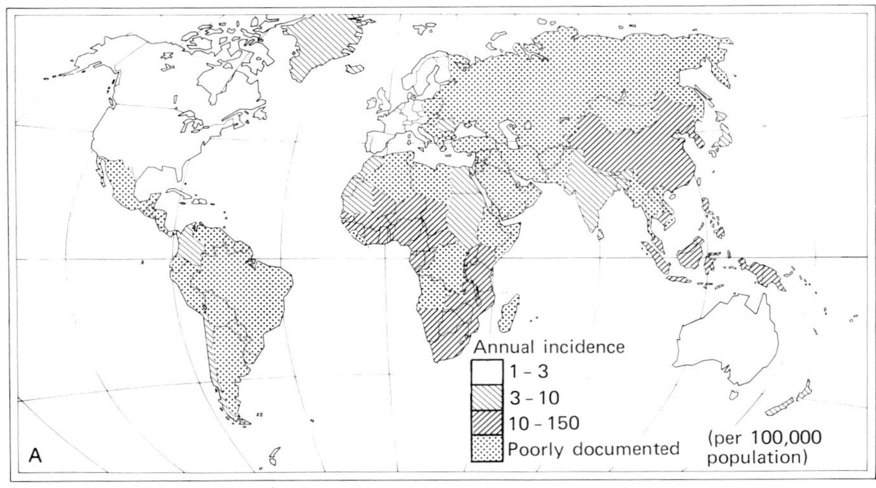

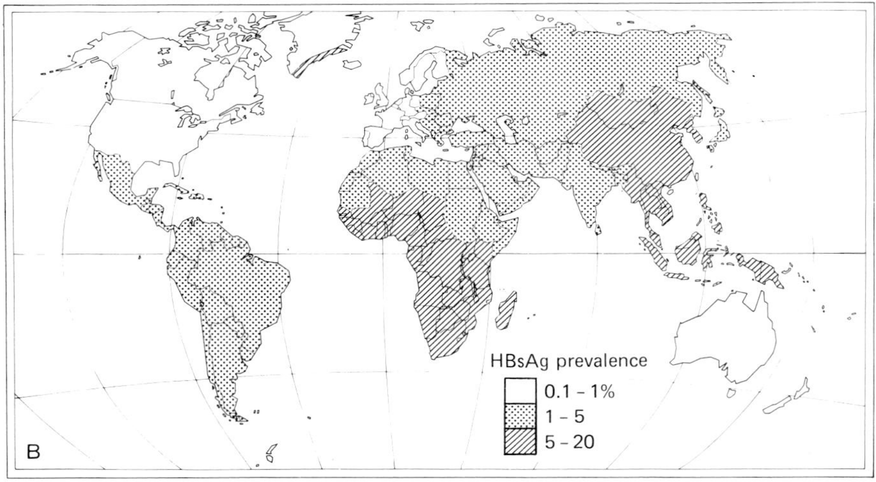

FIG. 8-3. *World distribution of primary hepatic cancer (A) compared with HBsAg carriage rate (B). [From P. Maupas and J. L. Melnick* Progr. Med. Virol. **27,** *1 (1981).]*

Hepatitis B Virus

Some 200 million people are chronic carriers of the hepatitis B virus. An estimated 250,000 people develop primary liver cancer (hepatocellular carcinoma, otherwise known as hepatoma) each year. The two cohorts closely correspond. Though primary liver cancer does not rank in the top 20 cancers in the Western world, where the hepatitis B carrier rate is 0.2–0.5%, it is one of the commonest cancers in much of Africa, China, southern Asia, and the Pacific Islands, where the rate of hepatitis B carriage is 5–20% (Fig. 8-3).

The great majority of hepatomas occur in HBsAg carriers. The viral genome is demonstrable (by hybridization with a radiolabeled cloned HBV DNA probe), integrated into the chromosomes of the cancer cells. The integrated HBV genomes often reveal evidence of major rearrangements, deletions, inversions, and duplications. Perhaps this explains why primary liver cancer cells and cultured hepatoma lines generally do not produce virions although they sometimes synthesize HBsAg. The monoclonal origin of the cancer and an etiological role for the virus is suggested by the fact that all cancer cells from a given patient display a unique pattern of chromosomal sites at which viral DNA is integrated.

A 3-year prospective study of 22,000 middle-aged Taiwanese, of whom 15% were HBsAg carriers, disclosed that hepatocellular carcinoma subsequently arose 223 times more commonly in HBsAg-positive than in negative individuals. Clearly then, the hepatitis B carrier state confers a high risk of liver cancer. As a recent WHO publication asserts, "HBV is second only to tobacco (cigarette smoking) among the known human carcinogens." The virus may be the principal or sole cause of the tumor, but this has not been proven. For example, aflatoxins, which are powerful liver carcinogens produced by fungi known to contaminate food in the parts of the world with high rates of liver cancer, may also be involved.

As in the case of EBV, definitive proof of an etiological role for HBV in liver cancer may have to await the results of long-term vaccine trials. The difference is that a satisfactory, if expensive, vaccine is already available (see Chapter 13), and such trials began in Africa and Asia in 1980. The possibility of preventing up to 250,000 cases of cancer a year (as well as much more morbidity from chronic hepatitis and cirrhosis) certainly justifies an immunization campaign on a vast scale in those populous areas of the world where HBsAg carriage is so prevalent.

ROLE OF VIRUSES IN HUMAN CANCER

From the foregoing it is evident that there is a *prima facie* case for incriminating certain viruses in certain malignant tumors of man. It is quite likely that the retrovirus HTLV-I causes adult T cell leukemia, that the herpesvirus EBV causes undifferentiated nasopharyngeal carcinoma and Burkitt's lymphoma, that the hepadnavirus HBV causes hepatocellular carcinoma, and that the papovavirus HPV causes epidermodysplasia verruciformis and perhaps cervical carcinoma.

Proving a causal connection is another matter altogether. Prospective studies such as those cited for EBV and HBV are much less subject to bias than those undertaken in retrospect. But since human transmission experiments are obviously out of the question, it is impossible to fulfil Koch's postulates in the conventional fashion. Certainly, the induction of malignant tumors in experimental animals by a human virus such as EBV provides strong support, but the more usual situation is that only a related host-specific virus of the same family does so (see hepadnaviruses, retroviruses, papillomaviruses). Transformation of cultured human cells is suggestive but not persuasive evidence of tumorigenicity *in vivo*. Ironically, unequivocal proof of a crucial etiological role for any particular virus in any particular cancer may have to await the demonstration that the cancer can be prevented by immunization with a viral vaccine, as has been so successfully accomplished with Marek's disease of chickens, and *Herpesvirus saimiri* lymphoma in monkeys. Furthermore, we may have to wait years for an answer, as the lack of any evidence for communicability of cancer, and the prospective studies with EBV and HBV indicate a very long "incubation period."

How should the search for new human cancer viruses be conducted? Obviously, the consistent finding of a particular virus in association with a particular tumor is a good start but there is reason to believe it will not usually be as simple as that. Most virus-induced cancer cells do not contain virus. The integrated viral genome is often defective, and even when it is intact it is usually fully or partially repressed. Hence the most generally useful approaches are (1) radiolabeled cloned DNA probes for integrated or episomal viral DNA, (2) fluorescent antibody screening for virus-coded T antigens, and (3) induction of virus production by cocultivation, cell fusion, irradiation, or chemical treatment. The discovery of cellular oncogenes has opened a Pandora's box of additional options. Now it is relevant to screen not only for viral oncogenes and their products but also for evidence of enhanced expression of cellular oncogenes.

Finding a consistent association between a virus and a tumor, even in a prospective study, tells us only that the virus constitutes a risk factor, i.e., predisposes to, thus increases the incidence of cancer. It may not be the only risk factor, or the most important. Several types of agent may be capable of initiating premalignant or malignant change, whether via mutation, translocation, insertional mutagenesis, or other mechanisms. Moreover, if progression to the full expression of cancer is a multistep process, cocarcinogens may be involved, or more than one class of oncogene may be necessary. Nevertheless, if the virus is an essential cocarcinogen, or the most important, intervention to prevent infection should significantly reduce the occurrence of the cancer.

FURTHER READING

Books

Epstein, M. A., and Achong, B. G. (1979). "The Epstein-Barr Virus." Springer Verlag, Berlin and New York.

Essex, M., Todaro, G., and zur Hausen, H., eds. (1980). "Viruses in Naturally Occurring Cancers." Vols. A and B. Cold Spring Harbor Lab. Cold Spring Harbor, New York.

Gallo, R. C., Essex, M. E., and Gross, L., eds. (1984). "Human T-Cell Leukemia-Lymphoma Viruses." Cold Spring Harbor Lab., Cold Spring Harbor, New York.

Klein, G. (1980). "Viral Oncology." Raven Press, New York.

Klein, G., ed. (1981–1985). "Advances in Viral Oncology." Vols. 1–5. Raven Press, New York.

Phillips, L., ed. (1983). "Viruses Associated with Human Cancer." Dekker, New York.

Rapp, F., ed. (1980). "Oncogenic Herpesviruses". Vols. I and II. CRC Press, Boca Raton, Florida.

Rigby, P. W. J., and Wilkie, N. M., eds. (1985). "Viruses and Cancer." *Symp. Soc. Gen. Microbiol. 37,* pp. 1–323.

Szmuness, W., Alter, H. J., and Maynard, J. E., eds. (1982). "Viral Hepatitis." Franklin Institute Press, Philadelphia, Pennsylvania.

Tooze, J., ed. (1980). "Molecular Biology of Tumor Viruses: DNA Tumor Viruses." Cold Spring Harbor Lab., Cold Spring Harbor, New York.

Vogt, P. K., ed. (1985). "Human T Cell Leukemia Viruses." Curr. Top. Microbiol. Immunol. Vol. 115. Springer-Verlag, Berlin and New York.

Weiss, R. A., Teich, N. M., Varmus, H. E., and Coffin, J. M., eds. (1984). "RNA Tumor Viruses." Vols. 1 and 2. 2nd ed., Cold Spring Harbor Lab., Cold Spring Harbor, New York.

Reviews

Bishop, J. M. (1982). Oncogenes. *Sci. Am.* **246,** 68.

Bishop, J. M. (1983). Cellular oncogenes and retroviruses. *Annu. Rev. Biochem.* **52,** 301.

Bishop, J. M. (1984). Exploring carcinogenesis with retroviruses. *Symp. Soc. Gen. Microbiol. 36,* Part I, 121–148.

Cairns, J. (1981). The origin of human cancers. *Nature (London)* **289,** 353.

Gallo, R. C. (1984). Human T-cell leukemia-lymphoma virus and T-cell malignancies in adults. *Cancer Surv.* **3,** 133.

Hinuma, Y. (1984). Retrovirus in adult T-cell leukemia. *Prog. Med. Virol.* **30,** 156.

Klein, G., and Klein, E. (1985). Evolution of tumours and the impact of molecular oncology. *Nature (London)* **315,** 190.

Land, H., Parada, L. F., and Weinberg, R. A. (1983). Cellular oncogenes and multistep carcinogenesis. *Science* **222,** 771.

Reitz, M. S., Kalyanaraman, V. S., Robert-Guroff, M., Popovic, M., Sarngadharan, M. G., Sarin, P. S., and Gallo, R. C. (1983). Human T-cell leukemia/lymphoma virus: The retrovirus of adult T-cell leukemia/lymphoma. *J. Infect. Dis.* **147,** 399.

Temin, H. M., and Baltimore, D. (1972). RNA-directed DNA synthesis and RNA tumor viruses. *Adv. Virus Res.* **17,** 129.

Varmus, H. E. (1983). Retroviruses. *In* "Transposable Elements" (J. Shapiro, ed.), pp. 411–503. Academic Press, New York.
Vogt, P. K. (1977). Genetics of RNA tumor viruses. "Comprehensive Virology" (H. Fraenkel-Conrat and R. R. Wagner, eds.) Vol. 9, pp. 341–455. Plenum, New York.
Vogt, P. K., and Koprowski, H., eds. (1983–1984). "Retroviruses, 1, 2, 3." Curr. Top. Microbiol. Immunol., Vols. 103, 107, and 112. Springer-Verlag, Berlin and New York.
Weinberg, R. A. (1980). Integrated genomes of animal viruses. *Annu. Rev. Biochem.* **49,** 197.
Weinberg, R. A. (1982). Oncogenes of spontaneous and chemically induced tumors. *Adv. Cancer Res.* **36,** 149.
Weiss, R. A. (1984). Viruses and human cancer. *Symp. Soc. Gen. Microbiol.* **36,** Part I, 211–240.
World Health Organization (1983). Prevention of liver cancer. *W.H.O. Tech. Rep. Ser.* **691.**
zur Hausen, H. (1980). Role of viruses in human tumors. *Adv. Cancer Res.* **33,** 77.
zur Hausen, H., Gissmann, L., and Schlehofer, J. R. (1984). Viruses in the etiology of human genital cancer. *Prog. Med. Virol.* **30,** 170.

CHAPTER 9

Epidemiology and Control of Viral Infections

Introduction 247
Transmission of Viruses 247
Survival of Viruses in Nature 252
Seasonal Variations in Disease Incidence 262
Epidemiological Importance of Immunity 264
Epidemiological Methods 266
Proving a Causal Relationship between Virus and Disease ... 272
Control and Eradication of Viral Diseases 273
Further Reading 280

INTRODUCTION

So far we have considered viruses as such (Chapter 1), in relation to the individual cells in which they multiply (Chapters 3 and 4), and in relation to the integrated assembly of cells that constitutes the animal body (Chapters 5, 6, 7, and 8). Viruses can survive in nature only if they are able to pass from one animal to another, whether of the same or another species. The study of the relationships of the various factors that determine the frequency and distribution of viral diseases in a community constitutes viral epidemiology.

TRANSMISSION OF VIRUSES

Transmission involves both the entry of viruses into the body and their shedding from the body surfaces: the skin and the mucosal sur-

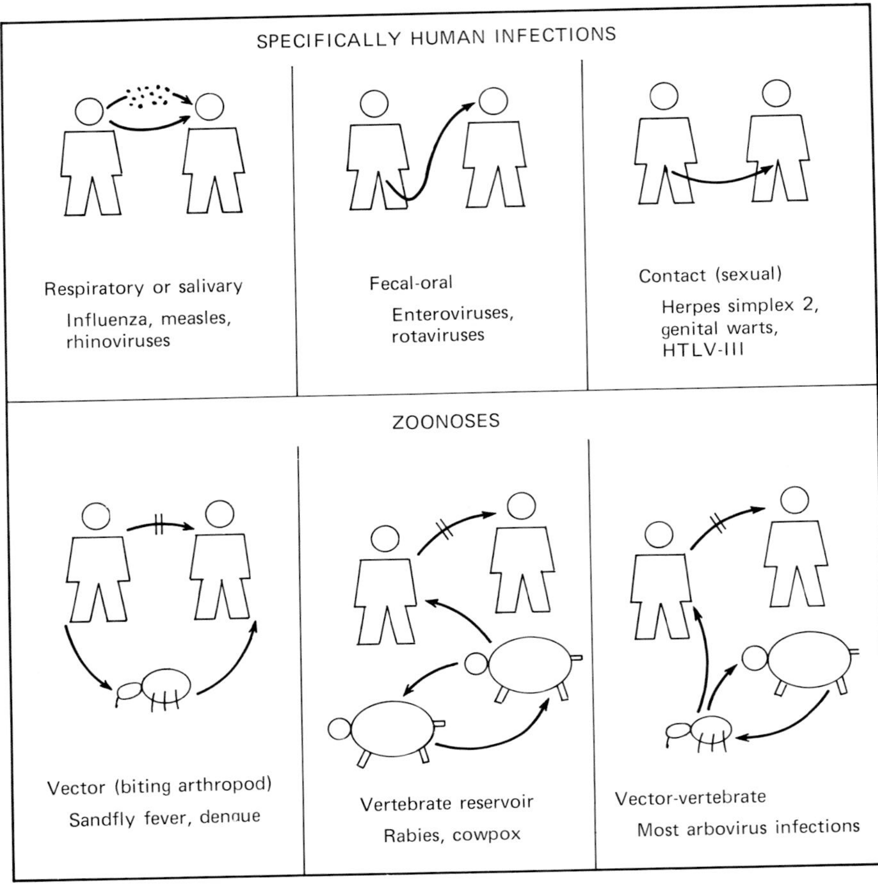

FIG. 9-1. *Types of transmission of human viral diseases. Respiratory and contact spread are not readily controlled. Fecal–oral spread can be controlled by public health measures. Zoonoses: infection of man sometimes may be controlled by controlling vectors or the vertebrate reservoir. (Modified from C. A. Mims, "The Pathogenesis of Infectious Disease," 2nd ed. Academic Press, London, 1982.)*

faces of the respiratory and intestinal tracts, the eye, and the genitourinary tract. It may be defined by a term that indicates the specific routes of exit and of entry (e.g., anal–oral), or more commonly by a general term that embraces both: respiratory, enteric, or contact (Fig. 9-1). Routes of entry and exit were discussed in Chapter 5. *Respiratory* and *enteric* routes are self-explanatory. *Contact infection,* however, is used to mean different things in different contexts. Some viruses are transferred by direct physical contact, e.g., during sexual intercourse. However, some epidemiologists subsume under the term "contact infection," infection at close range by large infectious droplets expelled during talking, coughing, and sneezing. In this chapter we shall use the term to include only physical contact, salivary exchange (e.g., kissing), and transfer of saliva or respiratory secretions via hands and fomites such as contaminated utensils, towels, handkerchiefs, toothbrushes, etc.

Another route of transmission, which bypasses the skin barrier, is by injection, either by man (hypodermic inoculation, transfusion, or transplantation) or by arthropods, which may act as either mechanical or biological vectors. Finally, certain viruses are transmitted congenitally, either across the placenta, in the egg, or as part of the germ plasm. Some viruses are transmitted by several routes; others are transmitted in nature exclusively by one route (Table 9-1).

TABLE 9-1
Common Routes of Transmission of Viruses of Man

VIRAL FAMILY	ROUTES
Hepadnaviridae	Contact (especially perinatal), injection
Papovaviridae	Contact
Adenoviridae	Respiratory, enteric
Herpesviridae	Contact (salivary) (EBV, CMV, herpes simplex 1)
	Contact (venereal) (herpes simplex 2)
	Respiratory (varicella)
	Contact, transfusion, transplacental (CMV)
Poxviridae	Contact (molluscum contagiosum, orf, cowpox)
	Arthropod (mechanical) (tanapox)
	Contact, respiratory (smallpox)
Parvoviridae	Respiratory, contact
Picornaviridae	Enteric, respiratory (*Enterovirus*)
	Respiratory, contact (*Rhinovirus*)
Caliciviridae	Enteric, contact
Togaviridae	Arthropod (*Alphavirus*)
	Respiratory, congenital (rubella)
Flaviviridae	Arthropod
Orthomyxoviridae	Respiratory
Paramyxoviridae	Respiratory, contact
Coronaviridae	Respiratory, contact
Arenaviridae	Contact, respiratory
Bunyaviridae	Arthropod, contact (Hantaan)
Rhabdoviridae	Animal bite (rabies)
Retroviridae	Contact (sexual) or injection (HTLV-III)
Filoviridae	Contact, injection (nosocomial)
Reoviridae	Enteric (*Rotavirus*)
	Arthropod (*Orbivirus*)

Sources of Infection

The sources from which susceptible persons acquire viral infections are influenced by a number of factors, including the route and duration of shedding, the resistance of different viral species to environmental conditions, and the existence of animal reservoirs, including arthropod vectors.

Respiratory Tract. A large number of viruses, which cause either localized disease of the respiratory tract or generalized infections, are expelled from the respiratory tract. Large droplets fall rapidly to the ground or may contaminate the hands or various fomites. Smaller droplets, with their relatively large surface-to-volume ratio, evaporate rapidly and are reduced to small droplet nuclei which do not fall to the ground but remain air borne indefinitely. In man, the relative importance of coughing, sneezing, talking, and breathing in producing infective aerosols depends on where in the respiratory tract the virus is multiplying. In general, little virus is shed by normal respiration or talking; coughs are important in spreading infections from the lower respiratory tract, and sneezes in infections of the upper respiratory tract.

Viruses responsible for the acute exanthemata of man (measles, rubella, and varicella) are not excreted from the respiratory tract until late in the incubation period, despite the fact that this is the site of their primary implantation. However, large amounts of virus are shed from the oropharyngeal lesions of the enanthems that are present at the end of the incubation period and during the first few days of illness. Shedding may also occur during convalescence or recurrently after that time, in the absence of symptoms, especially with various herpesviruses.

Viruses shed from the respiratory tract are in general highly susceptible to inactivation by the conditions that prevail in the environment. They produce infection under conditions in which there is close person-to-person contact, so that droplets or aerosols produced by infected persons are inhaled by susceptible persons soon after their production. Frequently, transfer occurs via the contamination of fingers or fomites (handkerchiefs, door handles, and the like) with infected droplets or oropharyngeal secretions. Fomites retain infectious virions for a short time only.

Enteric Tract. Enteric viruses are shed in the feces and are in general more resistant to inactivation by environmental conditions than are the respiratory viruses, especially when suspended in water, e.g., in water supplies contaminated with sewage. Thus, unlike the respiratory viruses, which usually spread only directly from infected to susceptible individuals, enteric viruses can persist for some time outside the body, especially as some, such as hepatitis A virus, rotaviruses, and caliciviruses, are relatively heat stable and may resist inactivation by ordinary levels of chlorination. Water-borne epidemics can occur. Epidemics of diarrhea and vomiting caused by caliciviruses have also been associated with the eating of raw shellfish; shellfish filter and concentrate viruses present in contaminated water.

Urogenital Tract. Urine, like feces, tends to contaminate food, drink, and living space. A number of viruses, e.g., mumps and measles, are shed in urine, and viruria is often prolonged in cytomegalovirus and polyomavirus infections and congenital rubella. Viruria is lifelong in arenavirus infections of rodents and constitutes the principal mode of contamination of the environment with these viruses.

Venereal transmission of viruses may occur because of lesions on the genitalia or because virions or infected cells are excreted in semen. With recent changes in sexual mores the incidence of genital herpes simplex type 2 and genital warts (due to human papilloma-viruses) has increased greatly. An extraordinary variety of interactions between oral, genital, and anal mucosae facilitates the transfer of numerous viruses between promiscuous male homosexuals, the most dramatic of which is the retrovirus HTLV-III, recently recognized to be responsible for the acquired immune deficiency syndrome (AIDS).

Infants may be infected during the birth process with herpes simplex type 2 virus present in the female genital tract; such neonatal infections are usually severe and often fatal. In some parts of the world newborn children are infected with hepatitis B virus during or shortly after birth, producing infections which may persist for life.

Blood. The blood of persons with a persistent viremia, such as occurs in hepatitis B and AIDS, for example, may be a source of infection by blood transfusion or the use of contaminated syringes. In diseases in which viremia is short-lived, virus in the blood is important for the transmission of infection mainly in those infections in which it is taken up by a susceptible arthropod vector. Arthropods may act merely as mechanical vectors, but usually the virus must multiply in the arthropod and spread to its salivary glands before transmission can occur.

Such *biological transmission* by arthropod vectors is the principal mode of transfer of a large number of viruses, which are called the arboviruses (*ar*thropod-*bo*rne). The arthropod vector acquires virus by feeding on the blood of a viremic person or animal. Multiplication of the ingested virus and its spread through the insect take some time, so that an interval of several days, called the *extrinsic incubation period,* elapses between the acquisition and transmission feeds. At the end of the extrinsic incubation period virions in the salivary gland of the vector are injected into the animal host during all subsequent blood feeds. Arthropod transmission provides a very effective way for a virus to cross species barriers, since the same arthropod may bite birds, reptiles, and mammals (including man) that rarely or never come into close contact in nature.

Animal Reservoirs. Vertebrates may constitute the source of infection of man in several circumstances. Some diseases are acquired from animals by direct physical contact (e.g., cowpox, orf); others (rabies and simian herpesvirus) are acquired by animal bite. Environmental contamination occurs via animal excretions in arenavirus infections and with Hantaan virus, but the most important animal source occurs if the animal harbors an arbovirus that can also infect man.

Excretion of Virus in the Absence of Disease

Virions are often shed by individuals who at the time show no symptoms or signs of disease. First, virus may be shed, sometimes in large amounts, during the last day or so of the incubation period, as in many respiratory infections. Second, virus may be shed from infected individuals who never develop disease; they are said to suffer from *subclinical, asymptomatic* or *inapparent* infections. Third, virus may be shed, continuously or intermittently, from *chronic carriers,* whose infection has persisted, often long after evidence of the original disease, if any, has disappeared. All three of these categories of shedders contribute to the maintenance and spread of viruses in the community.

SURVIVAL OF VIRUSES IN NATURE

It is a matter of common observation that the incidence of infectious diseases varies from time to time and place to place. Some are cosmopolitan and ubiquitous, others geographically localized. Some are endemic, others occur as sporadic cases or as epidemics.

Some of the factors that affect the survival of viruses in nature are set out in Table 9-2 and are discussed below. At the outset, there is an important distinction to be made between *infection* and *disease.* Survival of a virus in nature depends upon the maintenance of infection; the occurrence of disease is neither required nor necessarily advantageous.

TABLE 9-2
Factors Favoring the Survival of Viruses in Nature

FACTOR	EXAMPLE
High transmissibility but reduced virulence	Myxoma virus (rabbits)
Large urban populations	Measles virus
Persistent infections	Herpesviruses
	Hepatitis B virus
Congenital transmission	Retroviruses (birds and rodents)
Antigenic drift and shift	Influenza A virus
Animal reservoir(s)	
With arthropod transmission	Arboviruses
Without arthropod transmission	Arenaviruses

Inapparent Infections

Overall, inapparent infections are much more common than those that result in disease. Data on this and certain other epidemiological characteristics of some common human viral diseases are shown in Table 9-3.

The frequency of inapparent infections, in all except a minority of diseases, accounts for the difficulty of tracing case-to-case infection, even with the help of laboratory aids. Although clinical cases may be somewhat more productive sources of infectious virus than inapparent infections, in many viral diseases the latter are more numerous and do not restrict the movement of infected individuals, and thus they provide a major source of viral dissemination.

Transmissibility and Virulence

The best documented demonstration of how the virulence of a virus may directly affect the probability of transmission is myxomatosis of rabbits (see Chapter 4). Here mechanical transmission by biting arthropods is most effective when the diseased rabbit maintains highly infectious skin lesions for a long period. Very virulent strains of myxoma virus kill rabbits too quickly, whereas the lesions caused by attenuated strains heal too rapidly, so that viruses at either extreme of the virulence range cannot survive in nature.

In the past, because of medical intervention, a somewhat different relationship existed with the virulent and mild varieties of smallpox: variola major, with a case fatality rate in unvaccinated subjects of about

TABLE 9-3

Epidemiological Features of Some Common Human Viral Diseases

DISEASE	MODE OF TRANSMISSION	INCUBATION PERIOD[a] (DAYS)	PERIOD OF COMMUNICABILITY[b]	INCIDENCE OF INAPPARENT INFECTIONS[c]
Influenza	Respiratory	1–2	Short	Moderate
Common cold	Respiratory[d]	1–3	Short	Moderate
Bronchiolitis, croup	Respiratory[d]	3–5	Short	Moderate
A.R.D. (adenovirus)	Respiratory	5–7	Short	Moderate
Dengue	Mosquito bite	5–8	Short	Moderate
Herpes simplex	Contact	5–8	Long	Moderate
Enteroviruses	Enteric	6–12	Long	High
Poliomyelitis	Enteric	5–20	Long	High
Measles	Respiratory	9–12	Moderate	Low
Smallpox	Respiratory[d]	12–14	Moderate	Low
Chickenpox	Respiratory	13–17	Moderate	Moderate
Mumps	Respiratory[d]	16–20	Moderate	Moderate
Rubella	Respiratory	17–20	Moderate	Moderate
Mononucleosis	Contact	30–50	Long	High
Hepatitis A	Enteric	15–40	Long	High
Hepatitis B	Inoculation	50–150	Very long	High
Rabies	Animal bite	30–100	Nil	Nil
Warts	Contact	50–150	Long	Low
AIDS	Contact, inoculation	1–5 years	Very long	? Moderate

[a] Until first appearance of prodromal symptoms. Diagnostic signs, e.g., rash or paralysis may not appear until 2–4 days later.

[b] Most viral diseases are highly transmissible for a few days before symptoms appear, as well as after. Long, >10 days; short, <4 days.

[c] High, >90%; low, <10%.

[d] Also by contact.

25%, and variola minor, with a case fatality rate of about 1%. Where the population density was high and the health infrastructure weak, as on the Indian subcontinent, the more severe variola major prevailed, because more virus was excreted from these patients. In England and the United States, between the world wars, variola minor was the only endemic form of smallpox, largely because public health authorities acted vigorously to stamp out variola major if it was imported, but took little notice of variola minor.

Critical Community Size

Different specifically human viral infections (i.e., those in which there is no animal reservoir) can survive indefinitely in populations of very different sizes, depending on the persistence of immunity and the pattern of shedding. The principle can be exemplified by two common infections of children, measles and chickenpox.

Measles is a cosmopolitan disease that is characteristic in this respect of the generalized viral infections of childhood, like rubella, mumps, and poliomyelitis. It has long been a favorite disease for modeling epidemic behavior, because it is one of the few common diseases in which inapparent infections are very rare and diagnosis is easy. It also illustrates clearly the relationship of population size and dispersion to endemicity. Persistence of the virus in a community depends upon a continuous supply of susceptible subjects. With an incubation period of about 12 days, maximum viral excretion for the next 6 days and solid immunity to reinfection, between 20 and 30 susceptibles would need to be infected in series to maintain transmission for a year. Since nothing like such precise one-to-one transmission occurs, for a variety of reasons, many more than 30 susceptible persons are needed to maintain endemicity. Analyses of the incidence of measles in large cities and in island communities have shown that a population of about half a million persons is needed to ensure a large enough annual input of new susceptibles, provided by the annual incidence of births, to maintain measles as an endemic disease.

Because infection depends on close contact, the duration of epidemics of measles is correlated inversely with population density. If a population is dispersed over a large area, the rate of spread is reduced and the epidemic will last longer, so that the number of susceptible persons needed to maintain endemicity is reduced. On the other hand, in such a situation a break in the transmission cycle is much more likely. When a large proportion of the population is initially susceptible, the intensity of the epidemic builds up very quickly and attack rates are almost 100% (976 per 1000 exposed persons in southern Greenland in 1951, in an epidemic of measles which ran through the entire population in about 40 days). Such *virgin-soil epidemics* in isolated communities may have devastating consequences, due to lack of medical care and the disruption of social life, rather than to a higher level of genetic susceptibility.

The peak age incidence of measles depends upon local conditions of population density and the chance of exposure. In urban communities, epidemics occur every 2–3 years, exhausting most of the currently available susceptibles.On a continental scale epidemics used to occur annually in the United States until vaccination was introduced. Although newborns provide the input of susceptibles each year the age distribution of cases is primarily that of children just entering school, with a peak of secondary cases (family contacts) about 2 years old. The cyclic nature of measles outbreaks is determined by several variables, including the build-up of susceptibles, introduction of the virus, and environmental conditions which promote its spread. Both the seasonality of infectivity (see below) and the occurrence of school holidays affect the epidemic pattern.

Although chickenpox behaves in some ways just like measles, as an acute exanthem in which infection is followed by lifelong immunity to reinfection, it requires a dramatically smaller critical community size for indefinite persistence of the disease—less than 1000, compared with 500,000 for measles. This is explained by the fact that varicella virus, after being latent for many years, may be reactivated and cause zoster (see Chapter 7). Although zoster is not as infectious as chickenpox itself (secondary attack rates of 15%, compared with 70% for varicella), it can, in turn, produce chickenpox in susceptible children.

Persistent Infections

Viral persistence is often of trivial importance to the infected individual, but of considerable epidemiological importance because of continuous or recurrent viral shedding. Sometimes the reverse is true; persistent infection of the brain with measles virus is of no epidemiological significance but may have a severe effect on the patient, causing the lethal disease subacute sclerosing panencephalitis.

Infection with viruses of some groups is characteristically associated with long-continued shedding; with other viruses acute infections are

TABLE 9-4
Viral Infections of Man Associated with Persistent Viral Excretion

FAMILY	VIRUS	COMMENTS
Herpesviridae	EB virus Cytomegalovirus	Intermittent shedding in saliva, semen, milk
	Herpes simplex 1 and 2	Recurrent excretion in saliva, genital secretions and lesions for years
	Varicella	Recurs as zoster many years later
Adenoviridae	All adenoviruses	Intermittent excretion from throat and/or feces
Papovaviridae	BK and JC	In urine
Hepadnaviridae	Hepatitis B	Persistent viremia; may occur in semen, saliva
Arenaviridae	All arenaviruses	In rodents, intermittent shedding in urine
Togaviridae	Rubella	In urine, etc., of congenitally infected children
Retroviridae	HTLV-III	Persistent viremia; occurs in semen

associated with short periods of excretion, which ceases completely when recovery occurs. Table 9-4 lists some of the more important groups of viruses that are associated with persistent or recurrent shedding of virus in normal (not immunologically compromised) individuals. Most infections that are not listed have a brief period of transmissibility, measured in days or weeks, after which the subject ceases to excrete virus (see Table 9-3).

Vertical Transmission

The routes of transmission described so far apply to unrelated or related individuals of the same or different animal species. This is called *horizontal transmission,* to differentiate it from *vertical transmission* which refers to the transfer of virus from an individual of one generation to its offspring, then to the offspring's progeny, and so on, usually before birth. As outlined in Chapter 5, this may occur across the placenta, during development of the egg or via the germ-plasm.

The only proven examples of germ-plasm transmission of animal viruses are the oncoviruses (family *Retroviridae*), which exemplify the ultimate in biological adaptation. Although some are excreted in an infectious form and can be transmitted horizontally to other individuals (e.g., HTLV-I, the cause of adult T cell leukemia), most oncoviruses are also transmitted as part of the host's genome (see Chapter 8), in the form of cDNA. "Infection" of most animals with an exogenous oncovirus is usually a superinfection, since the DNA sequences of an endogenous oncovirus are already present. Transmission of avian lymphoid leukemia virus, for example, can occur by three pathways (see Chapter 7): horizontal contact transmission, congenital transmission of infectious virions, and through the germ line as the integrated provirus of an endogenous retrovirus.

More important in human viral diseases are those in which congenital infection occurs *in utero,* the most important examples being rubella and cytomegalovirus infection. Finally, "horizontal" transmission can occur from mother to offspring during the birth process (e.g., herpes simplex type 2) or immediately after birth (e.g., coxsackie B). Such *perinatal transmission* is epidemiologically important in cytomegalovirus and hepatitis B virus infections.

Multiple-Host Infections

Most human viral infections are maintained in nature within human populations. However, a number of viruses can spread naturally between different species of vertebrate host; the term *zoonosis* is used to

TABLE 9-5
Non-Arthropod-Borne Viral Zoonoses

VIRUS			
FAMILY	SPECIES	"RESERVOIR" HOST	MODE OF TRANSMISSION
Poxviridae	Cowpox	Cattle, ? rodents	Contact, through skin abrasions
	Monkeypox	? Monkey	? Contact, through skin abrasions
	Milker's nodes	Cattle	Contact, through skin abrasions
	Orf	Sheep, goats	Contact, through skin abrasions
	Tanapox	? Monkey	? Mosquito bite (mechanical)
Bunyaviridae	Hantaan	Rodents	Contact
Rhabdoviridae	Rabies	Carnivores, bats	Animal bite, respiratory
Filoviridae	Ebola, Marburg	?	Contact, iatrogenic (injection)
Orthomyxoviridae	Influenza A	Pigs, horses, birds	Respiratory (reassortment involved)
Arenaviridae	Junin	Rodents	Respiratory, contact
	LCM		
	Machupo		
	Lassa		

describe a multiple-host infection that is naturally transmissible from lower animals to man. Tables 9-5 and 9-6 list the viral zoonoses, the majority of which are caused by arboviruses (Table 9-6). A wide variety of *animal reservoir hosts* and *arthropod vectors* play a role in the maintenance of arboviruses in nature, and most human arbovirus infections originate from these reservoirs. Zoonoses listed in Table 9-5 are primarily infections of domestic or wild animals transmissible only under exceptional conditions to humans engaged in operations involving close contact with animals.

In the zoonoses, man is the *sentinel animal*, providing a warning of the presence of the "enemy" in the vicinity; parallel situations exist with domesticated or laboratory animals, which when appropriately exposed may acquire diseases not normally transmitted horizontally within the recipient (sentinel) species, e.g., monkeys acquire measles only from man.

Ecology of Arboviruses

There are many different patterns for the survival in nature of the arboviruses, whose ecology is complex because of their diverse vertebrate and invertebrate hosts. When arthropods are active, arboviruses multiply alternately in vertebrate and invertebrate hosts. A problem that has concerned many investigators is to understand what happens to the virus during the winter months in temperate climates, when the arthropod vectors are frequently inactive. An important mechanism of such

TABLE 9-6
Viral Zoonoses: Arboviruses

VIRUS			PRINCIPAL VERTEBRATE HOSTS	VECTOR
FAMILY	GENUS	SPECIES		
Togaviridae	*Alphavirus*	Chikungunya	Mammals	Mosquitoes
		Eastern equine encephalitis	Birds	
		O'nyong-nyong	Mammals	
		Ross River	Mammals	
		Venezuelan equine encephalitis	Mammals	
		Western equine encephalitis	Birds	
Flaviviridae	*Flavivirus*	Dengue	Primates	Mosquitoes
		Japanese encephalitis	Birds, pigs	
		Murray Valley encephalitis	Birds	
		Yellow fever	Primates	
		Kyasanur Forest disease	Primates	Ticks
		Louping ill	Mammals	
		Omsk hemorrhagic fever	Mammals	
		Powassan	Mammals	
		Tick-borne encephalitis	Mammals, birds	
Bunyaviridae	*Bunyavirus*	California encephalitis La Crosse Tahyna	Mammals	Mosquitoes
	Phlebovirus	Sandfly fever	Gerbils	Sandflies
		Rift Valley fever	Mammals	Mosquitoes
	Nairovirus	Crimean-Congo hemorrhagic fever	Mammals	Ticks
Reoviridae	*Orbivirus*	Colorado tick fever	Mammals	Ticks

"overwintering" is *transovarial transmission* from one generation of arthropods to the next. Long known to occur with the tick-borne flaviviruses, this type of vertical transmission has more recently been discovered to occur also with some mosquito-borne bunyaviruses and flaviviruses. Some species of bunyaviruses are found in high northern latitudes in which the mosquito breeding season is very short; some of the first mosquitoes to emerge each summer are infectious and the pool of virus is then amplified by transmission to small mammals.

Hibernating vertebrates may also play a role in the overwintering of arboviruses. In cool temperate climates, bats and some small rodents, as well as snakes and frogs, hibernate during the winter months and their body temperature falls. They can be infected with certain arboviruses, maintain an inapparent infection throughout a long period of hibernation, and become viremic when transferred to a warm environment.

Mosquito-Borne Encephalitis. Five different mosquito-borne flaviviruses, three alphaviruses, and two bunyaviruses have been incriminated as causes of encephalitis in man (see Chapter 29). All normally

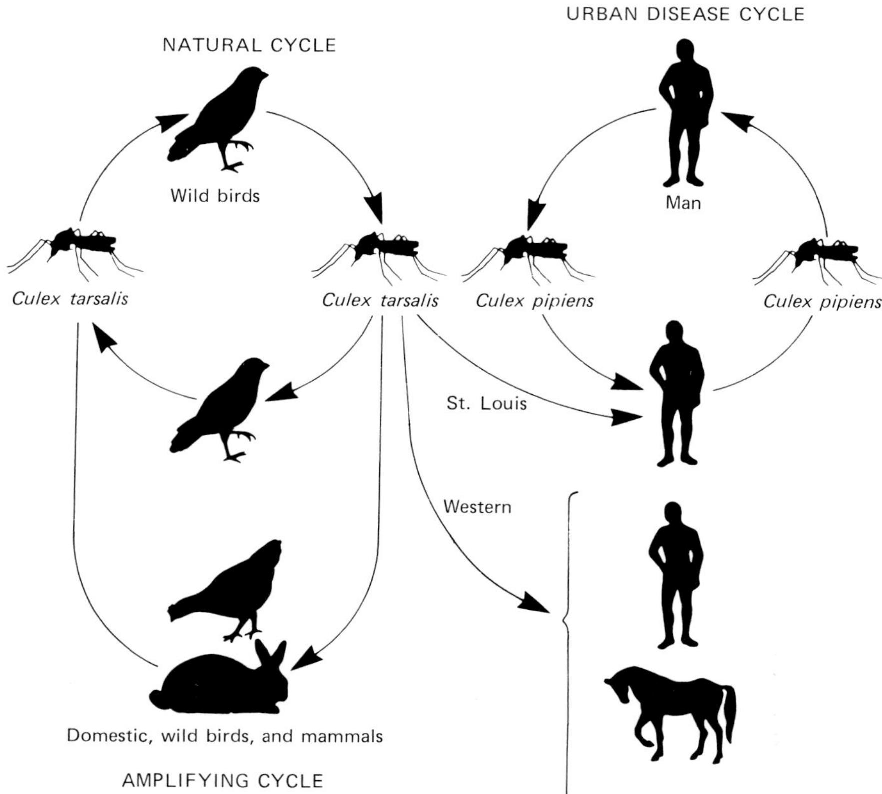

FIG. 9-2. *Sylvatic cycles of Western and St. Louis encephalitis viruses and the urban cycle of St. Louis encephalitis virus. The natural inapparent cycle is between* Culex tarsalis *and nestling and juvenile birds, but this cycle may be amplified by infection of domestic birds and wild and domestic mammals. Western encephalitis virus can multiply in the mosquito at cooler temperatures, so epidemic disease in horses and man may occur earlier in the summer and farther north into Canada. St. Louis encephalitis virus occurs primarily in the south-central United States and the Southwest and does not cause clinical encephalitis in horses. However, it can be transmitted directly from man to man by the city-breeding mosquito* Culex pipiens. *(Modified from R. T. Johnson, "Viral Infections of the Nervous System." Raven Press, New York, 1982.)*

cycle through birds or small mammals, man being only an incidental host. However, the best known flavivirus disease, yellow fever, can be maintained in a man–mosquito–man cycle. The flavivirus of St. Louis encephalitis, the most widespread arbovirus and commonest cause of human arboviral encephalitis in the United States, has more recently

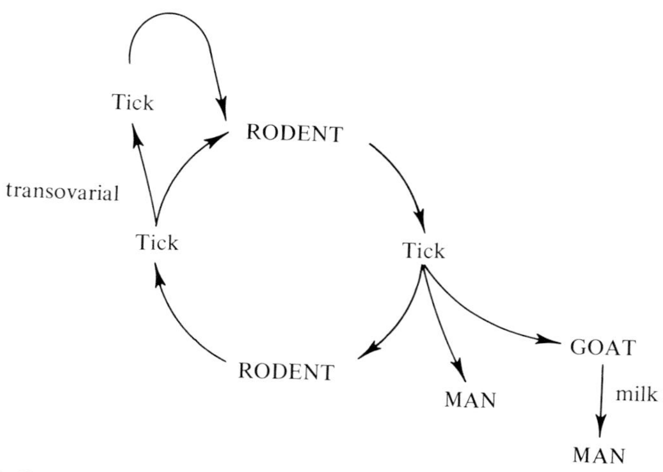

FIG. 9-3. *Central European tick-borne flavivirus infection. Congenital transmission in the tick occurs on a scale sufficient to maintain the virus in nature. The virus is transmitted by ticks to many other animals, including cows, goats, and man. Man may also be infected by ingesting milk from an infected goat.*

been shown to have the same capacity. Its natural cycle occurs between *Culex tarsalis* and nesting and juvenile birds, but when mosquitoes are numerous the cycle may be amplified by infection of domestic birds and wild and domestic mammals, causing encephalitis in a proportion of cases in which man is infected. In the eastern United States, *Culex pipiens,* which breeds in stagnant, polluted water, may set up urban St. Louis encephalitis, in a man–*C. pipiens*–man cycle (Fig. 9-2).

Western encephalitis virus, an alphavirus infection spread by *C. tarsalis,* normally circulates between small birds but may cause encephalitis in man and horses. Its sylvatic cycle can be amplified by extension to other hosts when seasonal conditions produce large mosquito populations, and disease in horses and man is then more common.

Tick-Borne Encephalitis. Central European tick-borne flavivirus encephalitis illustrates two important features not found in mosquito-transmitted flavivirus infections (Fig. 9-3): (1) transovarial infection of ticks is common enough to ensure its survival, independently of a cycle in vertebrates, and (2) transmission of this arbovirus from one vertebrate host to another can also occur by a mechanism not involving an invertebrate at all, namely, via milk. Tick-borne encephalitis is widespread in central Europe and the far east of the USSR. A large number of wild birds and mammals may be infected in nature; most investigators regard small rodents as the most important vertebrate reservoir. Man is an incidental host for tick-borne encephalitis virus but may become involved in two ways; he may enter natural foci of infection in pursuit of work or recreation and become infected by tick bite, or he may acquire infection from milk. Goats, cows, and sheep have been shown to excrete virus in their milk, but all milk-borne outbreaks have been associated with goat's milk. Adult and juvenile goats may acquire the virus by grazing in tick-infested pastures, and kids may be infected by drinking infected milk.

SEASONAL VARIATIONS IN DISEASE INCIDENCE

Many viral infections show pronounced seasonal variations in incidence. In temperate climates arbovirus infections transmitted by mosquitoes or sandflies occur mainly during the summer months, when vectors are numerous and active. Transmission by ticks or fleas does not show as pronounced a seasonal incidence, but human activities in places where tick bites could occur may be less frequent during the winter.

More interesting, but also more difficult to explain, are the variations in seasonal incidence of infections in which man is the only host animal. Table 9-7 shows the season of maximal incidence of several human respiratory, alimentary, and generalized infections. Most respiratory infections are most prevalent in winter or to a lesser extent in autumn, but adenovirus infections peak in the spring. Among enteric viruses, infections with enteroviruses are maximal in the summer, the Norwalk agent (a calicivirus) shows no regular seasonal pattern, and rotaviruses are more prevalent in winter.

In tropical countries, smallpox had a well marked seasonal maximum in the dry season, spread rapidly decreasing with the onset of the monsoon rains. It used to be primarily a winter–spring disease in Europe and North America. The acute exanthemata of childhood and mumps also show spring maxima in temperate climates, but infections with herpes simplex types 1 and 2, cytomegalovirus, and EB virus, which are contact and not respiratory infections, show no seasonal variations in incidence.

The patterns shown in Table 9-7 obtain, in general, in both the northern and southern hemispheres. Different factors probably affect seasonality in the tropics, the peak incidence of smallpox, measles, and chickenpox being late in the dry season, with an abrupt fall when the rainy season begins. On the other hand, in the tropics influenza and rhinovirus infections reach a peak during the rainy season. Both biological and sociological reasons have been suggested to explain these seasonal

TABLE 9-7
Season of Maximal Incidence of Specifically Human Viral Infections in Temperate Climates

INFECTION	WINTER	SPRING	SUMMER	AUTUMN	NONE IN PARTICULAR
Respiratory					
Adenoviruses		+			
Rhinoviruses		+		+	
Influenza	+				
Coronaviruses	+				
Respiratory syncytial virus	+				
Parainfluenza 1 and 2				+	
Parainfluenza 3					+
Enteric					
Enteroviruses			+		
Rotaviruses	+				
Caliciviruses (Norwalk)					+
Generalized					
Smallpox	+[a]				
Rubella		+			
Measles		+			
Mumps		+			
Varicella		+			
Hepatitis B					+
Herpes simplex 1 and 2					+
Cytomegalovirus					+
EB virus					+

[a] The dry season ("spring") in tropical countries.

variations. Biological reasons have included effects on both virus and host; measles, influenza, and vaccinia viruses survive in air better at low rather than high humidity, while polioviruses, rhinoviruses, and adenoviruses survive better at high humidity. All survive better, in aerosols, at lower temperatures. These situations correspond with conditions prevalent during the seasons when infections due to these viruses are most prevalent. It has also been suggested, without much supporting evidence, that there may be seasonal changes in the susceptibility of the host, perhaps associated with changes in nasal and oropharyngeal mucous membranes.

Biological effects are enhanced by social activities in different seasons. In places subject to monsoonal rains, the onset of the rains early in summer is accompanied by greatly reduced movement of people, both in their daily affairs and to fairs and festivals. The crowding into re-

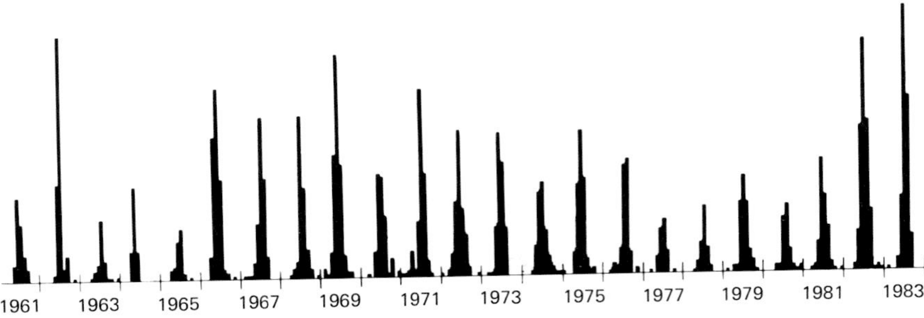

FIG. 9-4. *Epidemic occurrence of respiratory syncytial virus. The histogram shows the monthly isolations of the virus from patients admitted to Fairfield Hospital for Infectious Diseases, Melbourne, over the period indicated. (The Australian winter runs from June to August.) Note the regular annual winter epidemics, causing significant disease mainly in infants. (Data by courtesy of A. A. Ferris, F. Lewis, N. Lehmann, and I. D. Gust.)*

stricted areas and ill-ventilated vehicles and buildings that occurs in temperate climates in winter may promote the exchange of respiratory viruses, but experiences in the Arctic show that cold weather alone is not enough to cause "colds."

Annual winter outbreaks of severe respiratory syncytial virus infections in infants are a feature of western communities in both northern and southern hemispheres, the major impact of each epidemic lasting for only 2 or 3 months (Fig. 9-4). Likewise, in spite of seeding of virus throughout the community in the summer months, epidemic influenza occurs mainly during the winter. In contrast to the regularity of epidemics of respiratory syncytial virus infection, epidemics of influenza vary greatly in extent from year to year (Fig. 21-1).

Surveys at the Great Lakes Training Center in Illinois have shown that adenoviruses spread equally well among recruits throughout the year, but twice as many recruits developed feverish illness in winter as in summer.

EPIDEMIOLOGICAL IMPORTANCE OF IMMUNITY

The immune response to viral infections was described at length in Chapter 6. Immunity from prior infection or from vaccination plays such a vital role in the epidemiology of viral diseases that it will be reviewed briefly here, with examples of a generalized infection bestowing lifelong immunity (smallpox) and superficial infections attended by transient immunity (the common cold and influenza).

Generalized Infections

Although now extinct, smallpox is a good example of the way in which immunity due to prior infection, or following vaccination, influenced the spread of infection, its age, incidence, and its effects, at a population level. When it was first introduced into virgin populations, as in various parts of the Americas between the early sixteenth and the nineteenth centuries, the initial impact was devastating. Persons of all ages were susceptible, and the near simultaneous infection of most persons in most families totally disrupted social life. After this initial impact the situation was changed, for those who recovered were immune, and, in time, in dense populations, smallpox became a disease of childhood.

Vaccination with vaccinia virus gave protection without generalized disease, and further changed the epidemiology of smallpox, leading eventually to its global eradication. Before this occurred, American investigators evaluated all the factors that affected the epidemiology of smallpox in villages in Pakistan, and found that immunity, as determined by vaccination status, was overwhelmingly the most important factor in determining the occurrence of infection.

Exactly the same situation has been observed in other generalized viral infections. Immunity, even in the absence of the virus, appears to be lifelong, as evidenced by studies of measles and poliomyelitis in isolated populations. In a classical paper on measles in the Faroe Islands written in 1847, for example, Panum demonstrated the importance of respiratory droplet spread during the prodromal period, a constant incubation period of about 2 weeks, highest case fatality rates among infants and the elderly, an attack rate of almost 100% among exposed susceptibles, and persisting immunity conferred by an attack of measles experienced as long as 65 years earlier. Following the widespread use of an effective vaccine in the United States, the situation has now been reached in which essentially all cases of measles are in unimmunized immigrants and visitors.

Localized Infections

The common cold is a good example of an infection that is localized to the upper respiratory tract. A large number of serotypes of rhinoviruses and a few coronaviruses and enteroviruses can produce superficial infections of the mucous membrane of the upper respiratory tract. The seemingly endless succession of common colds suffered by urban communities reflects a series of minor epidemics, each due to one of these viruses.

Protection against reinfection is due mainly to antibodies in the nasal secretions, primarily IgA. Short-lived type-specific resistance does occur but there is no intertypic cross-immunity, hence the convalescent individual is still susceptible to all other rhinoviruses, coronaviruses, etc. Most persons contract two to four colds per year.

Epidemiological observations on isolated human communities illustrate the need for susceptible subjects or antigenically novel viral serotypes to maintain respiratory diseases in nature, and the importance of repeated (often subclinical) infections in maintaining herd immunity. Explorers, for example, are notably free of respiratory illness during their sojourn in the Arctic or Antarctica, despite the freezing weather, but invariably contract severe colds when they again establish contact with their fellow men.

In urban communities young children appear to be particularly important as the persons who introduce rhinoviruses into families, mainly because they bring back viruses from school and neighbors' children, and partly because they often shed larger amounts of virus than adults. A feature of rhinovirus infection that had not been observed previously emerged from family studies, namely prolonged shedding of virus for up to 3 weeks, i.e., long after the acute symptoms have subsided. The epidemiology of coronaviruses, another important cause of colds, is rather similar, but there are only a few serotypes of *Coronavirus* and they tend to produce colds in winter, and in adults rather than children.

In influenza A, as discussed in Chapter 4, respiratory tract immunity due to IgA selects for antigenic variation in the hemagglutinin (antigenic drift), leading to the emergence of new strains that can infect individuals immune to other strains of that subtype. At unpredictable intervals new subtypes enter and circulate in the human population (antigenic shift), probably because of genetic reassortment between influenza viruses of man and animals.

EPIDEMIOLOGICAL METHODS

Epidemiology could be said to stand at the top of the pyramid of virology, since it draws upon data from every other level. It is really a special branch of population biology and uses the methods and techniques of that disicpline, but there are some peculiar to the population biology of human viral infections.

Definitions

The terms incidence and prevalence are used to describe quantitative aspects of the occurrence of infections in populations. The *incidence* of infection (or of disease) is defined as the proportion of a population contracting that infection (or disease) during a specified period (usually a year), whereas *prevalence* refers to the proportion infected (or sick, or immune) at a specified point in time.

Comparisons of incidence and prevalence at different times and places are made by relating the appropriate numerator to a denominator which may be as general as the total population (of a city or country), or may be specified as the susceptible population at risk (*exposed* and *susceptible,* the latter usually being defined as those lacking antibodies). Incidence is useful in describing both acute and persistent conditions, prevalence only for states of relatively long duration such as immunity, persisting infection, or chronic disease. Thus, in influenza we speak of the incidence of disease and the prevalence of immunity (as determined by antibody surveys).

The *secondary attack rate,* applied to comparable relatively closed groups such as households or classrooms, is a useful measure of the "infectiousness" of viruses transmitted by aerosol or droplet spread. It is defined as the number of persons in contact with the primary or *index case* who become infected or ill within the maximum incubation period, as a percentage of the total number of "susceptibles" exposed to infection.

Deaths from a disease can be categorized in two ways: as a cause-specific *mortality rate* (the number of deaths from the disease in a given year, divided by the total population at mid-year), usually expressed as per 100,000; or the *case fatality rate* (the percentage of those with a particular disease who die from it).

All these rates may be affected by various attributes that distinguish one individual from another: age, sex, genetic constitution, immune status, nutrition, or various behavioral or socioeconomic parameters. The most widely applicable attribute is probably age, which may encompass (and therefore be confounded by) immune status as well as various physiological variables.

Seroepidemiology

The value of serological surveys of populations for understanding the epidemiology of viral infections has been implicit in earlier parts of this chapter. Serological epidemiology is useful in public health operations and in research. Because of the expense of collecting and properly storing sera (and if possible lymphocytes, in dimethyl sulfoxide), advantage is taken of a wide range of sources: planned population surveys, entrance examinations for military and other personnel, blood banks, hospitals, and public health laboratories. Such material can be used in several ways:

Estimates of Prevalence. Traditional surveillance is based on the reporting of clinical illness. Examination of representative sera by technically adequate measurements gives a much better index of the cumulative experience of the population with the disease(s) under study. By determining the presence of long-lived antibodies to selected viruses in various age groups of the population it is possible to determine how effectively endemic agents spread and the time of last appearance of agents such as the dengue viruses or polioviruses in communities where these diseases are currently absent.

Estimates of Incidence. The appearance of antibody to the virus under study in the second of two sequentially collected specimens indicates infection between the times of collection; a fourfold or greater rise in antibody titer (*seroconversion*) reflects infection. The presence of specific IgM (or, with some infections, significant titers of complement-fixing antibody) in a single serum specimen usually indicates the occurrence of recent infection. Correlation of the serological tests with clinical observations makes it possible to determine the ratio of clinical to subclinical infections.

Evaluation of Immunization Programs. A major follow-up to smallpox eradication has been the Expanded Immunization Programme of WHO, which seeks to ensure that all children in the world under the age of 5 years are immunized against diphtheria, pertussis, tetanus, tuberculosis, poliomyelitis, and measles. Serological surveys are the major tool used to determine how well such efforts have succeeded in various populations. In developed countries they are used to monitor these and other immunization programs and to determine strategies for the most efficient use of new viral vaccines.

Assessment of "New" Diseases or Viruses. Sometimes new viruses (or their antigens) are discovered for which no clinical manifestations have been recognized. Serological surveys can be used to determine their importance. The discoveries of Epstein–Barr virus and its role in glandular fever, and of hepatitis B virus, are classical examples of the application of this approach. It is also valuable in determining retrospectively the geographical and secular distribution of newly discovered agents, as demonstrated most recently with HTLV-III.

Molecular Epidemiology

This term has been applied to the use of molecular biological methods to enlarge our understanding of the epidemiology of viral infections.

Perhaps the best example is provided by investigations into influenza A, discussed at length in Chapter 4. The facility with which viral nucleic acids can now be sequenced has also permitted diagnostic laboratories to distinguish strains of virus containing nucelotide substitutions that are not reflected as antigenic changes demonstrable by serology. For instance, the genome of poliovirus recovered from paralyzed (or normal) people can be sequenced to determine whether it is a wild strain, an attenuated vaccine strain, or a vaccine strain that has reacquired neurovirulence as a result of subsequent mutations. Rather simpler techniques such as molecular hybridization and oligonucleotide mapping can also distinguish strains of virus within the same serotype. Such procedures have recently been applied to many viruses once thought to be invariate, e.g., dengue, rabies, and herpes simplex. This is more than just a "stamp collecting" exercise in describing finer and finer differences between strains from different parts of the world—it is a valuable epidemiological tool which can be exploited to trace the origin and course of epidemics and the progress of viral evolution.

Mathematical Modeling

From the time of William Farr in 1840, mathematicians have been interested in "epidemic curves" and secular trends in the incidence of infectious diseases. With the development of mathematical modeling techniques there has been a resurgence of interest in the population biology of infectious diseases and models are now being developed which allow estimates of critical community size for the maintenance of endemicity for several different kinds of infections, with short and long incubation periods, recurrent infectivity, and age-dependent pathogenicity.

Computer modeling provides useful insights into the effect of different vaccination regimes and different levels of acceptance of vaccination against rubella and measles on important complications of these two diseases, congenital rubella and measles encephalitis, respectively. One effect of vaccination programs is to raise the average age at which unimmunized individuals contract infection. Up to a certain level, increasing herd immunity may actually increase the risk in specially vulnerable (older) individuals, by increasing the age of first exposure to the virus. Thus, as discussed in Chapter 20, vaccination of infants against rubella may paradoxically increase the risk of the most important complication of the natural disease, namely congenital rubella, unless the acceptance rate is high (over 80%). A second conclusion from these studies is that because of the cyclic incidence of the viral exanthemata, with intervals between peaks that increase as vaccination coverage improves, the evaluation of vaccination programs that stop short of eradication must be carried out over a prolonged period of time.

Human Volunteer Studies

Epidemiological aspects of several specifically human diseases that have not been reproduced in other animals have been studied in human volunteers, e.g., early work with hepatitis viruses, rhinoviruses, and a range of other respiratory viruses (Plate 9-1). Many major discoveries that have led to the control of viral diseases were possible only with the use of human volunteers. It is important that careful consideration be given to any short- or long-term risks that may be involved, including

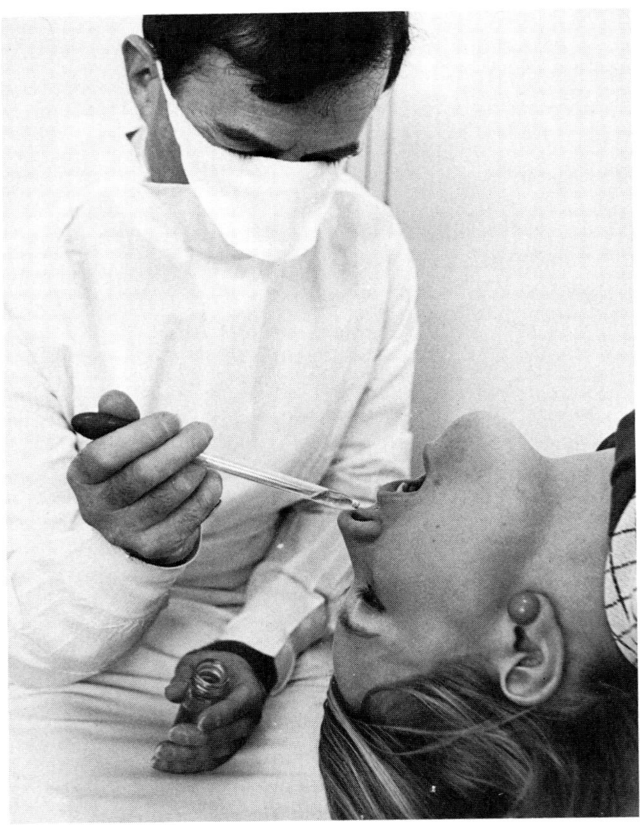

PLATE 9-1. *Intranasal inoculation of human volunteer with rhinovirus. (Courtesy Dr. D. A. J. Tyrrell.)*

the possibility of transferring other agents that may be present in the inoculum as contaminants.

Studies in Closed Communities

Long-term epidemiological study of selected institutional communities, such as the "Junior Village" in Washington, conducted with careful attention to the ethical problems involved, has provided knowledge that has greatly augmented our understanding of viral infections and accelerated progress toward the control of several of them.

Retrospective and Prospective Studies

In relation to attempts to determine etiology we can distinguish *retrospective* (case history) and *prospective* (cohort) studies. The causation of congenital defects by rubella virus provides examples of each kind of approach. Dr. N. M. Gregg, an ophthalmologist working in Sydney, Australia, was struck by the large number of cases of congenital cataract he saw in 1940–1941, and by the fact that many of the children also had cardiac defects. By interviewing the mothers he found that the great majority of them had experienced rubella early in the related pregnancy. The hypothesis that there was a causative relation between maternal rubella and congenital defects quickly received support from other retrospective studies, and prospective studies were then organized. To do this groups of pregnant women were sought who had experienced an acute exanthematous disease during pregnancy, and the subsequent occurrence of congenital defects in their children was compared with that in women who had not experienced such infections. Gregg's predictions were confirmed and the parameters defined more precisely.

Retrospective studies are valuable for providing clues but by their nature they are not conclusive, in that they lack suitable control groups; prospective studies are much more difficult to organize but can establish etiological relationships in a more convincing way.

Vaccine Trials

Having first been inoculated into experimental animals, then in individual human volunteers, then subjected to small-scale trials, often in institutions, viral vaccines must be tested on a very large scale before their worth and safety can be properly evaluated. Perhaps the best-known vaccine trial was the famous "Francis field trial" of the Salk inactivated poliovirus vaccine, carried out in 1954, but similar trials were necessary for live-virus poliovirus vaccines, for yellow fever vaccines before this, and for measles, mumps, and rubella vaccines since. There is no alternative way to evaluate a new vaccine or indeed a new drug, and the design of field and clinical trials has now been developed so that they yeild maximum information with minimum risk and cost.

Long-Term Family Studies

Another kind of investigation that provides the opportunity for both epidemiological studies and controlled trials of vaccines or therapeutic agents is the long-term family study. Because of the present advanced state of diagnostic and serological virology, such studies now yield a much greater array of valuable data than was possible a few years ago, but they are very expensive and require long-term dedication of both personnel and money.

PROVING A CAUSAL RELATIONSHIP BETWEEN VIRUS AND DISEASE

One of the great landmarks in the scientific study of infectious diseases was the development of what have come to be called the *Henle–Koch postulates* of causation. They were originally drawn up with bacteria and protozoa in mind, not viruses and were revised in 1937 by Rivers, who developed another set of criteria to cover viruses. With the advent of tissue culture for viral diagnosis a plethora of new viruses were discovered—"viruses in search of disease"—and Huebner further revised the Koch and Rivers postulates. Later the problem arose of determining whether viruses were causally involved in various chronic diseases and in cancers, a question that is still of major concern to medical virologists. Since the relevant disease cannot be reproduced by inoculation of experimental animals, scientists have to evaluate the probability of "guilt by association," a difficult procedure that relies on an epidemiological approach. Two tools are of central importance in assisting the epidemiologist: immunological investigations and demonstration of the presence of the viral genome in tumor cells by the use of nucleic acid probes.

The immunological criteria were first formulated by Evans in an assessment of the relationship of EB virus to infectious mononucleosis, at a time when there was no method of isolation of the virus. EBV antibodies were already present by the time the patients were first seen by the clinician. With ubiquitous viruses that might be expected sometimes to cause cancers or chronic neurological disorders the immunological criteria are difficult to apply. High viral antibody titers might precede the development of the disease in question and even play a role in its pathogenesis, they might accompany the onset of the disease due to a common immunological mechanism, or arise after its onset. As mentioned in Chapter 8, a large prospective study carried out on 45,000 children in an area of high Burkitt lymphoma incidence in Africa showed that (1) EBV infection preceded development of the tumors by 7–54 months; (2) exceptionally high EBV antibody titers often preceded the disease; and (3) antibody titers to other viruses were not elevated. In addition, it has been demonstrated that the EBV genome is always present in the cells of Burkitt lymphomas in African children, and a malignant lymphoma can be induced in certain primates with EBV or EBV-infected lymphocytes (see Chapter 8). There is thus a very strong causal association between EB virus and Burkitt lymphoma in African children.

Certain other associations of viruses with cancers and chronic disease are set out in Tables 8-1 and 29-3, respectively. It is clear that final proof of causation is very difficult, and special immunological criteria must be adopted, especially when the viruses concerned are ubiquitous, i.e., many, or even a majority, of the controls also have antibody. Even the presence of the viral genome in tumor cells is not a proof of its causative role, since herpesviruses in particular are notorious for their capacity for latency, usually in an episomal form; the virus or viral genome may merely be a "passenger." As with smoking and lung cancer, final "proof" may be difficult to achieve but epidemiological and laboratory studies may establish a very high probability of viral causation, perhaps as a cofactor, in some cancers and in some chronic diseases. The criteria set out in Table 9-8 may be viewed as useful guidelines for establishing causation for all kinds of diseases.

CONTROL AND ERADICATION OF VIRAL DISEASES

Nowhere in medicine is the adage "Prevention is better than cure" more appropriate than in viral diseases, for there is no effective treatment for most of them whereas there are several methods for their control or prevention. The ultimate step, which eliminates the need for control, is eradication, achieved on a global scale for smallpox in 1977. Within any country, control measures operate at various levels. Exotic diseases may be excluded by quarantine, a procedure that has more relevance nowadays in veterinary than in human medicine. Hygiene and sanitation are important methods of controlling enteric infections, and vector control may be useful for arbovirus diseases. However, the most generally useful control measure is the use of vaccines (see Chapter 10).

TABLE 9-8
Criteria for Causation: A Unified Concept[a]

1. *Prevalence* of the disease should be significantly higher in those exposed to the putative cause than in controls not so exposed[b]
2. *Exposure* to the putative cause should be present more commonly in those with the disease than in controls without the disease when all other risk factors are held constant
3. *Incidence* of the disease should be significantly higher in those exposed to the putative cause than in those not so exposed as shown in prospective studies
4. *Temporally,* the disease should *follow* exposure to the putative agent with a distribution of incubation periods on a bell-shaped curve
5. *A spectrum* of host responses should follow exposure to the putative agent along a logical biological gradient from mild to severe
6. *A measurable host response* following exposure to the putative cause should *regularly appear* in those lacking this before exposure (i.e., antibody, cancer cells) or should *increase* in magnitude if present before exposure; this pattern should not occur in persons not so exposed
7. *Experimental reproduction* of the disease should occur in higher incidence in animals or man appropriately exposed to the putative cause than in those not so exposed; this exposure may be deliberate in volunteers, experimentally induced in the laboratory, or demonstrated in a controlled regulation of natural exposure
8. *Elimination or modification* of the putative cause or of the vector carrying it should decrease the incidence of the disease (control of polluted water or smoke or removal of the specific agent)
9. *Prevention or modification* of the host's response on exposure to the putative cause should decrease or eliminate the disease (immunization, drug to lower cholesterol, specific lymphocyte transfer factor in cancer)
10. The whole thing should make biological and epidemiological sense

[a] From A. S. Evans, *Yale J. Biol. Med.* **49,** 175 (1976).
[b] The putative cause may exist in the external environment or in a defect in host response.

Quarantine

Originally introduced in the fifteenth century for the control of plague, quarantine of shipping was used by the English colonists in North America in 1647 to try to prevent the entry of yellow fever and smallpox. It proved very effective in keeping smallpox out of Australia in the nineteenth and early twentieth centuries, and in delaying the entry of pandemic influenza into that country in 1919. However, with the onset of air travel and the consequent arrival of passengers before the end of the incubation period, quarantine became much less effective. It was replaced, for smallpox, by the widespread requirement that international travelers had to have a valid certificate of smallpox vaccination, but this is no longer necessary. Currently, a similar provision operates for travelers who come from or pass through countries where yellow fever is endemic.

Hygiene and Sanitation

Viruses that infect the intestinal tract are shed in feces, and in many human communities there is a large-scale recycling of feces back into the mouth following fecal contamination of food or water. A more voluminous and more fluid output (diarrhea) increases the environmental contamination. Hands contaminated at the time of defecation and inadequately washed commonly transfer viruses directly or indirectly to food—a particular problem in those responsible for the preparation of meals to be ingested by others. Hygiene and sanitation have a profound effect on the incidence of enteric infections, both viral and bacterial. In most parts of the world, where reticulated sewerage systems are still unknown, sewage may seep into wells, streams or other drinking water supplies, particularly after heavy rains. Explosive outbreaks of hepatitis (A or non-A, non-B), poliomyelitis, or gastroenteritis occur from time to time even in sewered areas when sewerage mains burst or overflow to contaminate drinking water supplies.

Raw sewage contains 10^3–10^6 infectious virus particles per liter (typically 10^3–10^4 pfu per liter in Western countries). Titers drop 100-fold (typically to 10^1–10^2 pfu per liter) following treatment in a modern activated sludge plant, because virions adsorb to the solid waste which sediments as sludge. The primary sludge is generally subjected to anaerobic digestion, which reduces the titer of virus significantly. Some countries require that the treated sludge be inactivated by pasteurization prior to being discharged into rivers and lakes or utilized as landfill or fertilizer in agriculture.

In countries where wastewater has to be recycled for drinking and other domestic purposes, the treated effluent is further treated by coagulation with alum or ferric chloride, adsorption with activated carbon, and finally chlorination. Evidence that byproducts of chlorination are toxic for fish and may be carcinogenic for man has encouraged several countries to turn instead to ozonation. Ozone is a very effective oxidative disinfectant, for viruses as well as bacteria, provided that most of the organic matter, to which viruses adsorb, is removed first.

There is also a good case for chemical disinfection of recycled wastewater used for nondrinking purposes such as agricultural irrigation by sprinklers, public fountains, industrial cooling towers, etc., because such procedures disseminate viruses in aerosols, contaminate vegetables, and so on. However, the lability of viruses to heat, desiccation, and ultraviolet light ensures that the virions remaining in wastewater from which most of the solids have been removed will be inactivated without intervention within a few weeks or months, depending on environmental conditions. Even during a cold northern winter, the number of viable enteroviruses in standing water drops by about 1 log per month; during a hot dry summer the rate of decay is as high as 2 logs per week. Hence storage of the final effluent in an oxidative lagoon for 1–2 months is a cheap and effective way of inactivating viruses.

Hygienic measures have less effect on the incidence of respiratory infections, although washing of the hands is important in minimizing the risk of transfer of many respiratory viruses. In general, however, attempts to achieve "air sanitation" have failed and respiratory viral infections are probably more common now than they have ever been, because of growing populations and the constant and extensive movement of people, within cities, from rural to urban areas and internationally. For respiratory viruses, the human population of the world now constitutes a single ecosystem.

Vector Control

Control of the vertebrate reservoir host(s) being almost impossible to achieve, control of arbovirus infections may be approached by attacking the arthropod vectors, minimizing the opportunities for exposure to them, or enhancing human resistance by vaccination. Approaches to vector control are well illustrated by examining the methods advocated by WHO with respect to mosquitoes, namely elimination of breeding sites and destruction of mosquitoes or their larvae. The flight range of many vector mosquitoes is so limited that much can be achieved by concentrating on the immediate vicinity of human settlements, particularly in the case of anthropophilic species such as *Aedes aegypti*. Any still water constitutes a potential breeding site. Swamps and ditches can be drained, and water-collecting refuse such as discarded tires, tin cans, plastic containers, etc., should be destroyed. Larvicidal chemicals are placed in domestic water jars, and kerosine or diesel oil layered on the surface of nonpotable water. Biological control of mosquito larvae by fish or microorganisms, e.g., bacterial spores, is likely to be more widely exploited in the future.

The use of insecticides is a more controversial issue, because there are environmental objections and mosquitoes eventually develop resistance. Some countries have based their arbovirus control programs on aerial insecticide spraying but most retain this approach for emergency control aimed at rapid reduction of the adult female mosquito population in the face of an epidemic. Organophosphorus insecticides such as malathion or fenitrothion are delivered as an ultra-low-volume (short-acting) aerosol generated by spray machines mounted on backpacks, trucks, or low-flying aircraft.

Avoidance of exposure to the bite of arthropods is another obvious precautionary measure. Personal protection against mosquito bites can be achieved by the use of screens on doors and windows, nets over beds, protective clothing, especially at dusk, repellants, and so on.

With most arbovirus infections other than urban dengue and yellow fever the vectors breed over too wide a geographical area to make vector control feasible except on a local scale, usually in response to a threatened epidemic. Spraying of the luggage bays and passenger cabins of aircraft with insecticides reduces the chances of intercontinental transfer of exotic arthropods, whether infected or noninfected.

Change of Life Style

Certain life styles that have become common in western countries during the last few decades, such as the intravenous administration of addictive drugs and promiscuous male homosexuality, are associated with increased risks of infection with a variety of agents, notably hepatitis B and non-A, non-B, and HTLV-III. Observations on the incidence of gonorrhea suggest that fear of AIDS has substantially reduced the incidence and changed the nature of homosexual practices, but in general changes in life style designed to reduce the incidence of disease are difficult to achieve, as is evident with cigarette smoking and alcoholism.

Vaccination

Each of the foregoing methods of control of viral diseases is focused on reducing the chances of infection. The most generally effective method of control, immunization, is directed at making the individual resistant to infection, and if practised on a wide enough scale, in so reducing the likelihood of secondary infections that transmission is reduced or interrupted.

As outlined in Chapter 10, and discussed at length in the chapters of Part II, there are now effective vaccines for many common viral diseases. They are especially effective in diseases with a viremic phase, such as poliomyelitis, yellow fever, and the acute exanthemata. It has proved much more difficult to immunize effectively against infections of the alimentary or respiratory mucous membranes, or the skin. The dramatic effect of effective vaccination programs in reducing the incidence of

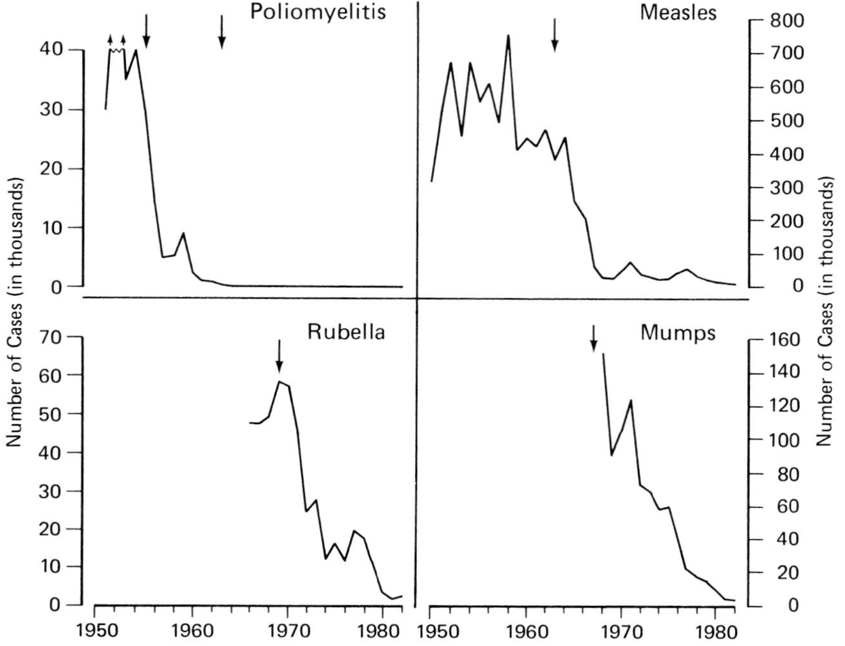

FIG. 9-5. *The fall in the incidence of poliomyelitis, measles, rubella, and mumps in the United States following the introduction of vaccination against these diseases (arrows). Inactivated poliovirus vaccine was introduced in 1954; live vaccine in 1963. The other three, measles (1963), mumps (1967), and rubella (1969), are all live vaccines (see Chapter 10). (Compiled from data kindly supplied by the Centers for Disease Control, United States.)*

poliomyelitis, measles, mumps, and rubella in the United States during the last two decades is illustrated in Fig. 9-5.

Eradication of Viral Diseases

Control, whether by vaccination alone, or by vaccination plus the various methods of hygiene and sanitation aimed at lessening the chance of infection, is an ongoing process, which must be maintained indefinitely. If a disease could be eradicated, so that the causative agent was no longer present anywhere except in microbiologically secure laboratories, control of that disease would no longer be required.

So far, global eradication has been achieved for only one disease, smallpox, the last endemic case of which occurred in Somalia in October 1977. It was achieved by an intensified effort that involved a high level of international cooperation and utilized a potent and very stable vaccine that was easy to administer. However, mass vaccination alone could not have achieved eradication of the disease from the densely populated tropical countries where it remained endemic in the 1970s, because it was impossible to achieve the necessary very high level of vaccine coverage. The effective strategy was to combine vaccination with *surveillance and containment*, by which cases were actively sought out, isolated, and their contacts vaccinated, first in the household and then in increasing distances from the index case.

The global smallpox eradication campaign was a highly cost-effective operation. The expenditure by the World Health Organization between 1967 and 1979 was $81 million, to which could be added about $32 million in bilateral aid contributions and some $200 million expenditure by the endemic countries involved in the campaign. Against this expenditure of about $313 million over the 11 years of the campaign could be set an *annual* global expenditure of about $1000 million, for vaccination, airport inspections, etc., made necessary by the existence of smallpox. This equation takes no account of the deaths, misery, and costs of smallpox itself, or of the complications of vaccination.

The achievement of global eradication of smallpox gave rise to discussions as to whether any other diseases could be eradicated worldwide. International conferences were held to discuss two other viral diseases; measles and poliomyelitis. The biological characteristics of the three diseases that affect the ease of eradication are set out in Table 9-9. These diseases share several essential characteristics: (1) no animal reservoir, (2) lack of recurrent infectivity, (3) one or few stable serotypes, and (4)

TABLE 9-9
Biological Characteristics Enhancing Chances of Eradication

CHARACTERISTIC	SMALLPOX	MEASLES	POLIOMYELITIS
No animal reservoir	+	+	+
No recurrent infectivity	+	+	+
Small number of serotypes	+	+	+
Vaccine			
Effective	+	+	+
Stable	++	+	±[a]; ++[b]
Number of doses	1	1	4[a]; 2[b]
No subclinical cases occur	+	+	–
Not infectious during prodromal stage	+	–	–
Early containment possible	+	–	–

[a] Live vaccine.
[b] Inactivated vaccine.

an effective vaccine. But some features of smallpox that were very important in the eradication of smallpox from tropical countries, notably the lack of infectivity in the prodromal stage, which made surveillance and containment possible, are lacking in measles, whereas the occurrence of a preponderance of subclinical infections in poliomyelitis renders its eradication much more difficult and would make certification of global eradication impossible.

In addition to the biological properties of smallpox being particularly favorable, there were strong financial incentives for the industrialized countries to promote the global eradication of smallpox, because of the costs associated with vaccination of international travelers, port inspections, etc. Finally, at the time global eradication of smallpox was proposed in 1958, all countries in Europe and North America had eliminated the disease, a situation not yet remotely in sight for any other disease.

FURTHER READING

Books

Anderson, R. M., and May, R. M. (1982). "Population Biology of Infectious Diseases," Life Sci. Res. Rep. No. 25. Springer-Verlag, Berlin and New York.

Bailey, N. T. J. (1975). "The Mathematical Theory of Infectious Diseases and its Applications," 2nd ed. Griffin, London.

Beran, G. W., ed. (1981). "Viral Zoonoses" (2 vols.), CRC Handb. Ser. Zoonoses, Sect. B. CRC Press, Boca Raton, Florida.

Berg, G., ed. (1983). "Viral Pollution of the Environment." CRC Press, Boca Raton, Florida.

Evans, A. S., ed. (1982). "Viral Infections of Humans. Epidemiology and Control," 2nd ed. Plenum Medical, New York.

Fenner, F., Henderson, D. A., Arita, I., Jezek, Z., and Ladnyi, I. D. (1987). "Smallpox and its Eradication." World Health Organ., Geneva.

Fox, J. P., and Hall, C. E. (1980). "Viruses in Families." PSG Publishing Co., Littleton, Massachusetts.

Melnick, J. L., ed. (1984). "Enteric Viruses in Water," Monogr. Virol. Vol. 15. Karger, Basel.

Paul, J. R., and White, C., eds. (1973). "Serological Epidemiology." Academic Press, New York.

Walsh, J. A., and Warren, K. S., eds. (1983). "The Great Neglected Diseases of Mankind: Strategies for Control." Univ. of Chicago Press, Chicago, Illinois.

World Health Organization (1982). Report of a WHO Expert Committee with the participation of FAO. Bacterial and viral zoonoses. *W. H. O. Tech. Rep. Ser.* **682.**

Reviews

Anderson, R. M., and May, R. M. (1983). Vaccination against rubella and measles: Quantitative investigations of different policies. *J. Hyg.* **90,** 259.

de-Thé, G. (1979). The epidemiology of Burkitt's lymphoma: Evidence for a causal association with Epstein–Barr virus. *Epidemiol. Rev.* **1,** 32.

Gwaltney, J. M., Jr., and Hendley, J. O. (1978). Rhinovirus transmission: One if by air, two if by hand. *Am. J. Epidemiol.* **107,** 357.

Hinman, A. R. (1982). World eradication of measles. *Rev. Infect. Dis.* **4,** 933.

Horstmann, D. M., Quinn, T. C., and Robbins, F. C., eds. (1984). International Symposium on Poliomyelitis Control. Rev. Infect. Dis. **6,** Suppl. 2, S301–S600.

Katz, S. L., Krugman, S., and Quinn, T. C., eds. (1983). Symposium on Measles Immunization. *Ref. Infect. Dis.* **5,** 389–625.

Mims, C. A. (1981). Vertical transmission of viruses. *Microbiol. Rev.* **41,** 267.

Panum, P. L. (1847). "Observations Made During the Epidemic of Measles on the Faroe Islands in the Year 1846." (Reprinted 1940, American Publishing Company, New York.)

Proceedings of the International Symposium on Measles Immunization, March 16–19, 1982. (1983). *Rev. Infect. Dis.* **5,** 389–625.

Szmuness, W. (1978). Hepatocellular carcinoma and the hepatitis B virus: Evidence for a causal association. *Prog. Med. Virol.* **24,** 40.

Tyrrell, D. A. J. (1981). Is it a virus? *Proc. R. Soc. London, Ser. B* **212,** 35.

World Health Organization Scientific Group (1979). Human viruses in water, wastewater and soil. *W. H. O. Tech. Rep. Ser.* **639.**

CHAPTER 10

Immunization against Viral Diseases

Introduction 283
Live Vaccines 284
Inactivated Vaccines 289
Purified, "Cloned," and Synthetic Proteins 290
Anti-Idiotypic Antibodies 293
"Live" or "Dead"? 294
Vaccination Policy 298
Passive Immunization 301
Further Reading 302

INTRODUCTION

Viral vaccines have traditionally been classified into two broad categories: "live" and "killed" ("inactivated") (Fig. 10-1). Most live vaccines are attenuated mutants, selected for their relative avirulence. They must nevertheless be capable of multiplying in the host and eliciting a natural type of immune response, for that is the rationale of their use. "Killed" vaccines, in contrast, are produced by inactivating, chemically or physically, the infectivity of the virulent wild virus, while retaining its immunogenicity. A refinement is to purify or synthesize the particular protein known to elicit neutralizing antibody. Recombinant DNA technology can be exploited to produce such proteins, either *in vitro*, or *in vivo* following administration of a live avirulent viral vector incorporating the gene for the protein in question.

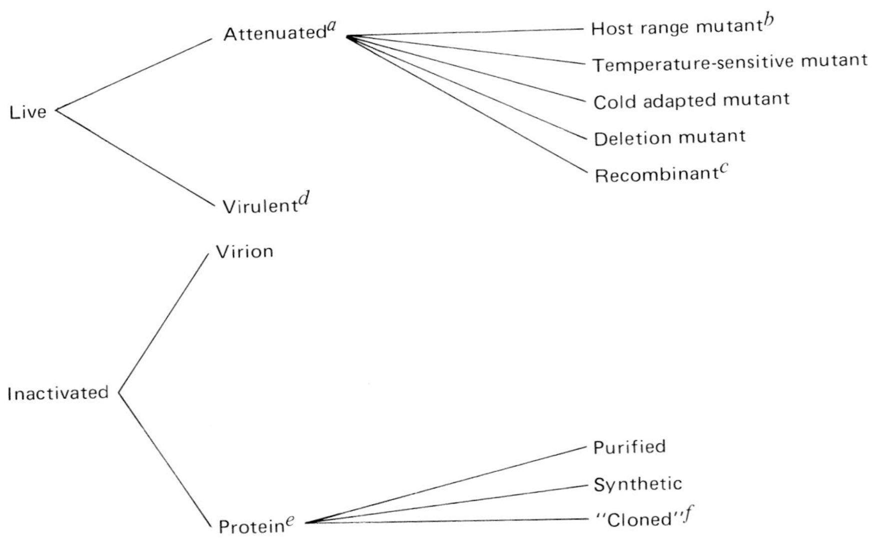

FIG. 10-1. *Types of viral vaccine.* [a]*Including genetic reassortants incorporating mutant genes.* [b]*Occurring naturally in a different animal species, or selected by passage in cultured cells.* [c]*Incorporating gene from another virus.* [d]*Delivered via an unnatural route or at a safe age (in veterinary medicine).* [e]*May be incorporated in liposome or adjuvant.* [f]*In prokaryotic or eukaryotic cells by recombinant DNA technology.*

LIVE VACCINES

The best of today's viral vaccines are live avirulent mutants. Several of them have been dramatically successful in reducing the incidence of important diseases of childhood (Fig. 9-5). One of them has eradicated a major disease, smallpox, from the earth (see Chapter 9). Some live vaccines are delivered via the natural route, e.g., by mouth or nose, but most are injected. The virus multiplies in the recipient, eliciting a lasting immune response, but causes little or no disease. In effect, a live vaccine produces a subclinical infection, which is "Nature's way" of immunizing.

Host Range Mutants

All of the live vaccines in common usage today (see Tables 10-6 and 10-7) have been derived empirically, by several dozen passages of the wild virus through one or more types of cell culture, sometimes after prior "adaptation" to laboratory animals or eggs. Accumulation of a large number of sequential mutations generally leads not only to more vigorous growth in that particular type of cultured cell, but also, quite fortuitously, to progressive loss of virulence for the original host. Avirulence is demonstrated initially in a convenient laboratory model, e.g., a primate, before being confirmed by clinical trial in human volunteers. Because of the requirement that the mutant must not be so attenuated that it fails to replicate satisfactorily, it is sometimes necessary to compromise with a vaccine that does in fact induce mild symptoms in a few recipients.

Such mutants have accumulated a large number of sequential mutations, e.g., the poliovirus type 1 strain used in the Sabin oral vaccine contains 57 separate base substitutions. It is still not known which of these is/are responsible for the loss of virulence, though it is clear that many of the changes are clustered in the region of the genome coding for the NH_2-terminal half of the viral capsid protein VP1. Virulent revertants from Sabin vaccine type 3 consistently contain a reversion to wild-type in a particular nucleotide in the 5′ noncoding region of the genome. For the great majority of viruses we have no idea which gene(s) confers virulence—in general, this vital property cannot be ascribed to any single gene (Chapter 4).

Moreover, in only a few cases has the avirulence of an attenuated vaccine been characterized in terms of its pathogenesis in the vaccinee. With many live respiratory viral vaccines, multiplication of the attenuated virus in the respiratory tract is severely restricted, i.e., the total yield of virus is greatly diminished. In other instances, however, e.g., live poliovaccine, the vaccine strain appears to have lost its capacity to infect the vulnerable target organ.

Despite the outstanding success of empirically derived live vaccines, research is proceeding apace on more calculated approaches to the design of live viruses of reduced virulence. These approaches, which have already yielded several promising experimental vaccines, are discussed below.

Temperature-Sensitive (*ts*) and Cold-Adapted (*ca*) Mutants

The observation that *ts* mutants generally display reduced virulence suggested that they might constitute very satisfactory live vaccines. Chanock and his colleagues vigorously pursued this possibility for the best part of a decade, producing *ts* vaccines against a number of respiratory viruses. The rationale was that *ts* mutants incapable of replication at the temperature of the lungs (37°C) would nevertheless multiply safely if instilled into the nose, where the temperature is consistently lower. Such proved to be the case but, unfortunately, even vaccines containing more than one *ts* mutation displayed a disturbing tendency to revert toward virulence during replication in man. Reversion was not attributable solely to simple back-mutation; the wild-type phenotype could be restored by a suppressor mutation in a gene quite distinct from that containing the *ts* lesion.

It transpires that several of the empirically derived host-range mutants that have been used for years as live vaccines are in fact also *ts*. These include the Sabin poliovirus strains, the Schwarz measles vaccine, and a number of live attenuated veterinary vaccines. The "ceiling" ("shutoff," "nonpermissive") temperature for several of these *ts* mutants must obviously be higher than 37°C, because the list includes vaccines that are administered systemically and are capable of multiplication at body temperature.

Concern has been expressed about the possibility that *ts* vaccines might be capable of setting up persistent infections in the brain or elsewhere. Such infections have been established experimentally in animal models using *ts* mutants of measles, and the measles virus recoverable from some cases of SSPE is *ts* (see Chapter 7). Yet the *ts* Schwarz mutant, which has been employed as the standard measles vaccine for two decades, has led, not to an increase, but to a marked decline in the incidence of SSPE.

Somewhat different from classical *ts* mutants are cold-adapted (*ca*) mutants, derived by adaptation of virus to grow at suboptimal temperature (25°C). A *ca* strain of influenza virus with mutations in every gene grows as well at 25° as at 33°C; it happens also to be *ts* in that it fails to grow at 38°C. So far, this type of *ca* vaccine has revealed less tendency to revert than the earlier *ts* vaccines.

Genetic Reassortment

In the case of viruses with segmented genomes, genetic reassortment can be exploited to introduce *ts*, *ca*, or other mutant genes from laboratory "master strains" into a recent isolate of wild virus to produce an avirulent vaccine strain relatively quickly. For example, the *ca* multiple mutant described above has been used as the master strain of influenza virus to produce a reassortant containing wild-type HA and NA genes plus six *ca* mutant genes. As this ingenious technique has now become standard practice in the manufacture of influenza vaccines, it will be discussed more fully in Chapter 21.

Deletion Mutants and Site-Directed Mutagenesis

In theory one might predict a negligible reversion rate in mutants from which parts of the genome had been deleted. Modern techniques of enzyme surgery permit the excision of a particular nucleotide sequence within a nonessential gene, ideally a gene conferring virulence, e.g., the thymidine kinase gene of herpes simplex virus which seems to be required for both pathogenicity and establishment of latency (see Chapter 11). There are no major impediments to the construction of deletion mutants from DNA viruses. In the case of nonsegmented ssRNA viruses of positive polarity, such as picornaviruses, cDNA can be transcribed using reverse transcriptase, manipulated by conventional recDNA techniques, and used to transfect mammalian cells which will then synthesize virions containing RNA with the prescribed deletion. Construction of deletion mutants from negative-stranded RNA viruses with a segmented genome poses special difficulties, which are nevertheless not insuperable.

Site-directed mutagenesis (see Chapter 4) now permits the introduction of prescribed nucleotide substitutions at will. Thus, when more becomes known about particular genes involved in virulence it will be possible to construct vaccines with any designated nucleotide sequence.

Recombinants Incorporating Cloned Genes from Other Viruses

Recombinant DNA technology has recently opened up another novel approach to vaccination which may prove to be of widespread applicability. The concept is to insert the gene for any given viral protein into the genome of an avirulent virus that can then be administered as a live vaccine. Cells in which the virus multiplies *in vivo* will produce the foreign viral protein, against which the body will then mount an immune response.

The prototype developed in 1983 was a live vaccinia virus incorporating the gene for hepatitis B surface antigen (HBsAg). In brief, the recombinant was constructed as follows. The gene for HBsAg was inserted, within a nonessential vaccinia viral gene (e.g., that for thymidine kinase, TK), into a bacterial plasmid. A strong vaccinia promoter was placed in the appropriate position upstream from the foreign (HBsAg) gene. Mammalian cells were then infected with wild-type vaccinia virus, and shortly thereafter transfected with the chimeric plasmid. Recombination occurred between the vaccinia DNA and the plasmid DNA. Appropriate selection procedures enabled recombinant vaccinia virions containing the HBsAg gene to be recovered. Following inoculation of such recombinants into cultured cells or animals, the virus multiplied, substantial quantities of HBsAg were secreted from infected cells, and the animals produced antibody to HBsAg. Furthermore, vaccinated chimpanzees were protected against challenge with hepatitis B virus. Analogous vaccinia–influenza, vaccinia-rabies and vaccinia–herpes simplex recombinants protected animals against influenza, rabies and herpes simplex, respectively.

TABLE 10-1
Live Vaccinia Virus as Vector for Cloned Viral Genes

Advantages
Large vaccinia genome can accommodate numerous foreign inserts
Single polyvalent vaccine protecting against several viral and nonviral diseases
Production cheap, simple; developing countries could manufacture
Relatively stable, even in tropics
Delivery (multiple puncture) by nonmedical personnel
Cell-mediated immunity also elicited, as foreign antigen expressed on surface of vaccinia virus-infected cells in association with class I MHC antigens
Disadvantages
Side effects: encephalitis, progressive vaccinia in immunodeficient recipients, eczema vaccinatum
Revaccination fails to take for at least 5 years

This dramatic demonstration of the power of recombinant DNA technology is a real *tour de force*. In theory at least, it should now be possible to insert the gene for any, or indeed several, foreign proteins into the DNA of an avirulent live virus, such as vaccinia. Thereby one could hope to vaccinate simultaneously against a whole range of infectious diseases, including not only those of viral etiology but also perhaps malaria and other scourges of mankind. The potential benefits, particularly to the Third World, are self-evident.

Vaccinia virus offers obvious advantages as a vector (Table 10-1). It has a large genome potentially capable of accommodating a number of foreign inserts (at least 22 kb) and still being packaged within the virion, and it has been used successfully as a live vaccine throughout the world for many years. It is reasonably, but not absolutely, safe—about 10 vaccinees in every million develop potentially lethal postvaccinial encephalitis. Perhaps the vaccinia genome can be manipulated to attenuate it further to lower the incidence of this serious side effect; even the loss of the TK gene may reduce virulence somewhat. Alternatively, other DNA viruses, albeit with smaller genomes, such as nononcogenic papovaviruses or adenoviruses could be harnessed as vectors. It may be advantageous to develop several alternative vectors because, once a particular virus has been used, it could not be reused as a live vaccine to immunize against another agent, for a few years at least, as it would not multiply satisfactorily. Alternatively, before substantial numbers of people are vaccinated with the chosen vector, careful thought could be given to the construction of a one-off vaccine incorporating genes for the key immunogenic proteins of most of the important human infectious diseases.

Live Virulent Viruses as Vaccines

A few vaccines still used in veterinary practice today consist of fully virulent live virus administered via an unnatural route, resulting in replication of the virus without producing disease, e.g., avian infectious laryngotracheitis virus scarified into the cloaca of chickens. Such an approach is generally considered to be far too dangerous to contemplate in man, but one interesting example, namely delivery of the respiratory pathogen, adenovirus, by mouth, enclosed in an "enteric" capsule, is discussed in Chapter 15. Another strategy occasionally pursued by veterinarians is to administer virulent live virus at a time of life when it will inflict little or no damage, e.g., to adult hens in order to protect their chicks, by maternal antibody, during the vulnerable first few weeks of life.

A rather different approach, pioneered by Jenner in his use of live cowpox virus to "vaccinate" against smallpox, involves the use of a naturally occurring virus of low virulence or a virus indigenous to another species of animal. It is sometimes possible to find a virus that is, on the one hand, nonvirulent for the host species one wishes to immunize, yet so closely related antigenically to the serotype(s) against which one wishes to protect that cross-immunity will result. One of many examples from veterinary medicine was described in Chapter 8, namely the use of virulent herpesvirus of turkeys to protect fowls against Marek's disease. Similarly, bovine rotavirus, administered by the natural (oral) route, protects calves and piglets against bovine, porcine, or human rotaviruses. It is not inconceivable that such an approach could be successfully exploited in man.

INACTIVATED VACCINES

Inactivated ("killed") vaccines are made from virulent virus by destroying its infectivity while retaining its immunogenicity. Being noninfectious, such vaccines are generally safe, but need to be injected in very large amounts to elicit an antibody response commensurate with that attainable by a much smaller dose of live attenuated vaccine. Normally, even the primary course comprises two or three injections, and further ("booster") doses may be required at intervals over the succeeding years to revive waning immunity.

The most commonly used inactivating agents are formaldehyde, β-propiolactone, and the ethylenimines. Formaldehyde is used to produce two human vaccines, influenza and "Salk" poliovaccine. It is not clear to what extent the abolition of viral infectivity is attributable to inactivation of nucleic acid (by reaction with the amino groups of nucleotides) and to what extent to denaturation of protein (by cross-linking). One aspect of the latter, known as "tanning," does produce some problems, including clumping and loss of immunogenicity. One of the advantages of β-propiolactone, now used in the manufacture of human vaccines against rabies, is that it is completely hydrolyzed within hours to a nontoxic degradation product found normally in the body. Ethylenimines have been widely used for veterinary vaccines, e.g., foot-and-mouth disease vaccines. Psoralens, now being used for the preparation of safe, noninfectious antigens for serology in diagnostic laboratories (see Chapter 12), might also have potential as agents for the production of inactivated vaccines.

Inactivated vaccines tend to be expensive because of the large numbers of virions that need to be injected. Recent efforts have been directed therefore toward scaling up the commercial production of virus, e.g., by using continuous cell lines capable of growing in suspension, or as monolayers coating "microcarrier" beads in fermentation tanks. Second, much higher standards are being applied to the purification and concentration of viruses, whether by zonal ultracentrifugation, gel filtration, ion-exchange chromatography, affinity chromatography using monoclonal antibodies, or a combination of such procedures. It is particularly important to remove aggregated virus prior to chemical inactivation; failure to do so led to a notorious tragedy with an early batch of Salk vaccine which contained residual live virus and was responsible for a number of cases of paralytic poliomyelitis in 1955.

PURIFIED, "CLONED," AND SYNTHETIC PROTEINS

Large doses of inactivated virions often produce a febrile response as well as a local reaction at the injection site, especially in young children. It was found quite empirically that the reactogenicity of inactivated influenza vaccines could be reduced somewhat by "splitting" the envelope of the virus with a lipid solvent. The logical extension of this compromise is to remove all nonessential components of the virion and inoculate only the relevant immunogen, namely the particular surface (envelope or outer capsid) protein against which neutralizing antibodies are directed. Such *"subunit" vaccines* are in use against influenza and hepatitis B and have been shown to be feasible against a wide range of other viruses, including rabies, herpes simplex, and adenoviruses.

Proteins from Genes Cloned by Recombinant DNA Technology

Recombinant DNA technology represents a means of producing cheaply on an industrial scale large amounts of viral protein that can be readily purified. Details of this technology were given in Chapter 4.

TABLE 10-2
Viral Proteins "Cloned" by Recombinant DNA Technology as Potential Vaccines

VIRUS	PROTEIN
Hepatitis B	HBsAg
Influenza	HA
Foot and mouth disease	VP1
Rabies	G
Herpes simplex	gD

Already, the complete genomes or selected genes of several mammalian viruses have been "cloned" in prokaryotic or eukaryotic cells. In many instances these genes are expressed, and in some the resultant protein is released into the supernatant. At the time of writing, antigenic surface proteins of a number of viruses had been produced by recombinant DNA technology and some of these were being seriously considered as potential vaccines (Table 10-2). The prospects appear brighter using eukaryotic cells such as yeasts or mammalian lines (e.g., CHO or COS) rather than bacteria as hosts for gene cloning, but persuasive verification of removal of all traces of DNA from the final purified product would be required before human cell lines would be licensed for this purpose.

Synthetic Peptides

The amino acid sequence of any viral protein can be read off the nucleotide sequence of the viral nucleic acid, or can be determined directly on the purified protein itself. The protein can then be synthesized chemically. Now that techniques are available for locating the antigenic sites on the surface of viral proteins (see page 162 and Fig. 4-5) it is possible to synthesize shorter peptides corresponding to the critical antigenic determinants to which neutralizing antibodies bind. Of particular interest is the possibility that relatively invariate ("conserved"), even buried, sequences, not normally immunogenic when presented *in situ* in the virion, may be capable in isolation of eliciting neutralizing antibodies which may then have the added advantage of cross-protection against heterologous serotypes.

This is an intriguing new approach which merits further research, but it must be emphasized that, because short synthetic peptides lack the tertiary conformation they assume in the intact virion, most antibodies raised against them are incapable of binding to virus, hence the neutralizing titer is generally orders of magnitude lower than that induced by inactivated vaccine, or even by the purified intact protein. Nevertheless, there are just one or two examples of synthetic viral peptides

TABLE 10-3
Synthetic Peptides as Potential Vaccines

Advantages
Short defined amino acid sequence representing protective epitope(s) only
Safe, nontoxic
Conserved sequence, normally nonimmunogenic, may be cross-protective
Priming with peptide may allow anamnestic response to challenge
Artificial constructs containing epitopes of more than one viral protein may be engineered
Disadvantages
Poorly immunogenic, hence adjuvant, carrier, and/or liposome essential
Most epitopes conformational and perhaps discontinuous, hence mimotopes may need to be constructed
May be too specific, not protecting against naturally occurring variants
Single-epitope vaccine will readily select single point mutants
Recipients lacking appropriate class II HLA antigen will fail to respond

that have been shown to elicit neutralizing antibodies (e.g., see hepatitis B, Chapter 13). Moreover, peptides incapable of inducing protective levels of antibody have been claimed to prime an animal to respond anamnestically when subsequently boosted with a subimmunizing dose of virus. However, most conformational determinants may be nonsequential (discontinuous), i.e., are composed of amino acids that are separated in the primary sequence but find themselves close together in the intact molecule. If so, new techniques of peptide chemistry may have to be invoked to construct "mimotopes" which mimic such discontinuous epitopes. Some additional observations are made in summary form in Table 10-3.

Immunopotentiation by Adjuvants, Liposomes, and Carriers

If isolated viral proteins—whether purified from virions, made by recombinant DNA technology, or chemically synthesized—are to be employed as vaccines, their immunogenicity will need to be enhanced by several orders of magnitude (Table 10-4). Such immunopotentiation could be achieved by coupling the protein to a suitable carrier, and/or incorporation in a liposome, or emulsification with an adjuvant.

Adjuvants are substances that, when administered with antigen, potentiate the humoral and/or cell-mediated immune response. The mechanism(s) of action of adjuvants is still unknown but may include (1) prolonged retention of antigen, (2) activation of macrophages, leading to secretion of lymphokines and attraction of lymphocytes, and (3) mitogenicity for lymphocytes. The most widely used adjuvants in man to date are aluminum phosphate and aluminum hydroxide gel, to which the immunogen is adsorbed, but the resulting immune response is not

TABLE 10-4
Defined Antigen Vaccines[a]

Advantages
Production and quality control simple
No nucleic acid, no extraneous proteins or lipid, hence less toxic
Safer in the case of viruses that are particularly dangerous (e.g., HTLV-III), or that may cause cancer (e.g., hepatitis B) or establish persistent infection (e.g., herpesviruses)
Feasible even if virus cannot be cultured (e.g., hepatitis B)
Disadvantages
May be less immunogenic than conventional inactivated whole-virus vaccines
Requires adjuvant or liposomes
Requires primary course of injections followed by booster(s)
Fails to elicit cell-mediated immunity

[a] (1) Purified protein; (2) protein "cloned" by recDNA technology (in bacteria, yeasts, or mammalian cells); and (3) synthetic peptide.

particularly prolonged, hence booster injections are required. There is a real need for new adjuvants, preferably chemically defined and of known mode of action.

Liposomes are artificial lipid membranes into which proteins can be incorporated. When purified viral envelope proteins are used, the resulting "virosomes" (or "immunosomes") somewhat resemble the original enveloped virion. This ingenious trick enables one not only to reconstitute virus-like structures lacking all nucleic acid and other extraneous material, but also to select nonpyrogenic lipids and to incorporate substances with adjuvant activity, thus regaining much of the immunogenicity lost when the viral protein was removed from its original milieu.

Carriers are proteins to which a smaller molecule (a "hapten") may be chemically coupled in order to render the latter more immunogenic. The potentiation of the immune response to the hapten is due to the helper T (T_h) cell response to the carrier. If such carrier proteins were to be employed for human vaccines, it would be important to ascertain that clinically significant hypersensitivity to that protein would not develop.

ANTI-IDIOTYPIC ANTIBODIES

The antigen-binding site (paratope) of every antibody molecule contains a unique amino acid sequence known as its idiotype (see page 153). Antibody can be raised against this idiotype and is known as *anti-idiotype (anti-Id)*. Because anti-idiotypic antibody is capable of binding to the same idiotope as binds a certain epitope on the original antigen, it may be surmised that anti-Id mimics the conformation of that epitope. If this were so, anti-Id raised against a neutralizing monoclonal antibody to a particular virus could conceivably be deployed as a vaccine. The antibodies it would elicit would bind to and neutralize the virus. In fact, anti-Id antibodies raised against antibodies to hepatitis B surface antigen or reovirus S1 capsid antigen do elicit an antiviral antibody response on injection into animals.

Moreover, anti-Id antibodies raised against the receptor on a T lymphocyte line specific for Sendai virus have been reported to elicit a cytotoxic T cell response in mice which protects them against viral challenge. It is still far from clear that this points the way to the vaccines of the future. First, the generality of these initial findings needs to be confirmed. Second, it is crucial to establish whether anti-Id antibody, when administered as an immunogen, elicits the same type of immune response as does antigen, i.e., that it achieves the right balance of humoral and cell-mediated immunity rather than, say, a suppressor T cell response. Third, since anti-Id for human use would need to be of human origin to avoid serum sickness, special precautions would need to be taken to ensure a DNA-free product if monoclonal anti-Id were to be produced from a human hybridoma.

"LIVE" OR "DEAD"?

Predating but brought to a head by the famous debate of the mid-1950s over the merits and disadvantages of Sabin versus Salk poliovaccines, the question of live versus dead viral vaccines remains an issue today. The major considerations are summarized in Table 10-5 and discussed below.

Immunological Considerations

The object of immunization is to protect against disease, not necessarily to prevent infection. If, as immunity wanes over the years following active immunization, or the weeks following passive immunization, infection with wild virus does occur, the infection is likely to be subclinical. Such an occurrence may have the beneficial effect of boosting immunity.

Acquired immunity to many respiratory viruses is so poor even following natural infection that it is perhaps unrealistic to expect any respiratory viral vaccine to be effective. In such cases, however, a reasonable objective may be to prevent serious (lower respiratory tract) disease.

It has been argued above that subclinical infection is Nature's way of immunizing, that by and large this is extremely effective, inducing lifelong immunity following systemic infection (see Chapter 6), and that live vaccines, preferably delivered via the natural route, are obviously

TABLE 10-5
Advantages and Disadvantages of Live versus Dead Vaccines

PROPERTY	LIVE	INACTIVATED
Route of administration	Natural[a] or injection	Injection
Dose of virus; cost	Low	High
Number of doses	Single[b]	Multiple
Need for adjuvant	No	Yes[c]
Duration of immunity	Many years	Generally less[d]
Antibody response	IgG; IgA[e]	IgG
Cell-mediated response	Good	Poor
Heat lability in tropics	Yes[f]	No
Interference	Occasional; OPV only[g]	No
Side effects	Occasional mild symptoms[h]	Occasional sore arm
Reversion to virulence	Rarely; OPV only[i]	No

[a] Oral or respiratory, in certain cases.

[b] Except oral poliovirus vaccine (OPV), to preempt possibility of interference. For some other live vaccines a single booster may be required by law (yellow fever) or desirable (rubella?) after a decade or so.

[c] But no satisfactory adjuvants yet licensed for human use.

[d] But satisfactory with some inactivated vaccines.

[e] IgA if delivered via oral or respiratory route. OPV can thereby prevent wild poliovirus from multiplying in the gut hence facilitate near eradication of the virus from the community.

[f] $MgCl_2$ and other stabilizers, plus maintenance of "cold chain" assist preservation.

[g] Especially in Third World.

[h] Especially rubella and measles.

[i] 10^{-6} vaccinees.

the nearest approach to this ideal. One of the great advantages of live oral poliovaccine (OPV) is that, by virtue of its replication in the alimentary tract, it leads to prolonged synthesis of local antibody of the IgA class which prevents the subsequent multiplication of wild poliovirus in the gut, thereby depriving it of hosts in which to circulate, hence rendering feasible the prospect of total eradication of the virus, as well as the disease, from the community. In general, IgA is considered to be the most (some would say the only) class of immunoglobulin relevant to prevention of viral infection of mucosal surfaces, such as the alimentary, respiratory, genitourinary, and ocular epithelia.

A second, perhaps even more crucial reason for favoring live vaccines is that they are evidently more effective in eliciting cell-mediated immunity. While it is clear that neutralizing antibody is the key to prevention of establishment of infection, there is good evidence that T lymphocytes, particularly T_c cells, play a crucial role in recovery from infection, once it has been established. The available evidence (see Chapter 6) indicates that T_c cells are activated not by soluble antigen, but by viral antigen presented on the surface of infected cells in association with MHC antigens. The differential response of T_s, T_h, T_d, and T_c cells to viral antigens presented in the form of live virus, inactivated virions, or soluble protein, respectively, merits fuller investigation.

A further major reason for favoring vaccines that elicit a T lymphocyte response is that T cells display broader cross-recognition of related viral strains than do B cells. Such cross-immunity would be advantageous where several viral serotypes circulate simultaneously or sequentially (antigenic drift), but is more relevant to recovery than to prevention of infection.

By contrast with the lasting immunity that follows systemic infections, immunity following localized infections of epithelial surfaces is relatively short-lived, and large numbers of respiratory and enteric viruses have evolved, presumably by antigenic drift. For both these reasons the prospects for development of truly successful vaccines against diseases such as the common cold are poor.

Safety

Properly prepared and tested, both live and inactivated vaccines are perfectly safe in immunocompetent people, as of course are purified, synthetic, or "cloned" proteins. Licensing authorities have become extremely vigilant and have insisted on rigorous tests for residual live virulent virus since a number of tragedies occurred in the pioneering days.

Inactivated or purified protein vaccines should be preferred in immunocompromised individuals, in whom live attenuated vaccines not uncommonly produce disease. Some virologists are also concerned about the wisdom of using live vaccines against herpesviruses because of the known propensity of members of this family to establish persistent infections and their association with cancer. The same would apply to any other virus with known or postulated oncogenic potential.

There are certain potential problems relating to safety which are unique to live vaccines and must be overcome for every product before it can be licensed for human use.

Underattenuation. Some live vaccines are not as avirulent as one would wish for in an ideal world. Rubella and measles vaccines, for example, produce some symptoms—in effect a mild case of the disease—in a minority of vaccinees. However, attempts to attenuate virulence further by additional passages in cultured cells have led to an accompanying decline in capacity of the virus to multiply in man, hence a corresponding loss of immunogenicity. Such trivial side effects as do occur with current human viral vaccines are of no real consequence and do not prove to be a significant disincentive to immunization, provided that parents are adequately informed in advance and perceive that the benefits of immunization are very much greater than the risks.

Genetic Instability. A quite different problem occurs in the case of vaccine strains with an inherent tendency to revert toward virulence during multiplication in the recipient or in contacts to whom the vaccine virus has spread. Most vaccine viruses are incapable of such spread, but in the case of those that do, such as oral poliovaccine, there are obviously greater opportunities for the accumulation of successive mutations leading gradually to a partial restoration of virulence. Vaccine-associated poliomyelitis does occur, but is exceedingly rare ($<10^{-6}$). No other human viral vaccine in use today displays any sign of genetic instability, although, as discussed above, experimental *ts* vaccines against some human respiratory viruses have proved to be unstable.

Contaminating Viruses. Cultured cells employed as a substrate for the growth of viruses for vaccine production may carry endogenous viruses. These may be overt or covert. Primary cell cultures, such as monkey kidney, are particularly liable to contain adventitious agents; over 75 simian viruses have been identified in this way, and some, such as *Herpesvirus simiae* ("B virus"), are lethal for man. Clearly, such contaminants are more of a worry in live vaccines, but even inactivated vaccines are not free from this risk, for the contaminant may be more resistant to the inactivating agent than the vaccine strain. For example, early batches of formaldehyde-inactivated, as well as live attenuated, poliovaccines were contaminated by live SV40 virus, which is oncogenic for baby rodents but not, fortunately, for humans!

The danger of contamination of live vaccines with adventitious viruses has triggered a great deal of debate about which types of cultured cell should be licensed for use as substrates for the growth of viruses for human vaccine production. If primary cell cultures are to remain legally acceptable for this purpose, the minimum requirement should be that the animals are bred in captivity, preferably under specific-pathogen-free conditions, e.g., closed colonies of monkeys or rabbits, or eggs from leukosis-free chickens, and that the cultured cells are rigorously scrutinized for all possible endogenous viruses. In general, however, primary cell cultures were replaced during the 1970s by diploid strains of human embryonic fibroblasts. Such strains can be subjected to comprehensive virological and karyological screening, certified as safe, then frozen for storage and distribution to vaccine manufacturers on request.

Recently, there has been a revival of discussion about the acceptability of heteroploid cell lines. Such "continuous" or "immortal" cell lines present great advantages to the vaccine manufacturers because of their infinite growth potential and ease of cultivation *in vitro,* including suspension culture. Several veterinary vaccine viruses are grown commercially in suspension culture in large fermentation tanks, e.g., foot-and-mouth disease virus in BHK-21 cells. In 1982, heteroploid cell lines were, for the first time, licensed for the production of human vaccines, provided that adequate techniques are used to demonstrate that the cells are nonmalignant and to exclude the presence of putative cancer viruses or oncogenes in the final product.

Interference

Live vaccines delivered by mouth or nose depend critically for their efficacy on multiplication in the alimentary or respiratory tract, respectively. Interference can occur between different live viruses contained in the vaccine, or between the vaccine virus and itinerant enteric or respiratory viruses that happen to be growing in the vaccinee at the time. OPV, for example, fails to "take" in substantial numbers of children in tropical countries where subclinical infection with any one of the numerous enteroviruses is the rule rather than the exception at any particular moment in time. The same phenomenon occurs with lower frequency in developed countries. This is the reason why OPV is routinely administered on three occasions, separated by at least 2 months. However, interference is not a problem with systemically administered polyvalent vaccines, live or inactivated. The measles–rubella–mumps live vaccine given subcutaneously to infants in the United States evidently elicits as good an antibody response against each of the three viruses as do the individual vaccines given separately.

Heat Lability

Live vaccines are vulnerable to inactivation by the high ambient temperatures encountered in the tropics. Since these countries are also, in the main, those with underdeveloped health services, formidable problems are encountered in maintaining refrigeration ("the cold chain") from manufacturer to the point of delivery, i.e., the child in some inaccessible rural village. To some extent the problem has been alleviated by the addition of stabilizing agents to the vaccines, and by packaging them lyophilized (freeze-dried), for reconstitution immediately before administration.

VACCINATION POLICY

This leads us into a discussion of the medical, political, economic, and sociological considerations which determine vaccination policy.

The aim of immunization is generally thought of as the protection of the vaccinee. This is usually so, but in the case of certain vaccines (e.g., rubella in women) the objective is to protect the vaccinee's offspring. Under other circumstances an important auxiliary objective may be to protect unvaccinated members of the community (e.g., against poliomyelitis), either by natural spread of live vaccine virus, or by reducing the circulation of wild virus.

Universal vaccination could lead to the eradication of certain human viruses, such as measles. However, the difficulties are much greater than those encountered with smallpox, partly because the diseases do not engender the same degree of fear, and partly because surveillance and containment would not work as effectively as they did with smallpox (see Table 9-9). Fear is the principal factor motivating people to seek or accept immunization for themselves and their children. Even in the case of a dreaded disease such as polio, it is difficult to maintain enthusiasm for a program of universal immunization after the disease has become very rare. Complacency has led to a limited resurgence of polio and measles in a number of countries with strong immunization programs. Continuation of routine immunization after the threat of disease has almost vanished but the virus has not been totally eradicated is doubly essential because reduction in circulation of wild virus in the community has left unimmunized people uniquely susceptible, by removing the protective effect of repeated subclinical infections. Highly organized and resolute health services are called for. Particular attention must be given to unimmunized pockets, e.g., urban ghettos, immigrants, and certain religious minorities. Legislation for compulsory immunization against particular diseases prior to crossing national borders or entering school is perhaps the most effective single measure.

Acceptability of a vaccine by the community is governed by a complex equation, balancing efficacy against safety, gain against risk, fear of the disease against fear of needles and side effects. If the disease is lethal or debilitating, e.g., poliomyelitis, the need for immunization should be clear to all; both the people and the vaccine-licensing authorities will accept a small risk of even quite serious consequences of vaccination gone wrong. If, on the other hand, the disease is perceived as trivial, no side effects will be countenanced. This is a major consideration with respiratory viral vaccines. Where equally satisfactory vaccines are available against a particular disease, considerations such as cost and ease of administration tip the balance, e.g., toward oral poliovaccine.

A significant impediment to comprehensive vaccine coverage of the community is the unnecessarily complicated immunization schedules officially recommended by some government health authorities. Most of the currently available vaccines, bacterial and viral, are aimed at preventing diseases the risk of which are greatest in early infancy, hence are given as soon as convenient, i.e., during the first 6 months of life for oral poliovaccine, or after antibody has completely disappeared (12–15 onths) in the case of systemically administered live vaccines for measles, mumps, and rubella (Table 10-6). Polyvalent vaccines, such as that available against measles–mumps–rubella, confer a major practical advantage in minimizing the number of visits the mother must make to the clinic. Similarly, there is no reason why at least some of the infant's OPV doses cannot be given simultaneously with "triple antigen" (diphtheria, pertussis, and tetanus). Live viral vaccines are generally the most convenient in that repeated boosters are not normally required.

Characteristics of the major human viral vaccines in common use are set out in Table 10-7. There are many more in various stages of development or clinical trial, e.g., against all of the herpesviruses, several respiratory viruses, and hepatitis A. Others are in use against viruses of

TABLE 10-6

Schedules for Immunization against Human Viral Diseases[a]

VACCINE	PRIMARY COURSE	SUBSEQUENT DOSES
Live vaccines		
Poliomyelitis	2–6 months of age, then 1 or 2 further doses at 2–3 month intervals[b]	School entry
Measles	12 months[c] (single dose)	—
Rubella	12 months[d] (single dose)	High school entry[d] (girls only)
Mumps	12 months (single dose)	—
Yellow fever	Before travel to endemic area	10 years
Inactivated vaccines		
Influenza	Autumn[e]	Annual booster
Rabies	After bite, then at 3, 7, 14, and 28 days	—
Hepatitis B	When at risk[f], then 1 and 6 months later	?

[a] Schedules vary from country to country. This table is to be taken only as a guide.

[b] Two or three doses spaced 2 months apart during first year of life, commencing between 2 and 6 months of age. A further dose at 15–18 months is not essential but should be given if there is any doubt about whether the first three doses were received.

[c] Given on or shortly after first birthday in most developed countries, but as early as 6–9 months in some tropical developing countries where measles death rate is high in the second 6 months of life.

[d] Current United States policy is to immunize males and females shortly after first birthday. Current United Kingdom policy is to immunize females only, on entry into high school. Recommended here is a combination of both.

[e] Vulnerable groups only, e.g., old respiratory invalids.

[f] Vulnerable groups only, e.g., family contacts and babies of carriers, staff of hemodialysis units, blood banks, laboratories, hospitals, institutions for mentally retarded, etc., immunosuppressed, drug addicts, male homosexuals.

TABLE 10-7
Viral Vaccines Recommended for Use in Man[a,b]

DISEASE	VACCINE STRAIN	CELL SUBSTRATE	ATTENUATION	INACTIVATION	ROUTE
Rabies	Pasteur	HEF	−	BPL or TBP	im or id
Yellow fever	17D	Chick embryo	+	—	sc
Poliomyelitis	Sabin 1,2,3 (OPV)	HEF	+	—	Oral
Measles	Schwarz	CEF	+	—	sc
Rubella	RA27/3 or Cendehill	HEF or RK	+	—	sc
Mumps	Jeryl Lynn	CEF	+	—	sc
Influenza	A/H3N2, B, (A/H1N1)	Chick embryo	−	BPL or formalin or HANA subunits	im
Hepatitis B	—[c]	—	−	HBsAg[c]	im

[a] A wide variety of different viral strains and cell substrates are used in different countries; the selection listed is not comprehensive.

[b] BPL, β-Propiolactone; CEF, chick embryo fibroblast cultures; HEF, diploid strain of human embryonic fibroblasts; id, intradermally; im, intramuscular; RK, rabbit kidney cultures; sc, subcutaneous; and TBP, tri-(*n*)-butyl phosphate.

[c] Hepatitis B surface antigen purified from serum of human donors, then formalin treated.

restricted geographical distribution, e.g., miscellaneous arboviruses. All will be discussed fully in the relevant chapters of Part II.

PASSIVE IMMUNIZATION

Instead of actively immunizing with viral vaccines it is possible to confer short-term protection by the intramuscular administration of antibody, as immune serum, or immunoglobulin (Ig) purified therefrom (usually designated *immune globulin,* IG). Human Ig is usually preferred, because heterologous protein provokes an immune response that can manifest itself as serum sickness or even anaphylaxis. Pooled normal human Ig can be relied upon to contain reasonably high concentrations of antibody against all the common viruses that cause systemic disease in man. Higher titers of course occur in "convalescent" Ig from donors who have recovered from infection with the virus in question, and such specific immune globulin is the preferred product if commercially available, e.g., for rabies, varicella-zoster, hepatitis B. Although passive immunization should be regarded only as an emergency procedure for the immediate protection of unimmunized individuals exposed to special risk, it is an important prophylactic measure against several human diseases, notably hepatitis A and B, rabies, varicella, and measles (see the relevant chapters in Part II).

Specific antibody can occasionally be used as therapy for an established viral disease, for instance, immune plasma reduces the mortality from infection with certain arenaviruses. Now that monoclonal antibodies of high specificity and high titers are readily obtainable, it is appropriate to reexamine the role of antibody in the treatment of established disease.

FURTHER READING

Books

Chanock, R. M., and Lerner, R. A., eds. (1984). "Modern Approaches to Vaccines." Cold Spring Harbor Lab., Cold Spring Harbor, New York.

Lerner, R. A., Chanock, R. M., and Brown, F., eds. (1985). "Vaccines 85." Cold Spring Harbor Lab., Cold Spring Harbor, New York.

Mims, C. A., and White, D. O. (1984). "Viral Pathogenesis and Immunology." Blackwell, Oxford.

Oxford, J., and Oberg, B. (1985). "Conquest of Viral Diseases." Elsevier, New York.

Quinnan, G. V., ed. (1985). "Vaccinia Viruses as Vectors for Vaccine Antigens." Elsevier, New York.

Reviews

Chanock, R. M. (1981). Strategy for development of respiratory and gastrointestinal tract viral vaccines in the 1980s. *J. Infect. Dis.* **143,** 364.

Edelman, R. (1980). Vaccine adjuvants. *Rev. Infect. Dis.* **2,** 370.

Gregoriadis, G. (1976). The carrier potential of liposomes in biology and medicine. *N. Engl. J. Med.* **295,** 704–710 and 765–770.

Hilleman, M. R. (1985). Newer directions in vaccine development and utilization. *J. Infect. Dis.* **151,** 407.

Kennedy, R. C., and Dreesman, G. R. (1985). Immunoglobulin idiotypes: Analysis of viral antigen–antibody systems. *Prog. Med. Virol.* **31,** 168.

Lerner, R. A. (1982). Tapping the immunological repertoire to produce antibodies of predetermined specificity. *Nature (London)* **299,** 592.

Mackett, M., Smith, G. L., and Moss, B. (1984). General method for production and selection of infectious vaccinia virus recombinants expressing foreign genes. *J. Virol.* **49,** 857.

Melnick, J. L. (1977). Viral vaccines. *Prog. Med. Virol.* **23,** 158.

Sabin, A. B. (1981). Immunization: Evaluation of some currently available and prospective vaccines. *JAMA, J. Am. Med. Assoc.* **246,** 236.

White, D. O. (1984). Antiviral chemotherapy, interferons and vaccines. *Monogr. Virol.* **16,** 1–112.

World Health Organization (1983). Viral vaccines and antiviral drugs. *W. H. O. Tech. Rep. Ser.* **693.**

CHAPTER 11

Chemotherapy of Viral Diseases

Introduction 303
Strategy for Development of Antiviral Agents 305
Clinical Application 308
Interferons 310
Nucleoside Analogs 314
Other Antiviral Agents 321
Other Possible Approaches 322
Further Reading 323

INTRODUCTION

Of the hundreds of antibiotics now available to fight bacteria not one has the slightest effect on any virus. Rickettsiae, chlamydiae, and mycoplasmas are susceptible to some of them, but all these we now recognize as being much more closely related to bacteria than to viruses. Of course, there are circumstances under which it may be appropriate to prescribe antibiotics in viral infections but they are confined to two general situations: (1) to play safe pending the laboratory identification of the etiological agent in the case of serious illnesses where there is real diagnostic doubt about the possibility of a bacterial cause, e.g., meningitis or pneumonia; and (2) to prevent serious bacterial superinfection of a viral disease, e.g., the secondary development of otitis media or pneumonia in a bronchiectatic child with measles, or bronchopneumonia in an elderly cardiac or pulmonary invalid with influenza. The widespread practice of prescribing antibiotics as a knee-jerk response to infection is bad medicine. Many infections are viral and most of these will resolve without treatment.

Why have we failed so comprehensively in the hunt for antiviral agents? The explanation is not so hard to find. Being obligate intracellular parasites, viruses are absolutely dependent on the metabolic pathways of the host cell for their replication. Hence most agents that block the multiplication of viruses are lethal to the cell. In recent years, however, we have come to know much more about the biochemistry of viral replication. This has led to a more rational approach to the search for antiviral chemotherapeutic agents.

Two recent events—the discovery of acyclovir and the production of interferon by recombinant DNA technology—have triggered a new sense of urgency in the research laboratories of the pharmaceutical industry and universities alike. While antiviral chemotherapy could be said to have been born in 1962 with the introduction of idoxuridine, it came of age 21 years later, in 1983, with the licensing of acyclovir. Progress during these two decades has been agonizingly slow but one senses that we have turned the corner at last. For the first time we can truly say that at least some specific antiviral agents are about to become standard weapons in the armamentarium of clinical practice. Admittedly they are as yet few in number, the principal ones being acyclovir, vidarabine, foscarnet, ribavirin, rimantadine, and interferon (Fig. 11-1), but they represent the first specific therapy for herpes simplex, varicella-zoster, hepatitis B, influenza, RSV bronchiolitis/pneumonia, and viral papillomas (Table 11-1).

TABLE 11-1
Antiviral Agents of Proven Efficacy in Man

AGENT	VIRUS
Interferon α, β, γ	Hepatitis B, papilloma, herpesviruses, respiratory viruses, cancer?
Nucleoside analogs	
Idoxuridine (IDU)	Herpes simplex
Trifluorothymidine (TFT)	Herpes simplex
Adenine arabinoside (ara-A, Vidarabine) and Ara-AMP	Herpes simplex; varicella-zoster; hepatitis B?
Acycloguanosine (Acyclovir)	Herpes simplex; varicella-zoster
Bromovinyldeoxyuridine (BVDU)	Herpes simplex; varicella-zoster
Ribavirin (Virazole)	Influenza?; respiratory syncytial virus?
Other antiviral agents	
Trisodium phosphonoformate (PFA, Foscarnet)	Herpes simplex
Amantadine and rimantadine	Influenza A

STRATEGY FOR DEVELOPMENT OF ANTIVIRAL AGENTS

We now recognize a considerable number of steps in the viral multiplication cycle which represent potential targets for selective attack. In particular, all virus-coded enzymes are theoretically vulnerable, as are all processes (enzymatic or nonenzymatic) that are more essential to the multiplication of the virus than to the survival of the cell. Obvious examples include (1) transcription of viral mRNA (or cDNA, in the case of the retroviruses) by the viral transcriptase, (2) replication of viral DNA or RNA by the virus-coded DNA polymerase or RNA-dependent RNA polymerase, and (3) posttranslational cleavage of protein(s) by virus-coded protease(s). Less obvious at first sight, but proven points of action of currently known antiviral agents are (4) penetration/uncoating, (5) polyadenylation, methylation, or capping of viral mRNA, (6) translation of viral mRNA into protein, and (7) assembly/maturation of the virion. In Table 11-2 we have tried to summarize the several points in the viral multiplication cycle at which antiviral chemotherapeutic agents are known to act.

The recent discovery of a new class of nucleoside analogs which selectively inhibit herpesvirus DNA synthesis has opened our eyes to the fact that virus-coded enzymes with a broader (or simply different) substrate specificity than their cellular counterparts can be exploited to convert an inactive precursor drug to an active antiviral agent. Because the viral enzyme occurs only in infected cells, such drugs are nontoxic for uninfected cells. Exploitation of this principle may revolutionize antiviral chemotherapy.

A logical approach to the discovery of new antiviral chemotherapeutic agents is to isolate or synthesize substances that might be predicted to serve as a substrate or as an inhibitor of a known virus-coded enzyme. Having found an agent that, in its native state or following modification by a virus-coded enzyme, displays a degree of specificity for a viral enzyme, synthetic pathway, or any other process integral to the viral replication cycle, analogs (congeners) of this prototype are then synthesized with a view to enhancing activity and/or selectivity. Inhibition of activity of a particular viral enzyme, e.g., DNA polymerase or reverse transcriptase, may be assayed directly *in vitro*, and techniques such as cryoenzymology, X-ray crystallography, and nuclear magnetic resonance harnessed to elucidate structure–function relationships. Recombinant DNA technology should facilitate and expedite the search, because it is now feasible to clone any viral gene, sequence it and/or its protein product, purify and crystallize the protein, solve its tertiary structure by X-ray crystallography, identify its active site(s), and design competitive inhibitors.

TABLE 11-2

Mechanism of Action of Antiviral Chemotherapeutic Agents

TARGET	DRUG	VIA[a]
Penetration (fusion)		
Cleavage of envelope protein (protease)	Diarylamidines	
	Virus-specific oligopeptides	
Uncoating	Arildone	
	Amantadine, rimantadine	
Primary transcription	Specific oligonucleotides	
RNA-dependent RNA polymerase	Ribavirin ⟶	RTP
Reverse transcriptase	Diarylamidines	
Processing of mRNA		
3′-Polyadenylation	Ara-A, Ara-AMP ⟶	Ara-ATP
5′-Capping (mRNA guanylyltransferase)	Ribavirin ⟶	RTP
5′-Capping (SAH hydrolase)	DHPA	
	Ara-A, Ara-AMP ⟶	Ara-ATP
5′-Capping (AdoMet and AdoHcy)	Interferons	
Translation		
mRNA degradation by activation of endoribonuclease	Interferons $\xrightarrow{\text{dsRNA}}$ activated 2-5A synthetase ⟶	2-5A ⟶ RNase L
Phosphorylation of initiation factor, eIF-2	Interferons $\xrightarrow{\text{dsRNA}}$	activated protein kinase

Replication of viral DNA			
DNA polymerase	Ara-A, Ara-AMP	→	Ara-ATP
	Acyclovir	viral dThd–dCyd kinase →	ACVTP
	BVDU	viral dThd kinase →	BVDUTP
	FIAC	viral dThd–dCyd kinase →	FIACTP
	Phosphonoformic acid		
IMP dehydrogenase	Ribavirin	→	RMP
Adenosine deaminase	DHPA		
Thymidylate synthase	Trifluorothymidine	→	TFTMP
Incorporation into DNA	Idoxuridine	→	IDUTP
	Ara-A, Ara-AMP	→	Ara-ATP
	Acyclovir	viral dThd–dCyd kinase →	ACVTP
	BVDU	viral dThd kinase →	BVDUTP
	FIAC	viral dThd–dCyd kinase →	FIACTP
Replication of vRNA	Enviroxime		
Processing of viral protein			
Posttranslational cleavage (protease(s)	Diarylamidines		

[a] MP, Monophosphate; TP, triphosphate.

The critical test of any putative chemotherapeutic agent is, of course, inhibition of viral multiplication. Suitable indicator viruses are assayed for plaque reduction in dilutions of the agent. Toxicity of the drug for human cells may be measured crudely by CPE (in the absence of virus) or, more sensitively, by reduction in cell plating efficiency or cell-doubling time. The *therapeutic index* may be defined as the ratio of the minimum cell-toxic dose to minimum virus-inhibitory dose. In general, only those agents displaying a therapeutic index of at least 10 and preferably 100–1000 are worth pursuing further. The literature is cluttered with worthless reports of "antiviral" agents that are lethal for cells. Indeed, many agents that look promising in cell culture also fall by the wayside at the next hurdle, namely the experimental animal.

Before embarking on clinical trials in man, the pharmacology and toxicology of the drug must be thoroughly investigated in experimental animals, including, desirably, primates. Ideally, the drug should be water soluble, chemically and metabolically stable, and moderately apolar, and must be taken into cells. Pharmacokinetic studies address such questions as the mechanism and rate of absorption following various routes of administration, tissue distribution, metabolism, detoxification, and excretion of the drug. Tests for acute toxicity encompass comprehensive clinical surveillance of all the body systems, biochemical tests (e.g., for liver and kidney function), hematological examination, tests for immunosuppression, and so on. Longer term investigations screen for chronic toxicity, allergenicity, mutagenicity, carcinogenicity, and teratogenicity.

CLINICAL APPLICATION

Route of Administration

The route of administration of an antiviral agent is a prime consideration in assessing its general acceptability. The oral route is naturally much the most convenient for the patient. Nasal drops or sprays may be acceptable for upper respiratory infections but can be irritating, while continuous delivery of aerosols through a face mask or oxygen tent is generally appropriate only for very sick hospitalized patients. Topical preparations (creams, ointments, etc.) are also satisfactory for localized superficial infections of skin or eye, but not for generalized skin or genital infections; penetration of drugs through the skin can be enhanced by mixing with substances such as polyethylene glycol. Parenteral administration, though unpopular with patients, may be required for serious systemic infections; intravenous infusion of course necessitates hospitalization. In the future we may see antiviral drugs conjugated to antiviral antibody, or incorporated into liposomes (see Chapter 10) coated with such antibody, to direct them to virus-infected cells.

Drug-Resistant Mutants

It is already clear that mutants resistant to many of the available antiviral drugs readily emerge *in vitro* and *in vivo*. These inevitable occurrences will have to be faced in much the same way as we have learned to live with the problem of antibiotic-resistant bacteria. Antiviral agents will need to be used only when absolutely necessary, but then applied in adequate dosage. Certain life-saving drugs may need to be retained for designated diseases only. Combined therapy, using agents with distinct modes of action, would be expected to reduce the probability of selection of resistant mutants. It might also minimize potential toxic side effects and, in certain instances, display synergism, as has been claimed for interferon + acyclovir, ara-A + acyclovir, and ribavirin + amantadine, for example. Mutants would not be expected to arise nearly as readily against drugs with multiple points of action; interferons are an obvious case in point.

Clinical Priorities

Diseases with a large number of different etiological agents are prime targets for antiviral chemotherapy, as are those against which no satisfactory vaccine is available; the common cold is an admirable example on both counts. Other upper respiratory infections, gastroenteritis, hepatitis, and infectious mononucleosis, must also be high on the list of priorities. Herpes zoster and recurrent episodes of herpes simplex are singularly troublesome, and effective chemotherapy would be a boon, although the latent infection itself is unlikely to be eradicated. Chronic and slow virus infections may be particularly amenable to antiviral chemotherapy, as might some other long drawn out diseases of currently unknown etiology such as certain cancers, autoimmune diseases, and degenerative conditions of the brain, should any of these, in the fullness of time, turn out to be of viral causation. Finally, we must not forget the rarer, lethal viral diseases, such as rabies, poliomyelitis, encephalitis, and the hemorrhagic fevers, for which successful chemotherapy would be life-saving.

Acute infections may need to be treated promptly, at the time of appearance of the first prodromal symptom, e.g., the paresthesia of recurrent herpes simplex, or the nasal congestion of the common cold or the exanthemata. Relatively broad-spectrum chemotherapeutic agents will be needed if treatment is to be commenced before a definitive diagnosis has been established. Otherwise, rapid diagnostic techniques such as enzyme immunoassay (EIA), immunofluorescence, or IgM serology (see Chapter 12) will be essential to select the appropriate drug for timely treatment.

Chemoprophylaxis may also have a role, not only in the prevention of complications, e.g., orchitis or meningitis in mumps, but also in protection against the spread of diseases such as hepatitis, mononucleosis, polio, influenza, measles, or rubella to unimmunized family contacts.

INTERFERONS

Ever since the discovery of interferons more than a quarter of a century ago it has been evident that they are, in theory at least, the ideal antiviral antibiotics. They are naturally occurring, relatively nontoxic, and display a broad spectrum of activity against essentially all viruses (for full description, see Chapter 6). Yet, clinical trials in man had been generally disappointing until techniques for the purification of human interferon were developed in the late 1970s and opened the way for trials which demonstrated that interferon at much higher dosage could be effective against at least some viral diseases.

Commercial Production of Interferons

Lymphocytes, NK cells, and macrophages, following stimulation by viruses, produce larger quantitites of interferon than any other cells. For many years, most of the world's supply of human interferon (Hu-IFN) came from one research laboratory in Finland, which produced a relatively crude, low-titer preparation of *leukocyte interferon* (Le-IFN), mainly IFN-α, by challenging "buffy coat" cells from human blood with Sendai virus. Greater yields of Hu-IFN-α have since been obtained by Wellcome Laboratories from continuous lines of *lymphoblastoid* cells, such as the B cell line, Namalwa, which is grown in suspension on an industrial scale. *Fibroblast interferon* (F-IFN), which is mainly IFN-β, is traditionally induced by the synthetic polynucleotide poly(I)·poly(C) in human fetal fibroblasts growing on the surface of tiny beads (microcarriers) in suspension. *Gamma interferon* (IFN-γ), previously known as interferon type II or immune interferon, has been produced, with considerable difficulty and low yield, by exposure of human lymphocytes to mitogens.

Recombinant Interferons. In 1980 a major breakthrough was achieved almost simultaneously by two new biotechnology companies, Biogen in

Switzerland, and Genentech in the United States, when the gene for a human IFN-α was cloned and expressed in *Escherichia coli.* Since then, the genes for virtually all known subtypes of human IFNs-α, -β, and -γ have been cloned in bacteria, yeasts, and/or mammalian cells. The yields of interferon so obtained are vastly greater than those from leukocyte, lymphoblastoid, or fibroblast cultures. Overall, it has been estimated that the cost of production of interferon should, theoretically at least, be reduced by up to 100-fold as a result of recombinant DNA technology.

Most of the cloned Hu-IFNs have now been purified and sequenced. Currently, there appear to be at least a dozen subtypes of Hu-IFN-α, one β, and one γ. The properties of these interferons were set out in Tables 6-5 and 6-6. The presumed mechanism(s) of their antiviral action was discussed in Chapter 6 and summarized in Fig. 6-5.

Assay of the antiviral and other biological properties of individual cloned Hu-IFN subtypes has begun. It is already clear that quite minor differences in primary amino acid sequence can markedly alter antiviral and antiproliferative spectra, as well as pharmacological properties, including stability. Hybrid interferons have been produced by cloning recombinant DNA molecules created by ligating segments derived from different IFN cDNAs that have been cleaved at common restriction endonuclease sites. Assay of such hybrid IFNs might shed light on structure–activity relationships and identify active sites for desirable antiviral, antiproliferative, and immunoregulatory functions. Thus, not only hybrid IFNs, but abbreviated IFNs might be manufactured by recDNA technology or by chemical synthesis.

Pharmacology

Hu-IFN-α is commonly administered intramuscularly (im), allowing serum levels of 50–500 units/ml to be maintained following the usual daily injection of 3×10^6–8×10^7 units. On the other hand F-IFN (Hu-IFN-β) has generally been given intravenously because most of it is not detectable in the circulation following im delivery; there is still some debate about whether this reflects destruction of IFN-β in muscle or binding by muscle. But the half-life of both IFN-α and -β is quite short following iv inoculation. Catabolism occurs mainly in the kidney and liver. Relatively little (approximately 3%) of any IFN type crosses the blood–brain barrier into the CSF, but it has been administered intrathecally. It is not effective by mouth. Following intranasal delivery it is rapidly swept away by ciliary movement of mucus. Recent successes with the treatment of cancer by local injection into the tumor itself suggest that this and other forms of topical delivery such as slow release from liposomes, or conjugation to an antibody or other targeting molecule, should be explored further.

Toxic side effects are regularly observed, and may be marked with doses in excess of 10^7 units per day, even when highly purified cloned IFN subtypes are employed. Fever regularly occurs at high dosage but lasts only a day or so. Severe fatigue is the most debilitating symptom and may be accompanied by malaise, anorexia, myalgia, headache, nausea, vomiting, weight loss, erythema and tenderness at the injection site, partial alopecia (reversible), dry mouth, reversible peripheral sensory neuropathy, or CNS signs. Various indicators of myelosuppression (granulocytopenia, thrombocytopenia, and leukopenia) and abnormal liver function tests, both reversible on cessation of therapy, are regularly observed if high-dose IFN administration is prolonged.

It used to be assumed that interferons were nonimmunogenic. Recently, however, anti-IFN antibodies have been detected in a small number of patients following the administration of interferon. Now it is apparent that this is not an uncommon development in patients receiving Hu-IFN, and, furthermore, that some patients, especially those suffering from autoimmune disease or cancer, can, less commonly, develop low levels of antibody to their own endogenous interferon. The short- or long-term clinical significance of such antibodies has yet to be evaluated.

Clinical Efficacy

Most of the clinical experience with interferon has been with HuLe-IFN (Hu-IFN-α), and to a lesser extent with HuF-IFN (Hu-IFN-β), of varying degrees of purity. Many of the trials have been uncontrolled, and some reports are little more than anecdotal. Still, well-planned, placebo-controlled, double-blind trials are now being conducted with purified cloned, as well as mixed interferons. Some modest successes are being obtained, the dosage generally required being in excess of 10^6 units per day (sometimes over 10^8 units per day).

Recent studies using *E. coli*-cloned Hu-IFN-α2 have confirmed earlier reports that colds can be prevented by intranasal installation of IFN in high dosage (total of 10^8 units) at frequent intervals for several days commencing prior to viral challenge. However, chemoprophylaxis is impracticable; what is required is a "cure" for the common cold. Interferon may have a role to play in the treatment of certain other viral diseases, notably those caused by the herpesviruses, hepatitis B virus, and papillomaviruses. Merigan and colleagues have reported that Hu-IFN limits the clinical progression of herpes simplex and herpes zoster in cancer patients, and there is also evidence that it can reduce the inci-

TABLE 11-3
Antiviral Activity of Interferons in Man

VIRUS	DISEASE	ROUTE[a]
Rhinovirus	Common cold	in
Herpes simplex	Keratitis	Ocular
	Herpes facialis, genitalis	im, topical
Varicella-zoster	Herpes zoster	im
	Varicella	
Cytomegalovirus	Activation in immunosuppressed	im
Hepatitis B	Chronic hepatitis or carrier	im,iv
Papilloma	Laryngeal papilloma	Local
	Warts (common or genital)	
Adenovirus	Keratoconjunctivitis?	Ocular

[a] in, Intranasal; im, intramuscular; iv, intravenous; local, injected into lesion.

dence of reactivation of cytomegalovirus in immunosuppressed patients (see Chapter 16). Prolonged treatment of chronic hepatitis cases with high doses of Hu-IFN-α or -β in alternation with adenine arabinoside monophosphate (see below) produced temporary remissions, but this apparently synergistic combination of drugs gave rise to an unacceptably high incidence of neurological side effects, including peripheral neuritis (Chapter 13). Juvenile laryngeal papilloma, a severe recurrent viral condition calling for repeated surgical removal, can be arrested by local injection of interferon, but the tumours recur when therapy is withdrawn. Hu-IFN-α and -β have also been used successfully to treat other types of human viral papilloma, including skin warts and genital warts (Chapter 14).

Table 11-3 lists those viral diseases against which human interferon appears to display at least some activity. Clinical trials against numerous other acute and chronic conditions of known or postulated viral etiology are underway.

In the late 1970s reports of partial regression of several types of cancer following interferon treatment precipitated a flurry of excitement, not only in cancer circles, but even on Wall Street where the stocks of the newly emerging recombinant DNA-based biotechnology companies skyrocketed overnight. However, subsequent more carefully controlled trials have given much less encouraging results.

It is too early to be dogmatic about the value of interferons in the treatment of viral infections or cancer. The euphoria of the late 1970s has been supplanted by a more cautious, balanced belief that interferons, while no panacea, may nevertheless come to occupy a limited place in both antiviral and anticancer chemotherapy. Presently, it appears that interferons have a demonstrable effect on a number of viral diseases, notably those due to papillomaviruses, herpesviruses, hepatitis B, and perhaps rhinoviruses, and that they may have a future against additional diseases if suitable protocols can be developed for timely delivery of the appropriate interferon by the appropriate route. On the other hand, results so far are not sufficiently impressive to conclude that interferon should be preferred over newer antiviral agents such as the second-generation nucleoside analogs to be discussed below.

The situation regarding cancer is more complex. As of now, it does seem that a small minority of patients with certain types of cancer may respond favorably to vigorous interferon therapy with some sort of temporary remission or partial regression. In several instances this is no more than can be achieved with other modern chemotherapeutic agents; in others, it may represent the best or only hope available at this juncture.

While some regard the whole interferon saga as a kind of bad dream from which we shall soon awaken to find that nothing is really there, others believe that we have barely scraped the surface so far. Much more work is now needed to test the effects of very high doses of pure preparations of each Hu-IFN subtype. Already it has become apparent that particular subtypes are considerably more potent than others, and/or "specialize" in different antiviral, antiproliferative, or immunomodulatory properties that could be exploited for the specific chemotherapy of particular viral, neoplastic, or indeed additional types of disease. Nor is it inconceivable that, by splicing DNAs of different IFN subtypes or synthesizing novel DNAs, more potent, more stable, less toxic IFNs with particular properties required for specific purposes may be tailor-made for the treatment of individual diseases. Reconstituted mixtures of particular cloned IFNs should also be more widely explored, for there is evidence to suggest that synergistic effects may be expected.

NUCLEOSIDE ANALOGS

Most of the successful antiviral agents described to date are nucleoside analogs (Table 11-4), most of which in turn are restricted in their antiviral activity to the herpesviruses, especially herpes simplex virus (HSV) and varicella-zoster virus (VZV). The prototype, *5-iodo-2'-deoxyuridine* (idoxuridine, IDU or "Stoxil", see Fig. 11-1) was shown over 20 years ago to be somewhat effective against herpes simplex viral infections of the eye when applied topically, before edema or scarring set in.

TABLE 11-4
Second Generation Nucleoside Analogs

ABBREVIATION	FULL NAME
Acyclovir	Acycloguanosine
BVDU	(*E*)-5-(2-Bromovinyl)-2′-deoxyuridine
FIAC	2′-Fluoro-5-iodoaracytosine
AIDDU	5′-Amino-5-iodo-2′,5′-dideoxyuridine
Ara-T	1-β-D-Arabinofuranosylthymine
DHPA	(*S*)-9-(2,3-Dihydroxypropyl)adenine
Ribavirin	1-β-D-Ribofuranosyl-1,2,4-triazole-3-carboxamide

Another halogenated nucleoside, *trifluorothymidine*, otherwise known as trifluridine, TFT, or F_3T, superseded IDU for the treatment of HSV keratitis in the 1970s. Both will be discussed with the herpesviruses (Chapter 16).

Adenine Arabinoside (Ara-A, Vidarabine) and Ara-AMP

The first agent to be licensed for the treatment of systemic herpesvirus infections in man was adenine arabinoside, familiarly known as ara-A or Vidarabine (see Fig. 11-1). Like IDU, ara-A, following phosphorylation by cellular enzymes to ara-ATP, blocks both viral and cellular DNA synthesis (Fig. 11-2). Hence, in common with many cytotoxic anticancer drugs, ara-A produces severe side effects on systemic administration and is used only for heroic treatment of potentially lethal herpesviral infections, notably HSV encephalitis, neonatal herpes, and disseminated herpes zoster, as well as hepatitis B.

Ara-A is relatively insoluble, hence must be administered in a large volume of fluid by slow intravenous (iv) infusion. Particular caution must therefore be exercised in patients with kidney or liver dysfunction, and cerebral edema is a distinct risk. Manifestations of neurotoxicity are frequently reported, especially when administered in conjunction with interferon. Furthermore, as might be expected with a drug that inhibits cellular DNA synthesis, toxic side effects referable to the gastrointestinal tract and bone marrow are not uncommon, particularly nausea, vomiting, diarrhea, thrombocytopenia, and leukopenia.

Adenine arabinoside monophosphate (ara-AMP) is much more soluble than ara-A, hence does not need to be delivered in large volumes of fluid, but can be administered intramuscularly or intravenously. Recent clinical trials indicate ara-AMP to be effective parenterally against HSV-

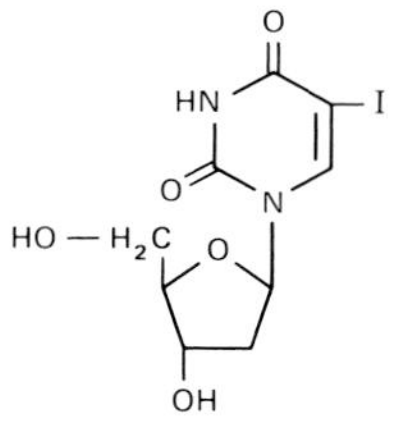

Idoxuridine (IDU)

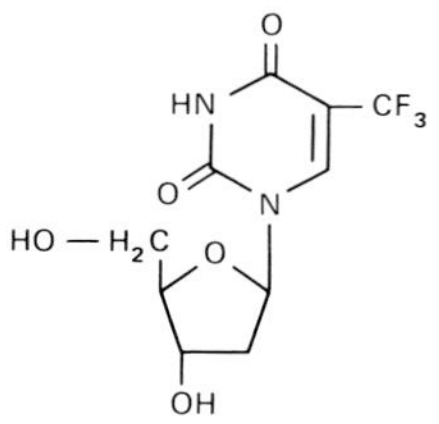

Trifluorothymidine (TFT)

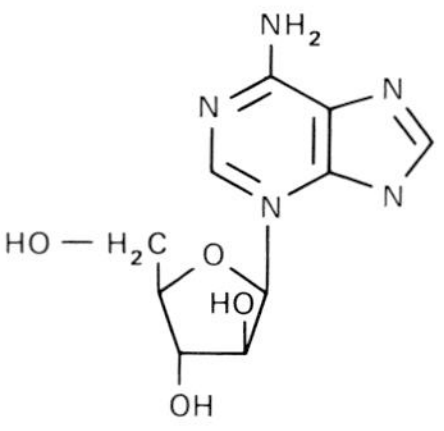

Adenine arabinoside (Ara-A)

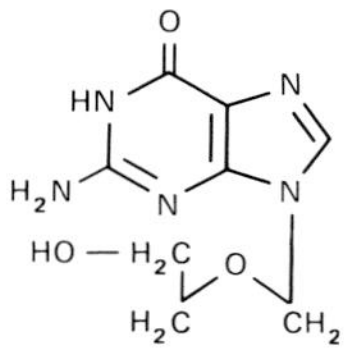

Acycloguanosine (Acyclovir)

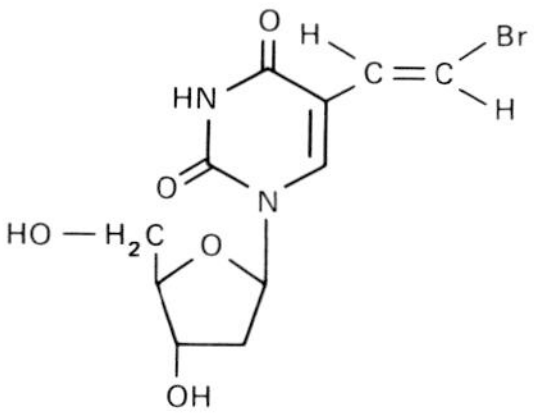

Bromovinyldeoxyuridine (BVDU)

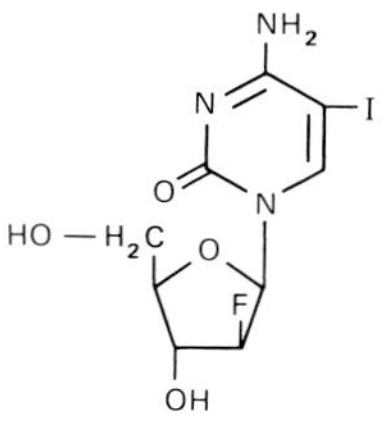

Fluoroiodoaracytosine (FIAC)

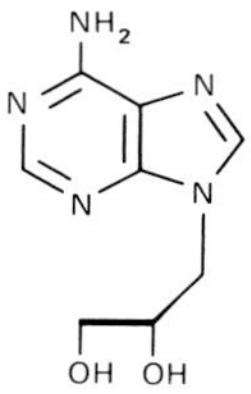

(*S*)-9-(2,3-Dihydroxypropyl)-adenine (DHPA)

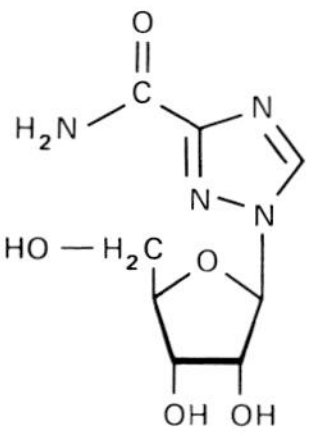

Ribavirin

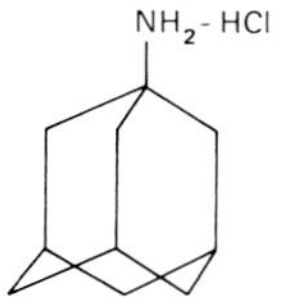

1-Adamantanamine hydrochloride (Amantadine)

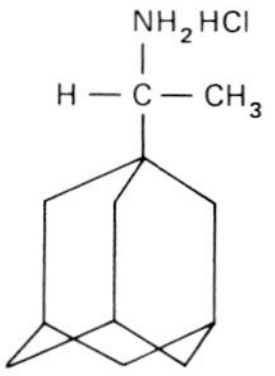

Rimantadine

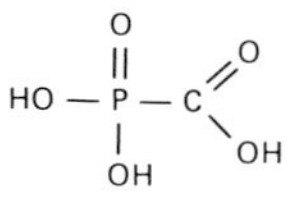

Phosphonoformic acid (PFA)

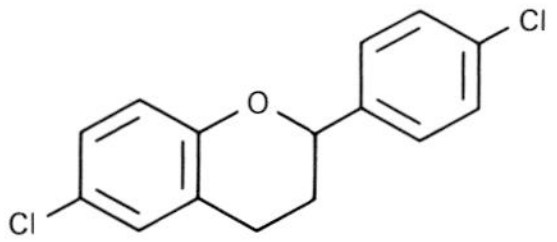

4′,6-Dichloroflavan

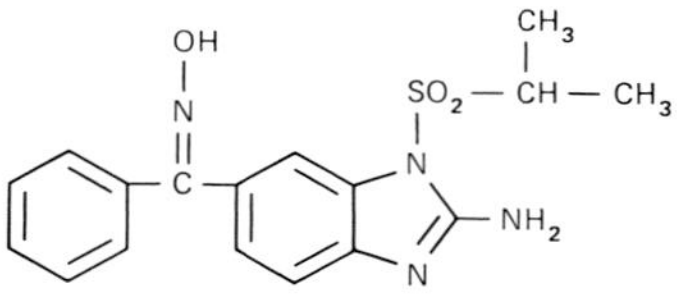

2-Amino-1-(isopropylsulfonyl)-6-benzimidazole phenyl ketone oxime (Enviroxime)

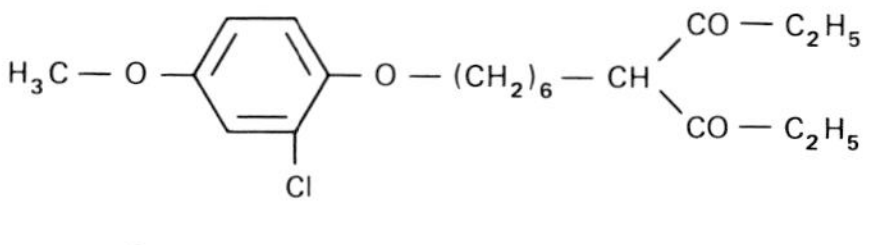

4-[6-(2-Chloro-4-methoxy) phenoxy]-hexyl-3, 5-heptanedione (Arildone)

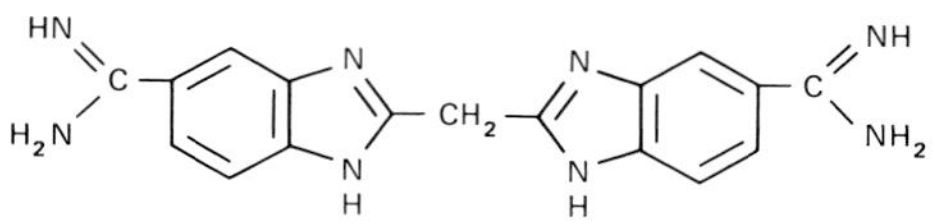

Bis(5-amidino-2-benzimidazolyl)methane

FIG. 11-1. *Antiviral chemotherapeutic agents.*

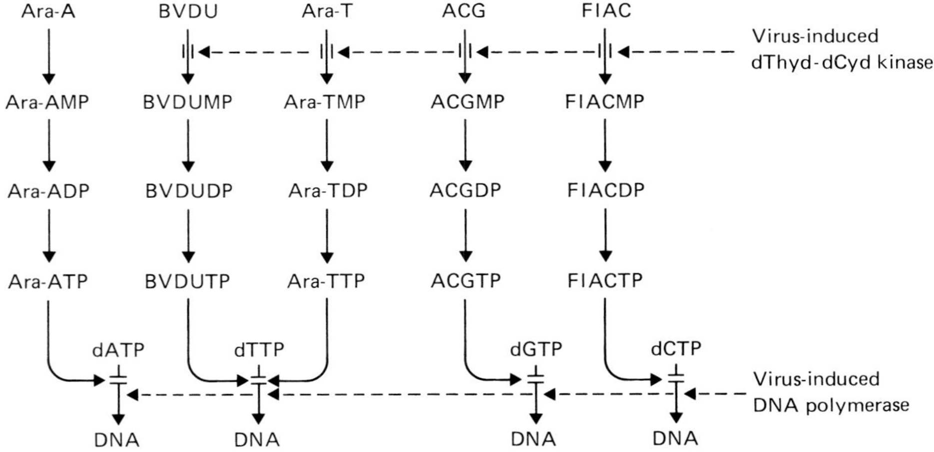

FIG. 11-2. *Mechanism of inhibition of herpesvirus multiplication by nucleoside analogs. [Reproduced in modified form from E. de Clercq,* Acta Microbiol. Acad. Sci. Hung. **28,** *289 (1981).]*

encephalitis, cutaneous herpes simplex and zoster in the immunosuppressed, and hepatitis B (see Chapters 16 and 13).

Acycloguanosine (Acyclovir) and Homologs

A major breakthrough in antiviral chemotherapy occurred in 1977 when Wellcome Laboratories developed an agent that depends upon a viral enzyme to convert it to its active form. Acycloguanosine, now commonly known as acyclovir, is a guanine derivative with an acyclic side chain, the full chemical name being 9-(2-hydroxyethoxymethyl) guanine (Fig. 11-1). Its unique advantage over earlier nucleoside derivatives is that it requires the herpesvirus-specified enzyme, deoxythymidine–deoxycytidine (dThd–dCyd) kinase to phosphorylate it intracellularly to acycloguanosine monophosphate; a cellular GMP kinase then completes the phosphorylation to the active agent, acycloguanosine triphosphate (Fig. 11-2). Acyloguanosine triphosphate also happens to inhibit the herpesvirus-specified DNA polymerase at least 10 times more effectively than it does cellular DNA polymerase. It acts as both inhibitor and substrate of the viral enzyme, competing with GTP and being incorporated into DNA, leading to chain termination because acyclovir lacks the 3′-hydroxyl group required for chain elongation. Since activation of the prodrug needs the viral dThd–dCyd kinase, acyclovir is essentially nontoxic to uninfected cells but is powerfully inhibitory to viral DNA synthesis in infected cells.

Herpes simplex viruses types 1 and 2 (HSV-1 and -2) are both very susceptible to acyclovir in cell culture; varicella-zoster virus (VZV) is susceptible at somewhat higher concentrations of the drug. The other human herpesviruses, EBV and cytomegalovirus (CMV), neither of which is known to possess a gene coding for thymidine kinase, are nevertheless susceptible to acyclovir, albeit only at very much greater doses. In the case of EB virus this has been attributed to a particularly high sensitivity of the viral DNA polymerase to the small amount of acyclo-GTP produced via the action of cellular enzymes. The relative sensitivity of different viruses seems to depend on a rather complex interplay of at least three variables: the efficiency of the virus-coded dThd–dCyd kinase (if any) in converting acyclovir to acyclo-GMP, the efficiency of cellular kinases in converting this intermediate to acyclo-GTP, and the susceptibility of the viral DNA polymerase to acyclo-GTP.

In man, acyclovir displays activity against herpes simplex and varicella-zoster (see Chapter 16). Applied topically to the eye, acyclovir is very effective against herpes simplex keratitis. Intravenous acyclovir protects transplant recipients against reactivation of their latent herpes simplex, and significantly inhibits the progress if delayed until lesions have already appeared. Acyclovir cream has been reported to shorten the healing time of recurrent herpes labialis ("fever blisters"). Furthermore, acyclovir is the first drug to offer some hope to genital herpes sufferers. Whether delivered topically, orally, or intravenously, acyclovir reduces the duration of virus shedding, local pain/itching, and lesion healing time, in primary herpes genitalis, though less persuasively in the (more short-lived) recurrent attacks. Unfortunately, there appears to be no significant reduction in the establishment of latency or in the frequency of subsequent recurrences. In the absence of any way of curing the latent state itself, the most attractive solution for herpes genitalis sufferers would undoubtedly be prevention of recurrences by chemoprophylaxis with acyclovir orally. Preliminary trials along these lines have been encouraging but careful longer term studies will be necessary before prolonged chemoprophylaxis could be recommended as a routine.

Herpes zoster also responds well to acyclovir, although there is no significant effect on that most distressing sequel, postherpetic neuralgia. Acyclovir has been disappointing against other herpesviruses. Most studies have revealed no effect against CMV (congenital CMID of infants, or CMV pneumonia following bone marrow transplantation), nor against EB virus. Hepatitis B also appears to be refractory.

Acyclovir (Zovirax) may be delivered by slow intravenous infusion or orally, or topically as an aqueous cream containing 5% acyclovir in propylene glycol. As anticipated from *in vitro* studies, the drug seems to be nontoxic. Rapid transient increases in blood urea and creatinine levels tend to occur. Acyclovir levels must be carefully monitored in patients with dehydration or renal impairment, as the drug, which is excreted unchanged through the kidneys, is rather insoluble, and crystalluria may occur. Elevated liver enzymes have been reported in immunocompromised subjects.

Acyclovir-resistant mutants of HSV readily develop in cell culture. The mutations can be located in either the gene coding for the viral thymidine kinase (TK) or that for DNA polymerase. There appear to be two kinds of TK^- mutant: those failing to induce appreciable levels of TK, and those in which the enzyme is produced in substantial amounts but has an altered substrate specificity such that it can no longer satisfactorily phosphorylate acyclovir. Acyclovir-resistant mutants of the TK^- class have been isolated from man following acyclovir therapy and demonstrated to be resistant also to BVDU (see below). It has been suggested that such TK^- mutants will not pose a serious threat because, in mice at least, they tend to display reduced pathogenicity and reduced capacity to establish latent infection in the trigeminal ganglion. Nevertheless, cross-resistance to other second-generation nucleoside analogs, arising in a single step, is likely to pose problems in the future.

Recently, derivatives of acyclovir, such as 9-[(2-hydroxy-1-(hydroxymethyl)ethoxy)methyl]guanine, otherwise known as 2′-nor-2′-deoxyguanosine, or 2′-NDG, or BIOLF-62, have been shown to be even more efficient substrates for HSV-1 thymidine kinase and GMP kinase than is acyclovir itself, and display greater activity against herpes simplex virus in cultured cells and mice. Deoxyacyclovir is better absorbed orally than acyclovir, while DHPG is more potent than acyclovir and is active against CMV and EBV in cultured cells. Clinical investigations have begun on a number of such analogs.

Bromovinyldeoxyuridine (BVDU) and Homologs

Following the example set by acycloguanosine, a wide range of other nucleoside analogs (Table 11-4) with similar modes of action (Fig. 11-2) have now been synthesized. Of these, one of the most promising is (*E*)-5-(2-bromovinyl)-2′-deoxyuridine (BVDU) (see Fig. 11-1), developed in Belgium. BVDU is particularly effective against VZV and HSV-1 in cultured cells but is, unfortunately, rapidly degraded *in vivo*. Yet more striking selectivity for VZV has been achieved with a BVDU derivative, [1-β-D-arabinofuranosyl-(*E*)-5-(2-bromovinyl)uracil], known as BVaraU, which inhibits the multiplication of VZV in cell culture at a minimum concentration of 1 ng/ml.

Recently, de Clercq has reported that 5-substituted 2′-deoxycytidines are more selective in their anti-HSV activity than the corresponding 5-substituted 2′-deoxyuridines, presumably because of the requirement that, following the phosphorylation of the drug by the HSV-coded enzyme, dThd–dCyd kinase, the dCMP analog must then be deaminated to the corresponding dUMP analog by a second (virus-coded?) enzyme, dCMP deaminase.

Fluoroiodoaracytosine (FIAC)

A new agent, 2′-fluoro-5-iodo-1-β-D-arabinofuranosylcytosine (FIAC) (Fig. 11-1), which also requires phosphorylation by the virus-induced enzyme, dThd–dCyd kinase, is very potent and displays a high chemotherapeutic index against HSV-1 and -2, VZV, and some strains of CMV in cultured cells. Clinical trials indicate efficacy of intravenous FIAC against herpes simplex and zoster. A related compound, 2′-fluoro-5-methyl-ara-U, known as FMAU, is more effective than FIAC against HSV-1 in mice.

S-Adenosylhomocysteine Analogs

Neplanocin A and (*S*)-9-(2,3-dihydroxypropyl)adenine (DHPA) (Fig. 11-1) are two of several compounds that have been shown to inhibit the cellular enzyme, SAH hydrolase, leading to the accumulation of *S*-adenosylhomocysteine (SAH), which inhibits methyltransfer from the methyl donor SAM (*S*-adenosylmethionine) to the 5′ cap of messenger RNA. Such agents would be expected to display a wide spectrum of activity against RNA and DNA viruses. It is not yet clear whether methylation of viral mRNA is inhibited preferentially, but DHPA has been reported to have some effect against rabies and rotavirus in animal models.

Ribavirin (Virazole)

A rather different nucleoside analog, known as ribavirin or Virazole, was first synthesized in 1972 but, despite extensive investigation, has still not been licensed for use in many countries, partly because of unpersuasive evidence of antiviral activity in man and partly because of indications of toxicity. At first sight, the drug would appear to have considerable potential, as it inhibits the growth of a wide spectrum of RNA and DNA viruses in cultured cells and experimental animals, by what appears to be a multipoint mechanism of action, quite unrelated to that of the nucleoside analogs we have been discussing so far. However, this early promise has not been matched by a comparable degree of efficacy in humans. Furthermore, following oral administration at the usual dosage of about 1 g per day, a substantial minority of recipients develop a reversible anemia with increased reticulocyte numbers and elevated serum bilirubin levels, while immunosuppressive and teratogenic effects have been demonstrated in animals. Nevertheless, recent reports of beneficial effects when ribavirin is delivered intravenously to patients with lassa fever, or as a continuous aerosol to patients with influenza or respiratory syncytial viral infections (see Chapters 21 and 22) merit further investigation.

OTHER ANTIVIRAL AGENTS

Trisodium Phosphonoformate (PFA, Foscarnet)

Trisodium phosphonoformate, known also as phosphonoformic acid (PFA) or Foscarnet (Fig. 11-1), inhibits the DNA polymerase of certain herpesviruses at 10–100 times lower concentration than that required to inhibit cellular DNA polymerase α. Clinical trials in man so far indicate that Foscarnet cream accelerates healing of recurrent facial or genital herpes (see Chapter 16). Some 30% of systemically available PFA is deposited in bone and cartilage. Although toxicity for man has not yet been reported, this will need to be carefully checked, in the long term as well as the short, especially before PFA is licensed for oral or systemic administration.

Amantadine and Rimantadine

It is now some 20 years since a simple three-ringed symmetrical amine known as 1-aminoadamantane hydrochloride, or more commonly as amantadine, or Symmetrel (Fig. 11-1) was synthesized and shown to inhibit the replication of influenza A viruses. The mode of action of the drug is still not completely understood, but it has been postulated that amantadine blocks uncoating by increasing the pH in the lysosome and so preventing the pH 5 mediated fusion with the viral envelope. Primary transcription of RNA is inhibited.

Therapeutically, amantadine is only marginally effective, but has been reported to reduce the severity of symptoms. Administered prophylactically, however, it can reduce the incidence of clinical influenza quite significantly (by up to 90% in some trials). Since, in practice, this demands the ingestion of 200 mg daily for 1–2 months from the commencement of an influenza epidemic in the community, it is important to consider the possible complications of such a prolonged annual chemoprophylactic regime.

Side effects commonly occur (7–33% of patients in various studies).

They relate mainly to the central nervous system—loss of concentration, insomnia, nervousness, lightheadedness, drowsiness, anxiety, confusion—but are generally reversible and mild unless the recommended dose is exceeded, or the patient has a history of mental illness or renal disease.

Rimantadine, or α-methyl-1-adamantanemethylamine hydrochloride (Fig. 11-1), found to be more active and less toxic in cultured cells, organ cultures, and experimental animals, has been extensively used in the USSR for several years. Since side effects are less frequent and less severe than with amantadine and efficacy is comparable, rimantadine appears to be the drug of choice. Like amantadine, rimantadine is given orally, but either drug can be delivered to hospitalized patients by aerosol spray, perhaps in conjunction with ribavirin.

OTHER POSSIBLE APPROACHES

Inhibitors of Reverse Transcriptase

The advent of AIDS has quickened interest in the search for inhibitors of reverse transcriptase. Numerous agents such as suramin and others displaying *in vitro* activity against retroviruses are currently undergoing clinical trial.

Virus-Specific Oligonucleotides

A short synthetic sequence of single-stranded DNA complementary to the 13 nucleotides of the 3′- and 5′-reiterated terminal sequences of Rous sarcoma virus RNA inhibits both virus production and cell transformation. On the other hand, an oligodeoxyribonucleotide complementary to the 12 nucleotides at the 3′ terminus of influenza viral (−)RNA failed to block transcription selectively. It remains to be seen whether this novel approach to antiviral chemotherapy has any general validity.

Virus-Specific Oligopeptides and Inhibitors of Proteases

Penetration of ortho- and paramyxoviruses by fusion requires prior cleavage of a viral envelope glycoprotein by a cellular protease. Synthetic oligopeptides with an amino acid sequence resembling the N-terminal sequence of the F_1 polypeptide which is exposed by cleavage of the paramyxovirus F (fusion) protein, or of the influenza HA_2 protein, specifically inhibit multiplication of the respective viruses in cell culture. Antiviral activity *in vivo* has yet to be demonstrated however.

Since cleavage of viral proteins by proteases, some of which are virus coded, is required at several stages in the multiplication cycle, including activation of some viral enzymes, posttranslational cleavage of the "polyprotein" product of polycistronic mRNA, and assembly of the virion, it may be possible to find or devise agents mimicking the cleavage site and thus specifically inhibiting the virus-coded protease.

Anti-Rhinovirus Compounds

Chemotherapy is the only practicable approach to the management of the common cold. Two compounds, enviroxime and 4′-6-dichloroflavan (Fig. 11-1), inhibit the replication of rhinoviruses in cultured cells but have been disappointing in clinical trials. A Japanese group has now described a naturally occurring compound, 4′-ethoxy-2′-hydroxy-4,6′-dimethoxychaleone (Ro 09-0410), which binds to a particular site on the surface of the rhinovirus particle *in vitro*. The recent X-ray crystallographic resolution of the three-dimensional structure of the rhinovirus capsid proteins may facilitate a systematic search for other such compounds.

FURTHER READING

BOOKS

Baron, S., Dianzani, F., and Stanton, G. J. (1982). "The Interferon System: Review to 1982," Parts 1 and 2, Tex. Rep. Biol. Med., Vol. 41. Univ. of Texas Medical Branch, Galveston.

Billiau, A., de Clercq, E., and Schellekens, H. (1985). "Antiviral Research," Suppl. 1, pp. 1–310. Elsevier, Amsterdam.

Came, P. E., and Caliguiri, L. A., eds. (1982). "Chemotherapy of Viral Infections," Hand. Exp. Pharmacol., Vol. 61. Springer-Verlag, Berlin and New York.

Collier, L. H., and Oxford, J., eds. (1981). "Developments in Antiviral Therapy." Academic Press, London.

de Clercq, E., and Walker, R. T., eds. (1984). "Targets for the Design of Antiviral Agents." Plenum, New York.

de Maeyer, E., and Schellekens, H., eds. (1983). "The Biology of the Interferon System 1983." Elsevier, Amsterdam.

Finter, N. B., ed. (1984–1985). "Interferon" Vols. 1–4. Elsevier, Amsterdam and New York.

Galasso, G. J., Merigan, T. C., and Buchanan, R. A., eds. (1984). "Antiviral Agents and Viral Diseases of Man," 2nd ed. Raven Press, New York.

Gauri, K. K., ed. (1981). "Antiviral Chemotherapy: Design of Inhibitors of Viral Functions." Academic Press, New York.

Gresser, I., ed. (1979–1984). "Interferon," Vols. 1–5. Academic Press, New York.

Merigan, T. C., and Friedman, R. M., eds. (1983). "Interferons," UCLA Symp. Mol. Cell. Biol., Vol. 25. Academic Press, New York.

Oxford, J., and Oberg, B. (1985). "Conquest of Viral Diseases." Elsevier, Amsterdam.

Shugar, D., ed. (1984–). "Viral Chemotherapy." Vols. 1–. Permagon, Oxford.

Stuart-Harris, C. H., and Oxford, J., eds. (1983). "Problems of Antiviral Therapy." Academic Press, London.

White, D. O. (1984). "Antiviral Chemotherapy, Interferons and Vaccines," Monog. Virol., Vol. 16. Karger, Basel.

REVIEWS

de Clercq, E. (1982). Specific targets for antiviral drugs. *Biochem. J.* **205,** 1.
Galasso, G. J. (1981). An assessment of antiviral drugs for the management of infectious diseases in humans. *Antiviral Res.* **1,** 73.
Pestka, S. (1983). The human interferons—from protein purification and sequence to cloning and expression in bacteria: Before, between, and beyond. *Arch. Biochem. Biophys.* **221,** 1.
World Health Organization (1982). Interferon therapy. *W. H. O. Tech. Rep. Ser.* **676.**
World Health Organization (1983). Viral vaccines and antiviral drugs. *W. H. O. Tech. Rep. Ser.* **693.**

CHAPTER 12

Laboratory Diagnosis of Viral Diseases

Introduction 326
Direct Identification of Virus, Antigen, or Nucleic Acid 328
  Electron Microscopy 328
  Radioimmunoassay 330
  Time-Resolved Fluoroimmunoassay 332
  Enzyme Immunoassay 332
  Immunodiffusion 334
  Reverse Passive Hemagglutination 336
  Immunofluorescence 336
  Immunoperoxidase Staining 338
  Nucleic Acid Hybridization: cDNA Probes 339
Virus Isolation 341
  Collection and Preparation of Specimens 341
  Inoculation and Maintenance of Cultures 346
  Recognition of Viral Growth 347
  Organ Cultures, Eggs, and Animals 348
  Identification of Viral Isolates by Serology 349
  Oligonucleotide Fingerprinting 354
  Interpretation 355
Measurement of Serum Antibodies 356
  IgM Serology 358
Further Reading 360

INTRODUCTION

It is fair to ask "why bother to establish a definitive laboratory diagnosis of a viral infection when almost no specific chemotherapy is available?" In any case, you may add, most viral diseases are so trivial (e.g., the common cold) or so obvious clinically (e.g., herpes simplex) that time and money should not be wasted on laboratory investigations. There is some basis for both of these assertions. The majority of viral infections can be handled without recourse to the laboratory. However, there are several important types of situation where a laboratory diagnosis is needed. These include the following:

1. Diseases caused by the few viruses for which antiviral chemotherapy is in fact already available, e.g., herpesviruses. More importantly, rapid advances in the development of new antiviral agents and in the extension of the known clinical applications of interferons will surely expand the range of viral diseases for which a precise diagnosis will become necessary. This will be so especially if new antiviral agents are relatively specific in their antiviral spectra. Furthermore, the anticipated emergence of drug-resistant mutants will before long necessitate drug-sensitivity tests following virus isolation.

2. Other diseases in which the management of the patient or the prognosis is influenced by the diagnosis. Specific treatment does not always comprise chemotherapy. For example, rabies is completely preventable by postexposure immunization (and is uniformly fatal without treatment). Abortion is recommended if rubella is diagnosed in the first trimester of pregnancy. Cesarean section may be the prudent course of action if a woman has genital herpes at the time of delivery. Special care and education are required for a baby with congenital defects attributable to rubella or cytomegalovirus. Passive/active immunization is needed for a baby born of an $HBeAg^+$ mother, and so forth.

In a negative way, establishment of a viral etiology may lead to a favorable prognosis and obviate the need for continued antibacterial chemotherapy, e.g., meningitis.

3. Infections that may demand public health measures to prevent spread to others. For instance, blood banks routinely screen for hepatitis B which may be present in blood donated by symptomless carriers. Herpes simplex type 2 is readily transmissible to sexual partners. Nosocomial infections (e.g., varicella or measles), often in epidemic form, may create havoc in a leukemia ward of a children's hospital, unless hyperimmune IgG is promptly administered to potential contacts following diagnosis of the sentinel case. Documentation of a novel strain

of influenza virus may herald the start of a major epidemic against which vulnerable older members of the community should be immunized. Positive identification of a particular arbovirus in a case of encephalitis enables the authorities to promulgate warnings and initiate appropriate antimosquito control measures. Introduction of a dangerous exotic disease, e.g., African hemorrhagic fever, demands containment and surveillance, and so on.

4. Surveillance of viral infections to determine their significance, natural history, and prevalence in the community, with a view to establishing priorities and means of control, and monitoring and evaluating immunization programs.

5. Continuous surveillance of the community for evidence of new epidemics, new diseases, new viruses, or new virus–disease associations. New viruses and new virus–disease associations continue to be discovered virtually every year. We need only make the point that over 90% of all the human viruses known today were completely unknown at the end of World War II. Opportunities are legion for astute clinicians as well as virologists and epidemiologists to be instrumental in such discoveries.

The traditional methods of laboratory diagnosis of viral infections have been (1) isolation of the virus, usually in cultured cells, followed by serological identification of the isolate; and (2) serological identification and quantitation of antibody in the patient's serum. While both these approaches (particularly the first) are still widely used, it is well recognized that both are unacceptably slow to provide an answer—a diagnosis taking more than a week is a diagnosis in retrospect. The severe limitations of such a delay are obvious: the diagnosis arrives too late to exert any influence on treatment, and the attending physician loses faith in the relevance of the laboratory. Therefore, the thrust of developments in diagnostic virology over the last decade has been toward rapid methods providing a definitive answer within 8–24 hours. Some of the first "same day" diagnostic techniques were limited in their application by such defects as low sensitivity, low specificity, and high cost, and demanded the attention of senior, experienced personnel and often expensive equipment. More recently, however, sensitive, reliable, cheap, and simple assays, which can be handled by a competent technician, have been developed. Standardized diagnostic reagents of much improved quality are now commercially available and a veritable cascade of "do-it-yourself" kits, often miniaturized, and sometimes incorporating semi-automated procedures has begun to flood the market. In the main, these are serological tests for viral antigen, or for antibody of the IgM class, which provide a diagnosis from a single specimen taken directly from the patient during the actue phase of the illness. Solid-phase radioimmunoassays (RIA) and enzyme immunoassays (EIA), in particular, have revolutionized diagnostic virology and are now the methods of choice.

DIRECT IDENTIFICATION OF VIRUS, ANTIGEN, OR NUCLEIC ACID

The last decade has witnessed the development of an impressive variety of rapid diagnostic methods involving the detection of viral antigens (or sometimes virions) in specimens taken directly from the patient without augmentation by cultivation *in vitro*. By and large, these are tests of high sensitivity, capable of detecting the very small concentrations of virus and viral antigens normally found in patients' feces or throats (approximately 1–10 μg of viral antigen/ml in nasopharyngeal aspirates and 100–1000 μg/g of feces).

Electron Microscopy

The morphology of most viruses is sufficiently characteristic to be able to place them into the correct family by electron microscopy. This was originally exploited to distinguish the herpesvirus of chickenpox from the poxvirus of smallpox in negatively stained vesicle fluids from patients with vesicular rashes; today it is more useful for identifying the viruses of orf, molluscum contagiosum and warts in skin lesions, polyomavirus or cytomegalovirus in urine, or herpes simplex virus in a brain biopsy. During the 1970s electron microscopy (EM) was the means to the discovery in feces of several new groups of viruses, previously noncultivable. The human rotaviruses, caliciviruses, astroviruses, hepatitis A, and previously unknown types of adenoviruses and coronaviruses were all first identified in this way (as were several previously unknown intestinal bacteriophages!). Indeed, it could be said that the major justification for the continued use of EM in viral diagnosis today is that this technique represents the best chance of unearthing further new viruses.

The biggest limitation of EM as a diagnostic tool is its low sensitivity. Because of the time required for even a skilled microscopist, occupying a very expensive machine, to scan the grid adequately, specimens should contain at least 10^6–10^7 virions/ml to be really satisfactory. Dirty specimens such as feces are usually first clarified by low-speed centrifugation. It was once considered desirable then to concentrate the virions by ultracentrifugation, preferably in an Airfuge, which is cheap, quick, and increases the sensitivity of detection by 2–3 logs. However, this has been rendered unnecessary by development of the agar-diffusion technique, in which salts and water from a drop of virus suspension hanging from the underside of a carbon-coated plastic ("Formvar") support film are allowed to diffuse downward into agar, leaving the concentrated virions on the film. The specimen is then stained with a negative stain, usually phosphotungstate, or sometimes uranyl acetate.

Immune Electron Microscopy. Definitive identification (and further concentration) of virions, may be achieved by immune (immuno)electron microscopy (IEM, Plate 12-1). Classically, the specimen was simply mixed with specific antibody and the virions were observed to be aggregated. The sensitivity of the method was greatly enhanced by the introduction of a solid-phase IEM technique in which capture antibody is attached to a staphylococcal protein A-coated grid.

As a result of the marked increase in sensitivity obtained by solid-phase IEM, it should now be practicable to extend this diagnostic pro-

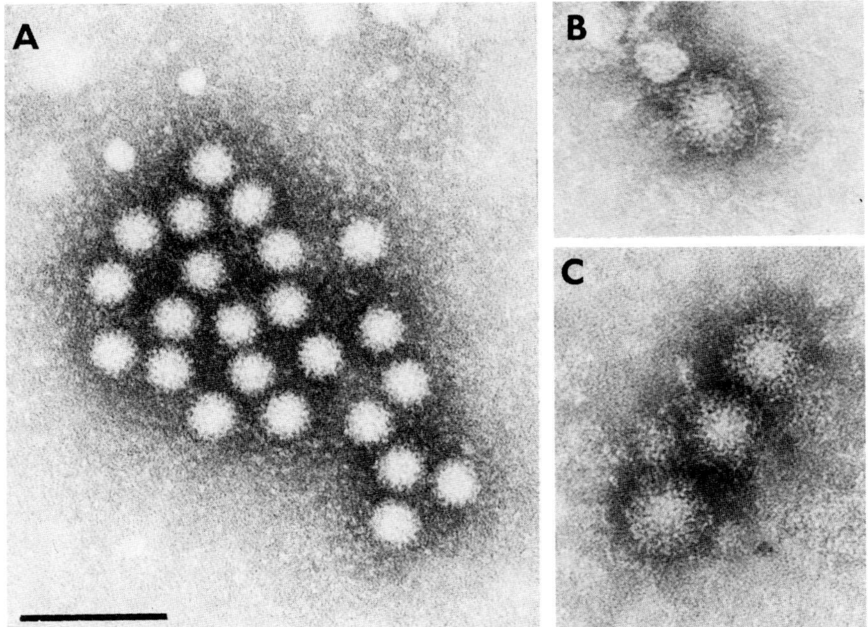

PLATE 12-1. *Immunoelectron microscopy. Norwalk agent from the feces of a patient with gastroenteritis; aggregation with convalescent serum containing a low (A) or high (B and C) titer of antibody against the virus. (Bar = 100 nm.) (Courtesy Dr. A. Z. Kapikian.)*

cedure to specimens other than feces, serum, and vesicle fluid. Cytomegalovirus has been detected in urine in this way. The major challenge, however, is detection of ortho- and paramyxoviruses in nasopharyngeal aspirates or throat washings.

Radioimmunoassay

Radioimmunoassay (RIA), which revolutionized the biochemical study of viral multiplication (see Chapter 3), has done the same to diagnostic virology. The exquisite sensitivity of the method enables as little as 1–10 ng of viral antigen/ml to be detected in specimens taken directly from the patient.

RIAs can be devised to detect antigen, or antibody of any designated class. The methodological permutations are almost limitless: a wide variety of direct, indirect, and reversed assays has been described. Most are solid-phase RIAs, i.e., the "capture" antibody or antigen is attached (by simple adsorption or by covalent bonding) to a solid substrate, typically polystyrene tubes or beads, or the wells of polyvinyl microtiter plates, so facilitating washing steps. In simplest form, the direct RIA, virus and soluble viral antigens from the specimen are allowed to adsorb to the capture antibody (bound to the solid phase), then, after washing, ^{125}I-labeled antiviral antibody is added as "detector/indicator" (Fig. 12-1, left). After a further washing step the beads or plate are counted in an appropriate spectrometer (gamma counter).

The indirect RIA is perhaps more widely used because of its greater sensitivity and avoidance of the need to iodinate more than one antibody. Here, the detector antibody is unlabeled and a further layer, ^{125}I-labeled anti-Ig, is added as indicator; of course, the antiviral antibodies constituting the capture and detector antibodies must be raised in different animal species (Fig. 12-1, right).

As an alternative to labeled anti-Ig as indicator, labeled staphylococcal protein A, is now widely used. Protein A binds to the Fc moiety of IgG (but not IgG_3, IgM, IgE, or some IgA) of most mammalian species (human > rabbit > mouse). It is readily conjugated to ^{125}I or enzymes without loss of binding activity. Hence this single reagent can be conveniently employed as indicator for an unlimited variety of detector antibodies, even from several mammalian species.

Even though backgrounds are lowered by minimizing nonspecific adsorption of protein to the plastic surface, e.g., by incorporating 1% bovine serum albumin (or fetal calf serum) and a wetting agent (e.g., 1–2% Tween 80) in the diluent, false positives may still be something of a problem. Positive (especially low titer) specimens may need to be re-

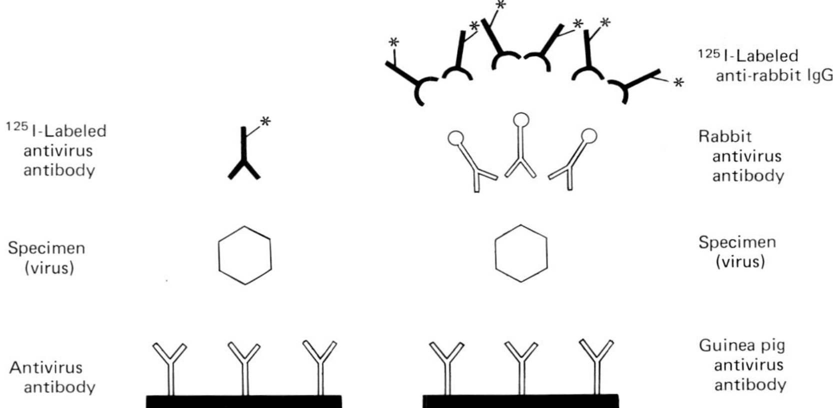

FIG. 12-1. *Radioimmunoassays for detection of virus and/or viral antigen. Left, direct. Right, indirect.*

tested against controls in which (1) preimmune (normal) serum from the same animal is used as "capture antibody"; (2) uninfected cell lysate is substituted as "antigen"; (3) hyperimmune capture antibody added again after virus attachment is demonstrated to block subsequent binding of detector antibody.

Increasingly, monoclonal antibodies will come to be used in serological assays such as RIAs and EIAs. Their obvious advantages are that they represent well-characterized, highly pure, monospecific antibody of a single class, recognizing only a single epitope, are by definition free of "natural" and other extraneous antibodies against host antigens or adventitious agents concurrently infecting the animal, and can be made available in large amounts as reference reagents. Their disadvantages are that they often bind with low affinity and they may in fact be too specific for certain purposes. For instance, only a high-affinity monoclone against a determinant common to all serotypes of a particular virus would be suitable for screening for that virus, otherwise variants could easily be missed. Even for the homologous strain, a monoclone may not be suitable for use in a "sandwich" assay if the antigen were in soluble monomeric form bearing only a single copy of the epitope in question. Therefore, there may be an advantage in employing different monoclones as capture and indicator antibodies, or even in reconstituting mixtures from selected monoclones. Using a monoclone as capture antibody should increase the sensitivity as well as the specificity of antigen detection compared with polyclonal sera in which most of the immunoglobulin bound to the plate is nonspecific. Polyclonal antibody may be more suitable as detector. Alternatively, an IgM monoclone can be used for capture, an IgG monoclone for detector, and a labeled anti-IgG or protein A as indicator, to eliminate false positives.

Originally introduced for the detection of HBsAg in serum, RIA is now widely used for the detection of viruses in feces and has recently been extended to CSF and nasopharyngeal aspirates. Respiratory mucus is solubilized with a mucolytic agent, and exfoliated cells disrupted by sonication, so releasing soluble antigens as well as virions. Sensitivity matches or exceeds the more time consuming and technically tricky immunofluorescence method (see below).

Time-Resolved Fluoroimmunoassay

Halonen and his colleagues in Finland recently described a new type of nonisotopic immunoassay, in which the indicator antibody is labeled, not with a radioisotope, but with a chelate of europium, which following excitation by very short pulses of light (1000 per second), emits fluorescence of high intensity; by introducing a delay before counting (in a fluorometer), background fluorescence, which is of short duration, is screened out. The method is as sensitive as RIA but has the advantages of a stable label and a very short counting time (1 second).

Enzyme Immunoassay

The enzyme-linked immunosorbent assay (ELISA), or enzyme immunoassay (EIA), represents a simple alternative to RIA, particularly for laboratories not equipped with a counter, which is expensive. EIA is essentially the same as an RIA except that the label on the indicator antibody is an enzyme and the read-out is based on the color change that follows the addition of a substrate appropriate to the enzyme. Figure 12-2 (left) shows a typical direct EIA.

Various enzymes have been used as labels, alkaline phosphatase, β-galactosidase, peroxidase, and urease being perhaps the best known. The enzyme is cross-linked to the antibody by glutaraldehyde, periodate, or a derivative of maleimide. An appropriate organic substrate is chosen for the particular enzyme, e.g., *O*-phenylenediamine (OPD) for horseradish peroxidase. Ideally, the substrate should be one that undergoes a dramatic color change, clearly recognizable by eye. Nevertheless, spectrophotometric estimation of the end point is generally more sensitive and quantitative.

Recent modifications of EIAs have been aimed at increasing sensitivity. High-energy substrates are now available which release fluores-

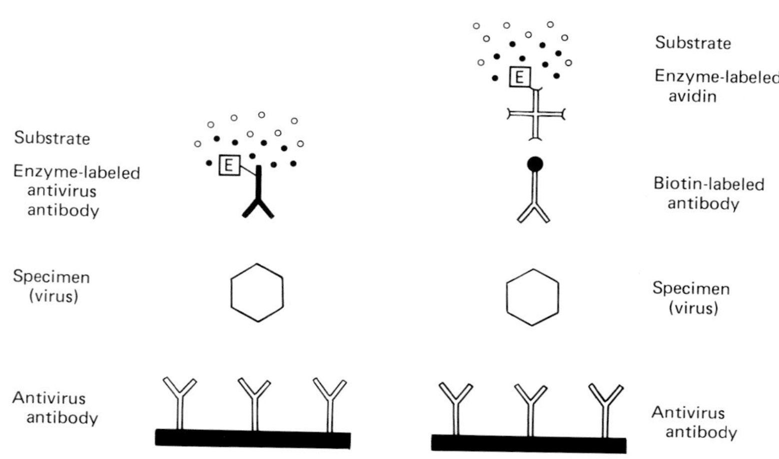

FIG. 12-2. *Enzyme immunoassays (EIA or ELISA) for detection of virus and/or viral antigen. Left, direct. Right, avidin-biotin.*

cent, chemiluminescent, or radioactive products that can be identified in very small amounts. For instance, the fluorogenic substrate of β-galactosidase, isoluminol, emits light which can be monitored in a scintillation counter.

A further refinement takes advantage of the extraordinarily high binding affinity of avidin for biotin. The antibody (or antigen) is conjugated with biotin, a low molecular weight marker which gives reproducible labeling and does not alter its antigen (or antibody)-binding capacity. The enzyme is coupled to avidin. Hence the antigen–antibody complex is recognized with exquisitely high sensitivity simply by adding enzyme-labeled avidin then substrate—no second antibody is needed (Fig. 12-2, right).

So-called "homogeneous" assays aim to save time by obviating the need for one rinse/incubation step, and can be conducted in solution. Such assays employ a very large substrate that is sterically hindered from binding to the enzyme when the latter is associated with the antigen– (enzyme-labeled) antibody complex, but can bind to free enzyme-labeled antibody. One simply assays for *unbound* enzyme-labeled antibody.

EIAs are emerging as the rapid diagnostic method of choice for most laboratories. Recent modifications such as the use of chemiluminescent substrates and magnifying systems, e.g., "sandwiches" encompassing complement or triple or quadruple antibody layers, have enhanced the sensitivity of EIAs to the point where they at least match that of RIAs and can now be used for diagnosing respiratory as well as intestinal infections.

Immunodiffusion

A serological technique that was once widely used for identification of viral antigen in vesicle fluids or serum is gel diffusion. Antigen and antibody are placed in separate wells cut in a thin layer of agar or agarose on a glass slide or petri dish. The reactants diffuse through the agar at a rate inversely related to their molecular weights. Where antigen and antibody meet in optimal proportions a sharp line of precipitate forms in the agar. If several antigens and their corresponding antibodies are present, as in the case of most unpurified preparations of virus and of unabsorbed antiviral sera, each antigen–antibody complex forms a discrete line. If two qualitatively identical preparations of antigen (or antiserum) are placed in adjacent wells and allowed to diffuse toward a common well of antiserum (or antigen) equidistant from them both, the corresponding pairs of precipitin lines will join exactly. This is called the "reaction of identity" (Plate 12-2) and is generally taken to indicate sharing of that particular antigen. This conclusion would be valid only if all the major proteins of the virus were present in soluble form and each carried only one major antigenic determinant. In practice, the crude viral antigen preparations used for ID usually include aggregates of different antigens bearing several distinct epitopes. Even homopolymers of any particular protein will diffuse at different rates according to their molecular weight and may produce several precipitin lines. Furthermore, ID is not a particularly sensitive test, i.e., rather high concentrations of both reagents are necessary to ensure the formation of a visible line, even with staining. Moreover, it does not lend itself readily to quantitation, i.e., it gives no measure of the titer of the unknown antibody or antigen.

Several refinements of ID have been introduced in recent years. In *single radial diffusion (SRD)*, one of the reactants is incorporated throughout the agarose gel, so that when the other diffuses out from the well, the antigen–antibody precipitate forms an opalescent halo, the diameter of which provides a degree of quantitation. In *counterimmunoelectrophoresis (CIEP)*, electrophoresis is used to increase the speed of diffusion of antigen and antibody through the gel. Though somewhat more sensitive than ID, it is not much used today.

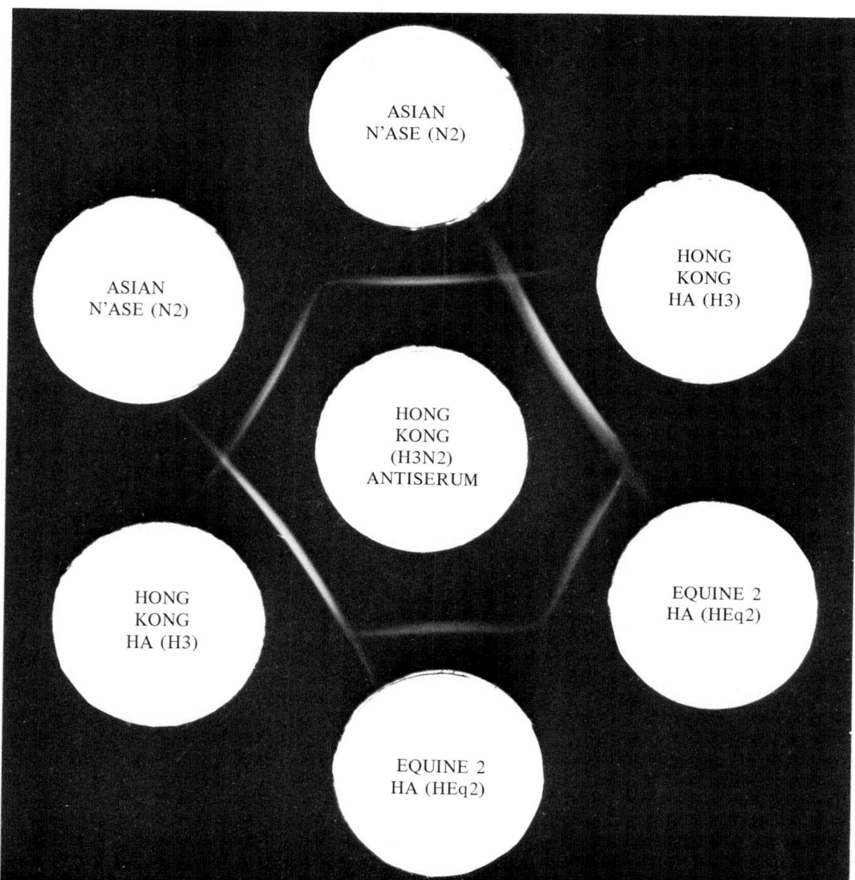

PLATE 12-2. *Immunodiffusion test. Example illustrates its use to analyze relationships between envelope antigens of influenza A virus. Center well: antiserum to Hong Kong influenza virus (H3N2). Peripheral wells, purified antigens: Hong Kong hemagglutinin (H3), Asian neuraminidase (N2), and Equine 2 hemagglutinin (HEq2). Antiserum to Hong Kong virus contains antibodies to all the antigens tested. Note (1) two pairs of antigens (N2 and HEq2) each show fusion of precipitin lines ("reaction of identity"), (2) neuraminidase N2 and hemagglutinin H3 show complete crossing over of precipitin lines ("reaction of complete nonidentity"), and (3) equine (HEq2) and Hong Kong (H3) hemagglutinins show partial fusion of lines ("reaction of partial identity") indicating serological cross-reactivity. (Courtesy Dr. R. G. Webster.)*

Reverse Passive Hemagglutination

Antibody may be coupled to erythrocytes (e.g., using chromic chloride); such red cells are agglutinable by viral antigen. This assay is quite sensitive and was used for some years to detect HBsAg in blood.

Immunofluorescence

If antibody is labeled with a fluorescent dye (fluorochrome), the antigen–antibody complex, when excited by light of short wavelength, e.g., ultraviolet or blue-violet light, emits light of a particular longer wavelength, which can be visualized as fluorescence in an ordinary microscope when light of all other wavelengths is filtered out. The sensitivity of the method is too low to detect complexes of fluorescent antibody with virions or soluble antigen (unless one uses a special "epifluorescence" microscope). Therefore, the antigen in the test generally takes the form of virus-infected cells. There are two main variants of the technique.

Direct immunofluorescence is conducted as follows. Viral antigen, e.g., in the form of an acetone-fixed, virus-infected, cell monolayer on a coverslip, is exposed to fluorescein-tagged antiviral serum (Fig. 12-3, left). Excess conjugated antibody is washed away and the cells are inspected microscopically using a powerful light source (e.g., tungsten–halogen, xenon, or mercury–vapor lamp) from which all but the light of short wavelength has been filtered out. Additional filters in the eyepieces, in turn, absorb all the blue and ultraviolet light, so that the specimen appears black except for those areas from which fluorescein is emitting greenish-yellow light (Plate 12-3).

Indirect immunofluorescence, sometimes known as the "sandwich" technique, differs in that the antiviral antibody is untagged, but fills the role of the "meat in the sandwich." It binds to antigen and is itself recognized by fluorescein-conjugated antiimmunoglobulin. For example, if the antiviral serum were made in a rabbit, then it would be appropriate to raise antibodies in a goat against normal rabbit IgG. The indirect technique has the advantages of greater sensitivity and, more important in diagnostic virology, of requiring only a single tagged reagent, e.g., fluorescein-conjugated (goat) anti-rabbit IgG, with which to test the interaction between any antigen and its corresponding rabbit antibody (Fig. 12-3, right).

There are several further variations on these themes. For example, complement can be included in the sandwich, and fluoresceinated anticomplement employed as indicator. The high affinity of avidin for biotin

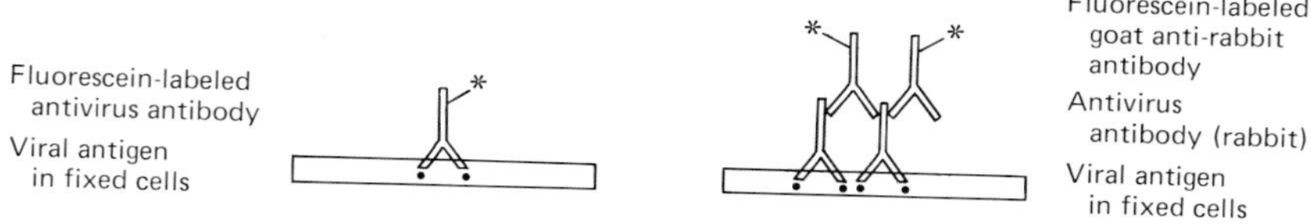

FIG. 12-3. *Immunofluorescence. Left, direct. Right, indirect.*

can also be exploited in immunofluorescence, by coupling biotin to antibody and fluorescein to avidin.

A particular problem confined to the herpesviruses, is that all members of this family induce the synthesis of "Fc receptors" in the membrane of infected cells. As Fc receptors bind normal IgG, unusually stringent controls are required if IF is to yield meaningful results with herpesviruses.

Considering that immunofluorescence has been known for over 40 years it is odd that this attractive technique has not before now become the standard approach to laboratory diagnosis of viral infection. No doubt this can be attributed to the cost, and to the technical difficulties involved in ensuring that the procedure is specific, as well as to the multitude of distinct serotypes responsible for most respiratory and enteric infections. Nevertheless, immunofluorescence has proved to be of great value in the identification of viral antigens in infected cells taken from patients with diseases known to have a relatively small number of possible etiological agents. There is no difficulty in removing partly detached infected cells from the mucous membrane of the upper respiratory tract, genital tract, eye, or from the skin, simply by swabbing or scraping the infected area with reasonable firmness. Cells are also present in mucus aspirated from the nasopharynx using a simple suction device (Plate 12-4E). Aspirated cells must be extensively and carefully washed by centrifugation to remove mucus before fixation and staining. Respiratory infections with paramyxoviruses, orthomyxoviruses, adenoviruses, and herpesviruses are particularly amenable to rapid diagnosis by this method.

Biopsy is sometimes justifiable to obtain infected tissue from very ill patients. Using such material, immunofluorescence has been applied, for example, to brain biopsies for the diagnosis of herpes simplex encephalitis or measles SSPE. At necropsy, immunofluorescence has for several years been the standard approach to the verification of rabies in the brain of animals trapped after biting man, or to the identification of a number of lethal viral infections of the human brain. The major viral

TABLE 12-1
Identification of Viral Antigen by Immunofluorescence

SPECIMEN	VIRUS
Brain (biopsy or PM)	Rabies
	Herpes simplex
	Measles (SSPE)
	Papovavirus (PML)
Corneal scraping	Herpes simplex
	Adenovirus
Vesicle scraping	Poxviruses
	Varicella-zoster
	Herpes simplex
Nasopharyngeal aspirate	Respiratory syncytial
	Parainfluenza
	Influenza
	Measles
	Adenovirus
Blood leukocytes	Arboviruses
	Retroviruses
	Many others
Heart (PM)	Coxsackieviruses
Liver (biopsy or PM)	Hepatitis A, B, and δ

diseases readily identifiable by immunofluorescence applied to fixed smears of cells taken directly from the patient are listed in Table 12-1.

Immunofluorescence has revolutionized the rapid diagnosis of respiratory viral infections, particularly in recent years since commercially available fluoresceinated reagents have become more reliable and most laboratories have learnt to conquer the problems of nonspecific fluorescence. Nevertheless, the method is technically demanding and time consuming and it remains to be seen whether IF will survive the competition from EIA and RIA.

A rather different application of immunofluorescence to laboratory diagnosis of viral infection is the identification of specific viral antigen in cell cultures inoculated a day or two earlier with material taken from the patient. This very sensitive probe reveals the growth of virus even in a very small number of infected cells, long before cytopathic effects become manifest (Plate 12-3).

Immunoperoxidase Staining

An alternative method of locating and identifying viral antigen in infected cells is to use antibody coupled to horseradish peroxidase; sub-

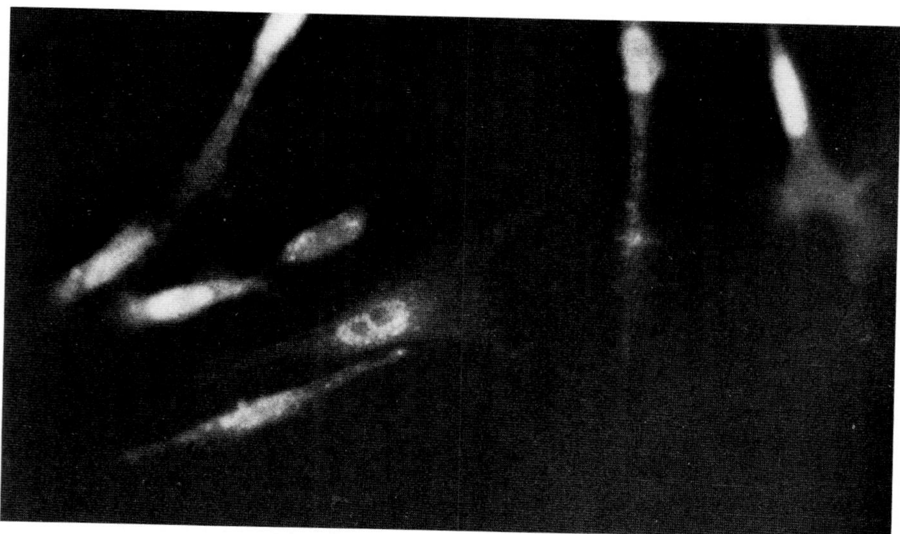

PLATE 12-3. *Immunofluorescence, used here for determining the site of assembly of components of influenza virus. Antibody against the NP (nucleoprotein) antigen shows nuclear accumulation of NP at 4 hours after infection of chick cells. Guinea pig antiserum to NP antigen, then fluorescein-conjugated rabbit anti-guinea pig IgG. (Courtesy Dr. N. J. Dimmock.)*

sequent addition of H_2O_2 together with a benzidine derivative forms a colored insoluble precipitate. The advantages of the method are that the preparations are permanent and only an ordinary light microscope is needed. The principal disadvantage is that endogenous peroxidase, present in a number of tissues, particularly leukocytes, produces false positives, but this problem can be circumvented by meticulous technique and adequate controls.

Nucleic Acid Hybridization: cDNA Probes

Nucleic acid hybridization is a technique that has been used for years as a research tool but only recently adapted to diagnosis. The principle is that, following "melting" of double-stranded DNA by heat or alkali treatment, the separated single strands will reanneal efficiently *in vitro* with one another, or competitively, with a similar or related complementary strand. If mixed with ^{32}P- or ^{35}S-labeled single strands of a related nucleic acid the extent and rate of hybridization provide a measure of the degree of homology (relatedness) between the two, since only those sequences that are complementary will hybridize. The reaction may be accomplished in solution, which is useful for determining the kinetics (rate) of annealing, or the stoichiometry of the reaction (concentration of the homologous sequence) from which can be calculated the percentage homology between the two sequences. However, the simplest technique is a two-phase system in which the denatured DNA is dried onto a nitrocellulose membrane; labeled single-stranded DNA (or RNA) is then hybridized to it *in situ* on the membrane. More recent refinements enable oligonucleotides derived by cleavage of DNA or RNA to be separated by electrophoresis on agarose or acrylamide gels, then transferred to nitrocellulose membrane sheets (Southern blotting) or diazobenzyloxymethyl (DBM) cellulose or paper ("Northern" blotting) and individually detected by radiolabeled nucleic aid probes. Alternatively, the homologous (i.e., basepaired) and nonhomologous (single-stranded) regions of the hybridized molecules can be visualized by electron microscopy—techniques known as heteroduplex mapping and R loop mapping. Moreover, nucleic acid hybridization using radiolabeled probes can be conducted *in situ* on fixed tissue sections or cultured cells and read by autoradiography. Alternatively the viral DNA probe can be labeled not with a radioisotope but with biotin, and its interaction with complementary nucleic acid revealed by techniques based on the powerful binding of biotin to avidin; the readout may be ELISA, immunofluorescence or immunoperoxidase staining of cells.

The principal applications of molecular hybridization to diagnosis are (1) detection of integrated or episomic sequences of viral DNA in persistently infected or malignant cells, (2) differentiation between related strains of virus to characterize geographical variants or to trace epidemiological connections, and (3) identification of DNA extracted from viruses in clinical specimens.

In situ hybridization in infected tissues is already firmly established as a technique for the investigation of persistent infections and malignant tumors involving hepatitis B, EBV, herpes simplex, and papillomaviruses (see Chapter 8).

Nucleic acid hybridization has not yet been widely exploited as a routine diagnostic tool in acute infections, but it has been shown to be feasible in the case of several herpesviruses, adenoviruses, hepatitis viruses and others. Using dot hybridization with autoradiography it yields results within 24–48 hours and is about as sensitive as EIA or RIA. It may have advantages over cell culture isolation in the case of viruses that are noncultivable, slowgrowing, dangerous, or nonviable as a result of suboptimal conditions of transport or storage. DNA is extracted from the specimen presumed to contain virus, immobilized on a nitrocellulose filter, and tested for hybridization to a ^{32}P-labeled probe consisting of either pure viral DNA (preferably cloned in bacteria to ensure freedom from contaminating mammalian DNA), or, for greater discrimination, cloned restriction endonuclease cleaved fragments, or even chemically synthesized oligonucleotide (DNA or RNA) sequences. The isotope ^{32}P has a short half-life and is somewhat hazardous for routine work, but the new method for labeling DNA with ^{35}S provides an excellent alternative. Biotinylated probes will also be used increasingly, particularly with an ELISA readout which may enable the test to be completed within an hour or so.

VIRUS ISOLATION

Despite this veritable explosion of new techniques for same-day diagnosis of viral disease by demonstration of virus, viral antigen or viral nucleic acid in specimens taken directly from the patient, it is still true to say that few of them achieve quite the sensitivity of virus isolation in cell culture. Theoretically at least, a single viable virion present in a specimen can be grown in cultured cells and expanded to a population of the order of a billion, which can then be identified by serology. It has been said that virus isolation remains the "gold standard" against which newer methods must be compared and, furthermore, is the only technique that can "detect the unexpected," i.e., identify a totally unforeseen virus, or even discover an entirely new agent. Accordingly, even those laboratories well equipped for rapid diagnosis often also inoculate cell cultures in an attempt to isolate the virus.

Collection and Preparation of Specimens

The chance of isolating a virus critically depends on the attention given by the attending physician to the collection of the specimen. Clearly, such a specimen must be taken from the right place at the right time. The right time is always as soon as possible after the patient first presents, because virus is usually present in maximum concentration at about the time symptoms first develop, then disappears during the ensuing few days (Fig. 5-5). Specimens taken as a last resort when days or weeks of empirically chosen antibacterial chemotherapy have failed are almost invariably a waste of effort.

The site from which the specimen is collected will depend on the pathogenesis of the particular infection. The most important specimen, routine in respiratory infections as well as in many generalized infections, is the throat swab; the back of the throat is firmly rubbed with a cotton-wool swab on the end of a wooden applicator. A nasal swab may also be useful. The best specimen to take from a young child is a nasopharyngeal aspirate; mucus is sucked from the back of the nose and throat into a mucus trap, using a vacuum pump (Plate 12-3E). The sec-

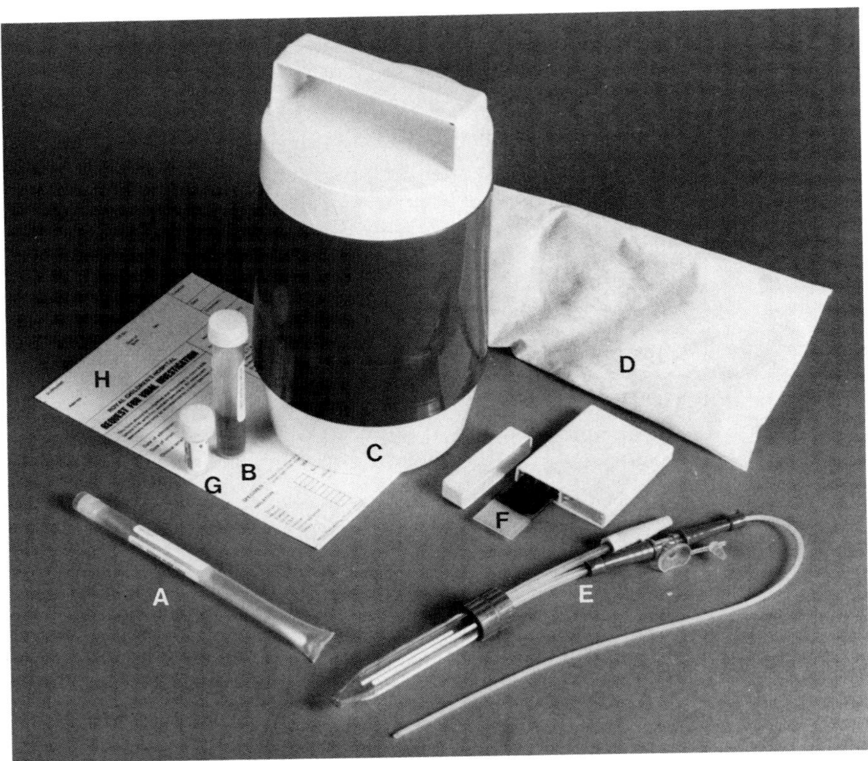

PLATE 12-4. *Basic equipment for collection of specimens for virus isolation. (A) Sterile swab. (B) Transport medium into which swab is broken off. (C) Vacuum flask or foam-plastic insulated container. (D) Refrigerant in flexible pack to be wrapped around specimen and placed in flask. (E) Apparatus for aspirating nasopharyngeal mucus by suction. (F) Slides for fixation of aspirated cells. (G) Anticoagulant for collection of blood (leukocytes for virus isolation and plasma for serology). (H) Requisition form. (Courtesy I. Jack.)*

ond important specimen, routine in enteric and many generalized infections, is feces; since about 4–8 g is desirable, a fecal swab or rectal swab is inadequate. Less commonly, other specimens may be appropriate. Some viruses responsible for systemic infections can be isolated from leukocytes; 10 ml of blood is withdrawn aseptically by venepuncture into a Vacutainer or a dry syringe containing heparin (in contrast to blood taken for serology, which should normally be allowed to clot). Scrapings or firm swabbings from skin (e.g., from the base of a vesicle or other lesion), or from other epithelial surfaces such as the eye or the genital tract are sometimes the most appropriate. Cerebrospinal fluid

(CSF) may yield virus in cases of meningitis. Biopsy or autopsy specimens may be taken by needle or knife from any part of the body for virus isolation, or snap-frozen for immunofluorescence. Obviously, tissue taken at autopsy or biopsy for the purpose of virus isolation must not be placed in formalin or any other fixative. In the case of many generalized viral diseases it may not be obvious what specimen to take. As a rough working rule it can be said that at a sufficiently early stage in the disease, virus can usually be isolated from a throat swab, or feces, or leukocytes from blood.

Because of the lability of many viruses, specimens must always be kept cold and moist. Immediately after collection the swab should be swirled around in a small screw-capped bottle containing about 2–5 ml of a *virus transport medium* (VTM). This medium consists of a buffered isotonic balanced salt solution, such as Hanks' BSS or tryptose phosphate broth, to which has been added protein (e.g., 0.5% gelatin or bovine serum albumen) or sucrose phosphate to protect the virus against inactivation, and antibiotics (e.g., gentamicin, or penicillin plus streptomycin, ± amphotericin B) to prevent microbial multiplication. (If it is at all probable that the specimen will also be used for attempted isolation of bacteria, rickettsiae, chlamydiae, or mycoplasmas, the collection medium must not contain antibiotics—the portion used for virus isolation can be treated with antibiotics later.) The swab stick is then broken off aseptically into the fluid, the cap is tightly fastened and secured with adherent tape to prevent leakage, the bottle is labeled with the patient's name, date of collection, and nature of specimen, then despatched immediately to the laboratory, accompanied by an informative clinical history and provisional diagnosis. If a transit time of more than an hour or so is anticipated the container should be sent refrigerated (but not frozen), e.g., packed with "cold dogs" or enclosed in a thermos-flask of ice (Plate 12-4). International or transcontinental transport of important specimens, particularly in hot weather, generally requires that the container be packed in "dry ice" (solid CO_2) to maintain the virus in a frozen state. Governmental and IATA regulations relating to the transport of biological materials require such precautions as double containers and absorbent padding in case of breakage.

For particularly labile agents such as respiratory syncytial virus or herpes simplex virus, it may even be desirable to inoculate the specimen directly from the patient into a monolayer cell culture. As the virus is promptly taken up by the cells, refrigeration is unnecessary; the specimen may be transported to the laboratory at ambient temperature.

On arrival in the laboratory the specimen is processed as soon as possible. If delays of more than a few hours are anticipated CSF and fecal samples are stored at −20° or −70°C. Before inoculation into cell

TABLE 12-2
Laboratory Hazards

HAZARD	CAUSE
Aerosol	Homogenization (e.g., of tissue in blender)
	Centrifugation
	Ultrasonic vibration
	Broken glassware
	Pipetting
Ingestion	Mouth pipetting
	Eating or smoking in laboratory
	Inadequate washing/disinfection of hands
Skin penetration	Needle-stick
	Hand cut by broken glassware
	Leaking container contaminating hands
	Pathologist handling infected organs
	Splash into eye
	Animal bite

culture, material is shaken from the swab into the VTM by hand or on an appropriate mechanical device. Feces may be dispersed with the help of a Vortex mixer or glass beads. Tissue specimens are homogenized in a high-speed blender or ground with a mortar and pestle. Cell debris and bacteria are deposited from the fluid by low-speed centrifugation, after which particularly "dirty" specimens like feces may be passed through a membrane filter. Some of the clarified fluid is then "snap-frozen" in thin-walled ampules and stored frozen for future reference. The remainder is inoculated into a range of cell cultures, and, rarely, into chick embryos or newborn mice. Methods of cultivation of viruses in these hosts have been discussed in Chapter 2.

This is an appropriate opportunity to say a little about safety precautions that must be taken in a virus diagnostic laboratory. While it has truthfully been observed that virology is one of the less hazardous human occupations, cf. building, mining, or driving a car, it is also true that several hundred deaths have been recorded in laboratory personnel over the years, particularly from togaviruses, arenaviruses, filoviruses, and other "Class 4" pathogens (known as Class C in the UK). Potentially hazardous procedures are set out in Table 12-2. In addition, "normal" serum, which may contain hepatitis B or non-A, non-B virus in high titer, has to be treated with caution at all times; the safest procedure is to regard all specimens as potentially infectious. Spills are disinfected with sodium hypochlorite, peracetic acid, or glutaraldehyde. Mouth-

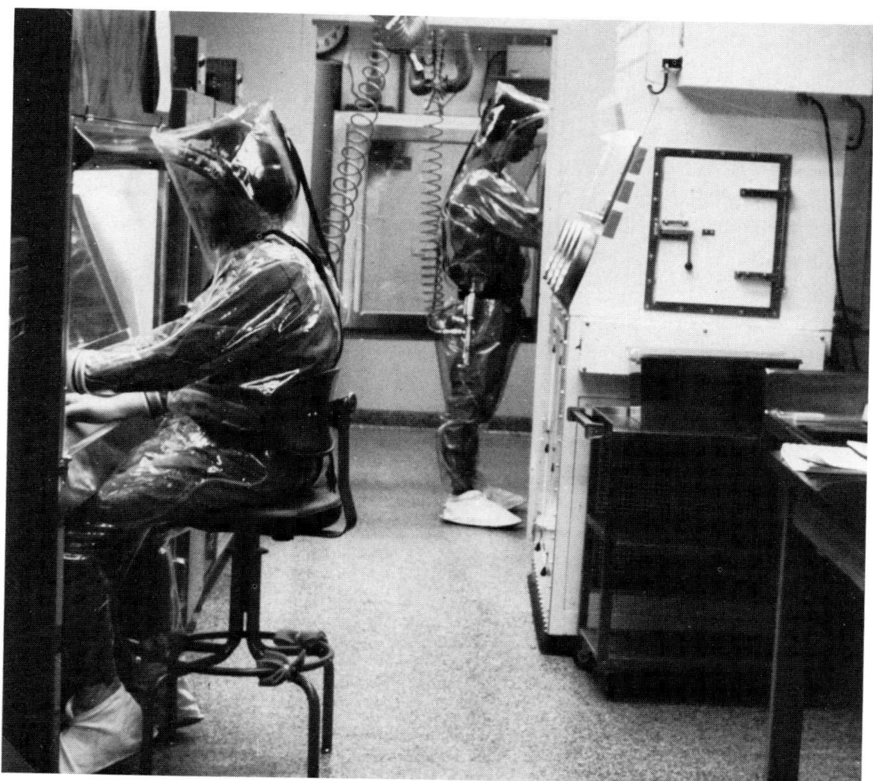

PLATE 12-5. *Maximum containment laboratory, Centers for Disease Control, Atlanta, Georgia. Workers are protected by positive-pressure suits with independent remote source of breathing air. Primary containment of aerosols is achieved by use of filtered vertical laminar-flow work stations. The ultracentrifuge is contained in an explosion-proof laminar flow hood. A full range of sophisticated equipment is available, and animals as large as monkeys can be studied. [From J. S. Mackenzie (ed.), "Viral Diseases in South-East Asia and the Western Pacific." Academic Press, Sydney, 1982, courtesy Dr. K. M. Johnson.]*

pipetting is banned. Gowns must be worn at all times, and gloves for anything dangerous. Various classes of safety cabinets are available for procedures of various degrees of hazard, and certain Class 4 viruses can be handled only in maximum-security laboratories (Plate 12-5). Personnel should be immunized against hepatitis B, poliomyelitis, and rubella (as well as against exotic agents in the special laboratories handling them). Limitations should be placed on the type of work undertaken by pregnant or immunosuppressed individuals.

Inoculation and Maintenance of Cultures

The choice of the most suitable host cell system will depend partly on availability, but mainly on the virus one expects to find, in the light of the patient's clinical history and the source of the specimen. The precise routine differs from one laboratory to another. There are essentially three different types of cell culture available. (1) Primary cultures, e.g., of human embryonic kidney (HEK) or monkey kidney (PMK) cells have a "broad viral spectrum" because they contain a variety of types of differentiated cells; many RNA viruses, particularly paramyxoviruses and some enteroviruses, are most easily isolated in such cultures. However, monkey kidneys are now very expensive in some countries and almost unobtainable in others, and they are frequently contaminated with adventitious simian viruses, while human embryonic tissue has become less readily available in many countries as a result of changes in the law, or in the frequency or surgical techniques of abortion. (2) Continuous human malignant epithelial cell lines, such as HEp-2 or HeLa, are useful but not essential, for the growth of adenoviruses, rhinoviruses, and respiratory syncytial viruses. (3) Diploid strains of human embryonic fibroblasts (known as HEF, or HDF, for human diploid fibroblasts) have a broad spectrum that overlaps with the other two; they are particularly valuable for the isolation of rhinoviruses, cytomegalovirus, varicella virus, and several enteroviruses, indeed are the most versatile of all the cultured cells available for viral isolation today. Some laboratories rely on well-known strains such as MRC-5 or WI-38, while others have derived their own diploid strain of fibroblasts from human embryonic lung (HEL) or kidney (HEK); the sensitivity of such cell strains to infection by individual viruses varies somewhat. It should also be appreciated that the cell line that is more sensitive to a particular virus (i.e., which will grow virus from the highest percentage of specimens) is usually but not always that in which CPE can be visualized earliest; both considerations are important.

In order to cover themselves against most contingencies some laboratories have a policy of inoculating a given specimen simultaneously into one type of primary culture, one diploid strain of human fibroblasts, and perhaps, in the case of throat but not fecal specimens, one malignant cell line. However, most laboratories today would tend to rely almost entirely on HEF (HDF) for routine isolations, and would reserve other cells for particularly fastidious viruses only, e.g., PMK for paramyxoviruses or orthomyxoviruses, the RD line of human rhabdomyosarcoma for coxsackie A and echoviruses, Vero or hamster kidney cells for arboviruses, RK-13 or BSC-1 for rubella virus, and so on.

Monolayer cultures for viral diagnostic purposes are generally grown in screw-capped glass tubes. Microtiter plates are very convenient for neutralization tests, where large numbers of cultures are required and economy of medium, equipment, and space is important, but the risk of cross-contamination argues against them for virus isolation. The inoculated cultures are held at the temperature and pH of the human body (37°C, pH 7.4), except in the case of the rhinoviruses and coronaviruses, which grow best at 33°C, the temperature encountered in the nasal mucosa. They may be slowly rotated on a roller drum, or kept stationary. If the pH of the maintenance medium falls below 7.0, it may be aseptically replaced with fresh medium. This is a necessary chore to maintain the cells in optimum health during the weeks required for the growth of certain very slowly replicating viruses, but the procedure must be conducted with great care to prevent cross-contamination of cultures with virus. Meanwhile, the cultures are observed three times a week for the development of CPE.

Recognition of Viral Growth

Although rapidly growing viruses such as poliovirus or herpes simplex virus can be relied upon to produce detectable CPE within a day or two, some of the notoriously slow agents such as cytomegalovirus, rubella, and some adenoviruses may not produce obvious CPE for 1–3 weeks. By this time, the uninoculated control cultures will often be showing nonspecific degeneration, so robbing the virologist of an appropriate standard of comparison. In such cases it may be necessary to subinoculate the cells and supernatant fluid from the infected culture into fresh monolayers ("blind passage"), after which clearcut CPE usually appears promptly.

When, at any stage, degenerative changes suggestive of viral multiplication become evident, the virologist has a number of courses open. The CPE is often sufficiently characteristic, even in the living unstained culture viewed *in situ* (Table 2-1), for the trained observer to be able to tender a provisional diagnosis to the physician immediately. Alternatively, it may be instructive to fix and stain the infected monolayer to identify the CPE, for which purpose it is usual to include a coverslip in one of the culture tubes. Inclusion bodies and multinucleate giant cells (syncytia) can be identified by staining with hematoxylin and eosin, or Giemsa stain; if present, these changes are usually sufficiently characteristic to enable the virologist to place the isolate at least within its correct family (Table 12-3). Coverslip cultures can also be stained with fluorescent antibody for positive identification. If none of these histo-

TABLE 12-3
Inclusion Bodies

SITE	STAINING	VIRUS[a]
Nuclear	Basophilic	Adenoviruses
	Acidophilic[b]	Herpesviruses
		Papovaviruses
Cytoplasmic	Acidophilic	Paramyxoviruses
		Reoviruses
		Rhabdoviruses
		Poxviruses
		(Togaviruses)
Nuclear and cytoplasmic	Acidophilic	Measles virus
		Cytomegalovirus
		(Orthomyxoviruses)

[a] The viruses in parentheses do not produce conspicuous inclusion bodies in many cell systems.
[b] Occasionally basophilic.

logical methods proves to be diagnostic, virus must be extracted from the culture for further examination.

Some viruses are relatively noncytocidal for cultured cells (see Chapter 5 and Table 2-1). Their growth in monolayer culture may sometimes be recognized by means of hemadsorption or interference (see Chapter 2). Most viruses that hemagglutinate will be amenable to test by hemadsorption; the growth of paramyxoviruses or orthomyxoviruses and, to a lesser extent, the togaviruses is routinely recognized in this way (Plate 2-2). Interference, on the other hand, is really a research tool rather than a routine diagnostic aid.

Organ Cultures, Eggs, and Animals

Some viruses, not readily recoverable in conventional monolayer cultures, can be grown in organ cultures. For example, certain strains of coronavirus can be cultivated only in rafts of intact fully differentiated human fetal tracheal (or nasal or laryngeal) epithelium (see Chapter 2). The agents responsible for certain persistent viral infections are recoverable only from explants of infected human tissue obtained by biopsy or autopsy. For instance, the etiological agents of slow infections of the brain, such as SSPE, may be isolated only following cocultivation of infected tissue with a permissive cell line (see Chapter 7).

Many viruses will grow satisfactorily in chick embryos or newborn mice (see Chapter 2) but neither animal is now commonly used in the diagnostic laboratory, because cell culture is generally the simpler option. The use of mice can be limited to the isolation of arboviruses, rabies virus, and some of the group A coxsackieviruses; suckling mice less than 24 hours old are injected with approximately 0.02 ml of virus suspension intracerebrally and 0.03 ml intraperitoneally (see Plate 19-4) then observed for up to 14 days for the development of pathognomonic signs before sacrificing for histological examination of affected organs. Chick embryos are still used, together with primate kidney cell cultures, for the isolation of influenza A viruses from man. Three to four days after amniotic and allantoic inoculation of 10-day-old embryonated hen's eggs, the fluids are tested from hemagglutination (see Chapter 21 for details). Smallpox virus used to be isolated on the chorioallantoic membrane and differentiated from vaccinia and other poxviruses by pock morphology (Chapter 17).

Larger animals are virtually never used for routine isolation of human viruses. Primates, especially chimpanzees, are employed by a small number of well-endowed research laboratories for attempted isolation of human viruses not yet cultivable in any other nonhuman host, e.g., hepatitis viruses (Chapters 13, 19, and 28) and various agents responsible for slow infections of the human brain (Chapters 7 and 28), as well as for experimental studies of viral pathogenesis and immunity, and for testing new viral vaccines (Chapter 10).

Identification of Viral Isolates by Serology

A virus newly isolated in cell culture, eggs, or suckling mice can usually be provisionally allocated to a particular family on the basis of the patient's clinical history, the laboratory host or the particular type of cell culture in which the virus was successfully grown, and the visible result of this growth (CPE, hemadsorption, hemagglutination, etc.). Final identification, however, rests on serological procedures. By using the new isolate as antigen against known antisera, e.g., in a complement fixation test, the virus can be placed into its correct family or genus. Having allocated it to a particular family (e.g., *Adenoviridae*), one can then go on to determine the species or serotype (e.g., adenovirus 5) by the more specific serological procedures of neutralization or hemagglutination inhibition. This sequential approach is applicable only to families that share a common family antigen, hence is of no help with picornaviruses, for example. The range of available serological techniques is now almost embarrassingly wide. Some are best suited to particular families of viruses. Each laboratory makes its own choice of favored procedures, based on considerations such as sensitivity, specificity, reproducibility, speed, convenience, and cost.

Immunofluorescence. The simplest way of identifying a newly isolated virus is by fluorescent antibody staining of the infected cell monolayer itself (Plate 12-3). This can provide an answer within an hour or so of recognizing the earliest suggestion of CPE. Immunofluorescence is best suited to the identification of monotypic genera, or to epidemic situations when a particular virus is suspected; otherwise replicate cultures must be screened with a range of antisera. The advantages and disadvantages of monoclonal antibodies, in comparison with polyclonal or "absorbed" sera, were discussed above in the context of radioimmunoassays. The arguments apply equally to other serological procedures, including immunofluorescence and neutralization.

Electron Microscopy. EM and IEM can also be very useful for the identification of cell culture isolates, e.g., enteroviruses during summer epidemics, if the facility is readily available.

Complement Fixation. As most readers will be familiar with the CF test, it will not be described here, but is outlined in the legend of Plate 12-6. Crude cell culture supernatants of course contain not only virions but the whole range of soluble antigens found in those virions. Since most of these are shared by many or all viruses within a particular genus or family, e.g., *Adenoviridae* (but not *Picornaviridae*), they will cross-react with antibodies in sera raised against any other member of that genus or family. In this sense, CF is not nearly as specific a test as say neutralization or HI (unless absorbed or monoclonal type-specific antibodies are used). Nevertheless, this very fact makes CF a suitable choice for preliminary screening of an isolate—to place it within the correct family or genus. *Immune adherence hemagglutination (IAHA),* which is basically a somewhat simplified version of the complement fixation test (see Table 12-5), can also be used, although it currently tends to be applied more often to the detection of antibody than of antigen.

Hemagglutination Inhibition. Antibody binds to viral hemagglutinin, thereby inhibiting hemagglutination (HA) when RBC of the appropriate species are added and incubated at the temperature and pH appropriate to that virus (Table 2-4, Plate 12-7). The HA titer of certain viruses, e.g., measles, rubella, may be increased by dissociation of the virions with detergents. Antisera may have to be pretreated to remove nonspecific inhibitors of HA [see Chapters 21 (influenza) and 20 (rubella) for detailed examples].

The HI test is sensitive and (except in the case of the togaviruses) highly specific, since it measures antibodies binding to the surface pro-

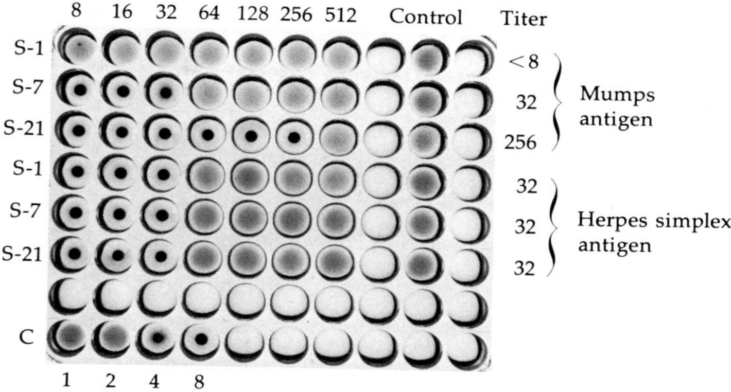

PLATE 12-6. *Complement fixation test. Titers are expressed as reciprocals of dilutions. The example illustrates the results of examination of three consecutive serum samples from a patient with aseptic meningitis without parotitis and complicated by herpes simplex infection of the lip (S-1, S-7, and S-21, serum taken, respectively, on days 1, 7, and 21 following admission to hospital). Following heating at 56°C for 30 minutes to inactivate complement, each serum was diluted in twofold steps from 1/8 to 1/512. A standard dose of antigen (inactivated mumps or herpes simplex virus) and complement (two hemolytic units) was added to each cup, and allowed to stand at 4°C overnight. Sheep erythrocytes, sensitized by addition of hemolysin (rabbit antiserum against sheep erythrocytes), were then added, and the plate incubated at 37°C for 45 minutes. Where complement has been fixed, there is no lysis, and the red blood cells have sedimented to the bottom of the well. Titration of the complement used in the test is shown in the lowest row (C). Interpretation: (1) The rising titer of antibody against mumps antigen indicates a diagnosis of mumps meningitis. (2) The unchanged titer of antibody against herpes simplex antigen indicates that the herpes labialis represented only a recrudescence of a previously existing latent infection. A small boost to the existing antibody titer might equally have been observed following such reactivation. In any event, since it is not a primary exogenous infection, it is unlikely to be responsible for the patient's meningitis. (Courtesy I. Jack.)*

tein most subject to antigenic change. Moreover, HI is simple, inexpensive, and rapid. It is therefore the serological procedure of choice for identifying isolates of hemagglutinating viruses.

Virus Neutralization. The infectivity of a virus is neutralized by specific antibody by mechanisms discussed in Chapter 6. For a virus neutralization test, serum must first be "inactivated" by heating at 56°C for 30 minutes to remove nonspecific inhibitors. Serum–virus mixtures are

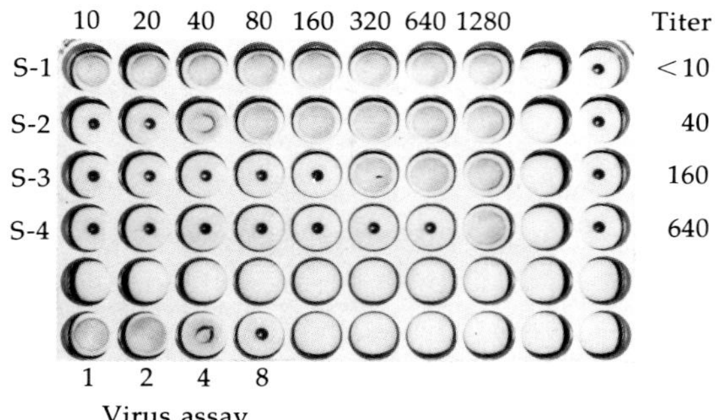

PLATE 12-7. *Hemagglutination inhibition test, used for titrating antibodies to the viral hemagglutinin. Titers are expressed as reciprocals of dilutions. In the example illustrated, an individual was immunized against the prevalent strain of influenza virus. Serum samples, S-1, S-2, S-3, and S-4 were taken, respectively, before immunization, 1 week after the first injection, 4 weeks after the first injection, and 4 weeks after the second injection. The sera were treated with periodate and heated at 56°C for 30 minutes to inactivate nonspecific inhibitors of hemagglutination, then diluted in twofold steps from 1/10 to 1/1280. Each cup then received four hemagglutinating (HA) units of the relevant strain of influenza virus, and a drop of red blood cells. Where enough antibody is present to coat the virions, hemagglutination has been inhibited, hence the erythrocytes settle to form a button on the bottom of the cup. On the other hand, where insufficient antibody is present, erythrocytes are agglutinated by virus and form a shield. The virus assay (bottom line) indicates that the viral hemagglutinin used gave partial agglutination (the end point) when diluted 1/4. Interpretation: The patient originally had no hemagglutination-inhibiting antibodies against this particular strain of influenza virus. One injection of vaccine produced some antibody; the second injection provided a useful booster response. (Courtesy I. Jack.)*

inoculated into appropriate cell cultures which are then incubated until the "virus only" controls develop CPE (Plate 12-8).

Neutralization tests can be exceedingly tedious in the case of viral families that contain large numbers of different serotypes, e.g., the picornaviruses and adenoviruses. To avoid the necessity to test the viral isolate against every single type-specific antiserum, "intersecting serum pools" can be employed. Up to a dozen antisera are combined to give one pool, while a second pool comprises a few of the same antisera together with several additional ones, and so on. With appropriate advice from a mathematician one can so construct the pools that an isolate may be positively identified by observing which particular pools neutralize it and which do not.

As discussed in Chapter 6, the sensitivity of neutralization tests can be

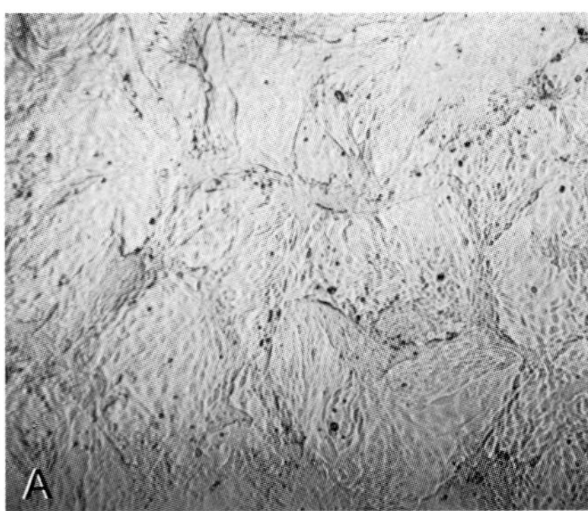

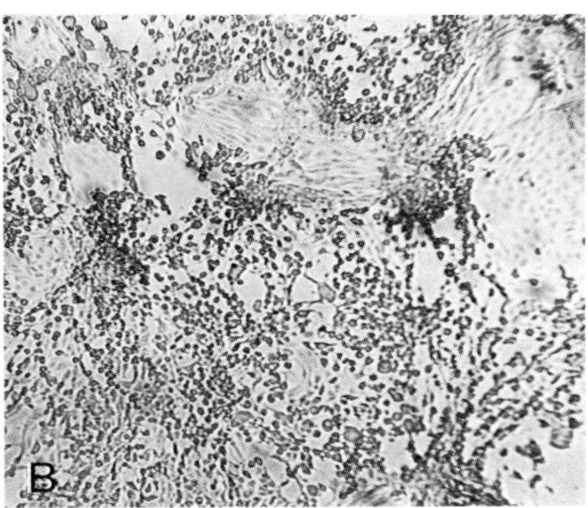

PLATE 12-8. *Virus neutralization test. A child developed meningitis during a summer epidemic of coxsackievirus B2. An enterovirus was isolated from the feces. One-hundred* $TCID_{50}$ *of this virus was incubated at 37°C for 60 minutes with a suitable dilution of "inactivated" (56°C, 30 minutes) anti-coxsackievirus B2 serum (a reference serum previously raised in a rabbit). The mixture was inoculated onto a monolayer of monkey kidney cells in a well of a microculture tray (A). Virus similarly incubated with normal rabbit serum was inoculated into well B. The cultures were incubated at 37°C for several days and inspected daily for CPE. (Unstained, magnification: ×23. Courtesy I. Jack.) Interpretation: The infectivity of this virus isolate has been neutralized by anti-coxsackievirus B2 serum (no CPE); the control culture (B) shows typical CPE.*

increased by the addition of complement, or antiimmunoglobulin. These technical tricks are sometimes employed to enhance neutralization of herpes simplex viruses and certain arboviruses.

In keeping with the general trend toward miniaturization, most neutralization tests today are conducted in disposable, nontoxic, sterile, plastic plates with flat-bottomed wells in which a cell monolayer can be established. Virus–antiserum mixtures can be added either to established monolayers, or simultaneously with the cell suspension. Whereas the standard neutralization test is read by inhibition of CPE, with some viruses the end point can be read simply by color—the so-called *metabolic inhibition* test; antibody, by neutralizing the infectivity of the virus, protects the cells against viral destruction, hence allows cellular metabolism to continue and the resulting acid turns the indicator yellow. In the *plaque reduction* test, which may also be conducted in microtrays, cell monolayers inoculated with virus–antiserum mixtures are overlaid with agar and incubated until countable plaques develop (Plate 2-3); the end point is taken to be the highest dilution of antiserum reducing the number of plaques by at least 50%.

If a newly isolated virus proves to be "untypable," i.e., not neutralizable by antisera against any of the known serotypes, it may indicate a mixed infection with two distinct agents, or aggregation of virions in the specimen. Aggregates can be removed by ultrafiltration or, in the case of nonenveloped viruses, dispersed with sodium deoxycholate, prior to repeating the neutralization test.

A refinement known as *kinetic neutralization* may be needed to differentiate antigenic variants, sometimes called "prime" strains, e.g., very similar arboviruses from different geographical localities (see Chapter 20), vaccine strains of poliovirus (Chapter 19), or herpes simplex type 2 (Chapter 16). The test determines the rate of neutralization of virus by a given concentration of serum, by measuring the amount of virus surviving after incubating the virus–antiserum mixture for various lengths of time.

Oligonucleotide Fingerprinting

For most routine diagnostic purposes it is often not necessary to "type" the isolate antigenically, even to the degree just described. On the other hand, there are certain situations when epidemiological information of vital importance to the public health can be obtained by going even further to characterize very subtle differences between "variants," "varieties," or subtypes within a given serotype (Table 12-4). Short of determining the complete nucleotide sequence of viral nucleic acid, the most useful method of doing this is by oligonucleotide mapping. Viral RNA is labeled with ^{32}P in culture; the labeled RNA is phenol extracted from purified virions, then digested with ribonuclease T1, and the resulting fragments separated by two-dimensional polyacrylamide gel electrophoresis, or by cellulose acetate electrophoresis followed by

TABLE 12-4
Techniques for Identification of Viruses at Various Taxonomic Levels

TAXONOMIC LEVEL	TECHNIQUES OF CHOICE
Family	CPE in cultured cells
	Electron microscopy
	Complement fixation
Species (type)	Neutralization
Subtype	Kinetic neutralization
	Monoclonal antibody serology
Variant	Nucleic acid hybridization
	Oligonucleotide mapping
Point mutation	Nucleic acid sequencing

DEAE-cellulose chromatography. Autoradiography reveals a "fingerprint" unique to that particular viral strain. Similarly, viral DNA can be digested with appropriately chosen restriction endonucleases, resolved by slab polyacrylamide gel electrophoresis and stained with ethidium bromide or silver.

The resolution of such fingerprints is such that different viral isolates are usually distinguishable unless they come from the same epidemic. Minor degrees of genetic drift, often too subtle to be reflected in serological differences, can be picked up in this way. Reference to a library of known fingerprints enables the investigator to trace the epidemiological origin of individual isolates with respect to time, geography, and even potentially to a particular person. For example, it is possible to trace the origin of a given strain of herpes simplex type 2 in a newborn baby to the nursery of the maternity hospital, or of HSV-2 in an adult to a sexual contact. Obviously, because such findings may have legal repercussions, they must be interpreted cautiously, with the limitations of current technology very much in mind. Other striking examples of the power of this technique in tracking the origin and course of outbreaks of poliomyelitis were mentioned in Chapter 9 and will be discussed further in Chapter 19.

Interpretation

The isolation and identification of a particular virus from a patient with a given disease is not necessarily meaningful in itself. Fortuitous subclinical infection with a virus unrelated to the illness in question is not uncommon. Koch's postulates are as apposite here as in any other microbiological context, but are not always easy to fulfil. In attempting to interpret the significance of any virus isolation one must be guided by the following considerations: (1) the site from which the virus was isolated, e.g., one would be quite confident about the etiological significance of rubella virus isolated from any organ of a congenitally deformed infant, or of mumps virus isolated from the CSF of a patient with meningitis, because these sites are usually sterile, i.e., they have no normal bacterial or viral flora. On the other hand, recovery of an echovirus from the feces or herpes simplex virus from the throat may not necessarily be significant, because such viruses are often associated with inapparent infections. Interpretation of the significance of the isolation in such instances will be facilitated by (2) isolation of the same virus from several cases of the same illness during an epidemic, and (3) knowledge that the virus and the disease in question are often causally associated.

TABLE 12-5
Serological Procedures Used in Virology

TECHNIQUE	PRINCIPLE
Virus neutralization	Antibody neutralizes infectivity of virion; CPE inhibition, plaque reduction, or protection of animals
Hemagglutination inhibition	Antibody inhibits viral hemagglutination
Enzyme immunoassay	Enzyme-labeled antibody (or antigen) binds to antigen (or antibody): substrate changes color
Radioimmunoassay	Radiolabeled antibody (or antigen) binds to antigen (or antibody), e.g., attached to solid phase
Time-resolved fluoroimmunoassay	Antibody labeled with europium chelate binds to antigen; fluorescence emitted after excitation by pulses of light
Immunofluorescence	Antibody labeled with fluorochrome binds to intracellular antigen; fluoresces by UV microscopy
Immunoperoxidase staining	Peroxidase-labeled antibody binds to intracellular antigen; colored precipitate on adding substrate
Immune electron microscopy	Antibody-aggregated virions visible by EM
Complement fixation	Antigen–antibody complex binds complement, which is thereafter unavailable for the lysis of hemolysin-sensitized sheep RBC
Immune adherence hemagglutination	Antigen–antibody complex binds C3 which agglutinates RBC
Immunodiffusion	Antibodies and soluble antigens produce visible lines of precipitate in a gel
Single radial diffusion	Antibody diffuses from well; forms zone of precipitate with antigen in gel
Single radial hemolysis	Antibody diffuses into gel; added complement lyses virus-coated RBC
Reverse passive hemagglutination	Antibody-coated RBC agglutinated by antigen
Latex agglutination	Antibody agglutinates antigen-coated latex particles

MEASUREMENT OF SERUM ANTIBODIES

We have discussed the application of several serological techniques to the identification of an unknown antigen, using antibodies of known specificity (for synopsis, see Table 12-5). Serological techniques may also be employed the other way around, to identify an unknown antibody using known antigens. *Paired sera* are taken from the patient, the first (*acute*) specimen as early as possible in the illness, the second (*convalescent*) specimen at least 1–2 weeks later. Blood should be collected in the absence of anticoagulants, given time to clot, and serum separated before freezing for storage. For most purposes several milliliters of blood should be taken, but for epidemiological surveys requiring only the spot-testing of large numbers of people for presence or absence of antibody to a given virus, a much smaller volume may suffice. A simple method of collecting a known volume of blood from a finger- (or ear-, or heel-) prick is to fill a circle of fixed diameter on a disc of filter paper by capillarity; such discs may be dried and stored conveniently at 4°C without appreciable denaturation of IgG (IgM is less stable).

Acute and convalescent specimens should be tested simultaneously. For cetain tests, e.g., neutralization, CF, and HI, the sera must first be *inactivated* by heating at 56°C for 30 minutes and sometimes treated by additional methods to destroy various types of nonspecific inhibitors. Then they are titrated for antibodies by any of the serological techniques described earlier or listed in Table 12-5.

For the safety of laboratory workers it is imperative to inactivate the infectivity of the more dangerous viruses employed as antigens in serological tests (other than neutralization), e.g., arenaviruses, rhabdoviruses, togaviruses. This can be done, without destroying antigenicity, by photodynamic inactivation with ultraviolet light in the presence of psoralen. In the future perhaps, purified viral proteins produced by recombinant DNA technology will be employed. This is desirable for another reason also. Finer specificity can be conferred on serological tests by using those particular viral proteins that characterize different serotypes.

A significant (four-fold or greater) rise in antibody titer between the two serum samples is indicative of recent infection. Unfortunately, because conventional serological techniques do not detect a rise during the first few days of the illness, they provide a diagnosis only in retrospect. However, there are a number of situations when finding antibody in a single specimen of serum can be diagnostic. The first and most important is when specific antibody of the IgM class is found; IgM serology is an important and growing field, discussed below. The second situation is much rarer, but interesting. When a single serum specimen in a seriously ill patient, e.g., a traveler recently returned from the tropics, is found to contain antibodies against a truly exotic virus, it is prudent to assume this agent to be the cause of the illness. For instance, a spot test is now available for the diagnosis of any one of the three lethal African hemorrhagic fevers: Lassa, Marburg, and Ebola; the patient's serum is spotted onto a cell monolayer previously infected with all three viruses (then fixed, and UV and γ irradiated for safety) and tested with fluorescein-labeled anti-human Ig. Third, examination of a single specimen of serum can determine the susceptibility of individuals at risk and enable appropriate measures to be taken, e.g., susceptibility to cytomegalovirus of a proposed recipient of an organ transplant, or susceptibility to varicella of children in a leukemia ward.

There are two further situations when serology can be of diagnostic importance despite the delay involved in waiting for a rising titer to make itself apparent. The first is when it is not practicable to attempt cultivation of viruses that are notoriously difficult to isolate, e.g., some togaviruses and bunyaviruses. The second is when the management of the case is not urgent, e.g., rubella in a pregnant woman with a rash; a clear demonstration of a rising antibody titer constitutes a strong indication for abortion.

In recent years a number of technological advances have been introduced to improve the sensitivity, reproducibility, and labor costs of serological procedures. First, *miniaturization:* most serological tests today are conducted not in test tubes but in wells of plastic trays, for convenience of handling, ease of replication, and conservation of reagents (100–200 μl in microtiter plates, 5–10 μl in Terasaki plates). Second, *automation:* instrumentation is now available for conducting tasks such as sampling, dispensing, diluting, transferring, rinsing, reading, printout, and computerized analysis of results. Finally, *do-it-yourself kits:* perhaps the best-known example being the *latex agglutination* test. Latex particles on a card have been presensitized with, for example, rubella antigen; patient's serum mixed with the latex for a few minutes will produce visible agglutination. The test can be refined to identify specific antibodies of the IgM class.

IgM Serology

In recent years serology has been revolutionized by the realization that a diagnosis can be made on a single acute-phase serum sample by demonstrating specific antibody of the IgM class. Because IgM antibodies appear early after infection but generally drop to very low levels or disappear altogether within 3 months, they are diagnostic of recent (or chronic) infection. They are also diagnostic of intrauterine infection if found in the newborn baby, because maternal IgM, unlike IgG, does not cross the placenta. All of the immunoassays described above can be readily adapted to this purpose—EIA, RIA, and IF are perhaps the most generally useful. A typical indirect RIA for virus-specific IgM is depicted in Fig. 12-4 (left).

A number of technical problems have been encountered with IgM assays. The first, namely lack of specificity of commercial anti-human IgM, is becoming less frequent now that reliable Ig-class-specific antiglobulins, sometimes monoclonal, are being marketed.

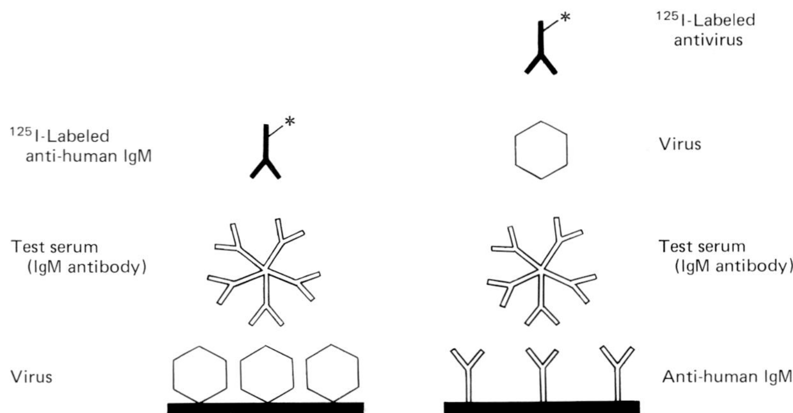

FIG. 12-4. *Radioimmunoassays for detection of specific antiviral antibodies of the IgM class. Left, indirect. Right, reverse.*

The second problem is that of the so-called *rheumatoid factor* (RF), which is antibody, mainly of the IgM class, directed against the constant-region domains of normal IgG. Though first described in rheumatoid arthritis and other "autoimmune" diseases, RF is in fact prevalent in many infectious diseases, and is found in the majority of neonates congenitally infected with any of the "TORCH" agents. This may in part be due to IgM antibodies made by the neonate against the mother's Ig allotypes. It may also be that anti-Ig has a function in recovery from infection, e.g., by augmentation of antibody neutralization titers. Be that as it may, RF creates false positives in IgM immunoassays, because it binds to antiviral (as well as normal) IgG in human serum, forming IgM–IgG complexes, which in turn bind the anti-μ employed in the assay as detector antibody; hence, antiviral IgG will register as if it were antiviral IgM. This is a particular problem in EIAs because, in the process of conjugation to enzyme, the indicator antibody is often aggregated, and RF binds much more avidly to aggregated than to native IgG.

There are several possible solutions to this problem. One can make the assumption that all human sera contain RF, or can screen for RF. In any event, to avoid RF, it is advisable to employ a *reverse(d)* immunoassay of the type illustrated in Fig. 12-4 (right) in the simplest form of which anti-human IgM is used as the capture antibody and labeled virus as the detector/indicator; more commonly, unlabeled virus is the detector, then labeled antiviral IgG, or perhaps better still, $F(ab')_2$ is added as indicator. Secondly, avian anti-human IgM can be used as capture antibody; only mammalian IgG is recognized by human RF. Third, monoclonal antibody of the IgM or IgA class can be used for capture. And finally, the test serum can be preabsorbed with heat-aggregated or glutaraldehyde-cross-linked human IgG, or with protein A-Sepharose. Protein A not only removes RF (by adsorbing the IgM–anti-IgG complexes via the Fc end of the IgG molecules) but also incidentally removes antiviral IgG, which otherwise tends to reduce the apparent titer of antiviral IgM by competing with it in the immunoassay. Only a minority of the antiviral IgM is lost on protein A. It is still necessary to use anti-human IgM as capture antibody in the assay itself, because antiviral antibodies of the IgG_3 subclass are not removed by protein A.

A quite different cause of false-positive IgM assays is attributable to "original antigenic sin" (see Chapters 6 and 21). Patients previously infected with a togavirus, flavivirus, paramyxovirus, enterovirus, or adenovirus may, during a subsequent infection with a quite different virus of the same family, produce low-titer but highly avid IgM cross-reacting with most or all viruses within that family. This trap does not apply to "monotypic" viruses such as rubella, measles, mumps, varicella, EBV, and hepatitis A. False-negative IgM results are occasionally encountered in infants, some respiratory infections, and reactivation of latent herpes simplex.

FURTHER READING

Books

Bachmann, P. A., ed. (1983). "New Developments in Diagnostic Virology," Curr. Top. Microbiol. Immunol. Vol. 104, Springer-Verlag, Berlin and New York.

Gardner, P. S., and McQuillin, J. (1980). "Rapid Virus Diagnosis: Application of Immunofluorescence," 2nd ed. Butterworth, London.

Grist, N. R., Bell, E. J., Follet, E. A., and Urquhart, G. E. D. (1983). "Diagnostic Methods in Clinical Virology," 4th ed. Blackwell, Oxford.

Howard, C. R., ed.(1982). "New Developments in Practical Virology," Lab. Res. Methods Biol. Med., Vol. 5. Alan R. Liss, Inc., New York.

Hsiung, G. D., and Green, R. H., eds. (1978). "CRC Handbook Series in Clinical Laboratory Science. Section H. Virology and Rickettsiology," Vol. 1, Parts 1 and 2. CRC Press, Boca Raton, Florida.

Langone, J. J., and van Vunakis, H., eds. (1980–1981). "Methods in Enzymology," Vols. 70, 73 and 74. Academic Press, New York.

Lennette, D. A., Specter, S., and Thompson, K. D., eds. (1979). "Diagnosis of Viral Infections: The Role of the Clinical Laboratory." University Park Press, Baltimore, Maryland.

Lennette, E. H., ed. (1985). "Laboratory Diagnosis of Viral Infections." Dekker, New York.

Lennette, E. H., and Schmidt, N. J., eds. (1979). "Diagnostic Procedures for Viral and Rickettsial Infections," 5th ed. Am. Public Health Assoc., Washington, D.C.

Lennette, E. H., Balows, A., Hausler, W. J., and Truant, J. P., eds. (1980). "Manual of Clinical Microbiology," 3rd ed. Am. Soc. Microbiol., Washington, D.C.

Reviews

Clewley, J. P., and Bishop, D. H. L. (1982). Oligonucleotide fingerprinting of viral genomes. *In* "New Developments in Practical Virology" (C. R. Howard, ed.), pp. 231–277. Alan R. Liss, Inc., New York.

Field, A. M. (1982). Diagnostic virology using electron microscopic techniques. *Adv. Virus. Res.* **27,** 1.

Flewett, T. H. (1980). Safety in the virology laboratory. *In* "Recent Advances in Clinical Virology" (A. P. Waterson, ed.), Vol. 2, pp. 169–187.

Gafre, G., and Milstein, C. (1981). Preparation of monoclonal antibodies: Strategies and procedures. *In* "Methods in Enzymology" (J. J. Langone and H. van Vunakis, eds.), Vol. 73, p. 1. Academic Press, New York.

Halonen, P., and Meurman, O. (1982). Radioimmunoassay in diagnostic virology. *In* "New Developments in Practical Virology" (C. R. Howard, ed.), pp. 83–124. Alan R. Liss, Inc., New York.

McIntosh, K., Wilfert, C., Chernesky, M., Plotkin, S., and Mattheis, M. J. (1978). Summary of a workshop on new and useful techniques in rapid viral diagnosis. *J. Infect. Dis.* **138,** 414.

McIntosh, K., Wilfert, C., Chernesky, M., Plotkin, S., and Mattheis, M. J. (1980). Summary of a workshop on new and useful techniques in rapid viral diagnosis. *J. Infect. Dis.* **142,** 793.

Madeley, C. R. (1977). "Guide to the Collection and Transport of Virological Specimens." World Health Organ., Geneva.

Meurman, O. (1983). Detection of antiviral IgM antibodies and its problems—a review. *Curr. Top. Microbiol. Immunol.* **104,** 101.

Richman, D., Schmidt, N., Plotkin, S., Yolken, R., Cherensky, M., McIntosh, K., and Mattheis, M. (1984). Summary of a workshop on new and useful methods in rapid viral diagnosis. *J. Infect. Dis.* **150,** 941.

Schmidt, N. J. (1972). Tissue culture in the laboratory diagnosis of viral infections. *Am. J. Clin. Pathol.* **57,** 820.

Schmidt, N. J. (1979). Laboratory diagnosis of viral infections. *In* "Antiviral Agents and Viral Diseases" (G. J. Galasso, T. C. Merigan and R. A. Buchanan, eds.) pp. 209–252. Raven Press, New York.

Schmidt, N. J., and Lennette, E. H. (1973). Advances in the serodiagnosis of viral infections. *Prog. Med. Virol.* **15,** 244.

Schmidt, N. J., Hamparian, V. V., Sather, G. E., and Wong. Y. W. (1978). Quality assurance practices for virology laboratories. *In* "Quality Assurance Practices in Health Laboratories" (S. L. Inhorn, ed.), pp. 1097–1144. Am. Public Health Assoc., Washington, D.C.

Sever, J. L. (1983). Automated systems in viral diagnosis. *Curr. Top. Microbiol. Immunol.* **104,** 57.

van Regenmortel, M. H. V. (1981). Serological methods in the identification and characterization of viruses. *In* "Comprehensive Virology" (H. Fraenkel-Conrat and R. R. Wagner, eds.), Vol. 17, p. 183. Plenum, New York.

World Health Organization (1981). Rapid laboratory techniques for the diagnosis of viral infections. *W. H. O. Tech. Rep. Ser.* **661.**

Yolken, R. H. (1982). Enzyme immunoassays for the detection of infectious antigens in body fluids: Current limitations and future prospects. *Rev. Infect. Dis.* **4,** 35.

Yolken, R. H. (1983). Use of monoclonal antibodies for viral diagnosis. *Curr. Top. Microbiol. Immunol.* **104,** 177.

PART II

Viruses of Man

CHAPTER 13
Hepadnaviruses

Introduction 365
Properties of *Hepadnaviridae* 366
Clinical Features of Hepatitis B 368
Pathogenesis and Immunity 369
Laboratory Diagnosis 371
Epidemiology 374
Control 376
Further Reading 380

INTRODUCTION

In 1963 Blumberg, a geneticist investigating hereditary factors in the sera of isolated racial groups, discovered an antigen in the serum of an Australian aborigine that reacted with sera from multiply transfused American hemophiliacs. In due course the antigen was demonstrated to be present on the surface of particles with three different morphological forms (Plate 13-1) and to be associated with the disease serum hepatitis (now known as hepatitis B). The 22-nm particles of "Australia antigen," subsequently renamed HBsAg, were found to be noninfectious, but the 42-nm particles were shown to be infectious virions capable of transmitting hepatitis to chimpanzees. The unique characteristics of these viruses led to their classification within a new family; the new name, *Hepadnaviridae,* adopted by the ICTV in 1984, reflects their association with hepatitis and their DNA genome (Table 13-1).

Hepatitis B is one of the world's major unconquered diseases. Some 200 million people are chronic carriers of the virus, and a significant minority of these go on to develop cirrhosis or cancer of the liver. Al-

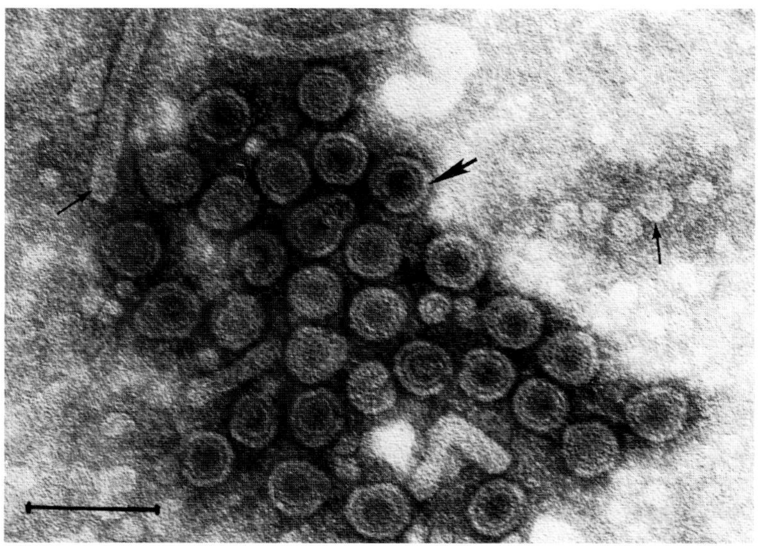

PLATE 13-1. Hepadnaviridae. *Negatively stained preparation of hepatitis B virions (Dane particles, large arrow) and accompanying HBsAg particles (small arrows). Bar = 100 nm. (Courtesy Dr. I. D. Gust and J. Marshall.)*

though the virus has yet to be cultivated *in vitro*, reliable diagnostic procedures and a much-needed vaccine are already available.

PROPERTIES OF *HEPADNAVIRIDAE*

The *Hepadnaviridae* family so far contains hepatitis viruses specific for man, woodchucks, ground squirrels, ducks, and snakes. We are concerned here only with the agent of human hepatitis B.

The icosahedral virion, sometimes called the Dane particle, is 42 nm in diameter and contains two concentric shells—an outer capsid and an inner core. The genome consists of a molecule of circular dsDNA, MW 1.6–2.0 million, of most unusual structure. One of the DNA strands is incomplete, leaving 15–50% of the molecule single-stranded, but this gap is repaired *in situ* in the virion by a DNA polymerase carried in the core. As far as we know, this situation is unique in virology. The complete (long) strand contains a discontinuity ("nick") at a unique site, and has a protein molecule covalently attached to its 5′ end.

The capsomers of the outer capsid of the virion are dimers consisting of one molecule each of the glycosylated and unglycosylated forms of

TABLE 13-1
Properties of Hepadnaviridae

Spherical virion, 42 nm
Outer icosahedral shell and inner core
dsDNA genome, circular, MW 1.6 million, negative strand nicked, positive strand variable length
DNA polymerase, protein kinase, and reverse transcriptase? in virion
DNA replicates via RNA intermediate?
Integration of viral DNA occurs

the major polypeptide of the virion. This is known as the hepatitis B surface antigen (HBsAg). The core of the virion is antigenically distinct (HBcAg); this polypeptide is phosphorylated by a protein kinase, present in the core. A third antigen, also detectable in the core, is known as HBeAg; it shares at least part of the amino acid sequence of HBcAg and is thought to be a monomeric form or breakdown product of HBcAg.

Virions present in the serum of chronic carriers are accompanied by a large excess—up to 10^{13} particles per ml—of the smaller, noninfectious particles of HBsAg (Plate 13-1). These are spheres or filaments, 22 nm in diameter, consisting solely of the HBsAg that makes up the outer capsid of the virion (together with some adsorbed albumin and other serum proteins). Though noninfectious, HBsAg particles are important for two reasons: (1) they serve as a diagnostic marker of infection, and (2) they can be harvested for use as a vaccine.

There are a number of subtypes of hepatitis B virus, defined by various combinations of antigenic determinants present on the HBsAg. All have the same group-specific determinant, *a*, but there are four major subtype-specific determinants, certain pairs of which (*d* and *y*; *r* and *w*) tend to behave as alleles, i.e., as mutually exclusive alternatives. Hence we have subtypes *adr, adw, ayr, ayw* with characteristic geographical distributions, though they often overlap. The situation has become complicated by the recent discovery of additional determinants, as well as by the finding that the so-called allelic pairs are not in fact always mutually exclusive. The precise molecular basis of these antigenic differences is now being elucidated by sequencing of the DNA and/or proteins of the various subtypes.

Hepatitis B virus infects man and certain other higher primates such as the chimpanzee and gibbon. It displays tropism only for the liver. Although there have been occasional reports of the cultivation of the virus in organ culture and even in continuous cell lines, there is still no system in which the virus can be consistently grown *in vitro*.

Viral Replication

As hepatitis B virus has not yet been cultured, little is known of its replication other than that HBcAg is found in liver cell nuclei and HBsAg in cytoplasm. However, there are some fascinating findings with duck hepatitis virus that, if found to be applicable to the *Hepadnaviridae* in general, would indicate that this family has evolved a most extraordinary procedure for replicating its genome. In brief, the complete (−) viral DNA strand appears to be transcribed to give a full-length (+)RNA copy. A reverse transcriptase, probably utilizing a protein primer, then transcribes a (−)DNA strand from the (+)RNA template, the latter being simultaneously degraded. Now the viral DNA polymerase is able to employ the (−)DNA strand as template for the transcription of (+)DNA. Newly synthesized DNA is packaged into virions before this last step is complete, so viral DNA is only partially double-stranded. Integration of viral DNA into cellular DNA is a regular occurrence and may lead to persistent infection or to liver cancer. The analogy with the retroviruses is fascinating.

CLINICAL FEATURES OF HEPATITIS B

The course of acute viral hepatitis is conventionally divided into three phases: (1) preicteric, (2) icteric, and (3) convalescent (Fig. 13-2). Following a long incubation period of 6 weeks to 6 months in the case of hepatitis B, the *preicteric (prodromal)* phase commences with malaise, lethargy, anorexia, and commonly nausea, vomiting, and abdominal pain in the right upper quandrant. A minority of patients (10–20%) develop at this time a type of serum sickness characterized by mild fever, urticarial rash, and arthralgia, resembling a benign, fleeting form of acute rheumatoid arthritis. Anything from 2 days to 2 weeks after the malaise begins, there commences the *icteric* phase, heralded by dark urine (bilirubinuria) closely followed by jaundice. On palpation, the liver is found to be only moderately enlarged and tender. Paradoxically, the patient feels better after jaundice develops. Prolonged jaundice implies cholestasis and may be accompanied by pruritis. The *convalescent* phase may be long drawn out, with malaise and fatigue lasting for weeks.

There are a number of possible outcomes (Fig. 13-1). In Western countries, most patients recover uneventfully but about 1% of hospitalized patients die of fulminant hepatitis. The case fatality rate (CFR) is related to the dose of virus received and to the underlying health of the patient; whereas the CFR in iv drug users does not significantly exceed that of hepatitis A, it can be up to 100 times higher (>10%) in posttransfusion

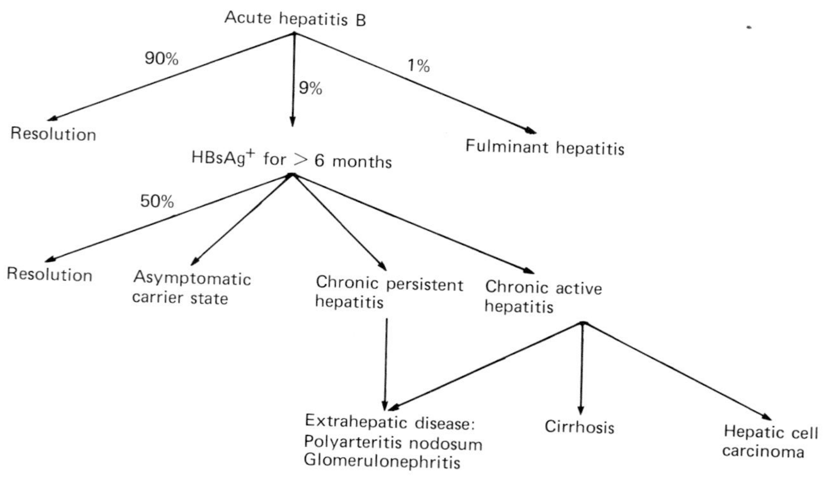

FIG. 13-1. *Clinical outcomes of acute hepatitis B.*

hepatitis B. Another 5–10% go on to one or another form of chronic infection, usually defined as HBs antigenemia persisting for at least 6 months. Hepatologists classify the spectrum of chronic conditions in various ways, e.g., into lobular, chronic persistent, and chronic active hepatitis, the latter being the most severe and often progressing to cirrhosis or to primary hepatocellular carcinoma. A proportion of those with chronic persistent or chronic active hepatitis develop manifestations of immune-complex disease, systemic necrotizing vasculitis (polyarteritis nodosa) and membranoproliferative glomerulonephritis being the two most common. Epidemiologically important also is the asymptomatic carrier state, in which liver damage does not progress but virus continues to be produced and to circulate in the bloodstream for years.

In the developing countries of Africa and Asia, where most of the world's hepatitis B occurs, the consequences of infection are much more serious, largely because the virus is generally acquired in the first few years of life, often perinatally. Of those infected at birth, 95% become chronic carriers and one in every three goes on to develop chronic active hepatitis, cirrhosis, or primary hepatocellular carcinoma in middle life.

PATHOGENESIS AND IMMUNITY

It is assumed, but not proved, that the hepatitis B virus reaches the liver via the bloodstream and multiplies principally or exclusively in hepatocytes. A prolonged incubation period of between 6 weeks and 6 months (average about 10 weeks) distinguishes the disease from hepatitis A. Liver biopsy at the height of hepatitis A or B reveals necrosis of liver parenchyma (Plate 13-2A). Rarely, the patient develops a fulminating hepatitis and dies within days from "acute yellow atrophy" (Plate 13-2B). Alternatively, subacute or chronic active hepatitis may develop. A proportion of these progress to cirrhosis or to carcinoma. More usually, however, regeneration of liver parenchyma is complete within 2–3 months.

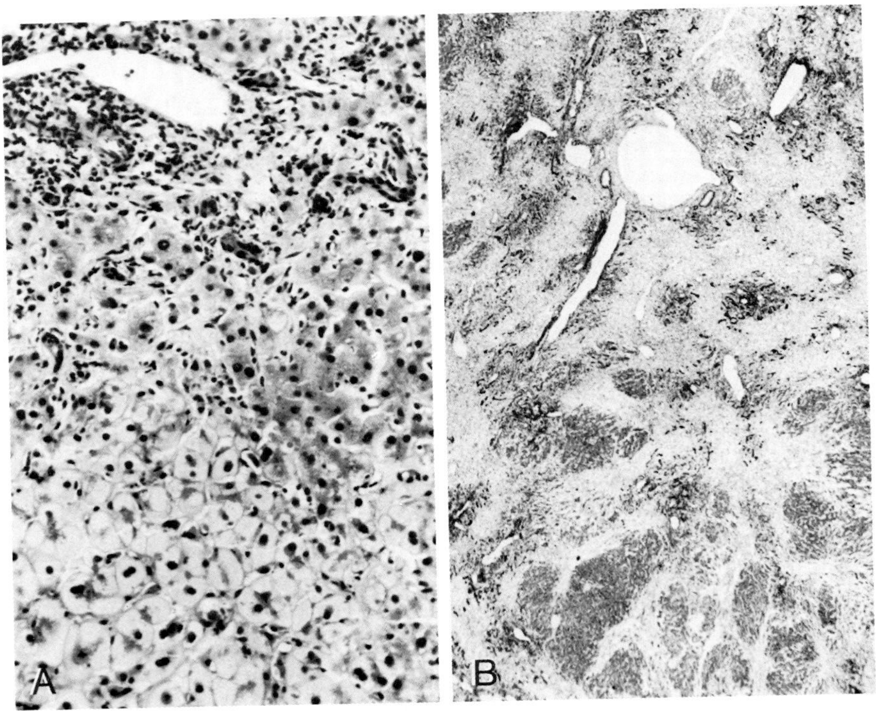

PLATE 13-2. *Viral hepatitis. (A) Liver biopsy 2 weeks after onset of acute hepatitis (H and E stain, ×125). Note focal necrosis within liver lobule with characteristic ballooning of hepatocytes (bottom left), biliary stasis (center right), and inflammatory cell infiltration around a widened portal tract (top left). (B) Liver at autopsy of patient who died 3.5 weeks after onset of viral hepatitis (H and E stain, ×12). Note massive necrosis of liver parenchyma and subsequent collapse leaving portal tracts with proliferating ductules and nodules of regenerating liver cells. (Courtesy Dr. P. S. Bhathal and I. Jack.)*

By immunofluorescence or electron microscopy HBcAg is demonstrable in the nuclei of affected hepatocytes, whereas HBsAg and virions are seen in the cytoplasm. By autoradiography using cDNA probes, HBV DNA is found in the cytoplasm of acutely infected cells but integrated into the chromosomes of chronically infected carriers. HBsAg, but not HBcAg, is found free in the serum of patients with acute or chronic active hepatitis. In the symptom-free carrier state, HBsAg persists, sometimes in concentrations up to 10^{13} particles per ml, for many years or even for life. The probability of becoming a hepatitis B carrier decreases markedly with age of acquisition of the infection. Perinatal infection of newborn infants leads regularly (95%) to persistent infection which can last for life, whereas only 5–10% of those first infected as adults become carriers (except for immunosuppressed individuals, most of whom become carriers).

Acquired immunity, evidently attributable principally to anti-HBs, is solid—perhaps lifelong—against exogenous reinfection with the homologous subtype. Cross-immunity, presumably due to antibodies directed against the common *a* determinant, confers protection against heterologous subtypes. Current research to measure the precise contribution of antibodies to the *d*, *r*, *w*, and *y* determinants in influencing the magnitude of protection may or may not indicate that the composition of future vaccines should be varied for use in various parts of the world.

Immune complexes are associated with both the "serum sickness" often seen during the prodrome of the acute illness and with the vasculitis or glomerulonephritis sometimes encountered in chronic hepatitis. A direct immunopathological role for these complexes, or indeed any other immunological phenomenon, has not yet been proved, although there is evidence that HBeAg–antibody complexes may induce membranous glomerulonephritis, especially in children.

The reader is referred back to Chapter 8 for a full discussion of the pathogenesis and epidemiology of primary hepatocellular carcinoma and its relationship to the hepatitis B carrier state.

LABORATORY DIAGNOSIS

Routine biochemical tests of liver function distinguish viral hepatitis from the many nonviral, e.g., obstructive or toxic, causes of jaundice. Characteristically, levels of serum transaminases (aminotransferases) are elevated markedly (5- to 100-fold) in acute symptomatic viral hepatitis, whether it be due to hepatitis A, B, or non-A, non-B. Alanine aminotransferase (ALT), otherwise known as serum glutamic-pyruvic transaminase (SGPT), and aspartate aminotransferase (AST), also known as serum glutamic-oxaloacetic transaminase (SGOT), rise together late in the incubation period to peak about the time jaundice appears; they gradually revert to normal over the ensuing 2 months in an uncomplicated case. Alkaline phosphatase and lactic dehydrogenase levels, on the other hand, are only slightly (1- to 3-fold) elevated. Serum bilirubin may rise, anything up to 25-fold, depending on the severity of the case, and may of course be close to normal in anicteric viral hepatitis.

Serology forms the basis of the diagnosis of hepatitis B, and of the differentiation of the various clinical forms of hepatitis B from one another. While many types of immunoassay have been successfully applied to HBV, the most widely used and most sensitive have been radioimmunoassay (RIA) and enzyme immunoassay (EIA). Five markers, all found in serum, are of particular diagnostic importance: hepatitis B surface antigen (HBsAg), HBeAg, antibody to HBsAg (anti-HBs), anti-HBe, and anti-HBc. (For some reason HBcAg is rarely found free in the serum, but only in the nuclei of hepatocytes.)

The typical *acute* infection with hepatitis B virus (Fig. 13-2A) is characterized by the appearance of HBsAg in the blood during the last month of the incubation period, rising to a peak shortly after symptoms develop, then gradually disappearing coincidentally with the fall in transaminase levels over the next 1–5 months. HBeAg (and viral DNA polymerase) appears at about the same time as HBsAg but disappears more abruptly when symptoms and enzyme levels peak. The earliest antibody to appear is anti-HBc; it becomes detectable even before symptoms develop, rises rapidly to high titer, and persists indefinitely. Anti-HBs, on the other hand, does not become detectable until HBsAg has totally disappeared and recovery is complete (usually after 8–15 months); indeed there is often a "window" in time during which neither HBsAg nor anti-HBs is demonstrable, and anti-HBc is the only positive marker of infection.

Chronic hepatitis B infection (Fig. 13-2B) is characterized by the persistence of HBsAg for at least 6 months, but often for years or even for life. As long as HBs antigenemia persists, no anti-HBs is ever found, but

FIG. 13-2. *The serologic events associated with the typical course of acute type B hepatitis (A), and the development of the chronic hepatitis B virus carrier state (B). Abbreviations: hepatitis B surface antigen, HBsAg; hepatitis B e antigen, HBeAg; alanine aminotransferase, ALT; antibody to HBsAg, anti-HBs; antibody to hepatitis B core antigen, anti-HBc; antibody to HBeAg, anti-HBe. (From J. H. Hoofnagle, 1981. Reproduced with permission, from the* Annual Review of Medicine, *Vol. 32, © 1981 by Annual Reviews Inc.)*

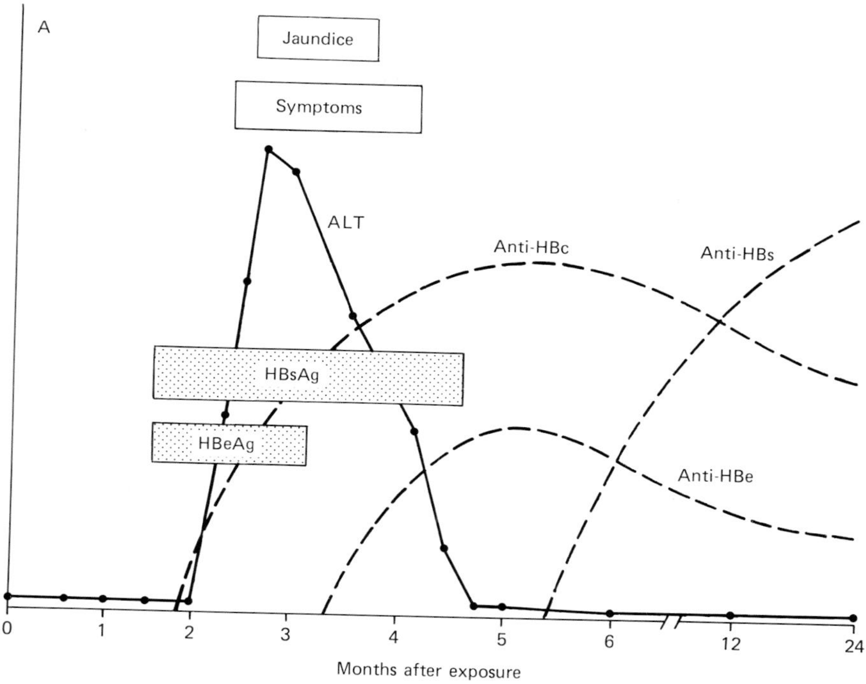

A
Jaundice
Symptoms
ALT
Anti-HBc
Anti-HBs
HBsAg
HBeAg
Anti-HBe
0
1
2
3
4
5
6
12
24
Months after exposure

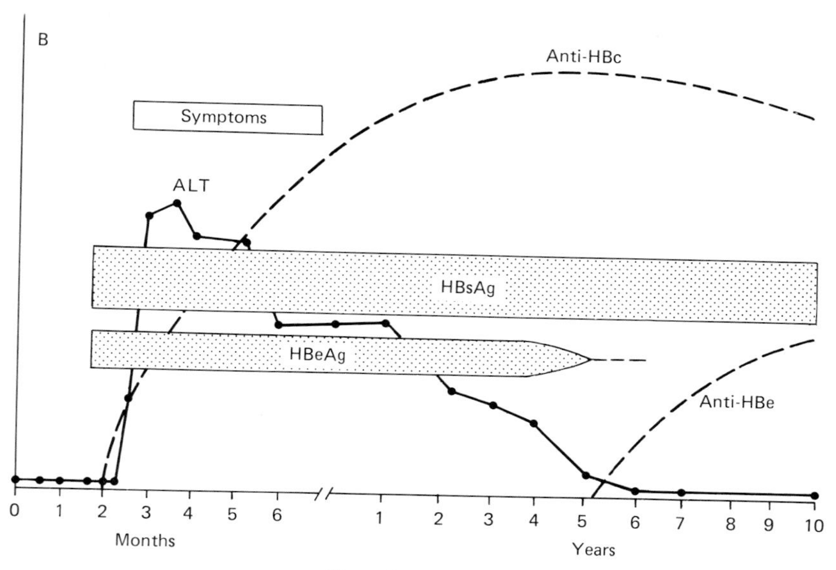
B
Anti-HBc
Symptoms
ALT
HBsAg
HBeAg
Anti-HBe
0
1
2
3
4
5
6
Months
1
2
3
4
5
6
7
8
9
10
Years
Time after exposure

anti-HBc rises to very high titer and persists for life in the normal way. This is the picture with the HBsAg *carrier state*.

Chronic active hepatitis is distinguished from the "healthy" or "asymptomatic" carrier state by the progression of liver damage, as indicated by continuing elevation of serum transaminase levels and histological evidence on liver biopsy. Persistence of HBeAg, HBV DNA polymerase, and Dane particles implies active viral multiplication, high infectivity, progressive liver damage, and poor prognosis. In contrast, anti-HBe, which develops only after HBeAg disappears and enzyme levels have returned to normal, indicates a long-standing asymptomatic carrier state of relatively lower infectivity.

Table 13-2 summarizes the patterns of serological markers that characterize the various outcomes of hepatitis B infection. Note that the key markers are HBsAg, anti-HBs, and anti-HBc; the pattern of these can distinguish most of the important situations. HBeAg and anti-HBe determinations are called for only if the patient is HBsAg$^+$.

Note that (1) the single most reliable marker of past or present HBV infection is anti-HBc; (2) the presence of HBsAg and anti-HBc IgM differentiates current from past infection; (3) persistence of HBeAg in chronic active hepatitis portends an unfavorable outcome; and (4) anti-HBs, which is the major protective antibody, appears only after HBsAg has vanished, hence is a reliable indicator of recovery and of immunity to reinfection.

EPIDEMIOLOGY

During World War II, United States troops destined for zones in which yellow fever was endemic were immunized with a live attenuated

TABLE 13-2
Serological Markers of Hepatitis B Infection

	SEROLOGICAL MARKER					
			ANTI-HBc			
CLINICAL CONDITION	HBsAg	ANTI-HBs	TOTAL	IgM	HBeAg	ANTI-HBe
Acute hepatitis B	+	−	+	+	+ → −	− → +
Asymptomatic carrier state	+[a]	−	+	±[b]	−	+
Past infection: immunity	−	+	+	−	−	+ → −
Past immunization	−	+	−	−	−	−

[a] Persisting for more than 6 months.
[b] Low titer.

yellow fever virus vaccine that had been "stabilized" with normal human serum. Many of these men contracted hepatitis, and the cause was traced to hepatitis B virus present in the serum. This unfortunate occurrence, together with a number of earlier disasters, e.g., transmission of hepatitis B to syphilis patients receiving regular injections of heavy metals from a communal syringe, led to the realization that hepatitis B virus is present in the blood of asymptomatic carriers and is readily transmitted via blood, particularly as a result of iatrogenic invasive procedures. Less than 1 μl of blood contaminating a syringe or needle, used by doctor, dentist, or drug addict can readily transmit hepatitis B from one individual to another. Even the amount of serum that oozes from skin wounds or sores can transmit hepatitis. Hard drug users ("mainliners") are particularly likely to contract the disease. Tattooing, acupuncture, and ear piercing without rigorous sterilization of equipment also constitute potential routes of transmission. Hepatitis B is an occupational risk in those medical and paramedical personnel frequently exposed to blood, e.g., dentists, surgeons, pathologists, mortuary attendants, technicians and scientists working in serology, hematology, biochemistry, and microbiology laboratories in hospitals or public health institutions, blood banks, or hemodialysis units. Until recently, patients receiving large or frequent transfusions of blood or blood products, e.g., hemophiliacs, open-heart surgery patients, renal dialysis patients, were particularly vulnerable to acquiring massive doses of virus. However, the incidence of posttransfusion hepatitis B has dropped dramatically since sensitive assays for HBsAg have been introduced as a routine screening procedure by blood banks throughout much of the world during the past decade or so.

In the developing countries, the prevalence of HBV infection is very much higher than in the West. Most adults in Asia, Africa, and Oceania have anti-HBs, indicative of past infection, whereas <5% ever become infected in Western Europe, North America, and Australia. More disturbing is the fact that the HBsAg carrier rate is up to 100 times higher in developing than in developed countries—frequently 10%, but even as high as 50% in some Pacific islands, compared with 0.1–1% in most of the countries of Western Europe, North America, and Australia. Indeed, there are estimated to be of the order of 200 million carriers of hepatitis B in the developing world today. Why the difference? There is no evidence to suggest a genetic predisposition to chronic HBV infection in particular racial groups; a newborn Australian baby, if infected with HBV, has the same chance of becoming a chronic carrier as does the baby of an infected Melanesian or Chinese. Even though such bloodletting procedures as tattooing, scarification and facial decoration, ear and nose piercing, and circumcision are an integral part of many tribal initiation ceremonies, it seems unlikely that they are important modes of transmission in these societies, because infection occurs at an age prior to the occurrence of these rituals. The infection seems to be acquired in early childhood, perinatally in many instances. Transmission via saliva or other types of close contact currently appears to be the most likely.

It is now clear that, in Western countries also, blood spread is not the only, or even the principal mode of transmission of the hepatitis B virus. Evidence is accruing which indicates that HBV can also be spread by close contact. Saliva contains the virus and may be the principal vehicle of transmission. Furthermore, hepatitis B is now recognized to be a sexually transmitted disease; semen and other genital secretions can transmit the virus to partners. Promiscuous male homosexuals almost invariably become infected. Perinatal transmission, though of minor importance to the overall spread of HBV in the West compared with its major significance in many developing countries, regularly occurs if the mother is HBeAg positive. Evidence that this is generally not through vertical transmission across the placenta comes from the finding that only about 6% of such infants are $HBsAg^+$ at birth whereas a majority are positive by one month of age and >90% by 3 months. Perinatal infection could occur via maternal blood contaminating the baby during parturition, or via saliva or other forms of contact during the early weeks of life. Transmission from carriers to other family members is also quite common.

CONTROL

Blood banks today routinely screen donors for HBsAg using a sensitive radioimmunoassay (or sometimes, reverse passive hemagglutination). This has almost eliminated posttransfusion hepatitis B.

Proper sterilization of equipment and careful aseptic technique by doctors, dentists, acupuncturists, tattooists, laboratory technicians, drug addicts, etc. constitutes another important approach to the control of hepatitis B. Disposable, rather than reusable equipment, particularly syringes and needles, should be preferred wherever feasible. Sterilization procedures must be sufficient to destroy hepatitis B virus, which is relatively hardy. Autoclaving or dry heat sterilization is perfectly adequate. Chemical disinfection is unreliable, but if it has to be employed for heat-labile materials or blood spills, sodium hypochlorite, peracetic acid, or glutaraldehyde are probably the most generally useful disinfectants. We still have no reliable technique for inactivating the virus in blood and heat-labile blood products.

Rigorous aseptic technique must be observed in all potentially dangerous situations, whether in the ward or the laboratory. For example, gloves should be worn by dentists, or by laboratory technicians handling blood specimens if there is a risk that the outside of the container may be contaminated with blood. Eating, smoking, and mouth pipetting must be prohibited in laboratories. Staff should be screened for HBsAg at regular intervals. Accidents, e.g., a spill of $HBsAg^+$ blood, must be reported and may need to be followed up by combined passive/active immunization of personnel, if they have not already been actively immunized with hepatitis B vaccine.

Passive Immunization

Hepatitis B immune globulin (HBIG), obtained by plasmapheresis of subjects with high titers of anti-HBs, is quite effective in prompt postexposure prophylaxis of hepatitis B, e.g., in people who have been exposed to infection with $HBsAg^+$ blood in hemodialysis units, blood banks, and serology laboratories, in newborn infants of HBsAg positive mothers, and in sexual contacts of persons with acute hepatitis B.

Vaccines

Paradoxically, we have an effective vaccine against hepatitis B even though the virus has not yet been cultured *in vitro.* Such exceedingly high concentrations of HBsAg can be found in the sera of carriers that a vaccine can be prepared simply by purifying the antigen directly from this source—with suitable precautions to inactivate any contaminating virus. The protocol for the preparation of the first hepatitis B vaccine licensed for human use is set out in Table 13-3.

In a controlled trial in male homosexuals the vaccine was demonstrated to produce no untoward side effects and no hepatitis, but induced antibodies and reduced the incidence of subsequent disease by over 95% in this high-risk population. A much lower degree of protection was achieved in a subsequent study in hemodialysis patients. A typical regimen is 3 doses (each 20 μg) of HBsAg, injected intramuscularly at intervals of 1 and 6 months; an occasional booster will probably also turn out to be necessary as antibody levels wane after a few years. The vaccine fails to eliminate the virus from carriers, but nor does it do them any harm, so immunization programs can be mounted without the necessity to screen potential recipients. The cost/benefit of prescreening depends on the cost of vaccine and the cost of testing for markers of current or past infection.

In an emergency, immediate protection can be conferred by giving immune human globulin at the same time as the first dose of vaccine.

TABLE 13-3
Steps in the Preparation of a Human Hepatitis B Vaccine[a]

1. Collect plasma by regular plasmapheresis of known high titer HBsAg human carriers negative for infectious virions (Dane particles), DNA polymerase, and HBeAg
2. Defibrinate with calcium
3. Concentrate HBsAg by precipitation with ammonium sulfate
4. Band by isopycnic (equilibrium) ultracentrifugation in sodium bromide
5. Reband by rate zonal centrifugation in sucrose gradient
6. Treat with pepsin, pH 2 to digest adsorbed plasma and liver proteins
7. Denature with 8 *M* urea, then renature (also to remove extraneous proteins)
8. Purify by gel (molecular sieve) filtration
9. Treat with formalin, 1 : 4000, 3 days to inactivate any residual infectious virus
10. Adsorb onto aluminium hydroxide as adjuvant (0.5 mg alum per 20 μg HBsAg) and add thimerosal 1 : 20,000 as preservative
11. Ampoule in 1 ml doses (20 μg HBsAg) and store at 4°C

[a] Reproduced in modified form from Hilleman *et al.* (1982).

Such combined active–passive immunization is advisable in the case of accidental exposure to contaminated blood, and in newborn babies of infected mothers.

This particular vaccine is very expensive and will always be in short supply. Therefore, a carefully considered strategy must be devised to meet the individual needs of each country. In the developed nations, such a strategy should have two principal objectives: (1) to reduce the pool of chronic carriers, by immunizing the babies of carrier mothers and the various other approaches mentioned above; and (2) to reduce the occurrence of clinical disease (icteric hepatitis B) in certain high-risk groups (Table 13-4). In the developing world, by way of contrast, the optimal strategy would be universal immunization of newborn babies, but this is out of the question with such a costly vaccine.

TABLE 13-4
Candidates for Hepatitis B Vaccine

Babies born of HBsAg-positive mothers
Family and sexual contacts of known chronic HBsAg carriers
Patients with diseases requiring frequent transfusions
Patients and staff of hemodialysis units
Medical and ancillary personnel in other high risk situations, e.g., blood banks, serology laboratories, dental surgery, surgery, pathology
Inmates and staff of homes for the mentally retarded and other custodial institutions
Immunosuppressed or cancer patients
Drug addicts
Male homosexuals and prostitutes

In the longer term it will be necessary to replace this type of vaccine with something much less expensive and much more widely available if we are to contemplate routine immunization of all infants in those regions of Africa, Asia, and Oceania where the infection is so prevalent. A conventional live attenuated or inactivated vaccine is likely to be highly satisfactory and should follow within a few years of the cultivation of the virus in cell culture. The cross-protection between subtypes reported with the existing HBsAg vaccines and with cross-challenge experiments in chimpanzees may be of sufficient magnitude to obviate the necessity to include more than one virus in such a vaccine.

Meanwhile, a succession of dramatic discoveries in the early 1980s has quickened interest in alternative approaches to vaccination (see Chapter 10).

1. Hepatitis B viral DNA has been cloned by recombinant DNA technology in bacteria, and subsequently in yeasts and mammalian cells from which HBsAg particles have been obtained. HBsAg vaccines prepared by cloning in yeast or in mammalian cells have been shown to protect chimpanzees against HBV challenge, and to be both safe and immunogenic in man. Prospects are that this will replace the current HBsAg vaccines by about 1987, but may not initially offer any substantial price advantage.

2. In a different approach, the HBsAg polypeptide, as well as shorter peptides representing important antigenic sites, have been synthesized chemically and demonstrated to have limited immunogenicity. While partial restoration of the tertiary structure of the critical peptide may improve its immunogenicity somewhat, such peptide vaccines will almost certainly require coupling to a carrier, incorporation into liposomes, and/or emulsification with an adjuvant (not yet available).

3. Perhaps the most exciting development of all is the recent demonstration that the gene for HBsAg can be incorporated into the genome of live vaccinia virus and that chimpanzees vaccinated with this recombinant are protected against challenge with hepatitis B (see Chapter 10). Such an approach, in which infants could be vaccinated with a multipurpose live recombinant vaccine incorporating the critical antigens of several major disease agents, may well be the answer to hepatitis B as well as to other serious problems in the populous Third World.

Chemotherapy

The results of clinical trials with interferons and adenine arabinoside monophosphate (ara-AMP) were discussed in Chapter 11. To recapitulate succinctly here, interferons in adequate dosage generally lead to a temporary decrease in serological markers of viral activity but to no long-term clinical improvement. Ara-AMP has shown limited promise However, in recent trials interferon and ara-AMP in combination or alternation proved to be neurotoxic, necessitating reevaluation of the whole program.

FURTHER READING

Books

Deinhardt, F., and Deinhardt, J. B., eds. (1983). "Viral Hepatitis: Laboratory and Clinical Science." Dekker, New York.

Gerety, R. J., ed. (1985). "Hepatitis B." Academic Press, New York.

Maupas, P., and Guesry, P., eds. (1981). "Hepatitis B Vaccine," INSERM Symp., Vol. 18. Elsevier/North-Holland Biomedical Press, Amsterdam.

Overby, L. R., Deinhardt, F., and Deinhardt, J., eds. (1983). "Viral Hepatitis." Dekker, New York.

Papaevangelou, G., ed. (1984). "Viral Hepatitis: Standardization in Immunoprophylaxis of Infections by Hepatitis Viruses." Karger, Basel.

Szmuness, W., Alter, H. J., and Maynard, J. E., eds. (1982). "Viral Hepatitis." Franklin Institute Press, Philadelphia, Pennsylvania.

Reviews

Centers for Disease Control (1982). Inactivated hepatitis B vaccine. *Morbid. Mortal. Wkly. Rep.* **31,** 317–328.

Hilleman, M. R., Buynak, E. B., McAleer, W. J., McLean, A. A., Provost, P. J., and Tytell, A. A. (1982). Hepatitis A and hepatitis B vaccines. *In* "Viral Hepatitis" (W. Szmuness, H. J. Alter, and J. E. Maynard, eds.) pp. 385–397. Franklin Institute Press, Philadelphia, Pennsylvania.

Hoofnagle, J. H. (1981). Serological markers of hepatitis B virus infection. *Annu. Rev. Med.* **32,** 1.

Marion, P. L., and Robinson, W. S. (1983). Hepadnaviruses: Hepatitis B and related viruses. *Curr. Top. Microbiol. Immunol.* **105,** 99.

Tiollais, P., Charnay, P., and Vyas, G. N. (1981). Biology of hepatitis B virus. *Science* **213,** 406.

CHAPTER 14

Papovaviruses

Introduction 381
Properties of *Papovaviridae* 381
Papillomaviruses 381
Polyomaviruses 386
Further Reading 388

INTRODUCTION

The main interest in papovaviruses lies in the possibility that they cause cancer in man. Members of the *Papillomavirus* genus are responsible for benign papillomas (warts) which occasionally turn malignant. In particular, the frequent association of genital warts with carcinoma of the cervix is currently attracting the attention of research workers. Members of the *Polyomavirus* genus, on the other hand, infect most people subclinically, persist for life, and are frequently reactivated by immunosuppression.

PROPERTIES OF *PAPOVAVIRIDAE*

The family *Papovaviridae* (Table 14-1) contains two genera, *Papillomavirus* and *Polyomavirus,* the former having a larger virion (Plate 14-1) and a larger genome. The characteristic twisted circular dsDNA molecule was shown in Plate 1-2.

PAPILLOMAVIRUSES

Species-specific papillomaviruses have been found in many animals. They all cause benign papillomas (warts), or more rarely fibromas, in the

TABLE 14-1
Properties of Papovaviridae

Icosahedral virion, 45[a] or 55[b] nm
dsDNA genome, circular, MW 3.4[a] or 5[b] million
Nuclear multiplication

[a] Polyomavirus.
[b] Papillomavirus.

skin or mucous membranes. There are several human papillomaviruses (HPV), each with a predilection for a particular site in the body. As none of these viruses has yet been grown in cultured cells, neutralization tests are unavailable, hence classification is based on differences in the genome itself. Molecular hybridization, heteroduplex mapping, restriction endonuclease mapping, and in some cases complete nucleotide sequencing have revealed differences in DNA sequence homology between human isolates of sufficient magnitude to separate them into over 2 dozen "types," as well as many more subtypes and variants. Paradoxically, we already know a great deal about the DNA of these viruses despite our inability to grow them *in vitro*. The parallel with the hepadnaviruses is interesting in this regard. The small circular dsDNA genome readily lends itself to cloning. Thus, ample quantities are available to molecular biologists for sequencing, for development of diagnostic

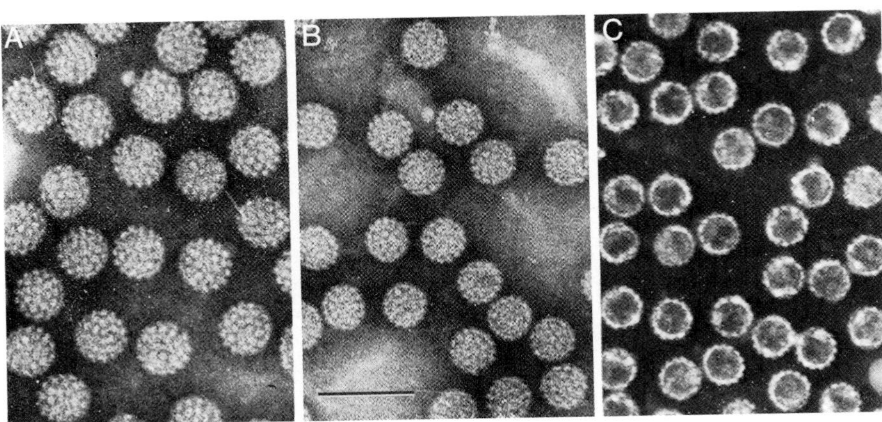

PLATE 14-1. Papovaviridae. *(A)* Papillomavirus *(human wart virus). (B)* Polyomavirus. *(C)* Polyomavirus, *empty virions. Bar = 100 nm. (Courtesy Dr. E. A. Follett.)*

probes, and for exploration of their putative role as carcinogens. Moreover, papilloma viral DNA itself is proving to be a cloning vector of great potential (see Chapter 4); bovine papillomavirus type 1 (BPV-1) DNA is now being widely exploited as a eukaryotic cloning vector for the manufacture of products such as interferons (see Chapter 11).

Relatively little is known about the proteins encoded by the genome, although electrophoretic profiles have been obtained for various types. There are, however, genus-specific structural proteins bearing epitopes common to papillomaviruses from all species. Accordingly, detergent-disrupted virions of, say, BPV-1, or antisera raised against these proteins, can be used as diagnostic probes in conventional serological tests for human papillomavirus infections.

Clinical Features

Virus-induced human papillomas may be classified into the following major categories:

1. *Common warts (verrucae vulgaris)* are raised lesions up to about 5 mm in diameter with a rough surface. They are found anywhere on the skin, but especially in localized crops on prominent regions subject to abrasion, such as the hands and knees, and are usually caused by HPV-2 (Table 14-2). HPV-7 is characteristically associated with "butchers' warts"—papillomatous lesions on the hand.
2. *Flat (plane) warts (verrucae plana)* are smaller, flatter, smoother, and more numerous, are especially seen on the face and knees of youngsters, and are commonly caused by types 3 and 10.
3. *Plantar and palmar warts (verrucae plantaris, or myrmecia)* are painful deep endophytic warts forced to grow inward by pressure on weight-bearing areas, especially the heel and sole of the foot. Types 1 and 4 are the etiological agents.

TABLE 14-2
Diseases Caused by Human Papillomaviruses

CLINICAL PRESENTATION	SITE	MALIGNANT CHANGE	COMMON TYPES
Common warts (*verrucae vulgaris*)	Skin	No	2,7
Flat warts (*verrucae plana*)	Skin	No	3,10
Plantar and palmar warts (*myrmecia*)	Sole, palm	No	1,4
Focal epithelial hyperplasia	Mouth	No	13
Epidermodysplasia verruciformis	Skin	Squamous carcinoma	5,8,9,12,14,15
Genital warts (*condyloma acuminata*)	Genitalia	Cervical carcinoma?	6,11,16,18,31
Laryngeal papillomatosis	Larynx, etc	Laryngeal carcinoma?	6,11

4. *Epidermodysplasia verruciformis (EV)* is a rare condition associated with profound depression of cell-mediated immunity. Flat warts are widely distributed over the skin. Occasional malignant change in a wart on an area of skin exposed to sunlight gives rise to squamous carcinoma, as was discussed in Chapter 8. Multiple HPV types can be isolated from a single EV patient but only types 5, 8, 9, 12, 14, and 15 seem to be involved in those tumors undergoing malignant change.

5. *Genital warts (condyloma acuminata),* caused by HPV-6, 11, 16, 18, and 31, are large moist soft pedunculated excrescences found in the male or female genital tract (Plate 14-2). Flat condylomas occurring inside the male urethra, or the female vagina or cervix may be an inconspicuous source of transmission to sexual partners. The association of condylomas with carcinoma of the cervix was discussed in Chapter 8.

6. *Laryngeal papillomas* are benign tumors seen primarily in young children. They grow rapidly to obstruct the airway, and continue to recur despite repeated surgical removal. If X irradiation is used to treat them, cancer may (rarely) develop years later. The fact that the types involved, HPV-6 and 11, are the same as those causing genital warts suggests, but does not prove, that they are acquired during parturition.

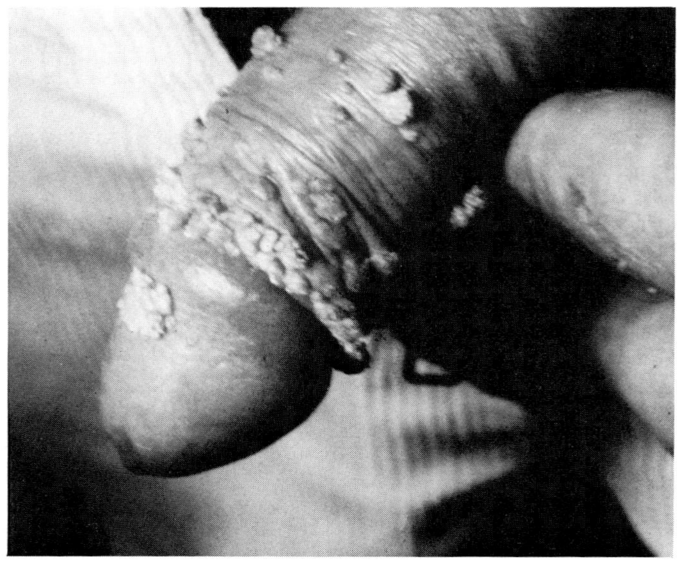

PLATE 14-2. *Genital warts* (condyloma acuminata). *(Courtesy Dr. D. Bradford.)*

Pathogenesis and Immunity

Human transmission experiments indicate an incubation period varying from 1 to 20 months. A wart is regarded by some as a benign tumor and by others as a simple hyperplasia. There is a marked proliferation of prickle cells in the Malpighian layer of the epidermis, resulting in a protruding papilloma, topped by extensive hyperkeratosis; the latter is a striking feature of older skin warts, but is not so prominent in genital warts. Large vacuolated cells are present in the stratum spinosum and the granular layer of young warts. Eosinophilic and basophilic intranuclear inclusions are also conspicuous. It is striking that, whereas viral DNA is present in basal cells, virions are found only in the keratinized end cells. Evidently, only dying epithelial cells that have progressed to this state of terminal differentiation are permissive. Presumably this is the reason why it has proved impossible to grow papillomaviruses in conventional dedifferentiated epithelial cell lines. It may also explain the failure of the immune system to reject the virus promptly.

Warts tend to disappear within a couple of years. When multiple, as they usually are, they all go together when they go—as mysteriously as they came. Attempts to ascribe this sudden regression to the immune response have not been altogether convincing. Specific IgM antibody is present continuously up until and during regression. Quite probably CMI is responsible; but we have no idea what triggers it, nor what delays it so. Following regression, IgG persists for years, providing immunity to reinfection.

Laboratory Diagnosis

Despite the fact that no method is yet available for culture of the virus, diagnosis of external skin or genital warts poses no great problem. Gross appearance and histology are usually pathognomonic. Internal condylomas can be visualized by colposcopy. Immunofluorescence or immunoperoxidase staining of fixed sections (or exfoliated cervical cells), using antisera raised against detergent-disrupted HPV-1 or BPV-1, identifies an antigen common to all members of the *Papillomavirus* genus. Electron microscopy reveals the characteristic virions, but is rarely needed. The use of *in situ* DNA hybridization to identify and locate viral DNA in papillomas, in premalignant cervical dysplasias and in carcinomas, was discussed in Chapters 8 and 12.

Epidemiology

Human papillomaviruses occur throughout the world. Skin warts are most prevalent in children in the second decade of life, genital warts in young adults, and laryngeal papillomas in young children. As far as we know, they are the only contagious tumors. Transmission of the viruses is by direct contact in the case of skin warts, venereally for condylomas, and perhaps perinatally for laryngeal warts. The prevalence of genital warts and of premalignant cervical dysplasia has increased dramatically in the past decade or so.

Treatment

The fact that warts regress spontaneously encourages the perpetuation of mythical cures ranging from hypnosis to Tom Sawyer's infallible dead-cat-in-the-cemetery-at-midnight cure. A comparable rate of success can be assured by letting nature take its course. Painful plantar warts or unattractive facial papillomas, etc. can be removed by surgery, cryotherapy, or caustic chemicals. Premalignant cervical dysplasia can also be destroyed by electrodiathermy, cryosurgery, or laser. As discussed in Chapter 11, there is evidence that interferons may cause viral papillomas to regress; some of the data with recurrent juvenile laryngeal papillomatosis are quite striking, but almost invariably the lesions reappear later.

POLYOMAVIRUSES

The prototype after which this genus was named is the polyoma virus of mice. Though causing only harmless inapparent infections in mice when spread by natural routes, this virus induces many different types of malignant tumor (poly-oma) when artifically injected into infant rodents, e.g., hamsters. Another *Polyomavirus,* simian virus 40 (SV40), infects monkeys subclinically but also induces tumors after inoculation into baby rodents. During the 1960s and 1970s these two viruses became the principal models for the biochemical investigation of virus-induced malignancy. The very latest techniques of molecular biology were brought to bear on the expression of the integrated viral genome and associated cellular changes in cultured fibroblasts transformed by these viruses *in vitro* (see Chapters 5 and 8). This research taught us an enormous amount about the mechanism of transformation by DNA viruses and about the biochemical and physiological alterations in transformed cells.

The last few years have seen the discovery of two, possibly three, polyomaviruses of man. BK virus was recovered from the urine of a renal transplant recipient, and JC virus from the brain of a patient with the rare demyelinating condition, progressive multifocal leukoencephalopathy (PML). More recently, humans have been found to have antibody to a lymphotropic polyomavirus recovered from lymph nodes of an African green monkey. Morphologically, all are typical polyomaviruses. Antigenically, they are distinct from one another, but minor cross-reactions between JC, BK, and SV40 are demonstrable in certain serological tests and the three code for similar T antigens, distinguishable by complement fixation but not by immunofluorescence. Like SV40, BK and JC are oncogenic in newborn hamsters and transform mammalian cells *in vitro,* but there is no convincing evidence to date that they cause cancer in man. Both are ubiquitous in man, producing inapparent infections that persist for many years in the urinary tract, and may be reactivated by immunosuppression. Epithelial cells in the lower urinary tract, e.g., ureter, contain characteristic inclusions in swollen, darkly staining nuclei.

BK Virus

BK virus has not been associated with any human disease. Yet it evidently infects man very readily. Worldwide, over 75% of people have antibody. Infection is presumably subclinical, usually occurring in the first 5 years of life. Thereafter it persists, possibly for life. The urinary tract appears to be an important site of persistence, for virus is shed into the urine from which it can be recovered intermittently, by cultivation in human embryonic kidney cells. Reactivation is triggered by immunosuppression, e.g., for renal transplantation, or during pregnancy.

JC Virus

JC virus has a similar natural history, although primary infection may occur somewhat later in childhood, and only about half the population has antibody. Again, virus is shed in urine sporadically throughout life and more frequently in pregnancy or renal transplantation. Unlike BK however, JC does seem to cause human disease. Virions are regularly found in oligodendrocytes in the brain of patients dying from progressive multifocal leukoencephalopathy. PML is seen mainly as a complication of advanced disseminated malignant conditions such as Hodgkin's disease or chronic lymphocytic leukemia, but also of certain immunodeficiency syndromes or immunosuppression for renal transplantation. Histologically, the disease is characterized by demyelination of neurons accompanied by proliferation of giant bizarre astrocytes. The nuclei of surrounding oligodendrocytes contain very large numbers of virions by electron microscopy, and antigens identifiable by immunofluorescence.

JC virus is more fastidious in cell culture than BK, growing poorly in human embryonic kidney or amnion but quite well in primary human fetal glial cells. Although JC virus induces brain tumors on artificial inoculation into adult owl monkeys, as well as sarcomas in baby hamsters, there is no direct evidence that this human papovavirus causes any cancer in man.

Lymphotropic Papovavirus

This simian polyomavirus grows (rather poorly) in B blasts, e.g., primate B lymphoblastoid cell lines such as TL-1 and BJA-B, the progeny consisting predominantly of DI particles. Although the virus has not yet been recovered from man or associated with any human disease at least 25% of humans have neutralizing antibody.

FURTHER READING

Books

Essex, M., Todaro, G., and zur Hausen, H., eds. (1980). "Viruses in Naturally Occurring Cancers." Cold Spring Harbor Lab., Cold Spring Harbor, New York.

Tooze, J., ed. (1980). "DNA Tumor Viruses." Cold Spring Harbor Lab., Cold Spring Harbor, New York.

Reviews

Howley, P. M. (1983). Papovaviruses—search for evidence of possible association with human cancer. *In* "Viruses Associated with Human Cancer" (L. Phillips, ed.), Dekker, New York.

Lancaster, W. D., and Olson, C. (1982). Animal papillomaviruses. *Microbiol. Rev.* **46,** 191.

Mounts, P., and Shah, K. V. (1984). Respiratory papillomatosis. Etiological relation to genital tract papillomatosis. *Prog. Med. Virol.* **29,** 90.

Takemoto, K. K. (1978). Human papovaviruses. *Int. Rev. Exp. Pathol.* **18,** 281.

zur Hausen, H., Gissmann, L., and Schlehofer, J. R. (1984). Viruses in the etiology of human genital cancer. *Prog. Med. Virol.* **30,** 170.

CHAPTER 15

Adenoviruses

Introduction 389
Properties of *Adenoviridae* 389
Clinical Features 393
Pathogenesis and Immunity 394
Laboratory Diagnosis 395
Epidemiology 398
Control 399
Further Reading 399

INTRODUCTION

In 1953 Rowe and colleagues, having observed that explant cultures of human adenoids spontaneously degenerated, isolated a new infectious agent which, fittingly, they named "adenovirus." Before long, it became evident not only that adenoviruses persist for years as latent infections of adenoids and tonsils, but also that they are a significant cause of disease in the respiratory tract and the eye. More recently certain species have been found to infect the genitourinary tract, while fastidious adenoviruses successfully cultured from feces have been associated with gastroenteritis.

PROPERTIES OF *ADENOVIRIDAE*

The *Adenoviridae* family is defined by the properties listed in Table 15-1. The virion is a perfect icosahedron (Plate 15-1B). The outer capsid, about 80 nm in diameter, is composed of two main types of capsomer: 240 "hexons" make up the 20 triangular faces of the icosahedron, while

TABLE 15-1
Properties of Adenoviridae

Icosahedral virion, 70–90 nm
dsDNA genome, linear, MW 20–25 million, terminal repeats, protein linked to 5′ termini, infectious
Nuclear replication

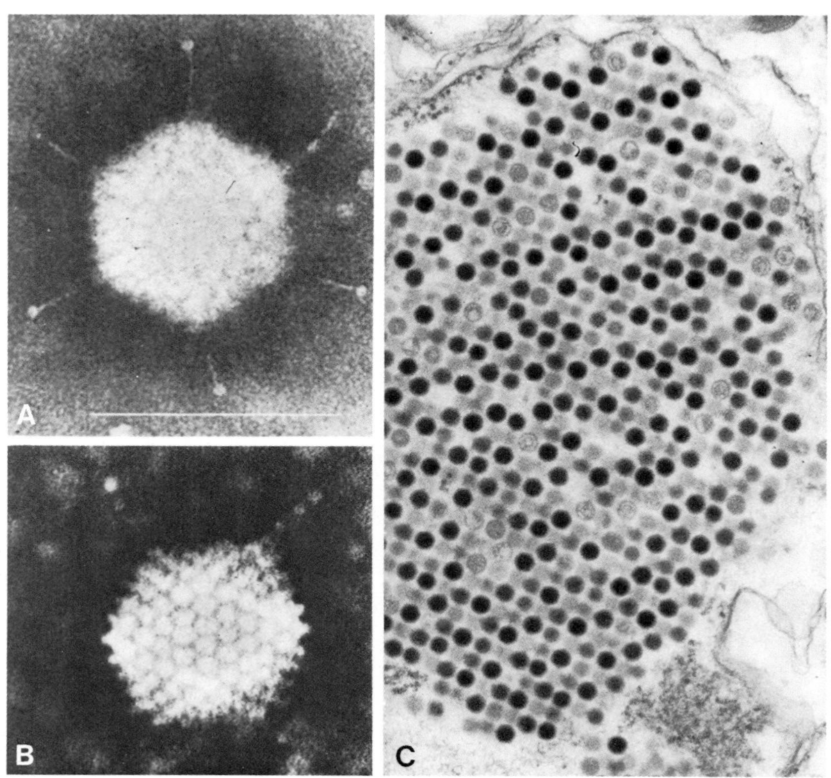

PLATE 15-1. Adenoviridae. *(A and B) Negatively stained preparations of* Mastadenovirus h5. *(A) Virion showing the fibers projecting from the vertices of the polyhedron. (B) Virion showing the icosahedral array of capsomers. Capsomers at the vertices (pentons) are surrounded by five nearest neighbors; all the others (hexons) by six. (C) Section showing crystalline array of mature virions of* Mastadenovirus h7 *in nucleus of a human fibroblast cell. Bars = 100 nm. [A and B from R. C. Valentine and H. G. Periera,* J. Mol. Biol. **13,** *13 (1965); C, courtesy Dr. A. K. Harrison.]*

12 "pentons" form the 12 vertices. From the base of each penton protrudes a "fiber," giving the virion the appearance of a communications satellite (Plate 15-1A). The genome is associated with an inner protein core, believed to have helical symmetry.

The genome consists of a single linear molecule of dsDNA of MW 20–25 × 10^6 with inverted terminal repetitions. This molecule, in association with the 55K protein covalently linked to each 5′ terminus, is infectious.

Classification

Mammalian and avian adenoviruses represent two distinct genera, designated *Mastadenovirus* and *Aviadenovirus,* respectively. In turn, the *Mastadenovirus* genus comprises numerous adenovirus species specific for particular mammalian hosts. Currently, 41 human adenoviruses are recognized and they fall into 6 groups as judged by genome homology, intragroup recombination potential, and several other criteria. There is keen debate about whether these should be regarded as 41 separate *species* (designated *adenovirus h1–h41*) falling into 6 subgenera (A–F), or as 41 *types* falling into 6 species. The chief argument for the latter is that the genomes of human adenoviruses from within each of the 6 groups generally display >90% homology (cf. <20% between members of different groups) and that such similar viruses, distinguishable mainly by serology, should be regarded only as serotypes, not species. Nevertheless, the former (Table 15-2) represents the official taxonomic position endorsed by the ICTV at the time of writing. Table 15-2 also shows how essentially the same grouping, with a few exceptions, is obtained by another criterion, namely the capacity of the virus to agglutinate erythrocytes from particular species of animals. Significantly, adenovirus species falling within a given subgenus tend to share common pathogenic and epidemiological characteristics (Table 15-3).

Particular capsid proteins carry antigenic determinants that are species-specific or cross-reactive between species (subgenus-specific or genus-specific). For example, the determinant α on the hexon is common to all human adenovirus species and can be used to characterize the *Mastadenovirus* genus serologically. Certain other determinants on the hexon and the fiber are shared between members of a particular subgenus. Species are distinguished by the ϵ determinant on the hexon (which elicits neutralizing antibody) and the γ determinant on the fiber (which elicits HI antibody). Hence, most species can be differentiated by neutralization or HI. However, certain adenoviruses, e.g., h40 and h41, though distinct species by lack of significant cross-reaction in neutralization tests, nevertheless cross-react by HI.

TABLE 15-2
Classification of Human Adenoviruses

SUBGENUS	SPECIES[a]	HEMAGGLUTINATION[b]		CPE
		SUBGROUP	RBC SPECIES	
A	12, 18, 31	IV	Nil[c]	Rounding
B	3, 7, 11, 14, 16, 21, 34, 35	I	Monkey	Aggregation
C	1, 2, 5, 6	III	Rat (incomplete)	Aggregation
D	8, 9, 10, 13, 15, 17, 19, 22, 23, 24, 26, 27, 29, 30, 32, 33, 36, 37, 38, 39	II	Rat	Rounding
	20, 25, 28		Atypical	
E	4	III	Rat (incomplete)	Aggregation
F	40, 41	III? IV?	Nil[c]	Rounding

[a] Written adenovirus h1, etc., but often abbreviated to Ad 1, etc.
[b] Hemagglutination subgroups reflect serological relationships and oncogenicity for baby rodents, with some exceptions.
[c] Hemagglutination of rat erythrocytes is detectable in the presence of heterotypic antiserum.

Restriction endonuclease analysis of viral DNA identifies the relationships between species and may indicate "clusters" of species within subgenera. Such restriction enzyme maps of DNA may also reveal the existence of types (or subtypes) within species, e.g., five (sub)types of the most pathogenic human species, Ad 7, and two of Ad 4.

TABLE 15-3
Diseases Produced by Human Adenoviruses

DISEASE	ADENOVIRUSES[a]	AGE
Acute febrile pharyngitis	**1, 2,** 3, **5,** 6, 7	Young children
Pharyngoconjunctival fever	1, 2, **3,** 4, 6, **7,** 14	Children
Acute respiratory disease	3, **4, 7,** 11, 14, 21	Military recruits
Pneumonia	1, 2, **3, 7,** 21	Young children
	4, 7	Military recruits
Epidemic keratoconjunctivitis	8, 11, 19, **37**	Any
Acute hemorrhagic cystitis	11, 21	Young children
Gastroenteritis	**40, 41**	Young children
Cervicitis, urethritis	37	Adults
Disseminated	4, 7, 34, 35	Immunocompromised

[a] Only the commonly occurring species are listed; the most common are in bold type.

Viral Multiplication

Adenoviruses apparently enter the cell by direct translocation across the plasma membrane. The pentons are removed in the cytoplasm and the core migrates to the nucleus. In the nucleus the genome is transcribed according to a complex program which was described in Chapter 3. RNA transcribed from five separate regions of the molecule, situated on both strands of the DNA, is spliced, then translated into about a dozen, mainly nonstructural, early proteins. Viral DNA replication, using a protein as primer, proceeds from both ends by a strand displacement mechanism. Following DNA replication, late transcription units are expressed, giving rise to structural proteins, which are made in considerable excess. Virions are assembled in the nucleus, where they form crystalline aggregates (Plate 15-1C). Cell shutdown develops progressively during the second half of the cycle.

CLINICAL FEATURES

Only about half of the known species of human adenovirus have been causally linked to disease (Table 15-3). Adenoviruses 1–7 are much the commonest species worldwide and are responsible for most cases of adenovirus-induced disease. Some 5% of acute respiratory illnesses in children under the age of 5 years (but less than 1% of that in adults) has been ascribed to adenoviruses. The recently cultivated "fastidious" species Ad 40 and 41 have been claimed to cause up to 10% of infantile gastroenteritis. Several of the lower numbered species as well as Ad 8, 19, and 37 are major causes of eye infections, while one of these (Ad 37) has recently been associated with genital disease.

Acute Febrile Pharyngitis is seen particularly in young children. The infant presents with a cough and a stuffy nose, the throat is inflamed, and cervical lymph nodes are enlarged. Adenoviruses 1, 2, 3, 5, 6, and 7 are usually responsible for these common sporadic infections which are relatively trivial except when otitis media or pneumonia supervenes.

Pharyngoconjunctival Fever (PCF), in contrast, tends to occur in outbreaks, e.g., at children's summer camps where "swimming pool conjunctivitis" may occur with or without pharyngitis, fever, and malaise. Adenoviruses 3 and 7 are commonly responsible. Ad 4a has caused a number of nosocomial outbreaks of conjunctivitis or PCF among hospital staff.

Acute Respiratory Disease (ARD) is the name given to a syndrome characterized by fever, pharyngitis, cervical adenitis, cough, and malaise which occurs in epidemic form when military recruits assemble in camps. Ad 4 and 7 are most often responsible.

Pneumonia, often severe and occasionally fatal, may develop in young children infected with any of the common species, but particularly Ad 7 and 3. ARD in military recruits also occasionally progresses to a pneumonitis. The more recently discovered species, Ad 34 and 35, isolated from renal transplant recipients, cause pneumonia and other life-threatening infections in immunosuppressed patients.

Epidemic Keratoconjunctivitis (EKC) is a severe eye infection, commencing as a follicular conjunctivitis and progressing to involve the cornea (keratitis). Often seen in industrial workers exposed to dust and trauma, the disease is familiarly known as "shipyard eye." The infection, which is highly contagious and often occurs in epidemic form, is caused by members of subgenus D. Ad 8 was the principal agent until 1973 when 19a took over. Then in 1976 Ad 37 suddenly appeared, spread worldwide, and is now the predominant cause of EKC.

Cervicitis and Urethritis are also common manifestations of infection with Ad 37, which was first identified in Rotterdam prostitutes in 1976.

Acute Hemorrhagic Cystitis is characterized by hematuria, dysuria, and frequency of micturition. It is seen mainly in young children, especially males. Ad 11 and 21 are the main species involved.

Gastroenteritis, especially in infants, has been associated with the most recently discovered adenovirus species, Ad 40 and 41. These fastidious "enteric" adenoviruses, previously visualized by electron microscopy in feces from children with or without diarrhea but regarded as "uncultivable," can now be grown in cultured cells. Their recovery from outbreaks of gastroenteritis, and significantly more frequently from symptomatic patients than from controls, points to an etiological role for them in this disease. However, further work is needed to establish the magnitude of their contribution, since one commonly encounters prolonged shedding of adenoviruses (especially of subgenera A, C, and F) in the feces of normal children.

Mesenteric Adenitis is a common, harmless condition of unknown etiology, presenting as vague abdominal pain which is sometimes mistaken for appendicitis. This is an occasional feature of adenovirus infections. More controversial, however, is the role of adenovirus in intussusception, a type of intestinal obstruction sometimes associated with mesenteric adenitis.

Generalized Infections culminating in meningoencephalitis, myocarditis, rashes, etc. have been reported, especially in infants or immunosuppressed patients. Adenoviruses 7, 34, and 35 are most frequently incriminated.

PATHOGENESIS AND IMMUNITY

Adenoviruses multiply initially in the pharynx, conjunctiva, or small intestine, and rarely spread beyond the draining cervical, preauricular, or mesenteric lymph nodes. As the disease process remains relatively localized, the incubation period is short (5–10 days). Most of the enteric infections and rather less than half of the respiratory infections are subclinical. Generalized infections are occasionally seen, particularly in immunosuppressed patients. Most deaths are caused by Ad 7, especially 7b. At autopsy, lungs, brain, kidney, and other organs reveal the characteristic basophilic nuclear inclusions.

Infection with the common endemic species 1, 2, and 5 persists asymptomatically for years in the child's tonsils and adenoids, and virus is shed continuously in the feces for many months after the initial infection, then intermittently for years thereafter. The state of the virus during this persistent infection is unclear; perhaps its replication is held in check by the antibody synthesized by these lymphoid organs. Fluctuation in shedding indicates that latent adenovirus infections can be reactivated. For example, infection with *Bordetella pertussis* can do so, and measles can actually be followed by adenovirus pneumonia. Subgenus B2 viruses are frequently shed in the urine of immunocompromised persons; Ad 34 and 35 were both originally isolated from renal transplant recipients and can be recovered quite commonly from AIDS patients. Ad 4 and 7 have also been associated with fatal pneumonia or other overwhelming infections in immunosuppressed patients.

In contrast to most respiratory viral infections, adenovirus infections lead to reasonably good immunity to reinfection with the same species, probably because of the extent of involvement with lymphoid cells in the alimentary tract and the regional lymph nodes. Maternal antibody generally protects infants under the age of 6 months against severe lower respiratory disease.

During the 1960s much attention was focused on the startling finding that adenoviruses of subgenus A, and to a lesser degree B, produce cancer following inoculation into baby rodents. Some of the resulting research on the mechanism of viral transformation of cultured cells contributed a great deal to our fundamental knowledge of the molecular biology of cancer (see Chapter 8 for details). However, retrospective and prospective epidemiological, serological, virological, and biochemical investigations have produced absolutely no evidence to suggest that adenoviruses are involved in human cancer.

LABORATORY DIAGNOSIS

Depending on the clinical presentation, appropriate specimens include feces, pharyngeal swab or aspirate, conjunctival swab, corneal scraping or tears, genital secretions, or urine.

Rapid diagnostic methods are not yet as widely used as they might be.

Indirect immunofluorescence can be employed to demonstrate common adenoviral antigens in cells from the throat, conjunctiva/cornea, or urine, following low-speed centrifugation then fixation of the pelleted cells. Enzyme immunoassay (or RIA) detects soluble viral antigen in feces or nasopharyngeal secretions. Species-specific (absorbed or monoclonal) antisera can then be used to identify the particular adenovirus concerned. Even the so-called "noncultivable" adenoviruses can be identified by this method, or by immunoelectron microscopy following detection by EM.

Isolation in cell culture is time consuming because many adenoviruses are very slow growing. Only human cells are satisfactory. Primary cultures of human embryonic kidney (HEK) are the most sensitive but are not available to most laboratories. Continuous malignant cell lines such as HeLa, HEp-2, or KB are generally susceptible and widely used, but suffer from the disadvantage that the monolayers tend to overgrow and slough off before viral CPE becomes apparent. Diploid lines of human embryonic fibroblasts (HDF), derived for example from lung or tonsil, are less sensitive and slower to develop CPE but hold up well during the prolonged maintenance period generally required. Only the fastidious enteric adenoviruses fail to grow satisfactorily in HEK or HDF; Ad 40 and 41 can be isolated with some difficulty in the Chang conjunctival cell line, in cynomolgus monkey kidney, or in Graham (293) cells. The 293 cell line was derived by transforming HEK with adenovirus 5 DNA; whether the integrated adenovirus 5 E1A and E1B genes code for a product that assists the fastidious adenoviruses to grow is not known.

Adenoviruses of subgenera B, C, and E often produce CPE within a week (4–6 days in malignant epithelial lines, 6–10 days in diploid fibroblasts). Classically, the pH of the medium falls and the cells become swollen, rounded, and refractile, and cluster together like a bunch of grapes. This aggregation is not seen with other adenoviruses, nor does the CPE develop so rapidly. Subgenus D viruses grow so slowly that CPE may sometimes not become evident for a month. Accordingly, even the uninfected control cultures tend to deteriorate (especially epithelial lines), hence the medium must be changed at least weekly, and one or more "blind passages" of frozen/thawed cells plus supernatant must be carried out if isolations are not to be missed. Basophilic (Feulgen-positive) intranuclear inclusions connected to the nuclear periphery by strands of chromatin are characteristic (Plate 15-2).

Confirmation of the isolate as an adenovirus can be made by immunofluorescence on a fixed cell monolayer a few days after infection. Alternatively, complement fixation can be used to identify interspecific

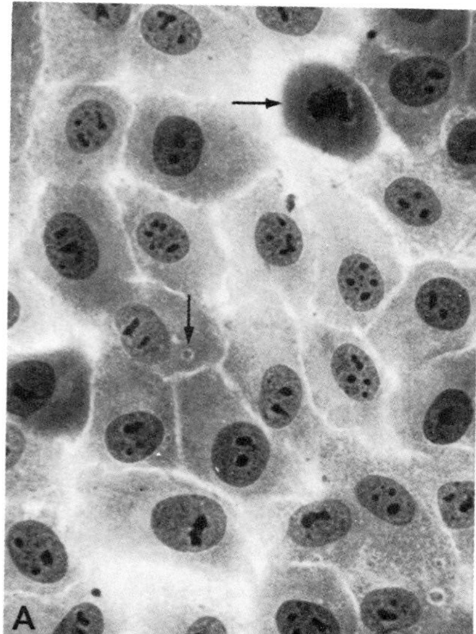

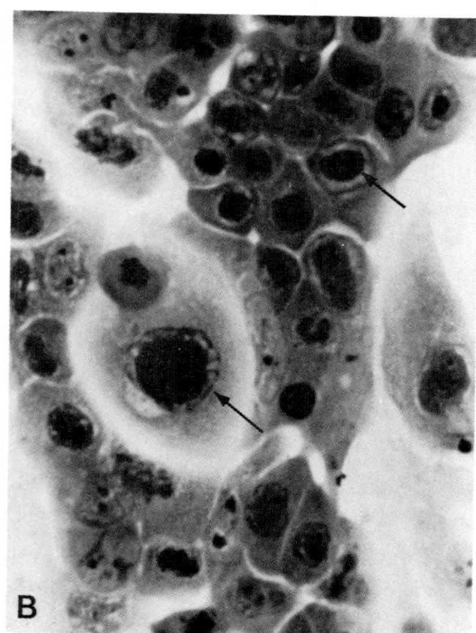

PLATE 15-2. *Cytopathic effects induced by adenoviruses (H and E stain, ×400). (A) Normal monolayer of HEp-2 cells. Horizontal arrow, cell in mitosis. Vertical arrow, phagocytosed cell debris, not to be confused with viral inclusion body. (B) Cytopathic effects induced by adenovirus in HEp-2 cells. Note distended cells containing basophilic intranuclear inclusions (arrows), which consist of masses of virions (see Plate 15-1C). Threads of chromatin sometimes radiate from the nuclear inclusions to the periphery of the nucleus. (Courtesy I. Jack.)*

determinants on antigens in the culture supernatant. If further characterization is desired, the first step is usually to test the isolate for hemagglutination using rat and rhesus monkey erythrocytes to place it in the correct subgenus (Table 15-2). Appropriate species-specific antisera can then be chosen to "type" the isolate by hemagglutination-inhibition (HI). Neutralization, being more cumbersome, is generally employed only to discriminate between those species known to cross-react in HI, e.g., Ad 40 and 41. Restriction endonuclease mapping or hybridization/electron microscopy of isolated DNA are research tools used to characterize possible new species or variants.

Interpretation of the significance of an isolate depends upon the origin of the specimen. Recovery of adenovirus from the eye, genital tract, lung, or brain is diagnostic; from the throat of a patient with respiratory disease, suggestive; and from the feces, ambiguous. This is because adenoviruses are, notoriously, shed in large numbers and for long periods in the feces of children with persistent infections, e.g., of tonsils and adenoids. Furthermore, recrudescent shedding may be precipitated by infection with a quite different agent.

Serology is rarely used directly to diagnose current infection. Epidemiological surveys for evidence of past infection usually rely on EIA or CF using the cross-reactive hexon antigen to detect adenovirus family antibodies (but CF antibodies may disappear after a few years), then proceed to HI to discriminate between species-specific antibodies.

EPIDEMIOLOGY

Although mainly associated with disease of the respiratory tract and eye, and clearly capable of respiratory transmission, adenoviruses spread principally via the enteric route. Long-term family studies have demonstrated that, following infection of young children, very large numbers of adenovirus particles are shed in feces over a period of several months and succeed in infecting about half of all susceptible members of the same family. Doubtless this explains the endemicity of the common members of subgenus C (1, 2, 5) and the fact that most children acquire one or more of these viruses by the age of 2 years. About half of such infections are subclinical, the others presenting as pharyngitis or pharygoconjunctival fever (PCF). The enteric adenoviruses of subgenus F (Ad 40 and 41), which have been associated with gastroenteritis, may spread exclusively via the fecal–oral route, and those of subgenus A (Ad 12, 18, 31) are generally isolated from the feces of infants.

Respiratory spread, via droplets or contact, occurs particularly in the case of the so-called "epidemic" species of subgenera B and E (notably 4 and 7) which are responsible for outbreaks of ARD in military recruits in late winter and spring.

Eye infections may be acquired in several ways, including transfer of respiratory secretions on fingers following infection via the respiratory or alimentary routes. However, two important direct routes of entry to the eye are well established. Outbreaks of "swimming pool conjunctivitis" (with or without PCF) often occur in children on summer camps. Furthermore, a number of major outbreaks of EKC have been traced to the surgeries of particular ophthalmologists whose aseptic technique leaves something to be desired. These iatrogenic infections are attributable to contaminated towels, ophthalmic solutions, and instruments, especially tonometers. Hand-to-eye transfer is also particularly important.

Adenoviruses can be excreted in urine, and the recent discovery of genital infections of male and female caused by adenovirus 37 indicates that venereal spread also occurs.

CONTROL

Fecal–oral spread of adenoviruses within families can be reduced by personal hygiene. Chlorination of swimming pools, drinking water, and wastewater largely removes the risk of outbreaks from these communal sources. Prevention of contact spread of eye infections by ophthalmologists demands that special attention be paid to segregation of EKC patients, hand washing, separate towels and ophthalmic solutions, and adequate sterilization of equipment. Similar precautions should also be taken to minimize the possibility of nosocomial outbreaks in wards in which a patient with a severe adenovirus infection is being nursed.

Vaccine

The regularity of outbreaks of ARD in United States military recruits in the 1960s prompted the development of a vaccine for protection of this particular population. The approach was novel. Live virulent virus is enclosed in a gelatin-coated capsule and given by mouth. In this way the virus bypasses the throat, in which it would normally cause disease, but is released in the intestine, where it grows without producing disease. When the two important species, 4 and 7, cultured in human fibroblasts, are combined in such a live vaccine, they grow with little or no mutual interference and produce highly effective immunity to challenge.

FURTHER READING

Books

Doerffler, W., ed. (1984). "The Molecular Biology of Adenoviruses." Vols. 1, 2, and 3. Curr. Top. Microbiol. Immunol., Vols. 109, 110, and 111. Springer-Verlag, Berlin and New York.

Ginsberg, H. S., ed. (1984). "The Adenoviruses." Plenum, New York.

Reviews

Flint, S. J., and Broker, T. R. (1980). Lytic infection by adenoviruses. *In* "Molecular Biology of Tumor Viruses: DNA Tumor Viruses" (J. Tooze, ed.), 2nd ed., Part 2, pp. 443–546. Cold Spring Harbor Lab., Cold Spring Harbor, New York.

Fox, J. F., Hall, C. E., and Cooney, M. K. (1977). The Seattle Virus Watch. VII. Observations of adenovirus infections. *Am. J. Epidemiol.* **105,** 362.

Schmitz, H., Wigand, R., and Heinrich, W. (1983). Worldwide epidemiology of human adenovirus infections. *Am. J. Epidemiol.* **117,** 455.

Wadell, G. (1984). Molecular epidemiology of human adenoviruses. *Curr. Top. Microbiol. Immunol.* **110,** 191.

Wigand, R., Bartha, A., Dreizin, R. S., Esche, H., Ginsberg, H. S., Green, M., Hierholzer, J. C., Kalter, S. S., McFerran, J. B., Pettersson, U., Russell, W. C., and Wadell, G. (1982). *Adenoviridae:* Second Report. *Intervirology* **18,** 169.

CHAPTER 16

Herpesviruses

Introduction 401
Properties of *Herpesviridae* 402
Herpes Simplex Viruses 404
Varicella-Zoster Virus 415
Cytomegalovirus (CMV) 419
EB Virus 426
Further Reading 430

INTRODUCTION

The *Herpesviridae* family contains five of the most important human pathogens. If anything their importance is increasing as a result of developments in modern medicine and social changes. It is also the family of viruses most amenable to chemotherapy.

The remarkable feature of all herpesviruses is their capacity to persist in their hosts indefinitely. Varicella (chickenpox) and herpes simplex viruses establish latent infections in neurons. Upon reactivation, the varicella virus precipitates an attack of herpes zoster ("shingles"), whereas herpes simplex type 1 causes recurrent attacks of "fever blisters" or "cold sores," and HSV-2 is mainly responsible for genital herpes, which has increased in incidence dramatically following the sexual revolution. Cytomegalovirus (CMV) and EB virus persist in lymphocytes. CMV is the major infectious cause of mental retardation and other congenital defects today. EB virus (EBV) is the etiological agent of infectious mononucleosis (glandular fever) and appears to be responsible also for two types of human cancer—one a carcinoma and the other a lymphoma. All five human herpesviruses are frequently reactivated following immunosuppressive therapy for organ transplantation or cancer.

PROPERTIES OF *HERPESVIRIDAE*

The herpesvirion has a complicated construction, with no less than two dozen distinct proteins arranged in four concentric layers. The DNA is wrapped around a fibrous spool-like core composed of histones, the fibers of which are anchored to the underside of the surrounding capsid. The capsid, or shell, is an icosahedron, 100 nm in diameter, composed of 162 capsomers—150 hollow hexamers and 12 pentamers (see Plate 16-1B). Surrounding the capsid is a layer of globular material, often asymmetrical and variable in quantity, known as the "tegument." This in turn is enclosed by a typical lipid envelope (Plate 16-1A). Because of the loosely fitting nature of the envelope the virion can range in overall diameter from 120 to 200 nm. One of the several viral glycoproteins projecting from the envelope possesses (perhaps fortuitiously) Fc receptor activity, i.e., it binds normal IgG.

The genome is a single linear molecule of dsDNA, MW 100–150 million, with remarkable arrangements of reiterated sequences. For instance, EB viral DNA displays (1) numerous reiterations of a single set of

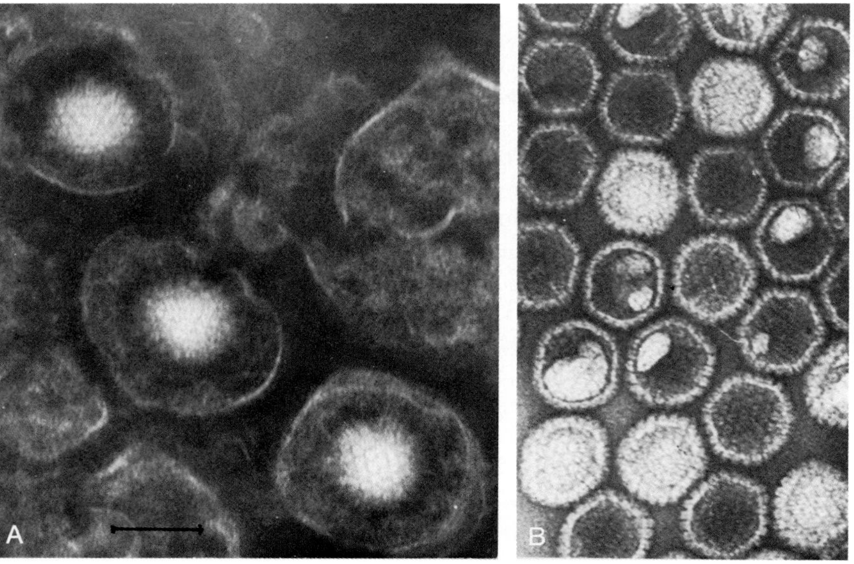

PLATE 16-1. Herpesviridae. *Negatively stained preparations of herpes simplex virus type 1. (A) Enveloped particles. (B) Icosahedral capsids with 162 capsomers (bar = 100 nm). (Courtesy Drs. E. L. Palmer and F. A. Murphy.)*

TABLE 16-1
Properties of Herpesviridae

Spherical virion, 120–200 nm
Envelope with glycoproteins, Fc receptors
Icosahedral capsid, 100 nm
Protein core on which DNA is wound
dsDNA genome, linear, MW 100–150 million, characteristic reiterated sequences
Nuclear replication, budding from nuclear membrane

sequences at both termini, and (2) internal tandem reiterations of other sequences. In contrast, herpes simplex DNA is characterized by (1) terminal reiterations, and (2) internal inverted reiterations of longer sequences from both termini. Furthermore, the two major coding sequences (L and S) that are flanked by these reiterated sequences may be inverted relative to one another, hence the genome occurs in four equimolar isomeric configurations (Fig. 1-3B; Table 16-1).

Viral Multiplication

The virion binds to specific cellular receptors via a particular envelope glycoprotein, fuses with the plasma membrane, and releases its nucleocapsid into the cytoplasm. The core moves into the nucleus, where viral DNA is transcribed by the cellular DNA-dependent RNA polymerase II. The "immediate-early" transcripts are spliced to produce individual mRNAs which represent mainly the terminal repeated sequences of the genome. Following the synthesis of immediate early ("alpha") proteins, the whole genome becomes available for transcription, yielding "early" mRNAs, which are translated into "beta" proteins. After viral DNA replication commences, the program of transcription switches once again, and the resulting "late" mRNAs, which also represent the whole genome, are translated into the "gamma" proteins. Over 50 virus-coded proteins are made during the cycle; many alpha and beta proteins are enzymes, whereas most of the gamma proteins are structural. Intricate controls must regulate expression at the level of translation.

Viral DNA is replicated in the nucleus utilizing a "rolling circle" mechanism; temporary circularization of the genome is brought about by virtue of the terminal repeated sequences. Newly synthesized DNA is spooled onto preformed immature nucleocapsids. Budding occurs through the inner lamella of the nuclear membrane; virions accumulate between the inner and outer lamellae and are transported via the endoplasmic reticulum to the cell surface.

Classification

Already over 80 herpesviruses are known. Ancient phylogenetically, they appear to have evolved with their hosts over millions of years and are now generally host specific. Unique herpesviruses are present in insects, reptiles, and amphibia as well as in virtually every species of bird and mammal that has been comprehensively explored. Well-known herpesviruses of veterinary importance include the viruses of infectious bovine rhinotracheitis, pseudorabies (of swine), and Marek's disease (of chickens). Various primates carry their own unique species of herpesviruses, some of which induce malignant lymphoma (see Chapter 8). There are 5 human herpesviruses, known as herpes simplex virus (HSV) types 1 and 2, varicella-zoster (V-Z) virus, EB virus, and cytomegalovirus (CMV).

The classification of viruses within the *Herpesviridae* family is a complex matter. Three subfamilies are recognized: the *Alphaherpesvirinae* are rapidly growing cytolytic viruses, the *Betaherpesvirinae* are slowly growing "cytomegalic" viruses, while the *Gammaherpesvirinae* are viruses that grow in lymphocytes and sometimes transform them to the malignant state (Table 16-2).

Serological relationships are complex. There are some shared antigens but different species have distinct envelope glycoproteins.

HERPES SIMPLEX VIRUSES

Clinical Features

In considering HSV infections one must distinguish between *primary* and *recurrent* infections (Table 16-3). Primary infections are often inapparent, but when clinically manifest, tend to be more severe than are subsequent recurrences at the same site. Nearly all recurrences of infection with the same type are reactivations of an endogenous latent infection. However some initial episodes of disease, e.g., genital herpes, may in fact occur in people with partial immunity from a previous heterologous (or less commonly, homologous) HSV infection; such cases are usually mild.

Gingivostomatitis, Pharyngotonsillitis, and Recurrent Herpes Labialis. Primary infection with HSV-1 most commonly involves the mouth and/or throat (Plate 16-2A and C). In young children the classical clinical presentation is gingivostomatitis. The mouth and gums become covered with vesicles which soon rupture to form ulcers. Though febrile, irritable, and in obvious pain with bleeding gums, the child recovers uneventfully.

TABLE 16-2
Classification of Herpesviridae

SUBFAMILY	BIOLOGICAL PROPERTIES	HUMAN VIRUSES	
		OFFICIAL NAME	POPULAR NAME
Alphaherpesvirinae	Fast-growing, cytolytic Latent in neurons	Human herpesvirus 1 Human herpesvirus 2 Human herpesvirus 3	Herpes simplex virus 1 Herpes simplex virus 2 Varicella-zoster virus
Betaherpesvirinae	Slow-growing, cytomegalic Latent in glands, kidneys	Human herpesvirus 5	Cytomegalovirus
Gammaherpesvirinae	Lymphoproliferative Latent in lymphocytes	Human herpesvirus 4	Epstein–Barr (EB) virus

In adolescents, primary infection more commonly presents as a pharyngitis or tonsillitis.

Despite recovery from primary oropharyngeal infection the individual retains HSV DNA in the trigeminal ganglion for life and may suffer recurrent attacks of herpes labialis (otherwise known as *herpes facialis, herpes simplex, fever blisters,* or *cold sores*) from time to time throughout the remainder of life. A brief prodromal period of hyperesthesia heralds the development of a cluster of vesicles, generally around the mucocutaneous junction on the lips (Plate 16-2B).

TABLE 16-3
Diseases Produced by Herpes Simplex Viruses

DISEASE	PRIMARY (P) OR RECURRENT (R)	AGE	FREQUENCY	SEVERITY	TYPE
Gingivostomatitis	P	Young children	Common	Mild	1
Pharyngotonsillitis	P	Adolescents	Common	Mild	1 > 2
Herpes labialis	R	Any	Common	Mild	1 > 2
Genital herpes	P, R	>15 years	Common	Mild to moderate	2 > 1
Keratoconjunctivitis	P, R	Any	Common	Moderate	1
Dermatitis[a]	P, R	Any	Rare	Mild	1, 2[b]
Encephalitis	P, R	Any	Rare	Severe[c]	1 > 2[d]
Neonatal herpes	P	Newborn	Rare	Severe[c]	2 > 1
Disseminated herpes	P, R	Any	Rare	Severe[c]	1 > 2

[a] Including HSV infection of burns, eczema herpeticum, etc.
[b] Skin above waist, 1 > 2; below waist, 2 > 1; arms, either.
[c] Often fatal.
[d] HSV-2 in neonates.

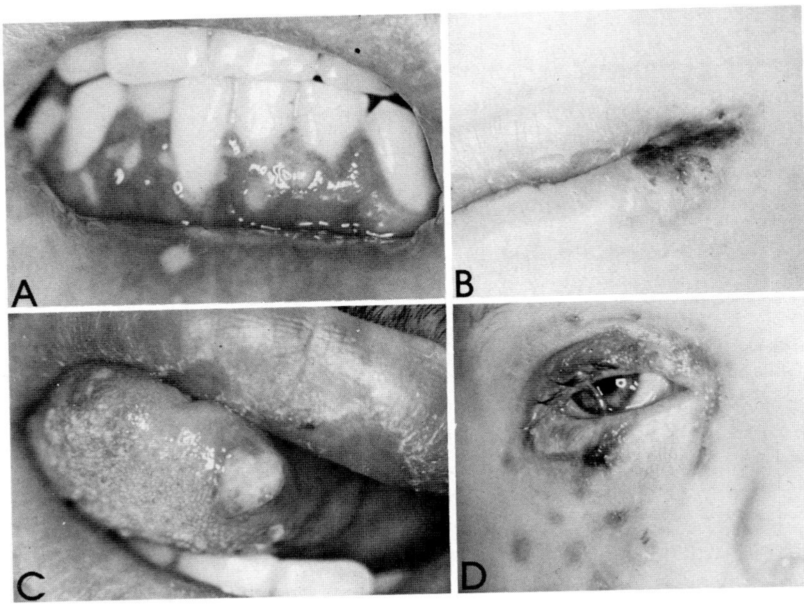

PLATE 16-2. *Herpes simplex. The lesions are vesicles, which rupture to become ulcers. (A), (B), (C), and (D) Lesions on gums, lips, tongue, and eye. (Courtesy Dr. J. Forbes, Fairfield Hospital for Infectious Diseases, Melbourne.)*

Genital Herpes. Because herpes genitalis is usually a sexually transmitted disease it is seen mainly in young adults, but is occasionally encountered following accidental inoculation at any age, or in young girls who have been victims of sexual abuse. Though the primary infection may be subclinical, severe disease often occurs, particularly in females. Ulcerating vesicular lesions develop on the vulva, vagina, cervix, urethra, and/or perineum in the female, penis in the male, or rectum and perianal region in male homosexuals (Plate 16-3). Local manifestations are pain, itching, redness, swelling, discharge, dysuria, and inguinal lymphadenopathy; systemic symptoms, notably fever and malaise, are often quite marked, especially in females. Spread may occur to the central nervous system. Initial disease is less severe in those with immunity resulting from previous subclinical infection with HSV-2, or from clinical or subclinical infection with HSV-1 ("initial disease, nonprimary infection").

About 80% of primary genital infections are caused by HSV-2, but an increasing minority are attributable to HSV-1. Some HSV-1 infections

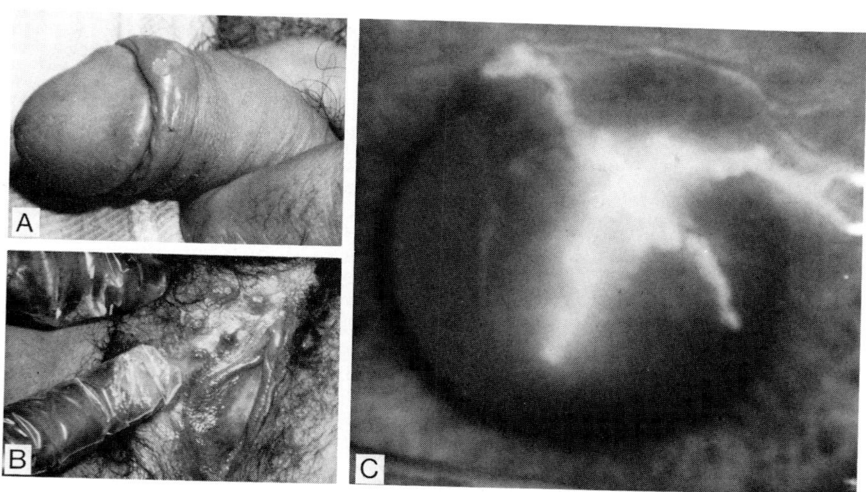

PLATE 16-3. *Herpes simplex viral diseases of the genitalia and eye. Genital herpes in the male (A) and female (B). Dendritic ulcer on cornea of the eye, stained with fluorescein (C). [Courtesy Dr. D. Bradford, Melbourne Communicable Diseases Clinic (A,B) and Dr. H. Maclean, University of Melbourne Department of Ophthalmology (C).]*

may arise by autoinoculation via fingers, but the majority are assumed to result from orogenital sexual practices. However, the vast majority of recurrences of genital herpes are found to be caused by HSV-2.

Recurrences tend to be more frequent following genital herpes than following oropharyngeal herpes. The average HSV-2 carrier suffers up to half a dozen every year, but some have to put up with one or more recurrences monthly throughout adult life. The resultant pain, humiliation, sexual frustration, and sense of guilt can have psychological as well as sexual consequences. It is some consolation that recurrences are clinically less severe than was the primary attack.

Keratoconjunctivitis. Primary infection of the eye (Plate 16-2D) may occur *de novo,* or result from autoinoculation. Involvement of the cornea (keratitis) often leads to a characteristic "dendritic ulcer" (Plate 16-3C) which may progress to involve the stroma beneath. The corneal scarring that follows repeated HSV infections may lead to blindness; such recurrences are usually unilateral.

Dermatitis. Primary and recurrent HSV infections may also involve any region of the skin. Rarely this occurs by direct traumatic contact, e.g., in wrestlers or rugby players—"*herpes gladiatorum*"! An occupational hazard of dentists, anaesthetists, and nurses is herpetic paronychia ("whitlow"), i.e., inflammation at the base of the fingernail.

Encephalitis and Meningitis. Though encephalitis is fortunately a rare manifestation of HSV infection, this virus is nevertheless the commonest endemic cause of the disease. The virus may spread to the brain during primary or, possibly, recurrent HSV infection, but vesicles are not usually present on the body surface. Type 1 is usually responsible except in the case of neonates. The temporal lobes are principally involved. The case fatality rate is 70%, and the majority of survivors suffer permanent neurological sequelae (see Chapter 29 for further details). A relatively benign meningitis is a much more common complication of both HSV-1 and -2 infections.

Neonatal Herpes. *Herpes neonatorum* is a serious disease acquired by babies whose mothers are infected with HSV. Although rare, it is increasing in incidence with the increase in genital herpes. Primary herpes genitalis carries a greater risk of intrapartum infection of the baby than does the recurrent form of genital herpes. The risk of serious disease is also greater in premature infants. Estimates of the overall risk of neonatal herpes if virus is present in the maternal genital tract at the time of delivery vary from 5 to 40%.

Neonatal herpes may present as (1) disseminated disease, with a case fatality rate of 60%, half of the survivors being left with permanent neurological or ocular sequelae; or (2) a more localized disease, principally affecting the brain (with high mortality), or mucocutaneous surfaces (nonfatal).

Disseminated Herpes in Compromised Hosts. Particularly at risk of potentially lethal disseminated HSV infections are patients who are already compromised by (1) congenital immunodeficiency disorders or malignancy, (2) immunosuppression, e.g., for organ transplantation, (3) severe malnutrition with or without concomitant measles, or (4) severe burns, eczema, or certain other skin conditions.

Pathogenesis and Immunity

The pathogenesis of HSV was described in Chapter 7 (Fig. 7-2, Table 7-2). It is not yet fully understood what regulates expression of the HSV genome during latency or reactivation in the neurons of the dorsal root ganglion in which it resides throughout the life of the carrier. Mutants with a defective thymidine kinase gene are inefficient in establishing latency. Evidence has been put forward to suggest that methylation of the viral DNA molecule might be involved in the maintenance of latency and that only a portion of the genome is transcribed. Nor is it apparent what the inducing stimuli—fever, emotional stress, sunlight, trauma, menstruation, and immunosuppression—have in common! Do they all induce hormone(s) that temporarily weaken one or more arms of the anti-HSV immune response? There is some evidence that nonspecific suppression of cell-mediated immunity may trigger reactivation of virus and recrudescence of disease but we know nothing about mechanisms or about whether there is a single common pathway of reactivation.

There is serological evidence of an epidemiological association between HSV-2 and carcinomas of the cervix and vulva. It has been postulated that HSV-2 may serve as an initiator, or cocarcinogen, with papillomavirus in the etiology of genital cancer, but the evidence is still only circumstantial (see Chapter 8).

Laboratory Diagnosis

Some of the clinical presentations of HSV, e.g., recurrent herpes labialis, are so characteristic that laboratory confirmation is not always required. Others are not nearly so clear-cut, particularly if vesicles are not visible, as in encephalitis, keratoconjunctivitis, or even those herpes genitalis infections that are confined to the cervix.

Rising concern about herpes genitalis has led to a situation in which genital swabs are now the most common specimen received by many virus diagnostic laboratories. Isolation in cell culture is the most sensitive procedure, providing that the specimen is taken early, placed in an appropriate transport medium, kept on ice, and transported to the laboratory without delay. Primary cultures of rabbit kidney or of guinea pig embryo are the most susceptible but are not widely available. Diploid strains of human fibroblasts are also very satisfactory, with the Vero cell line being rather less sensitive. Distinctive foci of swollen, rounded cells appear within 1–5 days (Plate 16-4A). However, the diagnosis can be made as early as 24 hours by immunofluorescent (IF) or immunoperoxidase staining of cultured cells. Sensitivity of IF can be augmented using biotin-labeled HSV–antibody and avidin–fluorescein conjugate.

Differentiation of HSV-1 from HSV-2 is achieved most directly by using the appropriate monoclonal antibodies for IF or EIA, or, rather more laboriously, for neutralization. Many laboratories take advantage of other biological properties that distinguish the two serotypes: (1) HSV-2 is 100 times more resistant than HSV-1 to bromovinyldeoxyuridine (BVDU), which may therefore be incorporated in the medium of half the cultures; (2) restriction endonuclease patterns are distinct;

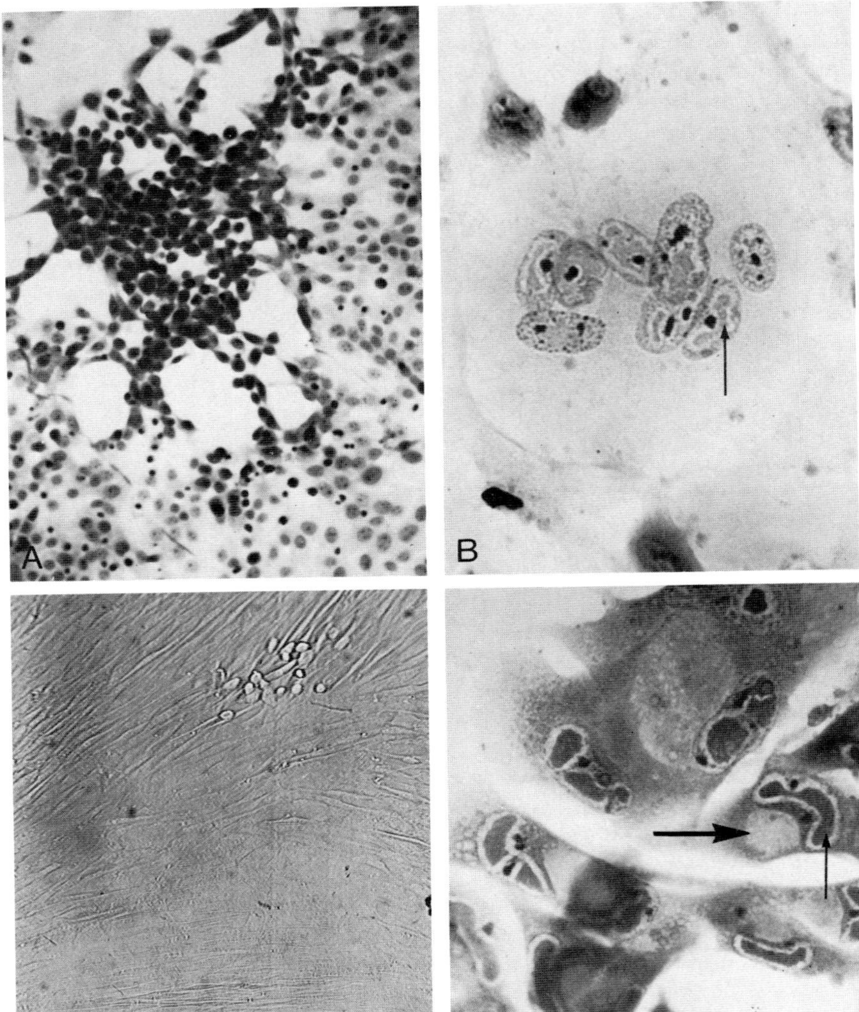
A
B
C
D

and (3) HSV-2 produces larger pocks on the chorioallantoic membrane of embryonated eggs and distinctive plaques in cultured cells. Restriction endonuclease maps or monoclonal antibodies can also be used to distinguish strains within each serotype.

Cell culture isolation is the method of choice for diagnosing HSV infections and, when monitored by IF after 24 hours, provides a rapid answer. Speed is important in a number of clinical situations, notably (1) the pregnant woman approaching term, who may be a candidate for cesarean section if found to be excreting HSV; (2) any patient who would benefit from specific chemotherapy, which should be commenced without delay. Hence there is a demand for even quicker tests that will provide a diagnosis on the same day. Currently the most reliable of these alternatives is the demonstration of HSV antigen in cells from cervix, cornea, brain biopsy, etc., as appropriate, by IF or immunoperoxidase staining (using type-specific monoclonal antibodies if desired). EIA or RIA is used to demonstrate HSV glycoprotein in CSF. Biotin–avidin amplified EIAs on detergent-solubilized cells/mucus are currently under development but require refinement to eliminate the possibility of false positives which may lead to unnecessary cesareans. Clinics without ready access to a virus diagnostic facility can diagnose genital herpes in females by Papanicolaou smear from the cervix; intranuclear inclusions in multinucleate giant cells are pathognomonic, but half the cases may be missed by this method. Hybridization to a labeled HSV DNA probe (see Chapter 12) is currently beyond the experience of most routine diagnostic laboratories but may one day be simplified.

Serum antibody determinations are not often required and suffer from the complication that HSV-1 and -2 share many antigens. Nevertheless, suitable EIAs have been developed using for capture, type-specific HSV glycoprotein (obtained by monoclonal antibody affinity chromatography). IgM serology is complicated by the fact that IgM antibodies may rise in recurrent as well as in primary infections; it may be of value, however, to supplement attempts at viral isolation in the diagnosis of

---

PLATE 16-4. *Cytopathic effects induced by herpesviruses. (A) Herpes simplex virus in HEp-2 cells (H and E stain, ×57). Note early focal CPE (top right). (B) Varicella-zoster virus in human kidney cells (H and E stain, ×228). Note multinucleated giant cell containing acidophilic intranuclear inclusions (arrow). (C) Cytomegalovirus in human fibroblasts (unstained, ×35). Note two foci of slowly developing CPE. (D) Cytomegalovirus in human fibroblasts (H and E stain, ×228). Note giant cells with acidophilic inclusions in the nuclei (small arrow) and cytoplasm (large arrow), the latter being characteristically large and round. (Courtesy I. Jack.)*

neonatal herpes, provided the standard precautions are taken to remove "rheumatoid factor."

Epidemiology

Herpes simplex viruses spread principally by contact with secretions. HSV-1 is shed in saliva, which may be communicated to others directly, e.g., by kissing, or indirectly, e.g., via contaminated hands, eating utensils, etc. Autoinoculation by fingers may spread virus to the eye or genital tract. HSV-2, on the other hand, is transmitted venereally, being known in some circles as "the virus of love." Neonates can be infected during passage through an infected birth canal.

As might be anticipated from what has been said, primary infection with HSV-1 tends to be acquired in childhood if domestic hygiene is suboptimal, but may be delayed until the teens in developed countries. In poor socioeconomic conditions about 80% of adults have antibody, contrasting with 40% of those in better circumstances. Overall, about one in every three adults suffers from recurrent herpes labialis. Virus is shed in saliva during these attacks, but may also be secreted in smaller amounts from time to time during asymptomatic intervals.

HSV-2, on the other hand, is only rarely encountered before puberty. Changing sexual mores during the 1960s and 1970s led to a dramatic increase in the prevalence of genital herpes in the United States and elsewhere. Estimates of the proportion of people who shed HSV-2 in their genital secretions vary greatly, depending on country, age, promiscuity, and so on—1–10% may be taken as an approximate guide but the figure is much higher in particular groups, e.g., prostitutes. The risk of contracting genital herpes from a sexual partner who is a known carrier is low during asymptomatic intervals but may exceed 50% over a lifetime of cohabitation.

Being endemic, HSV causes new primary infections sporadically the year round. Small nosocomial outbreaks sometimes occur in hospitals, e.g., in nurses responsible for the respiratory care of a baby with neonatal herpes.

Control

The very substantial risk of contracting genital herpes from a partner who is a known carrier can be reduced, but by no means avoided, by such measures as the use of condoms, and abstinence during and for several days after all recurrent attacks. Similarly, the risk of herpetic paronychia in dentists, and in nurses or anaesthetists handling oral or bronchial catheters or endoscopes, can be reduced by wearing gloves. Patients with active HSV, particularly heavy shedders such as babies with neonatal herpes or eczema herpeticum, should be isolated or treated with special caution.

The place of delivery by caesarean section in the prevention of neonatal herpes is controversial. If indeed the baby acquires the infection principally by passage through an infected birth canal (rather than transplacentally or nosocomially from someone other than the mother) it would seem logical to recommend caesarean section in the case of any woman with overt genital herpes at the time of parturition. The operation should be conducted promptly if the membranes have ruptured, otherwise there is a risk of ascending HSV infection. However, neonatal herpes is often acquired from a mother shedding HSV-2 asymptomatically. Accordingly, many authorities now recommend that women with a known history of genital herpes in themselves or their partner should be monitored for virus excretion during the last 6 weeks of pregnancy. The cost of routine screening of all pregnant women would be prohibitive. Moreover, if decisions were made on the basis of tests done earlier in pregnancy, many unnecessary cesareans would be conducted, as the frequency of HSV shedding falls off toward term.

Chemotherapy. The sequential development of the nucleoside analogs, iododeoxyuridine (IDU), trifluridine (TFT), and adenine arabinoside (ara-A) gave us the first chemotherapeutic agents against HSV. Now these in turn have been largely replaced by the "second generation" of nucleoside analogs (those requiring activation by a viral enzyme), the prototype of which is acycloguanosine (acyclovir). The chemistry, pharmacology, mechanism of action, and clinical usage of all these drugs against various HSV infections were fully discussed in Chapter 11. Only a few remarks need to be added here.

Acyclovir is clearly the drug of choice in 1985. Because of its proven efficacy and relative lack of toxicity it can be recommended with confidence for the treatment of all serious manifestations of HSV infection. Yet, because acyclovir-resistant mutants may be generated, it may be prudent policy not to squander it on more trivial infections. Slow intravenous infusion of acyclovir (10–15 mg/kg 8 hourly for about 1 week) is required for the treatment of life-threatening conditions such as HSV encephalitis, neonatal herpes, or overwhelming disseminated infections in immunocompromised patients. Topical application of 3% acyclovir ointment, commencing as early as possible, is effective against HSV keratoconjunctivitis, but so is 1% TFT solution, which should therefore be preferred, perhaps in conjunction with interferon, with which it is claimed to act synergistically. Cream containing 5% acyclovir in propylene glycol may have some effect if applied promptly to recurrent herpes labialis, but comparable results have been claimed for foscarnet (PFA) cream (see Chapter 11). Oral acyclovir (200 mg 5 times daily for 5–10 days) shortens the course of genital herpes; prompt self-medication inhibits the development of new lesions and reduces virus shedding. The provocative question of whether long-term chemoprophylaxis of recurrent herpes genitalis by oral acyclovir should be encouraged will be answered only after we have more data on long-term toxicity and on the epidemiological behavior and pathogenicity of acyclovir-resistant mutants. Meanwhile, some experts are advocating that iv acyclovir be used for very severe primary herpes genitalis meriting hospitalization and that other options, such as foscarnet, interferon, or nothing at all, be considered for milder primary and all recurrent cases.

The newer nucleoside analogs such as 2′-NDG, FIAC, and 5-substituted 2′-deoxycytidines (see Chapter 11) may emerge as useful alternatives to acyclovir, particularly if mutants do not display cross-resistance. The role of interferons (IFN) in the treatment of HSV infections is not as clear. IFN delivered parenterally appears to be beneficial in severe systemic infections, while topically administered IFN has been shown to hasten recovery from keratoconjunctivitis, and IFN cream applied prophylactically over a prolonged period has recently been claimed to diminish the frequency and length of attacks of recurrent herpes genitalis or labialis. It is also possible that additional subtypes of human IFN produced by genetic engineering may prove to be more potent and less expensive than the preparations tested so far.

Vaccines. The herpes genitalis epidemic has quickened interest in several experimental vaccines against HSV currently undergoing clinical trials. The facility with which herpesviruses establish latent infections and the association of HSV-2 with cervical cancer argue somewhat against a conventional live vaccine. Nevertheless work is proceeding on live attenuated HSV-1/HSV-2 recombinants and live deletion mutants of HSV-2 lacking a nonessential early gene. One "subunit" HSV-2 vaccine consists of viral glycoprotein adsorbed to alum adjuvant. Recombinant DNA technology has been used to clone the gene for the protective glycoprotein gD in *E. coli* and in mammalian cells; the resulting protein elicits antibodies that neutralize the infectivity of both HSV-1 and HSV-2. The gD gene has also been incorporated into the genome of vaccinia virus (refer Chapter 10). A short synthetic peptide embodying the critical epitope on gD has been shown to protect mice against HSV challenge. It remains to be seen which if any of these contenders wins the race for the first practicable HSV vaccine.

VARICELLA-ZOSTER VIRUS

Clinical Features

Varicella (Chickenpox). The rash of chickenpox appears suddenly, with or without prodromal fever and malaise. Erupting first on the trunk, then spreading centrifugally to the head and limbs, crops of vesicles progress successively to pustules then scabs. Though painless, the lesions are very itchy, tempting the child to scratch, which may lead to secondary bacterial infection and permanent scarring. Ulcerating vesicles on mucous membranes such as mouth and vulva are painful. The disease tends to be more severe in adults, with varicella pneumonia occurring more commonly than in children (Plate 29-2).

Neurological complications are uncommon, but potentially serious. In about 1 case in 1000, encephalitis develops a few days after the appearance of the rash; up to one-third of these patients are left with severe deficits or die. Rarer developments include the Guillain-Barré syndrome and Reye's syndrome (see Chapter 29).

Varicella is a particularly dangerous disease in the immunocompromised or in nonimmune neonates. Children with deficient cell-mediated immunity, whether congenital, or induced by malignancy (e.g., leukemia), anticancer therapy, or steroid therapy (e.g., for nephrotic syndrome) are especially vulnerable. The disease becomes disseminated, involving many organs in addition to the skin and lungs.

Varicella tends to be more serious in pregnant women and may affect the fetus. The only risk during the first half of pregnancy is that the fetus may very rarely develop the "congenital varicella syndrome" (cutaneous scarring, limb hypoplasia, and eye abnormalities) or may spontaneously abort. However, infections occurring in the few days immediately before or after parturition in a nonimmune woman can be particularly dangerous. As there is no maternal antibody to protect the baby, it may die from disseminated varicella-zoster.

Herpes Zoster. Herpes zoster results from endogenous reactivation of virus which has remained latent in a dorsal root ganglion ever since an exogenous attack of chickenpox many years earlier. Vesicles are confined to one or more dermatomes supplied by that particular sensory nerve(s) (*zoster* = girdle), usually on the trunk, less commonly on the face or neck. The accompanying pain is often very severe for up to a few weeks, but postherpetic neuralgia, which occurs in half of all patients over 60 years of age, may persist for many months. Motor paralysis and encephalomyelitis are rare complications. When the ophthalmic division of the trigeminal nerve is involved, the eye may be seriously affected in a number of ways (Plate 16-5). Disseminated (visceral) zoster is sometimes seen in cancer patients or those otherwise immunosuppressed.

Pathogenesis and Immunity

Varicella virus is believed to enter the body by the respiratory route but little is known for certain about its progression during the 2-week (10–20 days) incubation period. There is a cell-associated viremia. The rash itself results from multiplication of virus in epithelial cells of the skin. Prickle cells show ballooning and intranuclear inclusions, and virus is plentiful in vesicle fluid. In those rare instances when the patient dies from disseminated varicella or pneumonia, giant cells are found in affected organs. The immunity that follows varicella is lifelong. Second attacks of varicella are rare, although subclinical or very mild clinical reinfections can occur, particularly in immunocompromised individuals.

The pathogenesis of herpes zoster was described in Chapter 7 (Table 7-2, Fig. 7-2). The virus is thought to descend the sensory nerve by the same means as it ascended in the first instance, namely within the axon cylinder. In contrast with recurrent herpes simplex, second attacks of herpes zoster are exceptional; they do not necessarily occur on the same dermatome. Zoster affects about 1% of the 50–60 year age group annually, with the incidence climbing rapidly thereafter to the point where most 80 year olds have suffered an attack. The condition is particularly common in patients suffering from Hodgkin's disease, lymphatic leukemia or other malignancies, or following treatment with immunosuppressive drugs or irradiation of or injury to the spine. A sudden decline in the level of cell-mediated immunity is thought to precipitate an attack of herpes zoster, and the protracted course of the disease in older or immunocompromised patients is presumably due to their weaker CMI response.

Laboratory Diagnosis

The clinical picture of both varicella and herpes zoster is so distinctive that the laboratory is rarely called upon for assistance. Smears from the base of skin lesions, or sections from organs taken at autopsy, may be fixed and stained to reveal the characteristic intranuclear inclusions within multinucleated giant cells. These can be positively identified by fluorescent antibody staining.

The virus can be isolated from vesicle fluid, in cultures of human embryonic lung fibroblasts. Virus tends to remain cell associated, very little being released. Hence the CPE develops slowly in distinct foci over

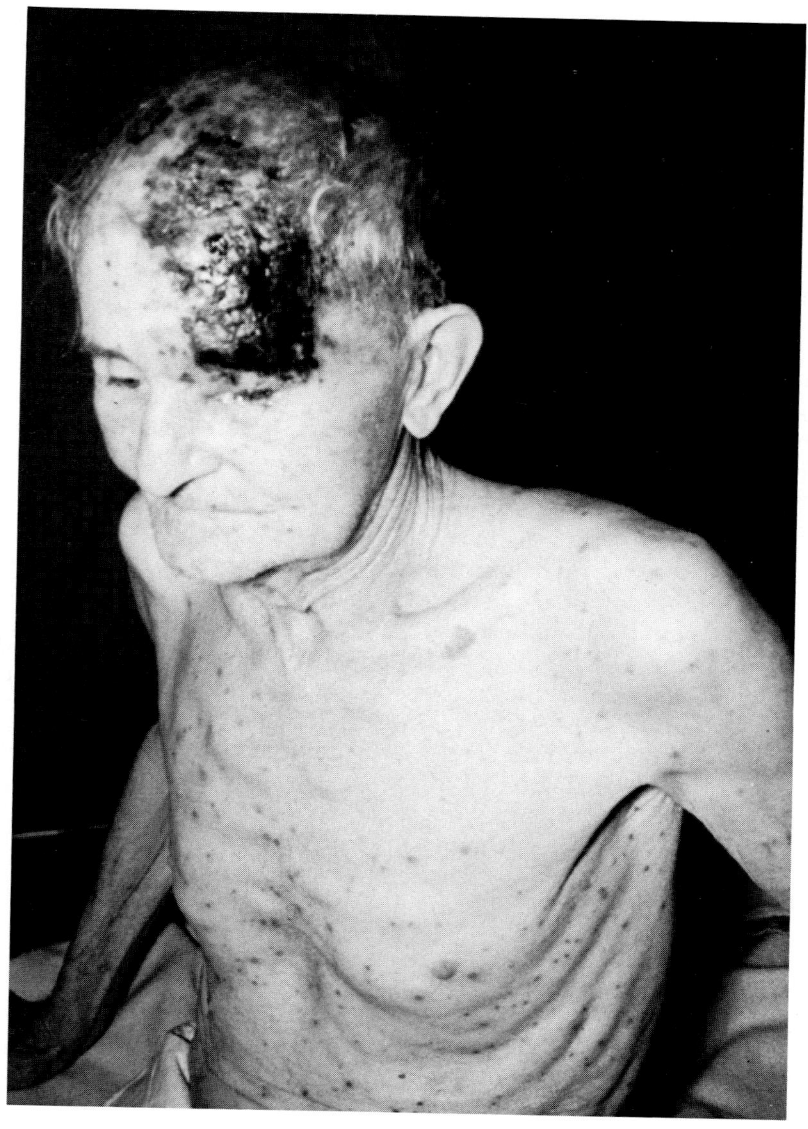

PLATE 16-5. *Ophthalmic herpes zoster. Note unilateral distribution of the lesions on the dermatome supplied by the ophthalmic division of the trigeminal nerve. In addition there are scattered lesions elsewhere on the skin (disseminated zoster), not an uncommon complication of the disease. (Courtesy Dr. J. Forbes.)*

a period of 2 or more weeks (Plate 16-4B). V-Z antigen can be demonstrated in nuclear inclusions by immunofluorescence before the end of the first week.

Recent infection can also be confirmed by detecting a rising titer of antibody, or by IgM serology, preferably EIA. Immune status, e.g., of potentially vulnerable leukemic children, can be determined by plaque-reduction neutralization, the low sensitivity of which may be enhanced by anti-human IgG. Specific tests for CMI are also desirable.

Epidemiology and Control

Varicella is endemic the year round but is most prevalent during late winter and spring. Most children become infected during their first years at school. Spread probably occurs by contact as well as via respiratory droplets. Children are contagious, and should be excluded from school, for as long as moist vesicles are present; 1 week is normally sufficient.

Chemotherapy. Varicella in the normal child is not generally a matter for great concern; management centers on prevention of itching, scratching, and secondary bacterial infection. Herpes zoster is treated with topical antibiotics plus powerful analgesics. Zoster or varicella in immunocompromised individuals may require more vigorous intervention. Several antiviral agents show promise (see Chapter 11). The first to be developed were ara-A and its more soluble form ara-AMP, both of which have been shown to accelerate healing of cutaneous lesions and to decrease visceral dissemination if iv administration is commenced promptly. Intravenous acyclovir appears to be at least as effective, and less toxic. Newer nucleoside analogs such as BVDU, FIAC, and 2′-NDG are currently under trial (see Chapter 11). Interferons, at very high dosage (3.5×10^7 units per day) have also been shown to limit somewhat the progression of zoster and varicella in immunocompromised patients.

ZIG. Passive immunization with zoster immune globulin (ZIG), obtained originally from convalescent zoster patients, has an important place in the prevention of varicella. ZIG should be administered to non-immune pregnant women who have come into close contact with a case of varicella within the preceding 3 days but is ineffectual if further delayed. If a pregnant woman has actually contracted varicella within the few days before (or very soon after) delivery, the probability of disseminated disease in the baby may be reduced by injecting both the mother (immediately) and the baby (at birth) with ZIG. ZIG is also indicated for immunocompromised patients who become exposed to the risk of infection—a typical crisis situation would be the occurrence of a case of varicella in the leukemia ward of a children's hospital.

Vaccine. Japanese and American workers have produced a live attenuated vaccine from the Oka strain, derived originally by serial passage in cultured human and guinea pig cells. A single injection induces protection in 95% of recipients. However, it is not yet clear how long this immunity lasts; detectable antibody sometimes disappears within a year or so and subsequent exogenous infections with wild varicella virus do occur. The vaccine induces fever and a papular rash occasionally in normal children, and more frequently in the principal target group for such a vaccine, namely immunocompromised children. For instance, a substantial minority of children with leukemia or nephrotic syndrome developed mild varicella following vaccination. Evidence so far indicates that the attenuated vaccine does not prevent herpes zoster but itself establishes latent infection in dorsal ganglia and leads to zoster in the years ahead. Whether such attacks are more or less frequent or severe than following natural varicella may not be known for certain for some considerable time. Meanwhile it is probably prudent not to recommend routine vaccination of normal children with this vaccine. It may come to have an important place in the protection of immunocompromised children and other risk groups.

CYTOMEGALOVIRUS (CMV)

Throughout much of the world CMV infection is acquired subclinically during childhood, but in some of the more affluent communities it tends to be delayed until an age when it is capable of doing considerably more damage. In particular, primary infections during pregnancy can lead to congenital abnormalities in the fetus. Moreover, iatrogenic infections may follow blood transfusion, or immunosuppression, e.g., for organ transplantation.

Clinical Features

Prenatal Infection ("Cytomegalic Inclusion Disease"). CMV infection during pregnancy is now recognized to be the major viral cause of congenital abnormalities in the newborn. About 1 baby in every 200 is born with asymptomatic CMV infection, and 1 in 2000 has signs of CID (Plate 16-6). The classical syndrome is not always seen in its entirety. The infant is small, with hepatosplenomegaly, jaundice, thrombocytopenia,

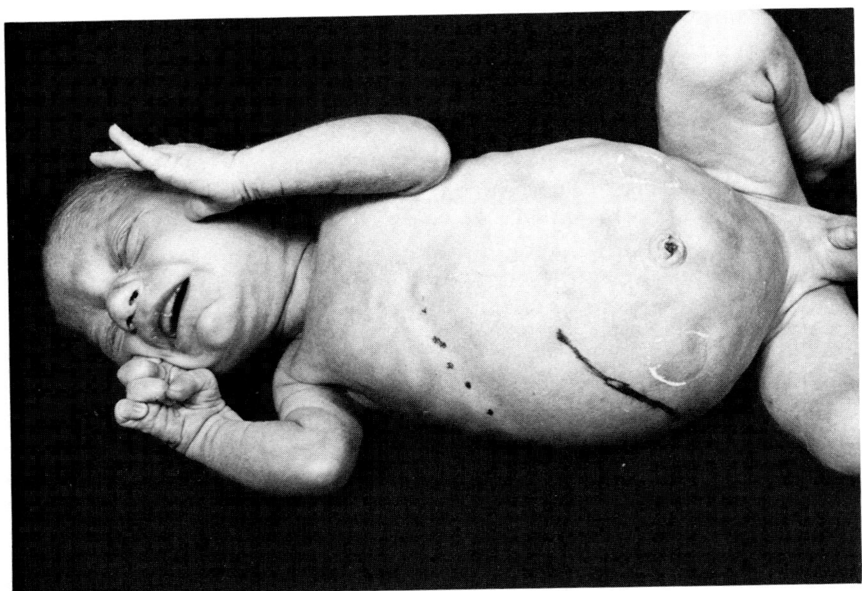

PLATE 16-6. *Congenital cytomegalic inclusion disease. Features are retardation of growth, microcephaly, thrombocytopenia, and hepatosplenomegaly. The edge of the liver has been marked. (Courtesy Dr. K. Hayes, Fairfield Hospital for Infectious Diseases, Melbourne.)*

and sometimes pneumonitis. Abnormalities of the brain (microcephaly, encephalitis) and eyes (chorioretinitis) lead to mental retardation, cerebral palsy, and impairment of hearing and sight. Socially and educationally important deficits such as hearing loss, subnormal IQ, epilepsy, and behavioral problems may not become apparent until as late as 4 years after birth.

Mononucleosis. Most infections acquired after birth, by whatever route, are subclinical. Not uncommonly however, a syndrome resembling EBV mononucleosis is seen, particularly in young adults and in recipients of blood transfusion. Typically, the patient presents with prolonged fever and malaise, and on examination is found to have splenomegaly, abnormal liver function tests, and lymphocytosis, often with "atypical lymphocytes" such as those observed with EBV (see below). In contrast to EBV mononucleosis however, pharyngitis and lymphadenopathy are uncommon, and heterophile antibody is absent.

Other Clinical Presentations. CMV infections may present in a wide variety of more serious forms, particularly in immunocompromised patients or premature infants. Disease may occur in almost any organ. The most important are interstitial pneumonia (seen especially in recipients of organ transplants and in premature babies following blood transfusion), hepatitis, chorioretinitis, arthritis, carditis, chronic gastrointestinal disease, and various CNS diseases, especially encephalitis and the Guillain-Barré syndrome.

Pathogenesis, Immunity, and Epidemiology

Once infected with CMV, an individual carries the virus for life and sheds it intermittently in saliva, urine, semen, cervical secretions, and/or breast milk. It is not known for certain which cells harbor the virus, nor what form it is in. The B lymphocyte is the most likely candidate, but there is some evidence incriminating monocytes, polymorphs, and epithelial cells. The genome of CMV has been recovered from oropharyngeal epithelium *in vivo.* Unlike the latent infections established by HSV and VZV, those of CMV are more akin to chronic infections, because virus can be readily recovered for long periods. Up to 10% of people may be found to be shedding virus at any time one cares to look, especially in young children. The intermittent nature of CMV shedding and the fluctuations observed in CF antibody levels suggest that asymptomatic exacerbations occur on several occasions throughout life. For example, reactivation occurs during pregnancy, though, oddly enough, cervical shedding seems to be suppressed early in pregnancy and rises markedly as term approaches. Hormonal factors may be at work here but immunosuppression is generally the most powerful trigger. CMV can be isolated from over 90% of patients profoundly immunosuppressed for organ transplantation. Indeed, the glomerulopathy induced by CMV adversely affects the survival of kidney grafts, while up to 25% of bone marrow transplant recipients die from CMV interstitial pneumonitis.

Cell-mediated immunity (CMI) appears to be responsible for controlling CMV and for limiting exacerbations. Natural killer (NK) and cytotoxic T (T_c) cells may both be involved. On the other hand, CMV itself is immunosuppressive, causing a general depression of CMI and an increase in suppressor T cells. There is evidence that the absence of CMV-specific T_c cells denotes a grave prognosis in bone marrow transplant recipients, and that possibly a CMI deficit in pregnancy prefaces the development of CID in the fetus.

Congenital infection of the fetus generally comes about by transplacental spread but may also occur as a result of ascending infection from the cervix. Approximately 1% of all babies become infected *in utero.* Of these, only about 1 in every 10 develops clinical features while another 1 in 10 appears normal at birth but reveals evidence of mental retardation or hearing loss during its early years. Whereas the majority of maternal infections are endogenous recurrences, these almost never lead to CID; most cases of frank CID result from primary infections occurring during the first 6 months of pregnancy. Hence CID tends to be a disease of affluence as >50% of women in developed countries, cf. <10% in Third World countries, are still seronegative as they enter their childbearing years. Primary infection during gestation carries a 40% risk of congenital infection and a 4% risk of clinical abnormalities in the baby. Higher titers of virus and of IgM antibody are found in babies from mothers with primary than with recurrent infection. Clearly, the preexisting immunity present in the latter confers considerable protection against disease in the fetus.

A minority of babies with clinical abnormalities at delivery are stillborn or die shortly after birth. Autopsy reveals fibrosis and calcification in the brain and liver. Typical cytomegalic cells may be found in numerous organs—as indeed they may be found in salivary glands or renal tubules of many normal children (Plate 16-7A). Progressive damage may occur not only throughout pregnancy but also after birth. The affected infant synthesizes specific IgM, and immune complexes are plentiful, but CMV-specific and nonspecific cell-mediated immune responses are markedly depressed.

Postnatal or natal (intrapartum) infection may occur in at least three different ways. About 10–20% of women shed CMV from the cervix at the time of delivery. The baby commonly becomes infected subclinically, but may rarely develop pneumonia or hepatitis. Second, 10–20% of nursing mothers shed CMV in their milk, hence infection may be acquired via breast-feeding. Such infections are subclinical. Perhaps this is a standard route of transmission of CMV in the developing countries, where asymptomatic infection in the first year or two of life is almost universal. But salivary and droplet spread may be the most important of all the known routes of spread. Via direct contact or contamination of hands, eating utensils, etc., oropharyngeal secretions are believed to constitute the principal vehicle of transmission, not only in childhood but again in adolescence. Kissing no doubt accounts in part for the sudden increase in CMV seropositivity from 10–15% to 30–50% between the ages of 15 and 30 in countries such as the United States. The other major route of transmission in this age group is sexual intercourse. CMV is shed intermittently both in cervical secretions and in semen. CMV infection is universal among promiscuous male homosexuals.

The two remaining mechanisms of acquiring CMV constitute special cases of iatrogenic infection—blood transfusion and organ transplanta-

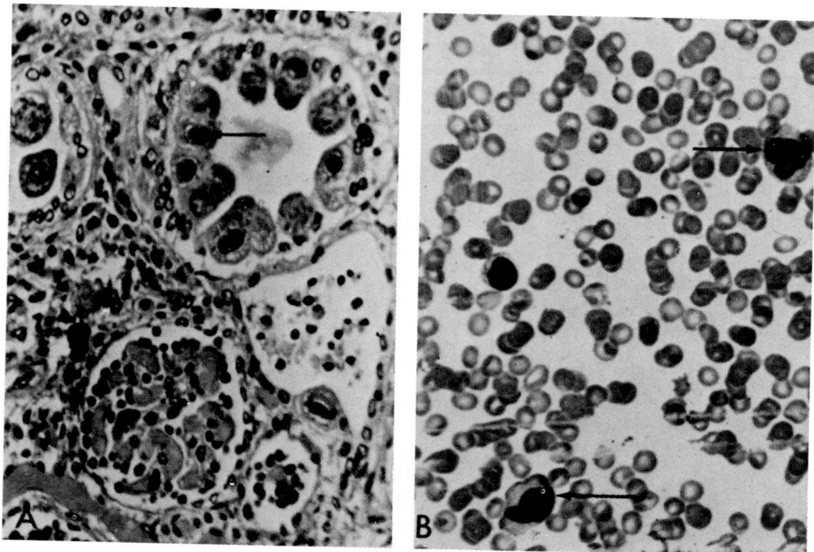

PLATE 16-7. *Histopathology of some diseases caused by herpesviruses. (A) Cytomegalic inclusion disease. Section of kidney (H and E stain, ×250). Arrow indicates one of several "cytomegalic" cells inside a renal tubule. The typical cytomegalic cell is greatly swollen, with an enlarged nucleus distended by a huge inclusion which is separated by a nonstaining halo from the nuclear membrane, giving the cell the appearance of an owl's eye. (B) Infectious mononucleosis (glandular fever). Smear of peripheral blood (Leishman stain, ×400). Note large "atypical lymphocytes" (arrowed). (Courtesy I. Jack.)*

tion. Almost all those who receive multiple transfusions of large volumes of blood develop CMV infection at some stage. Most such episodes are subclinical but the mononucleosis syndrome is not uncommon, and more serious manifestations of primary infection may occur in seronegative premature infants, pregnant women, and immunocompromised patients.

Disease in recipients of kidney, heart, or bone marrow transplants may also be of exogenous origin, introduced via the donated organ or via accompanying blood transfusion. Such an exogenous infection may be primary, or may be a reinfection with a different CMV strain. On the other hand, the profound immunosuppression demanded for organ transplantation, or indeed for other purposes such as for cancer therapy, is quite sufficient to reactivate previous infection.

The relationship between the circumstances of transmission of CMV and the commoner clinical outcomes is summarized in Table 16-4.

TABLE 16-4
Cytomegalovirus Infections

AGE	ROUTE	DISEASE
Prenatal	Transplacental	Cytomegalic inclusion disease
Perinatal	Cervical secretions	Pneumonia, hepatitis
	Breast milk	Nil
Any	Saliva	Mononucleosis
	Sexual intercourse	Nil
	Transfusion	Mononucleosis
	Immunosuppression	Pneumonia, hepatitis, retinitis, encephalitis, etc.

Laboratory Diagnosis

CMV may be grown from urine, saliva, milk, genital secretions, leukocytes, or various organs taken at autopsy. As cytomegaloviruses are host-specific, human fibroblasts must be used. Sensitivity can be increased by sedimenting virus onto the cells by low-speed centrifugation. The virus replicates very slowly, a single cycle having a latent period of 36–48 hours. Moreover, new virus tends to remain cell-associated, with spread occurring mainly to adjacent cells. Hence CPE may not become evident for up to 4 weeks, and the monolayers must be maintained in good condition for this length of time by weekly medium changes. Foci of swollen refractile cells with cytoplasmic granules are usually detectable after about 1–2 weeks (Plate 16-4C). When stained, these cells are found to contain a number of nuclei with pathognomonic large amphophilic "skein-like" nuclear inclusions and smooth round acidophilic cytoplasmic masses (Plate 16-4D). As early as 24–36 hours after inoculation the monolayer can be stained by the immunofluorescence (IF) or immunoperoxidase techniques using a monoclonal antibody against an early nuclear antigen.

There is only a single serotype of human CMV, but different strains can be distinguished by kinetic neutralization or restriction endonuclease mapping. The latter technique can be employed to identify the source of virus.

Laboratory confirmation of CID in a newborn infant is of great importance as, even in the absence of chemotherapy, it affects the medical and educational management of the child. The following diagnostic procedures are used.

1. *Virus Isolation*. This is the method of choice. Virus can be recovered from urine and saliva for years after birth but specimens should be taken

within the first 2 weeks of life to avoid confusion with natal or early postnatal infection, which is generally harmless.

2. *Exfoliative Cytology.* Cytomegalic cells (Plate 16-7A) may be identified in a centrifuged deposit of urine or saliva, or in biopsy or autopsy specimens. Immunofluorescence using monoclonal antibody identifies CMV antigen in the inclusions.

3. *Immunoassays and Nucleic Acid Hybridization.* EIA or RIA is now being used increasingly to identify CMV antigen in urine. Labeled nucleic acid probes (see Chapter 12) are more of an experimental tool so far.

4. *Anti-CMV IgM.* As IgM does not cross the placenta, finding CMV-specific IgM in the newborn baby is diagnostic of CID but it is not detected in a significant minority of cases. Traditionally, indirect immunofluorescence, or the rather simpler procedure of immune adherence hemagglutination (binding of antibody to antigen-coated sheep erythrocytes), has been used for this purpose. However, EIA and RIA lend themselves very well to IgM assays, providing adequate precautions are taken to remove IgG and rheumatoid factor, e.g., on protein A Sepharose. Anti-human IgM is preferred as the capture antibody (for details see Chapter 12 and Fig. 12-4).

Similar techniques are used to diagnose other CMV infections. IF, IHA, and EIA are also appropriate for determining immune status.

Control

Iatrogenic infection via transfusion or organ transplantation can be reduced by screening donor and recipient for evidence of CMV infection. The presence of antibody in the intended recipient indicates a significant degree of immunity, while antibody in the donor provides a warning that he is a carrier, hence his donation could be contaminated with the virus. Accordingly, the potentially dangerous combination is averted if seronegative recipients are given blood or organs taken only from seronegative donors. The amount of testing involved is not cost effective for routine blood transfusions but should be considered for seronegative premature infants, pregnant women, and immunocompromised individuals.

Chemotherapy. By and large, the nucleoside analogs which appear so promising against HSV and VZV have been very disappointing against CMV. Preliminary data with one or two of the newer agents such as 2′-NDG and DHPG (see Chapter 11) do look encouraging however. Interferon has been shown to reduce the incidence and clinical severity of CMV reactivation associated with renal transplantation.

Vaccines. Experimental live attenuated CMV vaccines have been developed by serial passage in human fetal fibroblasts. They induce a humoral and CMI response in normal recipients but a much reduced or delayed response in patients being prepared for renal transplantation. The vaccine failed to reduce the incidence of CMV shedding by such transplant recipients. The question of the safety of live CMV vaccines must be addressed, particularly in relation to adequacy of attenuation, establishment of persistence, subsequent reactivation, and possible oncogenicity.

EB VIRUS

Clinical Features

Infectious Mononucleosis (Glandular Fever). In the young, EB virus infections are asymptomatic or very mild. However, when infection is delayed until adolescence, as tends to happen in developed countries, there is a 50% chance that infectious mononucleosis (IM) will develop. Following an incubation period of 4–7 weeks the disease often begins insidiously with malaise and fatigue. The clinical presentation is protean, but three regular features are fever, sore throat, and enlarged cervical lymph glands. The fever is high and fluctuating, the pharyngitis is frequently severe with a white or grey malodorous exudate covering the tonsils and petechiae on the palate, and the lymphadenopathy seen first in the neck spreads to involve other lymph nodes as the disease develops. The spleen is often enlarged and liver function tests abnormal. Occasionally a rash may appear on the trunk. The disease usually lasts for 2–3 weeks, but convalescence may be very protracted ("prolonged atypical illness").

Neurological complications occur in <1% of cases, even in the absence of clinically evident IM. These include the Guillain-Barré syndrome, Bell's palsy, meningoencephalitis, and transverse myelitis (see Chapter 29). Other complications include autoimmune hemolytic anemia, thrombocytopenia, and carditis. Very rarely, death may occur following rupture of the spleen or respiratory obstruction.

X-Linked Lymphoproliferative (XLP) Syndrome. A rare fatal polyclonal B cell proliferative syndrome due to EBV infection has recently been observed in families with this particular X-linked recessive immunodeficiency to EBV. EBV infection in these boys is uncontrolled. Mortality in the acute stage of EBV infection is 70%, usually due to liver necrosis. The survivors usually develop common varied immunodeficiency, and some develop lymphoproliferative disorders.

Progressive Lymphoproliferative Disease. Progressive lymphoproliferative disease in children with primary immunodeficiencies, such as severe combined immunodeficiency, or in transplant recipients, other immunosuppressed patients, or homosexuals with AIDS, is a similar condition, thought to be associated in some cases with the use of cyclosporin A, which depresses the activity of T_c cells against EBV.

Burkitt's Lymphoma and Nasopharyngeal Carcinoma. The reader is referred back to Chapter 8 for a full discussion of these two highly malignant neoplasms and their relationship to EBV.

Pathogenesis and Immunity

The principal target of EBV is the B lymphocyte. Only the B cell is known to possess the receptor for EBV, which is associated with the C3d receptor. EBV proteins, such as the nuclear antigen EBNA, are found in the patient's B lymphocytes, but virus is produced only from occasional infected cells. The interaction is not cytocidal but transforming; B cells are actually induced to differentiate into plasma cells which secrete a wide variety of unrelated antibodies (*polyclonal activation*). These include the so-called *heterophil antibodies* which are the basis for some of the older diagnostic tests. The proliferation of these B cells, together with the T_c cells which attack and control them, accounts for the tonsillitis, hepatosplenomegaly, and widespread lymphadenopathy in IM. Normally, the B cell proliferation is kept in check by the cell-mediated immune response, perhaps antibody-dependent cell-mediated cytotoxicity but especially cytotoxic T cells which recognize EB membrane antigens on the B cell surface. These T blasts are among the "atypical lymphocytes" which characterize the disease, infectious mononucleosis. The number of B cells reaches its peak in the first week of the illness and returns to normal by 3 weeks; T cell numbers remain elevated for rather longer. When CMI is seriously compromised, as in XLP or in cyclosporin A-treated transplant recipients, progressive lymphoproliferative disease develops, often with a fatal outcome.

Infection with EBV induces lifelong immunity to exogenous reinfection. However, even fully immunocompetent individuals fail to reject the virus. The EBV genome persists in a tiny minority of B lymphocytes, mainly in the form of multiple copies of nonintegrated circularized DNA (episomes). The virus itself is not detectable in the lymphocytes of such asymptomatic carriers, nor indeed in acutely ill IM patients, but is revealed when they are cocultured *in vitro* with uninfected lymphocytes, which become transformed into a continuous line of lymphoblasts containing the EBV genome and antigens. Such immortalized EBV-induced lymphoblastoid cell lines (LCL) are comparable to those derived from Burkitt's lymphoma but without the karyological abnormalities (see Chapter 8). Some produce virus spontaneously but most are "non-producers," which yield virus only after induction by certain chemicals.

Presumably EBV must multiply productively in some other type of cell. The fact that severe pharyngitis is such an important feature of IM suggests that the virus may be cytolytic for certain epithelial cells at least. Furthermore, the high titer of cell-free virus secreted from Stensen's duct indicates that EBV multiplies in the parotid glands, where EBV DNA can also be found by hybridization with a radiolabeled probe. This is probably the source of virus in saliva, which is the principal vehicle of spread.

Laboratory Diagnosis

The clinical picture of IM is so variable, and often so vague, that laboratory confirmation is required. This rests on (1) differential white blood cell count, (2) heterophil antibodies, and (3) EBV-specific antibodies.

By the second week of the illness, white blood cells total 10,000–20,000/mm^3 or even higher. Lymphocytes plus monocytes account for 60–80% of this number. Of these, at least 10%, and generally more than 25%, are "atypical lymphocytes" (large pleomorphic blasts with deeply basophilic vacuolated cytoplasm and lobulated nuclei, Plate 16-7B), which persist for 2 weeks to several months.

Many years ago Paul and Bunnell made the empirical observation that sera from glandular fever patients agglutinate sheep erythrocytes. It is now clear that these agglutinins are just one of the heterophil IgM antibodies elicited by EBV infection. The Paul–Bunnell test is still a valid diagnostic aid. Sera are first absorbed with guinea pig kidney cells to remove Forssman antibody (and serum-sickness antibodies, which are rare these days). Commercially available spot test kits are convenient, although false positives may occur. The comparable horse RBC agglutinin test is more sensitive and remains positive for several months after IM. A third measure of heterophile antibodies is the ox red cell hemolysin test, which requires no serum absorption and is specific but remains positive for only a month or two.

EBV-specific antibodies might seem to offer a more reliable indicator of infection, but are not widely employed because the techniques of immunofluorescence (IF) required are still too cumbersome for the average laboratory. The single most useful antibody to assay is IgM against the EB viral capsid antigen VCA, because it develops to high titer early in the illness and declines rapidly over the next 3 months or so, disappearing within a year. Serum is absorbed with *Staphylococcus aureus* to remove IgG and RF, then tested by indirect IF on an EBV$^+$-LCL. Immune adherence hemagglutination is just as sensitive as IF and considerably simpler. Suitable EIAs are under development.

IgG (or total) antibodies against VCA represent the most convenient measure of immune status. Antibodies against the nuclear antigen EBNA are absent early in typical IM but persist thereafter indefinitely, hence can be helpful in differentiating reactivation from primary infection.

Isolation of virus is not used as a diagnostic procedure in IM, for two reasons. First, a high proportion of people are chronic excretors of EBV. Second, no known cell line is fully permissive for EBV. The accepted method for "growing" EBV from IM patients *in vitro* is to inoculate oropharyngeal secretions or peripheral blood leukocytes onto umbilical cord lymphocytes and to demonstrate the transformation of the latter to an LCL that can be stained successfully with fluoresceinated antibody against EBNA.

Epidemiology

Following IM, virus is shed in saliva for several months, commencing in the second week. Furthermore, some 15–20% of healthy seropositive young adults excrete EBV intermittently for years. The percentage of chronic shedders is even higher in developing countries, in young children, in early pregnancy, and in the immunocompromised.

In most developing countries EBV is ubiquitous. Essentially all infants become infected in the first 2 years or so of life, probably by salivary exchange, contamination of eating utensils, and perhaps also by respiratory aerosols. At this age almost all infections are subclinical; glandular fever is virutally unknown in the Third World.

By way of contrast, in countries with higher standards of living many reach adolescence before first encountering the virus. The intimate osculatory contact involved in kissing is the principal means of transmission, hence IM is a disease of 15–25 year olds. Some 80% of older adults have acquired infection and are permanently immune.

Conventional respiratory transmission by droplets does not appear to be significant. For example, casual roommates of IM patients are not at increased risk; closer contact seems to be required. Male homosexuals do have a very high rate of seropositivity. Blood transfusion can also rarely transmit the virus and mononucleosis, especially if the donor was incubating IM at the time of giving blood.

Control

Chemotherapy. Unfortunately none of the present armory of nucleoside analogs has been shown to have any effect on EBV infections.

Vaccines. If a successful vaccine could be developed against EBV it may be feasible to prevent not only IM but also nasopharyngeal carcinoma—the commonest cancer in the world's most populous nation (see Chapter 8). A conventional live or inactivated vaccine is a long way off since we cannot yet grow the virus productively in cultured cells. Nevertheless, some progress has been made toward a "subunit" vaccine consisting of the glycoprotein EBV membrane–antigen complex, which when incorporated into liposomes elicited EBV-neutralizing antibodies in marmosets. Recombinant DNA technology presents another possible approach.

FURTHER READING

Books

Epstein, M. A., and Achong, B. G., eds. (1979). "The Epstein-Barr Virus." Springer-Verlag, Berlin and New York.

Evans, A. S., ed. (1982). "Viral Infections of Humans: Epidemiology and Control," Chapters 8, 10, 13, and 23–26. Plenum, New York.

Galasso, G. J., Merigan, T. C., and Buchanan, R. A. eds. (1984). "Antiviral Agents and Viral Diseases of Man," 2nd ed., Chapters 6, 7, and 10–13. Raven Press, New York.

Glaser, R., and Gotlieb-Stematsky, T., eds. (1982). "Human Herpesvirus Infections: Clinical Aspects." Dekker, New York.

Ho, M. (1982). "Cytomegalovirus: Biology and Infection." Plenum, New York.

Nahmias, A. J., Dowdle, W. R., and Schinazi, R. F., eds. (1981). "The Human Herpesviruses." Am. Elsevier, New York.

Plotkin, S. A., Michelson, S., Pagano, J. S., and Rapp, F., eds. (1984). "CMV: Pathogenesis and Prevention of Human Infection." March of Dimes Birth Defects Foundation. Birth Defects: Original Article Series, Vol. 20, pp. 1–499. Alan R. Liss, Inc., New York.

Roizman, B., ed. (1982–1985). "The Herpesviruses," Vols. 1–4. Plenum, New York.

Reviews

Adler, S. P. (1983). Transfusion-associated cytomegalovirus infections. *Rev. Infect. Dis.* **5,** 977.

Enders, G. (1984). Varicella-zoster virus infection in pregnancy. *Prog. Med. Virol.* **29,** 166.

Epstein, M. A. (1982). Persistence of Epstein-Barr virus infection. *In* "Virus Persistence" (B. W. J. Mahy, A. C. Minson, and G. K. Darby, eds.), Symp. Soc. Gen. Microbiol., Vol. 33, pp. 169–184. Cambridge Univ. Press, London and New York.

Gershon, A. A. (1984). The success of varicella vaccine. *Pediat Infect. Dis.* **3,** 500.

Osborn, J. E. (1981). Cytomegalovirus: Pathogenicity, immunology, and vaccine initiatives. *J. Infect. Dis.* **143,** 618.

Rapp, F. (1980). Persistence and transmission of cytomegalovirus. *In* "Comprehensive

Virology'' (H. Fraenkel-Conrat and R. R. Wagner, eds.), Vol. 16, p. 193. Plenum, New York.

Roizman, B., Carmichael, L. E., Deinhardt, F., de-Thé, G., Nahmias, A. J., Plowright, W., Rapp, F., Sheldrick, P., Takahashi, M., and Wolf, K. (1981). Herpesviridae: Definition, provisional nomenclature, and taxonomy. *Intervirology* **16,** 201.

Stagno, S., Pass, R. F., Dworsky, M. E., Henderson, R. E., Moore, E. G., Watton, P. D., and Alford, C. A. (1982). Congenital cytomegalovirus infection: The relative importance of primary and recurrent maternal infection. *N. Engl. J. Med.* **306,** 945.

Takahashi, M. (1983). Chickenpox virus. *Adv. Virus Res.* **28,** 285.

Wildy, P., Field, H. J., and Nash, A. A. (1982). Classical herpes latency revisited. *In* ''Virus Persistence'' (B. W. J. Mahy, A. C. Minson, and G. K. Darby, eds.), Symp. Soc. Gen. Microbiol., Vol. 33, pp. 133–168. Cambridge Univ. Press, London and New York.

zur Hausen, H. (1983). Herpes simplex virus in human genital cancer. *Int. Rev. Exp. Pathol.* **25,** 307.

CHAPTER 17

Poxviruses

Introduction 433
Properties of *Poxviridae* 434
Laboratory Diagnosis 436
Smallpox 437
Vaccination with Vaccinia Virus 437
Human Monkeypox 440
Molluscum Contagiosum 441
Cowpox and Milker's Nodes 441
Orf 442
Yabapox and Tanapox 443
Further Reading 443

INTRODUCTION

The family *Poxviridae* is divided into eight genera on the basis of antigenic and morphological differences. Several poxviruses cause infections of man: smallpox (now extinct), monkeypox, molluscum contagiosum, cowpox, milker's nodes, orf, and tanapox (Table 17-2). Others cause economically important infections of domestic animals, and myxoma virus has been used with dramatic success in the biological control of a pest animal, the rabbit (see Chapter 4).

All diseases due to poxviruses are associated with skin lesions, which may be localized or may be part of a generalized rash, as in smallpox and monkeypox. Smallpox, one of the great plagues of mankind, which has played an important role in human history and a central role in the development of virology, was eradicated from the world in 1977.

PROPERTIES OF *POXVIRIDAE*

The poxviruses are the largest and most complex of all viruses. Using critical illumination, the virions are just large enough to be seen under a light microscope, but knowledge of their structure comes from electron microscopic studies. Plate 17-1A, C, and D illustrates the structure of the virion of vaccinia, which is similar to that of all the poxviruses that affect man except those belonging to the genus *Parapoxvirus* (Plate 17-1B). There is no nucleocapsid conforming to either of the two types of symmetry found in most other viruses (Chapter 1), hence it is sometimes called a "complex" virion. An "outer membrane" of tubular-shaped

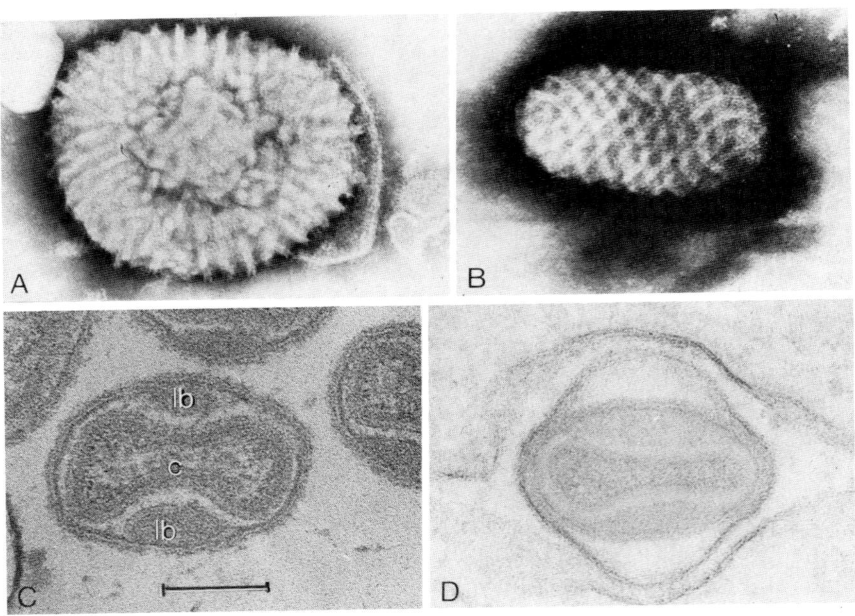

PLATE 17-1. Poxviridae. *(A) Negatively stained vaccinia virion, showing surface structure of rodlets or tubules characteristic of the outer membrane of the* Orthopoxvirus *genus. (B) Negatively stained orf virion, showing characteristic surface structure of the outer membrane of the* Parapoxvirus *genus. (C) Thin section of vaccinia virion in its narrow aspect, showing the biconcave core (c) and the two lateral bodies (lb). (D) Thin section of mature extracellular vaccinia virion lying between two cells. The virion is enclosed by an envelope originating from altered Golgi membranes. Bar = 100 nm. [A and D, from S. Dales,* J. Cell Biol. **18,** *51 (1963); B, from J. Nagington* et al., Virology **23,** *461 (1964); C, from B. G. T. Pogo and S. Dales,* Proc. Natl. Acad. Sci. U.S.A. **63,** *820 (1969).]*

lipoprotein subunits, arranged rather irregularly, encloses a dumbbell-shaped core and two "lateral bodies" of unknown nature. The core contains the viral DNA together with protein, but the detailed arrangement of these components has not been determined. Especially in particles released naturally from cells, rather than by cellular disruption, there is an envelope (Plate 17-1D) which contains cellular lipids and several virus-specified polypeptides including the hemagglutinin of orthopoxviruses.

The nucleic acid is dsDNA, MW from 85 million for parapoxviruses to 140 million for cowpox virus, with a low G + C content (35%). There are over 100 different polypeptides in the virion. The core proteins include a transcriptase and several other enzymes. The lipoprotein outer membrane of the virion is synthesized *de novo,* not derived by budding from cellular membranes; the envelope, when present, is derived from membranes of the Golgi apparatus but contains several virus-specified polypeptides (Table 17-1).

As would be expected from its many polypeptides, the virion contains numerous antigens recognizable by immunodiffusion. Most of these are common to all members of any one genus, although each species is characterized by certain specific polypeptides, while a few others appear to be shared by all poxviruses of vertebrates. There is extensive cross-neutralization and cross-protection between viruses belonging to the same genus, but none between viruses of different genera.

Viral Multiplication

Replication of vaccinia virus, the prototype poxvirus, occurs in the cytoplasm and can be demonstrated in enucleate cells. After fusion of the virion with the plasma membrane or endocytosis, the viral core is released into the cytoplasm. Transcription is initiated by the viral transcriptase and functional capped and polyadenylated mRNAs are produced within minutes after infection. Polypeptides produced by transla-

TABLE 17-1
Properties of Poxviridae

Brick-shaped virion, 300 × 240 × 100 nm[a]
Complex structure with core, lateral bodies, outer membrane, and envelope
dsDNA, linear, MW 110–140 million (*Orthopoxvirus*); 85 million (*Parapoxvirus*)
Transcriptase and several other enzymes in virion
Cytoplasmic multiplication

[a] Parapoxviruses: ovoid, 260 × 160 nm, with regular spiral arrangement of "tubule" on the outer membrane.

tion of these mRNAs complete the uncoating of the core, and transcription of about 100 genes, distributed throughout the genome, occurs before viral DNA synthesis begins. These early mRNAs are not produced by splicing or cleavage of polycistronic RNA precursors, but are monocistronic.

Early proteins include thymidine kinase, DNA polymerase, and several other enzymes, and viral DNA replication occurs within a discrete period 1.5 to 6 hours after infection. With the onset of DNA replication there is a dramatic shift in gene expression and almost the entire genome is transcribed, but transcripts from the early genes (i.e., those transcribed before DNA replication begins) are not translated.

The initial stages of virion formation occur in circumscribed areas of the cytoplasm ("viral factories"), the first morphologically defined structures being cupules which appear to be rendered rigid by an array of spicules. These develop into spherical immature particles and the spicules disappear. Subsequently the spherical bilayer of the immature particle becomes the outer membrane of the virion and the core and lateral bodies differentiate within it. Some of these mature particles move to the vicinity of the Golgi apparatus and acquire a double membrane, which includes several virus-specified polypeptides. Some of these particles are released from the cell with a single Golgi membrane that constitutes the viral envelope. However, most particles do not acquire an envelope and are released by cell disruption. Both enveloped and nonenveloped particles are infectious, but enveloped particles are more rapidly taken up by cells and appear to be important in the spread of virions throughout the body.

LABORATORY DIAGNOSIS

The morphology of the virions is so characteristic that electron microscopy is used to identify them in negatively stained vesicle fluid or biopsy material taken directly from skin lesions. All orthopoxvirus virions have the same appearance, which is also shared by the virions of molluscum contagiosum and tanapox, but milker's nodes and orf viruses can be distinguished by the distinctive appearance of the parapoxvirus virion.

The usual method of isolation of orthopoxviruses, which include smallpox, monkeypox, cowpox, and vaccinia viruses as human pathogens, is by inoculation of vesicle fluid or biopsy material on the chorioallantoic membrane (CAM) of chick embryos. The pocks of each species are characteristic (see Plate 2-4). The parapoxviruses, as well as molluscum contagiosum virus and tanapox virus do not grow on the

CAM. Orthopoxviruses grow well in cell culture, parapoxviruses and tanapox virus less readily, and molluscum contagiosum virus has not yet been cultivated.

Diagnosis of the species of orthopoxvirus, e.g., differentiation between variola and monkeypox viruses, or vaccinia and cowpox viruses, can be made by animal inoculation methods, of which the CAM and rabbit skin are the most useful. Each species of orthopoxvirus has a distinctive DNA map demonstrable by digestion with restriction endonucleases.

SMALLPOX

Smallpox is now an extinct disease, the last naturally occurring case in the world having been diagnosed in Somalia on October 26, 1977. Its eradication was a major triumph of preventive medicine, achieved through persistence in the face of major obstacles and through excellent international cooperation orchestrated by the World Health Organization. It came 176 years after Edward Jenner had proclaimed it as a possibility, following his discovery that cowpox protected against smallpox (see Chapter 9).

It is no longer necessary for medical students to learn about smallpox, but they should be aware of what this disfiguring and often lethal disease looked like (Plate 17-2), and of the fact that a positive laboratory diagnosis could be made by examination of vesicle fluid or scabs with the electron microscope and cultivation of the virus on the chorioallantoic membrane of chick embryos.

VACCINATION WITH VACCINIA VIRUS

With the eradication of smallpox, routine vaccination of the general public ceased to be necessary and the requirement that international travelers should have a valid vaccination certificate was abolished. However, the governments of several nations still (in 1984) require that military personnel should be vaccinated. Thus the possibility remains that nonvaccinated civilians in contact with recently vaccinated soldiers may be accidentally infected, which may have serious consequences in eczematous children. It is therefore necessary for medical practitioners to be aware of what a primary vaccinial lesion looks like and of the dangers of eczema vaccinatum.

Furthermore, recent studies have shown that foreign genes may be inserted into the genome of vaccinia virus in such a way that the pro-

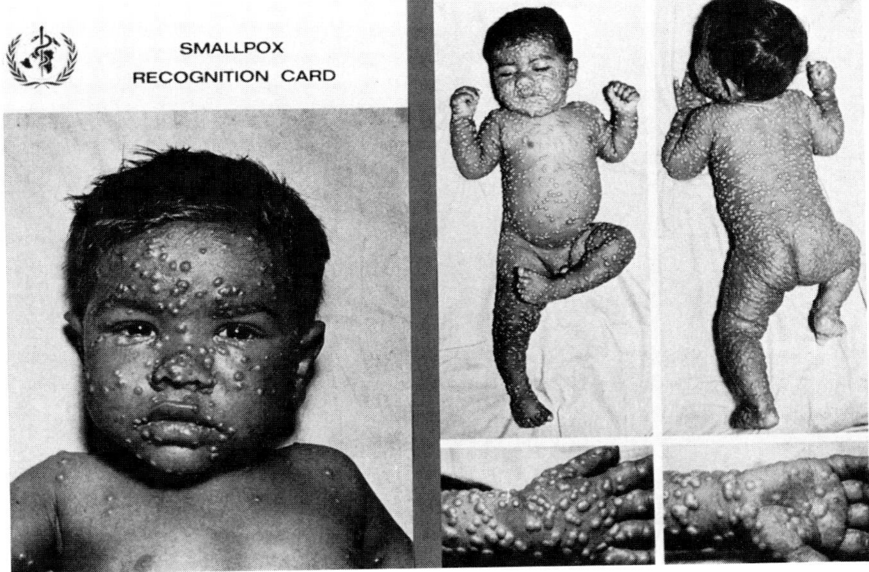

PLATE 17-2. *The rash of smallpox. "Smallpox recognition card" used during the Intensified Smallpox Eradication Programme of WHO, showing the distribution and nature of the rash of discrete ordinary-type smallpox in an unvaccinated child. (Courtesy World Health Organization.)*

teins for which they code are expressed when vaccination is performed (see Chapter 10). Such hybrid viruses have potential for the production of vaccines against several diseases of human and veterinary importance and could be particularly useful in developing countries, since freeze-dried smallpox vaccine is extremely stable and vaccination can be carried out by unskilled persons, without the need for a syringe and needle. Before such hybrid vaccines are licensed for use, it is likely that special attenuated strains will need to be developed to minimize the frequency of the rare but serious complications that were associated with smallpox vaccination.

Primary Vaccination

Vaccinia virus, employed as a live vaccine against smallpox, is a relatively avirulent virus of obscure origins, with properties resembling those of cowpox virus. Following reconstitution from the lyophilized

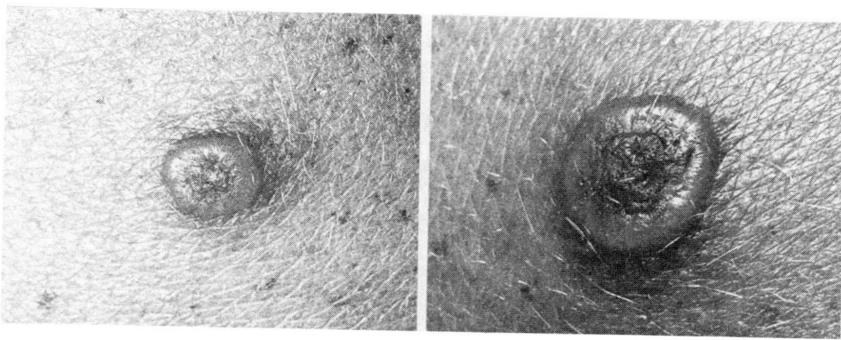

PLATE 17-3. *Primary response to vaccination in man: typical vesiculopustular response, maximal at 7–10 days. Left, at 7 days; right, at 11 days. (Courtesy Dr. J. R. L. Forsyth.)*

state, the virus is inoculated into the superficial layers of the skin of the upper arm by a "multiple puncture" technique.

Deliberate vaccination against smallpox is no longer carried out among the general public, but vaccinia accidentally acquired from vaccinated military personnel can occur in unvaccinated subjects and produce a skin lesion like that found in primary vaccination (Plate 17-3). A papule appears 3 to 5 days after inoculation or infection, and rapidly becomes a vesicle, which enlarges in size and develops a secondary erythema. The center of the vesicle usually becomes depressed. By the eighth or ninth day the contents become turbid and there is axillary lymphadenitis and fever. After the tenth day the pustule dries up and a scab forms which takes a week to separate, leaving a typical vaccination scar. Immunity to vaccinia and smallpox appears about the tenth day and persists, to a varying degree in different individuals, for several years.

Complications of Vaccination. The major complication that may occur in accidentally infected children is eczema vaccinatum, which can occur in eczematous subjects even if eczema is not active when they are infected. Large areas of skin may be infected, but healing usually occurs normally. Eczema vaccinatum is rarely fatal, but all cases should be treated with vaccinia-immune human γ-globulin.

Other very rare complications that were of considerable concern when vaccination used to be necessary were progressive vaccinia, which occurred only in persons with defective cell-mediated immunity, and postvaccinial encephalitis (Table 17-2).

TABLE 17-2
Diseases Produced in Humans by Poxviruses

GENUS	DISEASE	CLINICAL FEATURES
Orthopoxvirus	Smallpox (now extinct)	
	Variola major	Generalized infection with pustular rash; case fatality rate 10–25%
	Variola minor	Generalized infection with pustular rash; case fatality rate less than 1%
	Vaccination	Local pustule, slight malaise
	Complications of vaccination (rare)	Postvaccinial encephalitis; high mortality Progressive vaccinia; high mortality Eczema vaccinatum; low mortality Autoinoculation and generalized vaccinia; nonlethal
	Monkeypox	Generalized infection with pustular rash; case fatality rate 15%
	Cowpox	Localized ulcerating infection of skin, usually acquired from cows; nonlethal
Parapoxvirus	Milker's nodes	Trivial localized nodular infection of hands acquired from cows
	Orf	Localized papulovesicular lesion of skin acquired from sheep
Unclassified	Molluscum contagiosum	Multiple benign nodules in skin
	Yabapox	Localized skin tumors acquired from monkeys (in laboratory attendants)
	Tanapox	Localized skin lesions probably from arthropod bites; common in parts of Africa

HUMAN MONKEYPOX

Human infections with monkeypox virus, a species of *Orthopoxvirus*, were discovered in West and Central Africa in the early 1970s, after smallpox had been eradicated from the region. The signs and symptoms are very like those of smallpox, with a generalized pustular rash, fever, and toxemia. Many cases of human monkeypox have greatly enlarged submaxillary and inguinal lymph nodes, a feature not seen in smallpox.

Epidemiologically, monkeypox and smallpox are very different, since human monkeypox occurs as a rare zoonosis, only in villages in tropical rain forests in West and Central Africa, especially in Zaire. It is probably acquired by direct contact with wild animals killed for food, especially monkeys. A few cases of apparent person-to-person transmission have been recorded but the secondary attack rate is much lower than in smallpox. Up to December 1984, only 260 cases had been diagnosed. Vaccination with smallpox vaccine immunizes against monkeypox, but at present is not considered to be justified, since the disease is so rare.

MOLLUSCUM CONTAGIOSUM

This specifically human disease consists of multiple discrete nodules 2–5 mm in diameter, limited to the epidermis, and occurring anywhere on the body except on the soles and palms. The nodules are pearly white in color and painless. At the top of each lesion there is an opening through which a small white core can be seen. The disease may last for several months before recovery occurs.

Cells in the nodule are greatly hypertrophied and contain large hyaline acidophilic cytoplasmic masses called molluscum bodies. These consist of a spongy matrix divided into cavities in each of which are clustered masses of viral particles which have the same general structure as those of vaccinia virus.

The incubation period in human volunteers varies between 14 and 50 days. Attempts to transmit the infection to experimental animals have failed, and reported growth in cultured human cells has been hard to reproduce.

The disease is worldwide, but is much more common in some localities, for example, parts of Zaire and Papua New Guinea, than in others. It is probably transmitted through minor abrasions, and in developed countries it may be a sexually transmitted disease, or communal swimming pools may be a source of infection.

COWPOX AND MILKER'S NODES

Man may acquire two different poxvirus infections from cows, cowpox and milker's nodes, usually as lesions on the hands after milking (Plate 17-4). Recent work suggests that "cowpox" is primarily a disease of rodents that may occasionally spread to cows, humans, or animals, including large cats and elephants, in zoos and circuses. Cowpox virus has been found only in Europe and adjacent parts of Russia. It occurs in cattle as ulcers on the teats and the contiguous parts of the udder, and it is spread through herds by the process of milking.

The lesions in man usually appear on the hands and develop just like primary vaccinia (Plate 17-4A), although fever and constitutional symptoms may be more severe. Cowpox virus produces much more hemorrhagic lesions on the chorioallantoic membrane and in the skin of rabbits

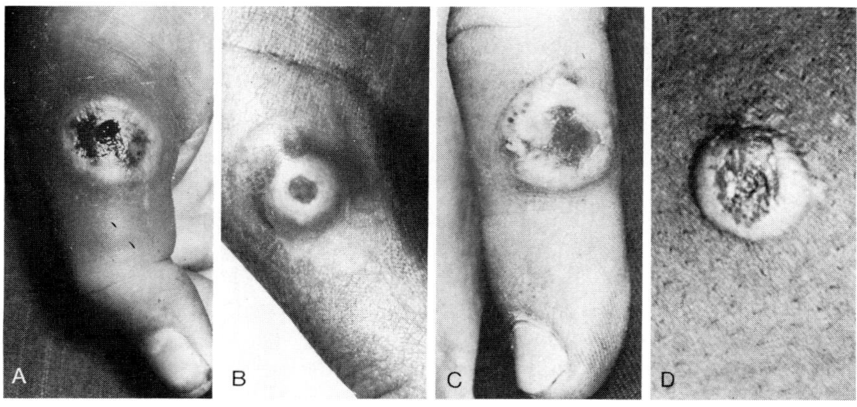

PLATE 17-4. *Localized zoonotic infections with poxviruses. (A and B) Lesions on hands acquired by milking infected cows: (A) cowpox, caused by an orthopoxvirus; (B) milker's nodes, caused by parapoxvirus; (C) orf, a parapoxvirus lesion acquired by handling sheep or goats suffering from contagious pustular dermatitis; (D) tanapox, lesion on arm of child in Zaire, probably transmitted mechanically by mosquitoes from an animal reservoir. (A and B, courtesy Dr. D. Baxby; C, courtesy Dr. J. Nagington; D, courtesy Dr. Z. Jezek.)*

than does vaccinia virus. Vaccination with vaccinia virus protects man against cowpox.

Milker's nodes (Plate 17-4B) also occur on the hands of man, derived from lesions on cows' teats. Unlike cowpox, it is a disease of cattle and occurs worldwide. In man the lesions are small nonulcerating nodules. They are caused by a poxvirus of the genus *Parapoxvirus* (Plate 17-1B). Immunity following infection of man does not last long and second attacks may occur at intervals of a few years. The disease is trivial and no measures for prevention or treatment are warranted.

ORF

Orf is an old Saxon term applied to the infection of man with the virus of contagious pustular dermatitis ("scabby mouth") of sheep. The disease of sheep, which occurs all over the world, is found particularly in lambs during spring and summer and consists of a papulovesicular eruption that is usually confined to the lips and surrounding skin. Infection of man occurs as a single lesion on the hand or forearm (Plate 17-4C) or occasionally on the face; a slowly developing papule becomes a flat vesicle and eventually heals without scarring. Orf is an occupational disease associated with handling of sheep.

The virus of orf (Plate 17-1B) belongs to the same genus (*Parapoxvirus*) as that of milker's nodes, but is a distinct species and occurs naturally among sheep and goats rather than cattle.

YABAPOX AND TANAPOX

Yaba poxvirus is enzootic in Africa and was discovered because it produced large benign tumors in Asian monkeys kept in a laboratory in West Africa. Subsequently cases occurred in primate colonies in the United States. Occasionally similar lesions have occurred in laboratory attendants handling affected monkeys.

Tanapox is a relatively common skin infection of humans in parts of Africa extending from eastern Kenya to Zaire. It appears to be spread mechanically by insect bites from a reservoir in wild animals (probably monkeys); occasionally multiple lesions occur. The skin lesion starts as a papule and progresses to an umbilicated vesicle (Plate 17-4D), but pustulation never occurs. There is usually a febrile illness lasting 3–4 days, sometimes with severe headache, backache, and prostration.

FURTHER READING

Books

Fenner, F., Dumbell, K. R., Dales, S., and Wittek, R. (1986). "The Orthopoxviruses." Academic Press, New York.

Fenner, F., Henderson, D. A., Arita, I., Jezek, Z., and Ladnyi, I. D. (1987). "Smallpox and its Eradication." World Health Organ., Geneva.

World Health Organization (1980). "Final Report of the Global Commission for the Certification of Smallpox Eradication," Hist. Int. Health No. 4. World Health Organ., Geneva.

Reviews

Benenson, A. S. (1982). Smallpox. *In* "Viral Infections of Humans: Epidemiology and Control" (A. S. Evans, ed.), 2nd ed., pp. 541–568. Plenum, New York.

Fenner, F. (1979). Portraits of viruses: The poxviruses. *Intervirology* **11,** 297.

Fenner, F. (1985). Poxviruses. *In* "Human Viral Diseases" (B. N. Fields, *et al.*, eds.), pp. 661–684. Raven Press, New York.

Moss, B. (1985). Replication of poxviruses. *In* "Human Viral Diseases" (B. N. Fields, *et al.*, eds.), pp. 685–703. Raven Press, New York.

CHAPTER 18

Parvoviruses

Introduction 445
Properties of *Parvoviridae* 445
Adeno-Associated Viruses 446
Parvovirus B19 447
Parvovirus RA-1 449
Further Reading 450

INTRODUCTION

The smallest of all viruses, the *Parvoviridae* (*parvus* = small), contain so little genetic information in their unique single-stranded DNA molecule that members of one genus, *Dependovirus,* depend on a helper virus for their replication. Their survival in nature is aided by persistent infections in which their genome becomes integrated with that of the cell.

A striking feature of the parvoviruses is their selective replication in dividing cells. No doubt this requirement accounts for their evident predilection for bone marrow, gut, and the developing embryo. Certain parvoviruses cause important diseases in animals (e.g., feline panleukopenia and enteritis, Aleutian disease of mink, congenital malformations in rats). One of the human parvoviruses has recently been shown to be the cause of "fifth disease" and to precipitate aplastic crises in patients with hemolytic anemia. Another has been associated with rheumatoid arthritis but its relevance is still being assessed.

PROPERTIES OF *PARVOVIRIDAE*

Only 18–26 nm in diameter, the *Parvoviridae* (Table 18-1) consist of a simple icosahedral shell surrounding a single-stranded DNA molecule

TABLE 18-1
Properties of Parvoviridae

Icosahedral virion, 18–26 nm
ssDNA genome, MW 1.5–2.0 million, negative polarity, some virions contain + strand
Multiply in nucleus of dividing cells
Dependovirus defective, requires helper
Persist via integrated DNA

of very limited coding potential (MW 1.5–2.0 million). The capsid is constructed from just 3 species of polypeptide, forming 32 capsomers. The ssDNA genome is generally of negative polarity. However, a most unusual situation obtains with the *Dependovirus* genus. Here, virions package a positive or a negative strand at random, even though presumably only one of them is infectious. This explains an observation that mystified early workers: following extraction from virions, *Dependovirus* DNA appears to be double-stranded because + and − strands from different virions, being complementary, spontaneously hybridize *in vitro*.

Viral Multiplication

Members of the genus *Parvovirus* multiply preferentially in actively dividing cells as they require at least one cellular function generated during late S or early G_2 of the mitotic cycle. Members of the *Dependovirus* genus are defective; they replicate only in cells coinfected with a helper virus–normally an adenovirus (Plate 18-1), occasionally a herpesvirus.

ADENO-ASSOCIATED VIRUSES

Five defective human "adeno-associated viruses" (AAV) of the *Dependovirus* genus have been described so far. AAV 1–4 were recovered from the throat or feces, and AAV-5 from a genital papilloma, by cultivation in the presence of an adenovirus as helper. None has yet been associated with any disease, but infection is common, as indicated by seroepidemiological surveys, and the genome persists indefinitely in infected cells.

The AAV genome is maintained in the form of dsDNA integrated into the cell's genome. Integration does not require the helper virus but replication does. Although AAV codes for all its own structural proteins, the helper codes for certain function(s) essential for the replication and

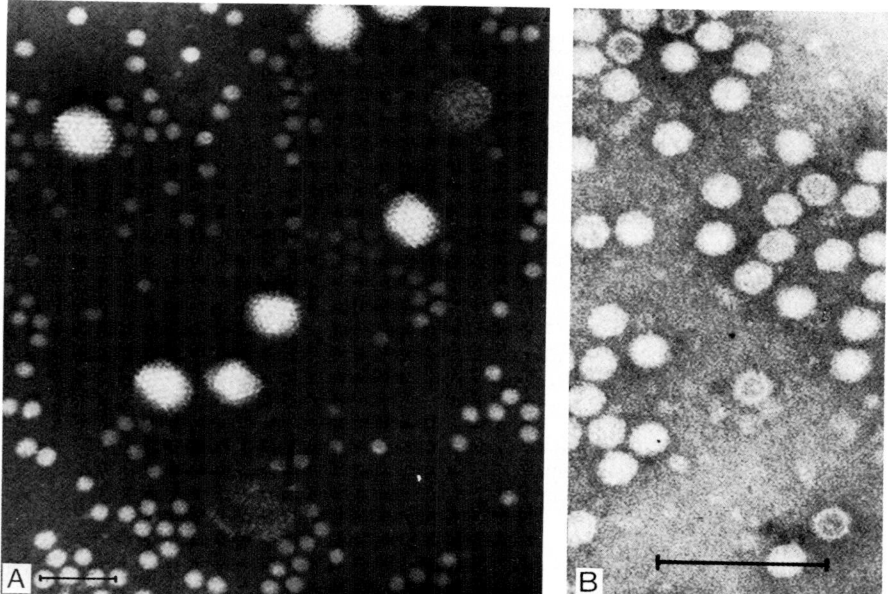

PLATE 18-1. Parvoviridae. *Negatively stained preparations. (A) Human* Dependovirus *(adeno-associated virus) particles together with their helper adenovirus. (B)* Parvovirus. *Bars = 100 nm. (A, courtesy Dr. H. D. Mayor; B, courtesy Dr. E. L. Palmer.)*

transcription of AAV DNA. Superinfection with adenovirus, even years later, leads to the production of both viruses in approximately equal amounts. Thus, explant cultures of apparently normal human tissues can occasionally be induced to produce AAV following infection with adenovirus. Certain herpesviruses also can serve as helpers, though generally with lower efficiency.

PARVOVIRUS B19

A parvovirus designated B19, originally isolated from the blood of healthy donors, has recently been identified as the cause of at least two quite distinct clinical syndromes (Table 18-2).

Properties of the Virus

While there is no evidence that B19 is defective and no putative helper virus has been found in association with it, formal proof that it is capable

TABLE 18-2
Human Infections Associated with Parvoviridae

GENUS	VIRUS	DISEASE
Dependovirus	AAV 1–5	Nil
Parvovirus	B19	Erythema infectiosum (fifth disease)
		Aplastic crisis in hemolytic anemia
	RA-1	Rheumatoid arthritis?

of autonomous replication must await its propagation in cultured cells. If so it will presumably be classified as a member of the *Parvovirus* genus, not as a *Dependovirus.* The ssDNA genome, which is of negative or positive polarity in different particles is, like that of other members of the *Parvoviridae,* a linear molecule of about 5.5 kb with terminal palindromes (hairpin termini), but displays more sequence homology with helper-independent parvoviruses of rodents than with the *Dependovirus* genus.

Clinical Features

"Fifth disease," also known as erythema infectiosum, is an innocuous contagious exanthem of childhood which has been well known to pediatricians for nearly a century though its cause remained a mystery. An erythematous rash on the face gives the child strikingly flushed cheeks. The rash spreads to involve the limbs and, as it fades, develops the characteristic appearance of fine lace. Mild URTI and/or fever are less usual. Arthralgia is seen only occasionally in children but is a regular feature in adult patients. Convincing evidence of seroconversion of virtually all the patients in several recent outbreaks, and induction of the disease in volunteers, has established parvovirus B19 as the etiological agent of fifth disease.

Almost concomitantly it was shown that parvovirus B19 is the principal cause of the well-known aplastic crises that complicate chronic hemolytic anemia. Typically, a child with preexisting sickle cell anemia or thalassemia experiences a sudden but transient drop in hemoglobin, lasting for a week or so.

Pathogenesis and Immunity

The nondefective parvoviruses multiply preferentially in rapidly dividing cells such as those of the bone marrow. Parvovirus B19 has been shown to affect cultured human erythroid progenitor cells. In volunteers, reticulocytes disappear from the blood and the hemoglobin level drops about 2 weeks after intranasal inoculation. Whereas this transient infection of red cell precursors in the bone marrow produces a barely detectable degree of anemia in normal healthy individuals, whose erythrocytes have a half-life of 120 days, it is a different story in hemolytic anemia where the red cell halflife is only a few days and the hemoglobin level is already low.

The rash of erythema infectiosum may be immunologically mediated for it appears a week after viremia has subsided. Immunity to reinfection appears to be solid.

In view of the teratogenicity of certain parvoviruses of other animals, prospective seroepidemiological surveys are currently being planned to assess the contribution, if any, of parvoviruses to congenital abnormalities in humans.

Laboratory Diagnosis

The virus has yet to be grown in conventional cell cultures. Nevertheless, immunoassays (e.g., RIA) using monoclonal antibody have been developed to demonstrate viral antigen in the blood during the viremic phase (i.e., the aplastic crisis in CHA) or to measure antibodies of the IgM class. Since the parvovirus B19 genome has been cloned and sequenced, the recDNA technology is available to produce antigen for use in such immunoassays.

Epidemiology

Studies in human volunteers have established that this parvovirus can be readily transmitted via the respiratory route. The incubation period appears to be just over 2 weeks, and many infections are subclinical. Children of primary school age seem to be most often affected. Outbreaks are common, especially in late winter and spring. Spread occurs readily within schools and families, with very high attack rates. A majority of adults have antibody.

PARVOVIRUS RA-1

Another putative parvovirus, RA-1, has recently been recovered from infant mice which had been inoculated with culture extracts from human fibroblasts that had been cocultivated with synovial cells from patients with rheumatoid arthritis (RA). Antisera raised against this virus react in an EIA with antigen extracted directly from RA synovial cells. At the time of writing it remains to be demonatrated that RA-1 is a parvovirus or that its relationship to RA is causal rather than casual.

FURTHER READING

Books

Berns, K. I., ed. (1983). "The Parvoviruses." Plenum, New York.

Ward, D. C., and Tattersall, P., eds. (1978). "Replication of Mammalian Parvoviruses." Cold Spring Harbor Lab., Cold Spring Harbor, New York.

Reviews

Anderson, M. J., and Pattison, J. R. (1984). The human parvovirus. *Arch. Virol.* **82,** 137.

Research Papers

Simpson, R. W., McGinty, L., Simon, L., Smith, C. A., Godzeski, C. W., and Boyd, R. J. (1984). Association of parvoviruses with rheumatoid arthritis of humans. *Science* **223,** 1425.

Summers, J., Jones, S. E., and Anderson, M. J. (1983). Characterization of the genome of the agent of erythrocyte aplasia permits its classification as a human parvovirus. *J. Gen. Virol.* **64,** 2527.

CHAPTER 19

Picornaviruses and Caliciviruses

Introduction 451
Properties of *Picornaviridae* 451
Polioviruses 454
Other Enteroviruses 461
Hepatitis A 468
Rhinoviruses 472
Caliciviridae 474
Further Reading 476

INTRODUCTION

The *Picornaviridae* comprise perhaps the largest of all the families of viruses. The fact that almost 200 host-specific picornaviruses have been identified in man alone suggests that thousands of others occur in other animals yet to be comprehensively searched. Some 69 enteroviruses, including the polioviruses and hepatitis A virus, inhabit the enteric tract, while well over 100 rhinoviruses cause common colds. Though the largest family of RNA viruses, they are the smallest in size (*pico* = very small, *RNA* viruses). Only slightly larger are the *Caliciviridae,* which have recently been accorded the status of a separate family. They cause gastroenteritis in man.

PROPERTIES OF *PICORNAVIRIDAE*

The picornavirion is a naked icosahedron only 25–30 nm in diameter (Plate 19-1). The capsid, appearing smooth and round in outline, is

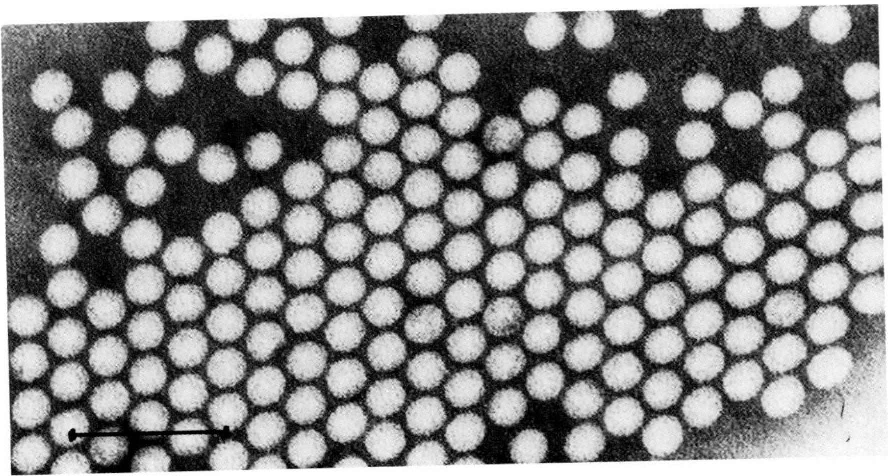

PLATE 19-1. Picornaviridae. *Negatively stained poliovirions. Bar = 100 nm. (Courtesy Dr. J. J. Esposito.)*

constructed from 60 "protomers," each comprising a single molecule of each of four polypeptides, VP1, 2, 3, 4 (see Fig. 3-9).

The genome is a single linear molecule of single-stranded RNA of positive polarity, MW 2.5×10^6, polyadenylated at its 3′ end, with a protein, VPg, covalently linked to its 5′ end. Being a single molecule of + sense it is infectious in its own right (Table 19-1).

Classification

The *Picornaviridae* family is divided into four genera, only two of which, *Enterovirus* and *Rhinovirus*, contain human pathogens. The enteroviruses, as might be expected, are remarkably resistant to the conditions prevailing in the gut—acidic pH, proteolytic enzymes, and bile salts. The rhinoviruses, which multiply in the nose, are acid labile; they also have a substantially greater buoyant density by CsCl equilibrium gradient centrifugation (1.40, cf. 1.34).

The *Enterovirus* genus contains 69 viruses known to infect man. Their nomenclature is a mess (Table 19-2). First, they are known as "types," but should more properly be regarded as species because most of them share no antigens, hence can be distinguished by complement fixation or immunodiffusion as well as neutralization tests. Second, they were historically allocated to three groups—polioviruses, coxsackieviruses, and echoviruses—on the basis of criteria which are now regarded as

TABLE 19-1
Properties of Picornaviridae

Spherical virion, 25–30 nm
Naked icosahedral capsid
ssRNA genome, linear, positive polarity, MW 2.5 million, poly(A) at 3′ end, protein VPg at 5′ end, infectious
Cytoplasmic replication
vRNA translated into polyprotein which is then cleaved

trivial. Coxsackieviruses are pathogenic for infant mice, indeed were originally isolated in this host, from inhabitants of the village of Coxsackie in the United States. Echoviruses (*e*nteric *c*ytopathogenic *h*uman *o*rphan viruses), isolated in cell culture from the feces of asymptomatic people (hence "orphan" viruses, with no apparent parent disease), are not pathogenic for suckling mice. Because this distinction between coxsackieviruses and echoviruses is neither absolute nor especially important, it was abandoned some years ago and all enteroviruses identified since that time have simply been allocated an enterovirus number, beginning with enterovirus 68. As is the wont of committees, however, ICTV bowed to the preference of workers in the field to retain the existing nomenclature for all viruses named prior to that date. Thus we have to live for the moment with the compromise you see in Table 19-2. As if this were not enough, note also that human coxsackievirus A23 does not exist, nor do echoviruses 10 and 28, all three having been reclassified. Furthermore, many hepatitis specialists feel somewhat peeved that the very important and clinically un-enterovirus-like virus they have known for years as hepatitis A now finds itself an also-ran with the inglorious title of enterovirus 72.

TABLE 19-2
Classification of Human Picornaviruses

GENUS	PROPERTIES	SPECIES
Enterovirus	pH 3 stable Buoyant density 1.34	Polioviruses 1–3 Coxsackieviruses A1–24 (no A23), B1–6 Echoviruses 1–34 (no 10 or 28) Enteroviruses 68–72
Rhinovirus	pH 3 labile Buoyant density 1.40	Rhinoviruses 1–113

Viral Multiplication

Quite early in the study of viral replication, poliovirus became the preferred model for the analysis of RNA viral multiplication. The work of Baltimore in particular has provided us with a complete description of the poliovirus genome, and of the mechanism of RNA replication, post-translational proteolytic processing of polyproteins, and morphogenesis of simple icosahedral virions. These processes were described fully in Chapter 3 (see also Figures 3-4, 3-8, and 3-9).

POLIOVIRUSES

Poliomyelitis was once one of the most feared of all viral diseases; its tragic legacy of paralysis and deformity was a familiar sight a generation or more ago. Today, by contrast, few medical students have seen a case, such has been the impact of the Salk and Sabin vaccines. The foundation for these great developments was laid by Enders, Weller, and Robbins in 1949, when they demonstrated the growth of poliovirus in cultures of nonneural cells. From this fundamental discovery flowed all subsequent work dependent on viral multiplication in cultured cell monolayers. Enders and his colleagues were rewarded with the Nobel Prize. A new era in virology had begun.

Clinical Features of Poliomyelitis

It is important to realize that paralysis is a relatively infrequent complication of an otherwise trivial infection. Of those infections that become clinically manifest at all, most take the form of a minor illness ("abortive poliomyelitis"), characterized by fever, malaise, and sore throat, with or without headache and vomiting which may indicate some degree of aseptic meningitis. However, in about 1% of cases muscle pain and stiffness herald the rapid development of flaccid paralysis. In bulbar poliomyelitis death may result from respiratory or cardiac failure (Plate 19-2). Otherwise, some degree of recovery of motor function may occur over the next few months, but paralysis remaining at the end of that time is permanent.

Pathogenesis and Immunity

Following ingestion, poliovirus multiplies first in the pharynx and small intestine. It is not clear whether the mucosa itself is involved, but the lymphoid tissue (tonsils and Peyer's patches) certainly is. Spread to the draining lymph nodes leads to a viremia, enabling the virus to be-

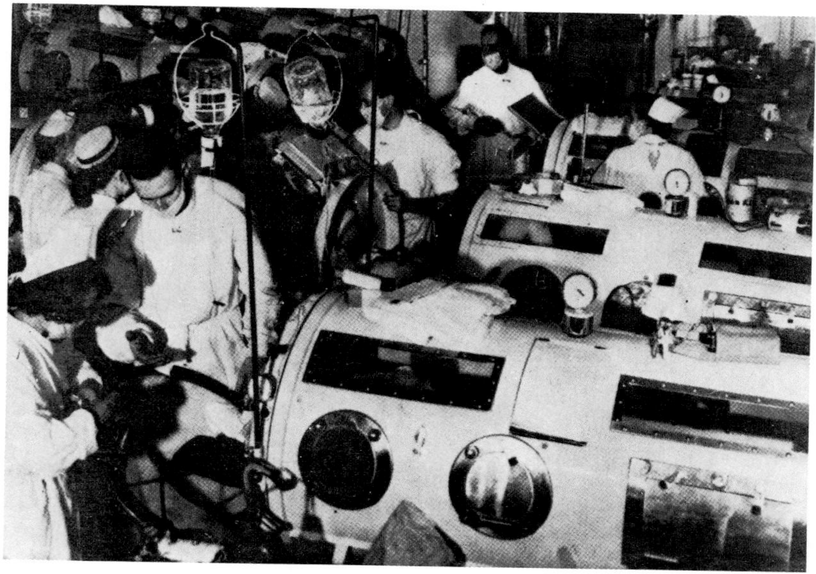

PLATE 19-2. *Totally paralyzed poliomyelitis patients in mechanical respirators during the last epidemic in the United States before the advent of universal immunization (1955). [From L. Weinstein,* J. Inf. Dis. **129,** *480 (1974).]*

come disseminated throughout the body. It is only in the occasional case that the central nervous system becomes involved. Virus is carried via the bloodstream to the anterior horn cells of the spinal cord and motor cortex of the brain. The resulting lesions are widely distributed, but variation in severity gives rise to a spectrum of clinical presentations, with spinal poliomyelitis the most common and the bulbar form less so. The incubation period of paralytic poliomyelitis averages 1–2 weeks, with outer limits of 3 days to 1 month. Acquired immunity is permanent but monotypic.

Laboratory Diagnosis

Virus is readily isolated from feces but not from CSF. Any type of human or simian cell culture is satisfactory, the virus growing so rapidly that cell destruction is usually complete within a few days. Early changes include cell retraction, increased refractivity, cytoplasmic granularity, and nuclear pyknosis (Plate 19-3). The isolate is identified by neutralization tests.

The ubiquity of attenuated vaccine strains poses a difficult problem for

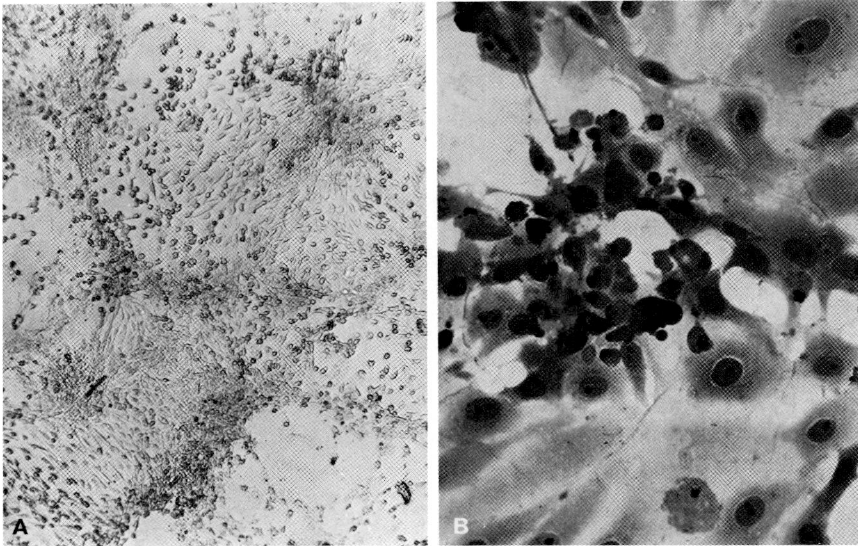

PLATE 19-3. *CPE induced by enterovirus in primary monkey kidney cell culture. (A) Unstained, low power. Note rapidly developing generalized cell destruction. (B) H and E stain, higher power. Note disintegrating pyknotic cells. (Courtesy I. Jack.)*

diagnostic laboratories today. Numerous tests have been employed to differentiate wild-type virulent polioviruses from oral poliovaccine (OPV) strains and there is some confusion about their validity. Only one test unequivocally determines whether a field strain is neurovirulent, namely intraspinal inoculation of monkeys, inducing paralysis. On the other hand, if one wants to know whether the antigenic characteristics of the isolate resemble the wild-type or the vaccine, the appropriate test is a microneutralization assay using absorbed strain-specific antisera. But if the object is to decide whether a "vaccine-associated" case of paralysis is caused by a "vaccine-like" or by a wild strain (see below), the most sensitive test is nucleic acid sequencing, or more simply, oligonucleotide fingerprinting of the viral RNA. The maps (fingerprints) of vaccine strains remain fairly constant; even following reversion to neurovirulence, generally only one or two oligonucleotides will have altered.

Epidemiology

Being enteric in their habitat, polioviruses spread mainly via the fecal–oral route. Direct fecal contamination of hands, then food or eating utensils, is probably responsible for most case-to-case spread. Explosive epidemics have been known to result from contamination of water supplies by sewage. In the tropics the disease is endemic throughout the year; in temperate countries before the introduction of vaccination it classically occurred in summer epidemics. The chain of infection is rarely obvious, because the vast majority of infections are inapparent.

Worldwide, at least 85% of paralytic cases are caused by poliovirus type 1, but the rare "vaccine-associated" cases (see below) are generally due to reversion in the type 3 or type 2 OPV strains. Oligonucleotide fingerprinting and sequencing, which have been used to track the course of epidemics as well as to identify the origin of individual cases, reveal that mutations are continually occurring as the virus moves from one intestine to another.

Two major developments have greatly changed the epidemiology of poliomyelitis over the years. The first of these was the introduction of modern standards of hygiene and sanitation to the more advanced countries of the world, which, paradoxically, had the effect of increasing the incidence of paralytic polio in older children and adults. The reduction in the spread of viruses by the fecal–oral route limited the circulation of polioviruses and the incidence of infection in the community as a whole. As a result, most people no longer had acquired immunity by the time they reached adolescence. The consequence was a shift in the age incidence of paralytic poliomyelitis to include young adults, making the old term "infantile paralysis" something of a misnomer in Western nations. Primary infection of adults is, for reasons still unknown, much more likely to result in severe paralytic disease than is primary infection of young children. This was exemplified most strikingly in "virgin soil" epidemics occurring in isolated communities with no prior experience of the virus; most of the deaths occurred in adults. It was also commonplace during epidemics in Western countries for a parent to become paralyzed after acquiring poliovirus from his or her asymptomatically infected child.

The second major influence on the epidemiology of poliomyelitis has become apparent only during the last three decades. In those countries in which a vigorous policy of immunization with OPV has been successfully pursued, not only has the disease been abolished, but the wild virus itself has almost disappeared. In North America, Europe, and Australasia the incidence of paralytic poliomyelitis has been reduced 1000-fold since 1955, and wild polioviruses are rarely found in human sewage, but vaccine strains are ubiquitous.

In the tropical countries of Africa and Asia immunization has generally not met with the success anticipated and in some countries the incidence of poliomyelitis actually seems to be increasing. The problems, and current attempts to overcome them, are discussed below.

Control

In the late 1940s, the US National Foundation for Infantile Paralysis organized a nationwide doorknock appeal, imaginatively labeled "The March of Dimes." The response was overwhelming and the Foundation set about sponsoring a massive research drive on several fronts, in an attempt to turn Enders' recent discovery to advantage by developing a poliomyelitis vaccine. Salk was commissioned to work toward an inactivated vaccine and Sabin toward a live one. The formalin-inactivated (Salk) poliovaccine (IPV) was the first to be licensed and was enthusiastically embraced in North America, Europe, and Australia during the mid-1950s. Sweden, the country in which paralytic poliomyelitis first became apparent in epidemic form and which for many years continued to have the highest rate of poliomyelitis in the world, eradicated the disease by the use of Salk vaccine. Meanwhile, however, the live attenuated oral poliovaccine (OPV) of Sabin was adopted in the USSR and Eastern Europe where it was demonstrated to be so successful that the whole world except Sweden, Finland, Iceland, and Holland now uses OPV because of its greater convenience and lower cost. The several advantages of such living vaccines over their inactivated counterparts were set out in detail in Chapter 10. The virtual abolition of polio in much of the Western World since the introduction of poliovaccine (see Fig. 9-5) represents one of the truly great achievements of medical science.

OPV is a "trivalent" vaccine, consisting of attenuated strains of all three types, derived empirically by serial passage in primary cultures of monkey kidney which have been carefully monitored to exclude simian viruses. Because of the diminishing availability of monkeys and the difficulty of ensuring that even laboratory-bred animals are virus free, some manufacturers have moved to human diploid fibroblasts as a substrate for the production of poliovaccine (see Chapter 10). As discussed in Chapters 4 and 10, the vaccine strains have acquired literally dozens of mutations during serial passage in cultured cells. Attenuation is confirmed by absence of neurovirulence for monkeys. The three serotypes are pooled in carefully adjusted proportions to balance numbers against growth rate, hence minimizing the probability of mutual interference. In the presence of molar $MgCl_2$ to protect the virus against heat inactivation the vaccine is stable for about a year under proper conditions of refrigeration (but for only a few days at 37°C). It can be administered simply, by unqualified personnel. OPV is usually given in the first 6 months of life, in 3 doses, spaced 2 months apart. Theoretically, one successful "take" would suffice, but concurrent infection with another enterovirus may interfere with the replication of the vaccine viruses, as commonly occurs in the developing countries of the tropics. Earlier concern that breast-feeding may represent a contraindication has not been substantiated—though colostrum contains moderate titers of maternal IgA, milk itself does not contain enough antibody to neutralize the vaccine virus. Ideally, the children should receive a booster dose of OPV at 1 year, then on entering school, again on entering high school, and as adults if traveling to a developing country.

Two further significant advantages of OPV over IPV follow from the fact that it multiplies in the alimentary tract. First, this subclinical infection elicits the prolonged synthesis not only of IgG antibodies which protect the individual against paralytic poliomyelitis (as does IPV) by intercepting wild polioviruses during their viremic phase, but also of IgA. These "copro-antibodies" prevent primary implantation of wild virus in the gut and hence diminish very considerably the circulation of virulent viruses in the community. Indeed, wild polioviruses have now virtually disappeared from countries such as the United States; the only strains that can be isolated from sewage are "vaccine-like." This striking replacement of wild virus by vaccine strains is facilitated by the fact that the latter are excreted in the feces and spread to nonimmune contacts.

A consequence of the spread of vaccine virus to contacts is that it provides greater opportunities for selection of mutants displaying varying degrees of reversion toward virulence. Very rarely (about once per million vaccinees) the vaccinee or a family contact develops poliomyelitis. Indeed, such has been the success of OPV in the United States that about half of the miniscule number of cases of paralytic polio that still occur are "vaccine associated." Because OPV-associated polio is 10,000 times more common in those who are immunocompromised in some way, only IPV should be used in these children. Furthermore, because there are still significant numbers of parents who have never received poliovaccine, there are good arguments for immunizing nonimmune family contacts prior to or simultaneously with their infants.

In the more advanced nations the disease has been effectively conquered but it remains imperative that very high levels of immunization are maintained in spite of the lack of perceived threat. Pockets of unimmunized people, e.g., immigrants concentrated in urban ghettos, are particularly vulnerable to imported viruses. Similarly, outbreaks occurred among the Arab population of Gaza and the West Bank of the Jordan during the 1970s. Another salutory instance was the outbreak of poliomyelitis in 1978 in Holland which was confined entirely to members of a particular religious minority who had refused immunization. The epidemic spread to Canada and then to the United States where it affected only unimmunized members of the Amish community, causing many cases of paralysis.

IPV of greater potency is now being produced as a result of a number of technological improvements. Unfortunately, human diploid cell strains do not grow sufficiently well to make them a commercial proposition for production of the very large amounts of virus required for IPV, but nonmalignant, aneuploid monkey kidney cell lines such as Vero have recently been approved for this purpose. They are grown on the surface of myriads of DEAE-Sephadex beads known as "microcarriers" suspended in large fermentation tanks. Poliovirus grown in such cells is purified by gel filtration and ion-exchange chromatography, affinity chromatography on Sepharose-immobilized antibodies, and/or zonal ultracentrifugation. The importance of removing aggregates of virions before inactivation with formaldehyde was established many years ago following the 1955 disaster in which clumped virus escaped inactivation and paralyzed numerous children in the United States. Two doses of the newer preparations of IPV appear to confer protection equivalent to the four previously required; this has encouraged the proposal that IPV replace OPV, particularly in the Third World where unsatisfactory seroconversion rates are frequently reported even after the full course of three doses of OPV. Since two injections of "triple vaccine" (against diphtheria, pertussis, and tetanus) is already the standard regimen in most countries, it would not be too difficult to add IPV to the cocktail to make a "quadruple" vaccine.

In stark contrast to the dramatic success of the polio vaccination programs of the wealthy temperate countries, relatively little impact has been made on the disease throughout most of the Third World. The reasons for this are complex but are related to the lack of adequate health service infrastructure. OPV is simply not delivered satisfactorily to the children, particularly in inaccessible rural villages. In many developing countries it is still true to say that most children never receive a full course of OPV. Even when they do, there are doubts about whether the degree of refrigeration during storage and transport in the hot tropical climate has been adequate to retain the viability of the virus. Maintenance of the cold chain is typical of the type of problem to which much greater attention is now being paid. That polio can indeed be, to all intents, eradicated by comprehensive coverage with OPV in developing countries has been amply demonstrated by the superb record of Chile, Brazil, and Cuba, to name three particularly impressive examples. Eradication of poliomyelitis from the Third World is a major objective of WHO's Expanded Programme of Immunization (EPI).

OTHER ENTEROVIRUSES

Clinical Features

Most enterovirus infections are subclinical, particularly in young children. Nevertheless, they can cause a wide range of clinical syndromes involving many of the body systems (Table 19-3). It should be noted that each syndrome can be caused by several viruses, and that each virus can cause several syndromes even during the same epidemic. Rashes, URTI, and "undifferentiated summer febrile illnesses" are common. Moreover, enteroviruses are the commonest cause of meningitis, albeit a relatively innocuous form. In general, coxsackieviruses tend to be more pathogenic than echoviruses; they cause a number of diseases rarely seen with echoviruses, e.g., carditis, pleurodynia, herpangina, hand-foot-and-mouth disease, and occasionally paralysis. The recently discovered enterovirus 70 has been responsible for pandemics of hemorrhagic conjunctivitis, while enterovirus 71 causes encephalitis, and enterovirus 72 hepatitis A.

TABLE 19-3
Diseases Caused by Enteroviruses

SYNDROME	VIRUSES[a]
Neurological	
Meningitis	*Many enteroviruses*
Paralysis	*Polio 1, 2, 3; Entero 70, 71;* Cox A7
Cardiac and muscular	
Myocarditis	*Coxsackie B;* some Cox A and echos
Pleurodynia	*Coxsackie B*
Skin and mucosae	
Herpangina	*Coxsackie A*
Hand-foot-and-mouth disease	*Cox A9, 16;* Entero 71; others
Maculopapular exanthem	*Echo 9,* 16; many other enteroviruses
Respiratory	
Colds	*Cox A21, 24; Echo 11, 20;* Cox B; others
Ocular	
Acute hemorrhagic conjunctivitis	*Entero 70; Cox A24*
Abdominal	
Hepatitis	*Enterovirus 72*
Pancreatitis	Coxsackie B
Gastroenteritis	Various?
Hemolytic-uremic syndrome	Various?

[a] The commonest causal agents are italicized.

Neurological Disease. *Meningitis* is more commonly caused by enteroviruses than by all the bacteria combined, but fortunately this "aseptic" meningitis is not nearly as serious (see Chapter 29). Numerous enteroviruses have been implicated from time to time. The most commonly involved are echoviruses 3, 4, 6, 9, 11, 14, 16, 18, 19, 30, 33, coxsackie A7, 9, coxsackie B1–6, enterovirus 71, and (in developing countries) poliovirus 1–3.

Paralysis is usually poliomyelitis, but has very rarely been associated with coxsackie A7. In recent years outbreaks of enterovirus 71 have been marked by meningitis, some encephalitis, and many cases of paralysis with quite a number of deaths. During the 1969–1971 enterovirus 70 pandemic, acute hemorrhagic conjunctivitis was complicated in a small minority of cases by radiculomyelitis, which manifested itself as an acute flaccid lower motor neuron paralysis, resembling polio but often reversible.

Cardiac and Muscular Disease. *Myocarditis* (Plate 19–4C) is a grave consequence of perinatal infection with coxsackie B, or more rarely other enteroviruses. Classically, a neonate suddenly becomes dyspneic and cyanosed within the first week of life; examination reveals tachycardia and ECG abnormalities, sometimes accompanied by jaundice, and/or pulmonary hemorrhage, meningitis, or other signs of an overwhelming systemic infection. There is a high mortality.

Recently, it has become apparent that carditis is quite a common manifestation of coxsackie B virus infections at any age. Though mortality is low except in neonates, permanent heart damage may result, leading eventually to cardiomegaly, constrictive pericarditis, or congestive heart failure.

Pleurodynia, also known as Bornholm disease or epidemic myalgia, is basically a myositis and is characterized by paroxysms of stabbing pain in the muscles of the chest and abdomen. Seen mainly in older children and young adults, sometimes in epidemic form, it is caused principally by coxsackie B viruses. Despite the severity of the pain, recovery is invariable.

Enanthems and Exanthems. *Rashes* are a common manifestation of enteroviral infections. Generally transient, always inconsequential, their principal significance is as an index of an epidemic which may have more serious consequences for other patients. Differential diagnosis from other rashes, viral or otherwise, is quite difficult. For example, the fine maculopapular ("rubelliform") rash of the very common echovirus 9 may easily be mistaken for rubella. The "roseoliform" rash of echovirus 16 ("Boston exanthem") appears only as the fever is declining.

Many other enteroviruses can also produce rubella-like or measles-like rashes with fever and sometimes pharyngitis, especially in children.

Vesicular ("herpetiform") rashes are less common but very striking. The quaint name of *"hand-foot-and-mouth disease"* is given to a syndrome characterized by ulcerating vesicles on these three sites. Coxsackie A9 and 16 and the coxsackie A-like enterovirus 71 are most frequently responsible. The notorious foot-and-mouth disease of cattle is caused by an unrelated picornavirus which is not transmissible to man.

Herpangina is a severe febrile pharyngitis characterized by vesicles (vesicular pharyngitis) or nodules (lymphonodular pharyngitis), principally on the soft palate. Despite the confusing name, it has nothing to do with herpesviruses but is one of the common presentations of coxsackie A virus infections in children.

Respiratory Disease. Enteroviruses are not major causes of respiratory disease but can produce a range of febrile colds and sore throats during summer epidemics. These come under the umbrella label of "summer grippe" or "undifferentiated febrile illness." Coxsackie A21 and 24, and echo 11 and 20 are prevalent respiratory pathogens; coxsackie B and certain echoviruses are less common.

Ocular Disease. *Acute hemorrhagic conjunctivitis (AHC)* burst upon the world as a new disease in 1969. Extremely contagious, with an incubation period of only 24 hours, the disease began in West Africa and swept across Asia to Japan, infecting some 50 million people within 2 years. As its name implies, this conjunctivitis was often accompanied by subconjunctival hemorrhages and sometimes transient involvement of the cornea (keratitis). Mainly adults were affected. Resolution usually occurred within a week or two, uncomplicated by anything other than secondary bacterial infection. However, a small minority of cases, particularly in India, were complicated by neurological sequelae, notably radiculomyelitis, discussed above. In 1981 another major epidemic of AHC arose in Brazil and swept north through Central America and the Caribbean to the Southeastern United States and the Pacific Islands. The cause of the original pandemic was a previously unknown agent now called enterovirus 70, but some recent epidemics have been caused by a variant of coxsackie A24.

Abdominal Disease. Oddly enough, the enteroviruses do not induce much disease referable to the alimentary tract itself, despite the fact that the GIT is their primary, and usually their only site of replication. While certain enteroviruses have been isolated more regularly from children with *gastroenteritis* than from controls, the possibility that they may be irrelevant passengers has generally not been excluded. Immunocompromised individuals may be unusually susceptible, as evidenced by the death of several bone marrow transplant recipients during a nosocomial outbreak of diarrhea associated with coxsackievirus A1. Caliciviruses, now regarded as taxonomically distinct from enteroviruses, are, however, a major cause of gastroenteritis (see below). Bloody diarrhea is often associated with the *hemolytic-uremic syndrome* (HUS) (hemolytic anemia plus impaired renal function) from which enteroviruses have frequently been isolated; coxsackie A4, various coxsackie B, and echo 11 viruses have been incriminated and the frequency of multiple infection with two different enteroviruses has encouraged the proposition that HUS may be due to a Shwartzman reaction. Many organs, including liver and pancreas, may be involved in the overwhelming infections of neonates by coxsackie B, or occasionally other enteroviruses. Enterovirus 72 is a major cause of hepatitis (see below) but other enteroviruses are only rarely incriminated in infections of this organ. There is some provocative speculation that coxsackie B viruses, particularly B4, may sometimes precipitate juvenile-onset diabetes mellitus.

Pathogenesis and Immunity

Most enteroviruses grow well in both the throat and the intestinal tract, but are shed in the feces for much longer than in respiratory secretions. Dissemination via the blood stream is doubtless the route of spread to the wide range of target organs susceptible to attack, particularly by coxsackieviruses. Perinatal infections may arise from transplacental transmission or by fecal contamination of the perineum at the time of birth.

The incubation period is a week or so in the case of systemic enterovirus infections but may be as short as a day or two for conjunctival or respiratory disease.

Immunity is type specific and long lasting. Perinatal infections tend to produce no disease or mild respiratory or gastrointestinal symptoms in neonates with maternal antibody, but overwhelming disseminated disease in those without antibody.

Laboratory Diagnosis

Enteroviruses are of course most readily isolated from feces, but they can also be recovered from the throat. In the case of certain diseases they may be found in the eye, vesicle fluid, urine, CSF, or organs such as heart or brain at autopsy. Enzyme immunoassays (EIA) have been developed to detect enteroviruses in feces, especially coxsackieviruses, which tend to be shed in large numbers for long periods. Immunofluorescence (IF) has been employed to identify antigen in leukocytes from CSF of meningitis patients but is cumbersome because the lack of any common enterovirus antigen requires the use of pools of numerous type-specific antisera. Most enteroviruses are readily isolated in cell culture and this currently remains the diagnostic method of choice.

Classically, primary cultures of monkey kidney (PMK) or human embryonic kidney (HEK) have been standard, but their diminishing availability has led to their replacement in most laboratories by continuous lines. The buffalo (African) green monkey kidney line (BGM) is a satisfactory substrate for coxsackie B and polioviruses. A human rhabdomyosarcoma line (RD) supports the growth of many coxsackie A viruses previously regarded as uncultivable in anything except infant mice (and is also susceptible to polioviruses and some echoviruses). Human diploid lung fibroblasts (HDF) are sensitive to echoviruses, as are PMK and HEK. Enterovirus 70 is fastidious but can be isolated with difficulty in cultures of human conjunctiva/cornea or HDF.

Cytopathic effects resemble those described above for poliovirus (Plate 19-3) but develop more slowly, commencing with foci of rounded refractile cells which then lyse and fall off the glass. The higher numbered serotypes are generally slower growers and/or produce only incomplete CPE, so blind passage may be necessary to reveal their presence. Unfortunately, there is no single CF antigen common to all enteroviruses, hence provisional allocation to the *Picornaviridae* family is generally made on the basis of CPE, or if still in doubt, EM. Neither is HI a very useful method for identifying the isolated virus as only about one-third of enteroviruses hemagglutinate. EIA is feasible but neutralization is more reliable for typing. This laborious procedure can be short-cut by the use of an "intersecting" series of "polyvalent" serum pools. Each pool contains horse antiserum against say 10 human enterovirus serotypes; the pools are constructed to ensure that antiserum to any given serotype is present in several and absent from several others, hence the isolate can be positively identified by scrutinizing the pattern of pools that neutralize. Aggregated virions may escape neutralization hence appear untypable; this problem can be overcome by dispersing clumps with chloroform, or by using a plaque reduction assay.

The use of the infant mouse is diminishing but a few coxsackie A viruses, e.g., types 1, 19, 22, can still be grown only in this way. Newborn mice (less than 24 hours of age) are inoculated intraperitoneally and intracerebrally. BALB/c mice are more sensitive than outbred Swiss mice. Coxsackie A produces a fulminating generalized myositis, resulting in flaccid paralysis (Plate 19-4, A) and death within a week. In

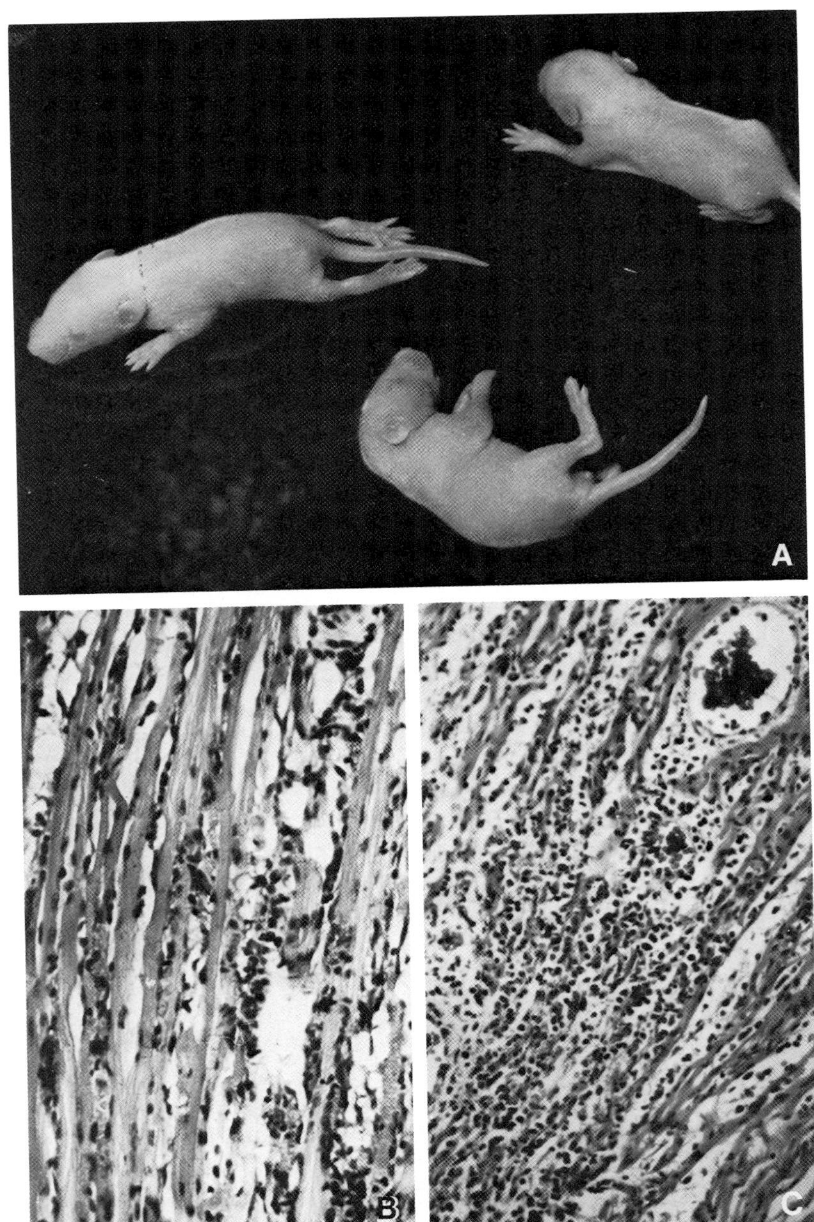

contrast, coxsackie B more slowly induces scattered focal myositis (Plate 19-4, B), brown fat necrosis, and often pancreatitis, hepatitis, and encephalitis, the latter resulting in spastic paralysis.

IgM serology is increasingly being used to diagnose enterovirus infections. For example, in coxsackie B carditis or pleurodynia, virus-specific IgM can be identified by EIA any time up to 2 months after symptoms appeared.

Epidemiology

Enteroviruses are transmitted mainly via the fecal–oral route, spreading rapidly and efficiently within families. Echoviruses appear in the alimentary tract of most infants shortly after birth. In underdeveloped countries with low standards of hygiene and sanitation an enterovirus may be recovered from the feces of as many as 80% of children at any time. In contrast, the New York "Virus Watch" program revealed an enterovirus carriage rate of only 2.4%, while comparable studies gave similar figures in other northern United States cities but substantially higher in the South.

Droplet spread also occurs, more so with the coxsackieviruses, and may be more relevant to the acquisition of the upper respiratory infections for which enteroviruses are quite often responsible. Acute hemorrhagic conjunctivitis is highly contagious in the crowded coastal cities of Asia and Africa, spreading by contact, with an incubation period of only 24 hours. Over half a million cases occurred in Bombay alone during the 1971 pandemic.

The commonest enteroviruses worldwide are echo 6, 9, and 11, coxsackie A9 and 16, and coxsackie B2-5. Outbreaks of these serotypes, and others, often reach epidemic proportions, peaking in late summer/early autumn. A different type tends to become prevalent the following year as immunity develops. Nevertheless, distinct serotypes do cocirculate to some degree, mainly among nonimmune infants.

Young children constitute the principal target and reservoir of enteroviruses. It is significant that the prevalence of enteroviruses tends to increase at the time children return to school after vacations. Having

---

PLATE 19-4. *Coxsackieviruses. (A) Infant mice 4 days after inoculation with coxsackievirus group A. The mice are moribund, with flaccid paralysis, particularly noticeable in the hind legs. (B) Section of muscle from infant mouse infected with coxsackievirus group B (H and E stain, ×250). Note focal myositis. (C) Section of heart from fatal case of coxsackievirus group B-induced myocarditis in a newborn baby (H and E stain, ×100.) (Courtesy I. Jack.)*

acquired the virus from other youngsters they then bring it home to the parents and siblings. In general, coxsackieviruses are more highly transmissible than echoviruses, probably because they are shed for longer in both feces and respiratory secretions. The probability of transmission to a nonimmune sibling was demonstrated in one study to be 76% for coxsackieviruses and 43% for echoviruses. In young children most infections are inapparent, or mild undifferentiated fevers, rashes, or URTI. Disproportionately, the severe diseases—carditis, meningitis, encephalitis—are seen in older children and adults, or in neonates. Coxsackie B viruses are particularly dangerous in neonates, but other enteroviruses, e.g., echovirus 11, have also been incriminated in fulminating infections of the newborn and in nosocomial infections in hospital nurseries.

HEPATITIS A

Properties of the Virus

The etiological agent of hepatitis A was discovered in 1973 by Feinstone and his colleagues at the United States National Institutes of Health who observed a 27-nm icosahedral virion in patients' feces by immunoelectron microscopy (Plate 19-5). Subsequent biophysical and biochemical studies in Australia and Germany led the ICTV to classify the agent as *enterovirus type 72*. Although this decision is regarded by some in the field as premature (even demeaning!) it cannot be denied that the virus closely resembles poliovirus and other enteroviruses in both structure and chemistry. The genome consists of a single molecule of ssRNA, MW 2.5×10^6, of positive polarity, polyadenylated at its 3′ end, VPg at its 5′ end, and fully infectious in its own right. The capsid is in the form of a rhombic triacontahedron or a pentakis dodecahedron, with 32 capsomers, composed of one molecule each of polypeptides VP1, 2, 3, 4, which have similar molecular weights to those of other known enteroviruses. Hepatitis A virus is somewhat less acid stable than some enteroviruses, but is considerably more heat stable, resisting 60°C for 4 hours. There is only one known serotype.

The virus produces hepatitis in marmosets and chimpanzees and can now be isolated in primary or low-passage cultures of African green monkey kidney cells, albeit with some difficulty. The virus generally grows only very slowly, without CPE, producing a steady-state infection. On serial passage, however, mutants are selected which grow more quickly, produce CPE, and shed virus into the supernatant fluid. Thereafter, the virus can be grown to reasonable titers in continuous monkey kidney cell lines such as BS-C-1, FRhK-4, and FRhK-6, and also in human diploid lung fibroblasts (HDLF). Clearly, cell culture isolation cannot yet be regarded as a routine procedure for the diagnosis of hepa-

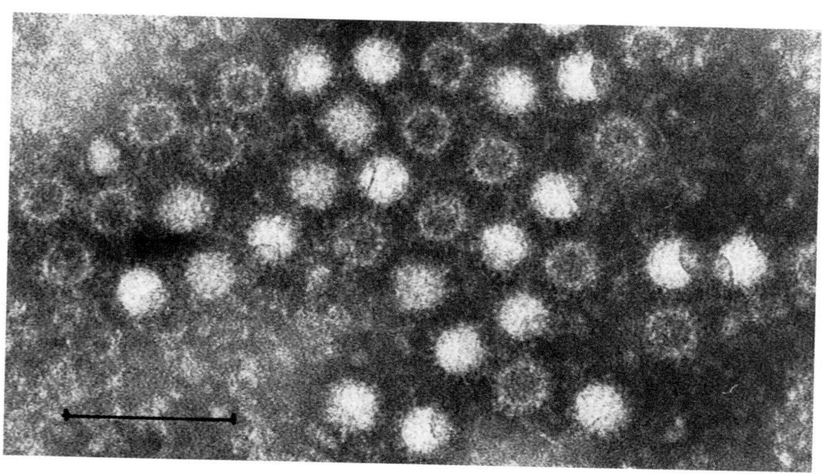

PLATE 19-5. *Hepatitis A virus (*Enterovirus 72*). Negatively stained preparation of virions clumped by antibody as seen by immunoelectron microscopy. Bar = 100 nm. (Courtesy Drs. J. Marshall and I. D. Gust.)*

titis A, but fortunately a satisfactory alternative, in IgM serology, is available (see below). The ice has been broken and we can look forward to improvements in cell culture systems that should pave the way to a vaccine.

Clinical Features, Pathogenesis, and Immunity

It is widely assumed that hepatitis A virus, being an enterovirus entering the body by ingestion, multiplies first in intestinal epithelium then spreads, presumably via the bloodstream, to infect parenchymal cells in the liver, but none of these assumptions has been formally proven. Virus is detectable in the feces (and sometimes in blood) during the week prior to the appearance of the cardinal sign, dark urine, and disappears after serum transaminase levels reach their peak. The incubation period of hepatitis A (once called "infectious" hepatitis) is about 4 weeks (limits: 2–6 weeks), substantially shorter than that of hepatitis B ("serum" hepatitis).

The clinical features of hepatitis A closely resemble those already described for hepatitis B (Chapter 13). The onset tends to be more abrupt, fever is more common, but the constitutional symptoms of malaise, anorexia, nausea, and lethargy tend to be somewhat less debilitating and less prolonged. Most infections occur in children and are anicteric, but in western countries an increasing proportion of clinical cases are in young adults. Severity increases with age. Mortality is lower (0.1–0.2% of hospitalized cases) than with hepatitis B, immune-complex mediated serum sickness is unusual, and in striking contrast to hepatitis B, no chronic carrier state develops following hepatitis A.

As there is only one known serotype of hepatitis A virus and infection leads to lifelong immunity, second attacks are unknown.

Laboratory Diagnosis

Markedly elevated serum aminotransferase levels distinguish viral from nonviral hepatitis, but do not discriminate between viral hepatitis A, B, and non-A non-B. A single serological marker, anti-HAV IgM, is absolutely diagnostic of hepatitis A. RIA or EIA is the method of choice for detecting the IgM antibody, which is demonstrable from the time symptoms and signs appear until about 4–6 months later.

Epidemiology

Like poliovirus, hepatitis A virus is spread via the fecal–oral route. As might be expected therefore, the disease is hyperendemic in the developing countries of Asia, Africa, and Central/South America where overcrowding, inadequate sanitation, and poor hygiene are rife. Infection, usually subclinical, is acquired in early childhood, so that virtually all adults have protective antibody. Most of the clinical cases are seen in children, young adults, and visitors from the West. Contaminated food and water are major vehicles of spread, but direct person-to-person contact spread is also very important. Major common-source outbreaks may occur, particularly when wells or other communal water supplies become polluted with sewage.

In developed countries by contrast, the epidemiological picture is one of declining endemicity. The infection rate is gradually declining, such that only a minority of the population have antibody. Since, as with polio, primary infections tend to be more severe with increasing age, the peak incidence of disease is in the 15 to 30 age group. Typically, a child contracts the infection at its school or creche and about a month later members of its family may come down with the disease. Infection tends to be more prevalent in lower socioeconomic groups and unsewered areas, as well as in particular minority groups such as sewer workers, male homosexuals, and returning travelers. Outbreaks frequently originate in communal living establishments with marginal standards of hygiene or special problems, such as children's day-care centers, homes for the mentally retarded, mental hospitals, prisons, army camps, and so forth. Infected handlers of food (especially uncooked or inadequately heated food) represent a particular danger in restaurants, etc. Sewage contamination of water supplies, swimming areas, or shellfish farms can also lead to explosive outbreaks. Special problems arise in times of war or natural disaster.

Control

As hepatitis A is transmitted exclusively via the fecal–oral route, control rests on heightened standards of public and personal hygiene. Reticulated drinking water supplies and efficient modern methods of collection, treatment, and disposal of sewage (see Chapter 9) should be the objective of every municipal government. Where this is impracticable, e.g., in remote rural areas, particular attention needs to be given to the siting, construction, maintenance, and operation of communal drinking water supplies such as wells; for example, these should not be downhill from pit latrines because of the risk of seepage, particularly after heavy rains. Public bathing and the cultivation of molluscs for human consumption should not be permitted near sewerage outlets.

Those employed in the dispensing of food should be subject to special scrutiny and required to observe high standards of hygiene, especially handwashing after defecation. In so far as children often contract the infection at school, attention should be given to proper instruction of children in this matter. Especially difficult problems occur in day-care centers and homes for mentally retarded children. Routine precautions include separate diaper-changing areas, and chemical disinfection of fecally contaminated surfaces and hands.

Passive immunization against hepatitis A is a well established procedure which has been in use for many years. For instance, during wars or other operations involving the mass movement of personnel into endemic areas of the tropics, it is standard to immunize all the visitors with normal human immunoglobulin on arrival and every 4–6 months thereafter. Normal Ig (0.02 ml/kg) is also used to protect family and institutional contacts following outbreaks in creches, schools, and other institutions.

Development of a vaccine against hepatitis A is obviously a top priority. Virus isolated in fetal rhesus monkey kidney cell culture has been adapted to human embryonic lung fibroblasts, though yields are low. Serially passaged virus of greatly reduced virulence which retains immunogenicity for chimpanzees has now been derived and could constitute the starting point for the development of a live attenuated or inactivated vaccine for use in man. An experimental live vaccine is currently undergoing trials in the United States Army.

RHINOVIRUSES

In this era of organ transplantation, genetic engineering, and other dramatic demonstrations of the wonders of medical science, the man-in-the-street perceives a certain irony in the inability of modern medicine to make the slightest impact on that most trivial of all man's ailments, the common cold. An even greater irony is the fact that years of searching for the elusive common cold virus have led us finally not to one, but to over 100 separate rhinoviruses. Moreover, it is now evident that rhinoviruses are by no means the only cause of this syndrome. A common cold vaccine is further away than it ever was.

Clinical Features

We are all too familiar with the symptomatology of the common cold—profuse watery nasal discharge (rhinorrhea) and congestion, sneezing, and quite often a headache, mildly sore throat, and/or cough. There is little or no fever. Resolution generally occurs within a week but sinusitis or otitis media may supervene, particularly if secondary bacterial infection occurs. Rhinoviruses may also precipitate wheezing in asthmatics, or exacerbate chronic bronchitis, especially in smokers.

Pathogenesis and Immunity

Consistent with their predilection for replication at 33°C, rhinoviruses usually remain localized to the upper respiratory tract. Inflammation, edema, and copious exudation begin after a very short incubation period (2–3 days) and last for just a few days. Interferons may play a role in terminating the illness and in conferring transient resistance to infection with other viruses, but this has yet to be proven. Acquired immunity is type-specific and correlates with the level of locally synthesized specific IgA rather than with that of IgG in serum transudate.

Laboratory Diagnosis

One would hardly ask the laboratory to assist in the diagnosis of the common cold, except perhaps in the course of epidemiological research. Nevertheless, rhinoviruses can be isolated, albeit with some difficulty. Interestingly, these fastidious viruses were first grown in cell culture by mimicking the conditions prevailing in the nasal mucosa, namely 33°C (not 37°C), pH 7.0 (instead of 7.4), and adequate oxygenation (achieved by continuously rolling the culture tubes). Now, most known rhinoviruses can be recovered in particular strains of human cells demonstrated to be highly susceptible, notably the WI-38 strain of diploid lung fibroblasts or the so-called M strain of HeLa. Foci of refractile granular cells develop slowly, producing microplaques. Organ cultures of human embryonic nasal or tracheal epithelium were originally used to isolate certain strains but this is hardly a routine diagnostic procedure.

Epidemiology

As discussed in Chapter 9 the average person contracts two to four colds per year, by contact with respiratory secretions. Children tend to shed larger numbers of virions for longer periods and suffer more colds than do adults. Rhinovirus colds occur throughout the year, with peaks early in the autumn and spring. Three or four rhinovirus serotypes circulate among the community simultaneously, one sometimes predominating; then a new set moves in after a year or so as immunity develops.

The fact that many colds occur during winter and the "changes of seasons" has encouraged the popular myth that one is most vulnerable if exposed to cold wet weather. The lie was given to this old wives' tale by an experiment conducted years ago on honeymooning couples offered free accommodation at the Common Cold Research Establishment in England (Plate 9-1). Moreover, colds are rare in the Arctic and Antarctic. It seems more likely that cold weather or abrupt changes of temperature and/or humidity may (1) alter the susceptibility of the nasal mucosa, and (2) cause people to coop themselves up together in ill-ventilated, overheated rooms where they can swap parasites more effectively.

Control

Clearly, the diversity of serotypes displaying little or no cross-protection rules out any possibility of an effective vaccine. Certain commercially available "Christmas stocking" vaccines directed against miscellaneous secondary bacterial invaders are also worthless.

Antiviral chemotherapy, or possibly chemoprophylaxis, seems to be the only hope. A great deal of effort is currently going into the pursuit of such a money-spinner, but none has yet reached the marketplace. The claims for vitamin C have been discredited. Enviroxime proved to be ineffective by nasal spray and produced unacceptable side effects when delivered orally. Interferon may have a marginal effect but only when administered prophylactically; this is impracticable. Certain flavan derivatives that bind to rhinoviruses and inhibit their replication are currently undergoing clinical trials (see Chapter 11).

CALICIVIRIDAE

The caliciviruses have been judged by ICTV to be sufficiently distinct from the picornaviruses to be accorded the status of a separate family (Table 19-4). They cause a variety of syndromes in animals, including a vesicular exanthem in swine and pneumonia in cats. Human caliciviruses have been observed only in feces and are responsible for a significant amount of gastroenteritis. Various currently uncultivable viruses known as the Norwalk group resemble caliciviruses by electron microscopy and certain biochemical criteria, but their taxonomic status has yet to be determined. For the purposes of discussion they will be considered here to be caliciviruses.

Properties of *Caliciviridae*

The calicivirus particle is slightly larger than that of the typical picornavirus and its icosahedral capsid is constructed from not four but only one type of polypeptide. The family derives its name from the 32 cup-shaped (*calix* = cup) surface depressions that give the capsid its unique outline (Plate 19-6A). The particle has a buoyant density of about 1.38 in cesium chloride, and is relatively resistant to heat and acid. The genome resembles that of the picornaviruses, but the strategy of its expression is different. For example, the structural protein is translated from a subgenomic mRNA rather than from full-length viral RNA.

Several viruses make up the so-called Norwalk group of human enteric agents. They are smaller (25–32 nm) than the prototype animal caliciviruses but otherwise similar in so far as they have been studied to date. The agents known as Norwalk, Hawaii, Ditchling, Snow Mountain, and Marin County are serologically distinct from one another, and the Parramatta and Cockle agents are different from Norwalk at least. Other viruses morphologically resembling caliciviruses, as well as various even less well defined "small round viruses" in the 27–40 nm range,

TABLE 19-4
Properties of Caliciviridae

Spherical virion, 35–40 nm[a]
Icosahedral capsid with 32 cup-shaped depressions
ssRNA genome, positive polarity, MW 2.6 million, poly(A) at 3′ end, protein at 5′ end, infectious
Cytoplasmic replication
Subgenomic mRNA for capsid protein

[a] Norwalk group of agents 25–32 nm.

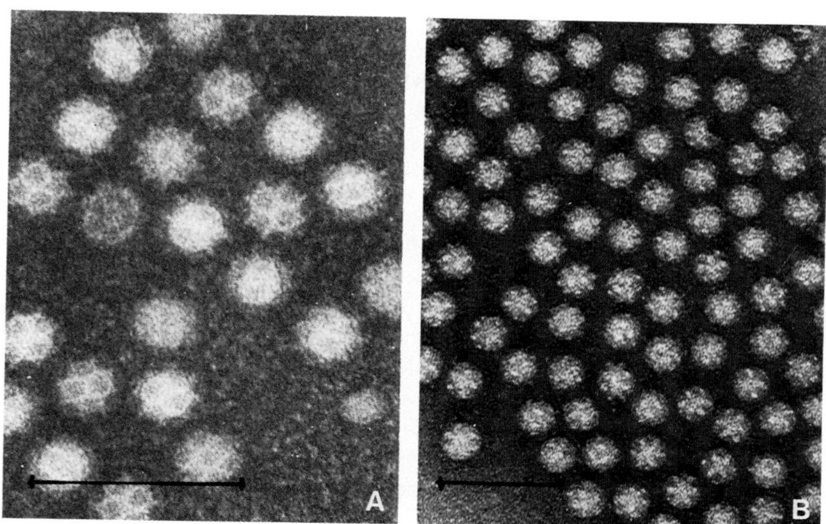

PLATE 19-6. *Negatively stained electron micrographs of virions of* Caliciviridae *(A) and astrovirus (B). Bars = 100 nm. [Courtesy Drs. M. Szymanski and P. J. Middleton (A), and Dr. D. Snodgrass (B).]*

have been visualized by EM in feces from infants with diarrhea but not yet characterized serologically or biochemically.

To complicate the issue even further, 28-nm particles called *astroviruses* have been described in feces from infants and from lambs and calves with diarrhea. The negative stain fills the hollows on the surface the particles, leaving a remarkable star-shaped unstained center, hence the name (Plate 19-6B). The taxonomic position of astroviruses, particularly in relation to caliciviruses and picornaviruses, remains unsettled, as does their role if any, in human gastroenteritis.

Clinical Features

The disease caused by the Norwalk agent has been most comprehensively studied, in volunteers as well as in natural outbreaks. Diarrhea, nausea, vomiting, and abdominal cramps are regular features; headache, myalgia, and low fever are variable. Adults are affected more often than children.

Laboratory Diagnosis

All these viruses are known only by their electron microscopic appearance in feces, and EM, or immuno-EM, remains the principal method of rapid diagnosis, although it is not always positive. RIAs have been developed for detection of antigen in feces, or of antibody in serum. Serotypes are distinguished by IEM or RIA, but are still inadequately characterized. Human caliciviruses have not yet been cultured, but there is a report of propagation of human astrovirus in HEK cultures.

Pathogenesis and Immunity

The incubation period of Norwalk gastroenteritis is only 2–3 days, and may be as short as 24 hours in a common-source outbreak. As in other viral enteritis, the tips of the villi slough off but rapidly regenerate. Immunity is poor. Indeed, volunteers with preexisting antibody proved to be even more susceptible to challenge with Norwalk virus than did "nonimmune" controls. This demonstrates (1) that prior infection confers no significant immunity, and (2) that unknown genetic or long-term physiological factors may predispose certain individuals but not others to gastroenteritis following exposure to virus.

Epidemiology

The Norwalk group of agents is uncommon in children, but about half of all adults reveal serological evidence of infection. Common-source outbreaks frequently occur via fecal contamination of food, e.g., salad, in restaurants, schools, camps, cruise ships, etc. Several major outbreaks have been traced to consumption of uncooked or partially cooked oysters, cockles, clams, or other shellfish that have been harvested from sewage-polluted estuaries. Others have been attributed to discharge of sewage into drinking water supplies, or into pools and lakes in which people swim. Secondary spread occurs from person to person within households.

Some of the other agents may have different epidemiological patterns. For example, a calicivirus epidemic described in Japan occurred in young children, in winter. Astroviruses also infect young children, being found quite commonly in asymptomatic infants as well as in outbreaks of gastroenteritis in children or adults; transmission studies in volunteers indicated that the virus has relatively low pathogenicity for adults. Much remains to be learned about this confusing constellation of small round enteric viruses.

FURTHER READING

Picornaviruses

Fox, J. R., and Hall, C. E. (1980). "Viruses in Families." PSG Publishing, Littleton, Massachusetts.

Melnick, J. L. (1983). Portraits of viruses: The picornaviruses. *Intervirology* **20,** 61.

Pérez-Bercoff, R., ed. (1979). "The Molecular Biology of Picornaviruses." Plenum, New York.

Rueckert, R. R. (1976). On the structure and morphogenesis of picornaviruses. *In* "Comprehensive Virology" (H. Fraenkel-Conrat and R. R. Wagner, eds.), Vol. 6, p. 131. Plenum, New York.

Enteroviruses

Grist, N. R., Bell, E. J., and Assaad, F. (1978). Enteroviruses in human disease. *Prog. Med. Virol.* **24,** 114.

Hennessen, W., and van Wezel, A. L., eds. (1981). "Reassessment of Inactivated Poliomyelitis Vaccine," Dev. Biol. Stand., Vol. 47. Karger, Basel.

Melnick, J. L. (1978). Advantages and disadvantages of killed and live poliomyelitis vaccines. *Bull. W. H. O.* **56,** 21.

Melnick, J. L., ed. (1984). "Enteric Viruses in Water," Monogr. Virol., Vol. 15. Karger, Basel.

Sabin, A. B. (1980). Vaccination against poliomyelitis in economically underdeveloped countries. *Bull. W. H. O.* **58,** 141.

Sabin, A. B. (1981). Paralytic poliomyelitis: Old dogmas and new perspectives. *Rev. Infect. Dis.* **3,** 543.

Yin-Murphy, M. (1984). Acute hemorrhagic conjunctivitis. *Prog. Med. Virol.* **29,** 23.

Hepatitis A

Deinhardt, F., and Deinhardt, J., eds. (1983). "Viral Hepatitis: Laboratory and Clinical Science." Dekker, New York.

Deinhardt, F., and Gust, I. D. (1982). Viral hepatitis. *Bull. W. H. O.* **60,** 661.

Francis, D. P., and Maynard, J. E. (1979). The transmission and outcome of hepatitis A, B, and non-A, non-B. *Epidemiol. Rev.* **1,** 17.

Gerety, R. J., ed. (1984). "Hepatitis A." Academic Press, New York.

Szmuness, W., Alter, H. J., and Maynard, J. M., eds. (1982). "Viral Hepatitis." Franklin Institute Press, Philadelphia, Pennsylvania.

Rhinoviruses

Gwaltney, J. M. (1982). Rhinoviruses. *In* "Viral Infections of Humans: Epidemiology and Control" (A. S. Evans, ed.), 2nd ed., pp. 491–517. Plenum, New York.

Macnaughton, M. R. (1982). The structure and replication of rhinoviruses. *Curr. Top. Microbiol. Immunol.* **97,** 1.

Caliciviruses

Kapikian, A. Z., Greenberg, H. B., Wyatt, R. G., Kalica, A. R., and Chanock, R. M. (1982). The Norwalk group of viruses—agents associated with epidemic viral gastroenteritis. *In* "Virus Infections of the Gastrointestinal Tract" (D. A. J. Tyrrell and A. Z. Kapikian, eds.), pp. 147–178. Dekker, New York.

Kaplan, J. E., Gary, G. W., Baron, R. C., Singh, N., Schonberger, L. B., Feldman, R., and Greenberg, H. B. (1982). Epidemiology of Norwalk gastroenteritis and the role of Norwalk virus in outbreaks of acute nonbacterial gastroenteritis. *Ann. Intern. Med.* **96,** 756.

Schaffer, F. L. (1979). Caliciviruses. *In* "Comprehensive Virology" (H. Fraenkel-Conrat and R. R. Wagner, eds.), Vol. 14, pp. 249–284. Plenum, New York.
Schaffer, F. L., Bachrach, H. L., Brown, F., Gillespie, J. H., Burroughs, J. N., Madin, S. H., Madeley, C. R., Povey, R. C., Scott, F., Smith, A. W., and Studdert, M. J. (1980). Caliciviridae. *Interviology* **14,** 1.

CHAPTER 20

Togaviruses and Flaviviruses

Introduction 479
Properties of *Togaviridae* and *Flaviviridae* 481
Clinical Features in Man 484
Pathogenesis and Immunity 487
Laboratory Diagnosis 488
Epidemiology 490
Control 498
Rubella 499
Further Reading 507

INTRODUCTION

Because most members of the *Togaviridae* and *Flaviviridae* families are arthropod-borne, this chapter is mainly about arboviruses. Many of the principles enunciated here apply just as appropriately to Chapter 25, in which we discuss the other major family of arboviruses, the *Bunyaviridae*.

An *arbovirus* may be defined as a virus that multiplies in a hematophagous (blood-feeding) arthropod and is transmitted by bite to a vertebrate, in which it also multiplies, producing a viremia of sufficient magnitude to infect another blood-feeding arthropod and so perpetuate the cycle. The arthropod is usually a mosquito, tick, sandfly, or midge. This *invertebrate host* is known as the *vector*. The principal *vertebrate host*—usually a mammal or bird—is known as the *reservoir*. In its natural environment an arbovirus alternates between its invertebrate vector and its vertebrate hosts. This *cycle* (see Figs. 9-2 and 9-3) can be bypassed by vertical transmission of the virus from the arthropod directly to its offspring, a mechanism of particular importance for "overwintering" in temperate climates (see Chapter 9). Man generally plays no role in the natural history of the arbovirus and is irrelevant to its survival. But he may become accidentally involved when he intrudes upon the natural ecosystem (e.g., by penetrating a jungle or a swamp), or when he changes it (e.g., by clearing a forest or irrigating a desert). The consequences may be dire—even lethal—for, unlike the natural reservoir host, man has not evolved in company with the virus over millions of years and has acquired no genetic resistance. In a few arbovirus infections, such as dengue and urban yellow fever, man may serve as the reservoir, so that a man–arthropod–man cycle is established independently of any animal reservoir host.

There are over 400 known or presumed arboviruses, most of them discovered in mosquitoes or ticks during the halcyon days of the Rockefeller Foundation's Virus Program through the 1950s and 1960s. About 80 of them are capable of infecting man. Half of these, mainly togaviruses and flaviviruses, induce significant disease. Some arboviral diseases, the encephalitides and hemorrhagic fevers, rank among the most lethal of all human infections.

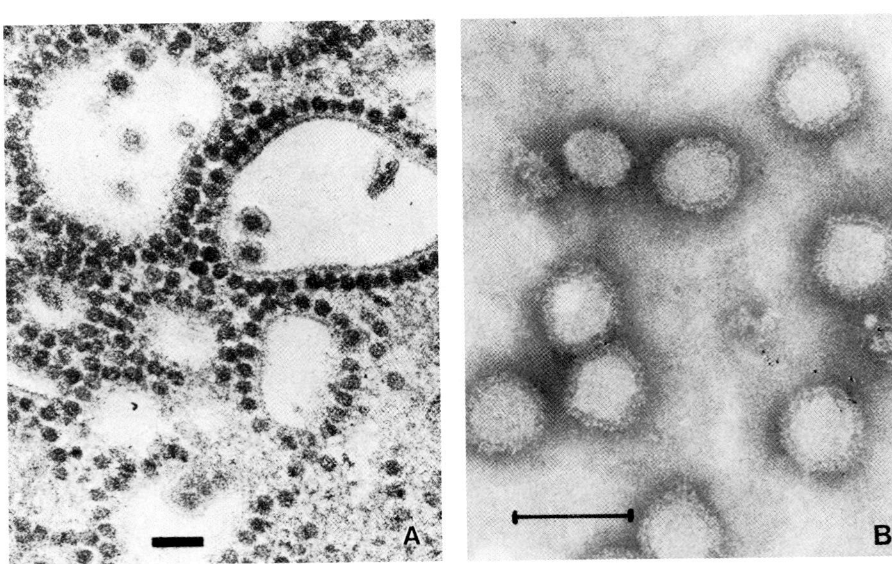

PLATE 20-1. Togaviridae. *(A) Alphavirus nucleocapsids acquiring an envelope as they bud into cytoplasmic vesicles. (B) Negatively stained virions of rubella virus. Bars = 100 nm. [Courtesy Dr. I. H. Holmes (A) and Dr. C-H. von Bonsdorff (B).]*

PROPERTIES OF *TOGAVIRIDAE* AND *FLAVIVIRIDAE*

The flaviviruses have always been regarded as a genus within the *Togaviridae* family but in 1984 were provisionally reclassified as a distinct family, the *Flaviviridae,* mainly on the basis of gene sequence and a different strategy of expression of the genome. Nevertheless, the two are considered here in a single chapter because they resemble one another so closely, not only in their structure but especially in their ecology.

The virion comprises an icosahedral core enclosed within a tightly fitting envelope (*toga* = cloak), total diameter 60–70 nm (*Togaviridae,* Plate 20-1) or 40–50 nm (*Flaviviridae,* Plate 20-2). The envelope of togaviruses generally contains two species of glycoprotein and that of flaviviruses one (Tables 20-1 and 20-2). These peplomers bind erythrocytes, hence the virions hemagglutinate.

The genome is a single linear molecule of single-stranded RNA of positive polarity, MW 4×10^6, which is infectious. The 5′ end is capped, but the 3′ end is polyadenylated only in the case of the togaviruses.

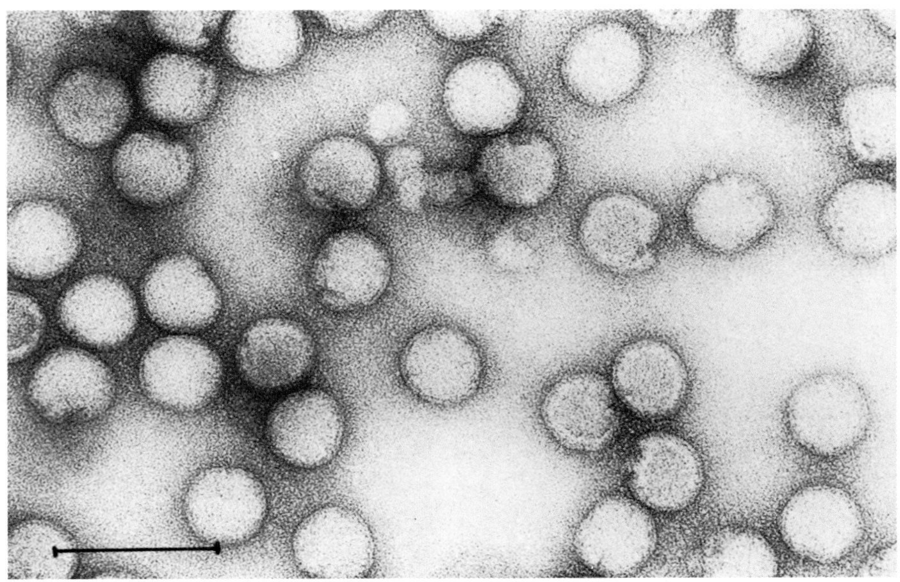

PLATE 20-2. Flaviviridae. *Negatively stained virions of tick-borne encephalitis virus. Bar = 100 nm. (Courtesy Drs. W. Tuma, F. X. Heinz, and C. Kunz.)*

TABLE 20-1
Properties of Togaviridae

Spherical virion, enveloped, total diameter 60–70 nm
Two envelope glycoproteins
Icosahedral capsid, 28–35 nm
ssRNA genome, linear, positive polarity, MW 4 million, 5′ end capped, 3′ end polyadenylated, infectious genes for nonstructural proteins located at 5′ end
Cytoplasmic replication, budding from plasma membrane
Full-length and subgenomic mRNA; posttranslational cleavage of polyproteins
Genus *Alphavirus* is arthropod-borne; *Rubivirus, Pestivirus* and *Arterivirus* are not

Viral Multiplication

Togaviridae. The alphaviruses, Sindbis and Semliki Forest virus (SFV), and *ts* mutants derived from them (see Chapter 4), have been intensively investigated as models for the study of viral replication and a great deal is known about them. In Chapter 3 we described the uptake of a togavirus by receptor-mediated endocytosis and its transport in a coated vesicle to a lysosome where, at pH 5, it is uncoated (see also Plate 3-2). The viral (+)RNA molecule thus released into the cytoplasm functions as mRNA, the 5′ two-thirds of which is translated into one major polyprotein and one minor polypeptide. While still nascent on the ribosome, the polyprotein is cleaved to produce four nonstructural polypeptides which form the RNA-dependent RNA polymerase. This enzyme is now available to transcribe full-length cRNA (minus sense), from which in turn two types of plus-sense mRNA are copied: full-length (= vRNA) and subgenomic (corresponding to the 3′ one-third of vRNA). The latter is translated into another polyprotein which carries protease activity and cleaves itself to yield all the viral structural poly-

TABLE 20-2
Properties of Flaviviridae

Spherical virion, enveloped, total diameter 40–50 nm
One envelope glycoprotein
Icosahedral capsid, 25–30 nm
ssRNA genome, linear, positive polarity, MW 4 million, 5′ end capped, but 3′ end not polyadenylated, infectious, genes for structural proteins located at 5′ end
Cytoplasmic replication, maturation within cytoplasmic vesicles
No subgenomic mRNA, no polyprotein precursor yet detected
One genus, *Flavivirus,* all members of which are arthropod-borne

peptides (capsid protein plus two to three membrane proteins). As previously described, the membrane proteins are glycosylated in stepwise fashion as they progress from the rough endoplasmic reticulum through the Golgi complex to the plasma membrane from which the completed virion buds (see Fig. 3-10 and Plate 3-4A).

Flaviviridae. Despite the superficial similarity of their genomes, flaviviruses differ significantly from alphaviruses in both gene sequence and replication strategy. In contrast with the alphaviruses, the genes coding for the structural proteins are located at the 5′ end of the flavivirus genome. Moreover, no subgenomic mRNA has been detected, nor has any polyprotein precursor. Taken at face value, this apparent contradiction would indicate that there is neither posttranscriptional cleavage of polycistronic RNA nor posttranslational cleavage of polyprotein. This has led to the hypothesis that internal initiation of translation is possible at various points along the polycistronic vRNA molecule—an occurrence that would be unique among mammalian viruses. Another significant difference is that those flaviviruses examined so far mature within vesicles of the endoplasmic reticulum; budding has never been demonstrated unequivocally.

Classification

The *Togaviridae* family is subdivided into four genera: *Alphavirus, Rubivirus, Pestivirus,* and *Arterivirus.* Only the alphaviruses are arthropod-borne. Rubella, a very important human non-arthropod-borne togavirus, is the only species within the *Rubivirus* genus, and will be discussed at the end of this chapter. The *Pestivirus* genus contains no human pathogens but several viruses causing important diseases of livestock, notably bovine virus diarrhea and swine fever, while the *Arterivirus* genus contains the virus of equine arteritis.

The *Alphavirus* genus currently contains over two dozen viruses, most (perhaps all) being mosquito-borne arboviruses. About a dozen of them have been shown to cause disease in man, seven of which have produced significant epidemics (Table 20-3). All alphaviruses share a common "group" antigen (the nucleoprotein of the capsid) which is demonstrable by CF. Species differ in "type-specific" antigenic determinants on their envelope glycoproteins as recognized by HI and neutralization tests. The latter proteins also carry cross-reactive determinants which are shared by certain species comprising a given "complex" (= "subgroup" or subgenus); six such "complexes" have been described within the *Alphavirus* genus. It also seems likely that some "species" will subsequently turn out to be more appropriately regarded as serotypes.

TABLE 20-3
Human Alphaviruses

VIRUS	DISTRIBUTION	INVERTEBRATE VECTOR	VERTEBRATE RESERVOIR	DISEASE IN MAN
Chikungunya	Asia, Africa	Mosquito	Monkeys?, man	Fever, arthritis, hemorrhagic fever
O'nyong-nyong	Africa	Mosquito	Monkeys?, man	Fever, arthritis
Ross River	Australia, Pacific	Mosquito	Marsupials?	Arthritis
Sindbis	Africa, Asia, Australia	Mosquito	Birds	Fever
Eastern equine encephalitis	Americas	Mosquito	Birds	Encephalitis
Western equine encephalitis	Americas	Mosquito	Birds	Encephalitis
Venezuelan equine encephalitis	Americas	Mosquito	Rodents, horses	Encephalitis

The *Flaviviridae* family comprises a single genus, *Flavivirus,* containing over 60 serologically related viruses, which can be assigned to a small number of subgroups. Some are transmitted by mosquitoes and others by ticks. At least a dozen of them are significant causes of human disease (Table 20-4).

CLINICAL FEATURES IN MAN

Fortunately, the great majority of arbovirus infections in man are asymptomatic, the only evidence of infection being seroconversion to a state of lifelong immunity. Furthermore, even the clinical manifestations of many of the arboviruses are trivial, taking the form of a nondescript fever which is difficult or impossible to distinguish from other tropical fevers. Yet two or three dozen of these viruses are capable of inducing serious disease and several cause potentially lethal encephalitis or hemorrhagic fever. The most important togaviruses and flaviviruses are listed in Tables 20-3 snd 20-4. Although the syndromes they can cause are diverse, they can conveniently be classified into three main categories: (1) fever/rash/arthralgia, (2) hemorrhagic fever, and (3) encephalitis.

Fever/Rash/Arthralgia

This is the most common presentation of arboviral disease and dengue fever, the most widespread of all the arboviruses, is the prototype.

TABLE 20-4
Human Flaviviruses

VIRUS	DISTRIBUTION	INVERTEBRATE VECTOR	VERTEBRATE RESERVOIR	DISEASE IN MAN
Yellow fever	Africa, South America	Mosquito	Monkeys, man	Hepatitis, hemorrhagic fever
Dengue 1–4	Asia, Pacific, Caribbean, Africa	Mosquito	Monkeys?, man	Fever, hemorrhagic fever
Japanese encephalitis	East and SE Asia	Mosquito	Birds, pigs	Encephalitis
Murray Valley encephalitis	Australia, New Guinea	Mosquito	Birds	Encephalitis
West Nile fever	Africa, Europe, Middle East	Mosquito	Birds	Fever, encephalitis
St. Louis encephalitis	Americas	Mosquito	Birds	Encephalitis
Rocio	Brazil	Mosquito?	Birds?	Encephalitis
Tick-borne encephalitis	USSR, Europe Europe	Tick	Rodents, birds, goats, cattle	Encephalitis
Louping ill	Britain	Tick	Rodents, sheep	Encephalitis
Powassan	North America, USSR	Tick	Rodents	Encephalitis
Omsk hemorrhagic fever	USSR	Tick	Rodents	Hemorrhagic fever
Kyasanur Forest disease	India	Tick	Rodents, monkeys	Hemorrhagic fever

There is a sudden onset of fever, chills, headache (especially retroorbital), conjunctivitis, and excruciating pains in the back, muscles, and joints, which gives the disease its popular name of "break-bone fever." The fever often falls and rises again a few days later, producing the characteristic "saddleback" temperature chart. Lymphadenopathy and leukopenia are usually demonstrable. A scarlatiniform or maculopapular rash appears on the third or fourth day and fades 2–3 days later. Convalescence is slow but certain.

Arthritis is a more prominent feature in the similar syndromes produced by the togaviruses, Ross River, chikungunya, o'nyong-nyong, Mayaro, and Sindbis. Indeed, chikungunya and o'nyong-nyong are colorful African terms evoking the agony of the affected joints! Nevertheless, the arthritis is usually short lived.

The fever–myositis syndrome, without rash or arthritis, is also caused by other important arboviruses to be discussed in Chapters 25 (Bunyaviruses) and 27 (Reoviruses).

Hemorrhagic Fever

As the name implies, the striking feature of this potentially lethal syndrome is hemorrhage, which makes itself apparent as petechiae and ecchymoses on the skin and mucous membranes (Plate 25-2) with bleeding from any or all of the body orifices. Examination of the blood reveals thrombocytopenia, and usually leukopenia. In fatal cases the patient collapses abruptly from hypotensive shock. Indeed, in the lethal dengue shock syndrome the child may die even before hemorrhagic signs have become manifest.

The hemorrhagic fevers (Chapter 29) are a diverse group of diseases caused by several very different agents and differing in their pathogenesis (see below). In addition to the flaviviruses, dengue, yellow fever, Kyasanur Forest disease, and Omsk hemorrhagic fever, and the alphavirus, chikungunya, several bunyaviruses, filoviruses, and arenaviruses, some of which are not arboviruses, cause clinically similar hemorrhagic fevers, which will be discussed in Chapters 24, 25, and 26.

Yellow fever is a special case in that the liver is involved and the patient jaundiced, hence the name. In other respects it is a kind of hemorrhagic fever. Indeed its rather revolting nickname, "black vomit," refers to the massive gastrointestinal hemorrhages that characterize this disease. Both the blood pressure and the pulse rate drop, proteinuria and oliguria signal kidney failure, jaundice intensifies, and the patient may die. Fortunately, most cases are much milder than this, presenting only with fever, chills, headache, backache, myalgia, and vomiting; the overall case fatality rate has ranged from 2 to 50% in different epidemics.

Encephalitis

This devastating disease begins innocently enough with the usual arboviral syndrome of fever, chills, headache, widespread aches, and vomiting. However, within a day or two the patient develops drowsiness, often accompanied by neck rigidity indicative of meningeal involvement, and may progress to confusion, paralysis, convulsions, and coma. Case fatality rates average about 10–20%, but are considerably higher with Eastern equine encephalitis. Survivors are often left with permanent neurological sequelae such as mental retardation, epilepsy, paralysis, deafness, and blindness.

Even the "nonencephalitogenic" arboviruses have encephalitogenic potential. This is apparent not only from the neurological damage they produce in newborn mice but also from rare cases of encephalitis in man following natural infection or after their use many years ago as oncolytic agents in the experimental treatment of human cancer.

PATHOGENESIS AND IMMUNITY

The infected female mosquito introduces its proboscis directly into a capillary beneath the skin and injects saliva which contains the arbovirus. Viral multiplication ensues in vascular endothelium and in both blood monocytes and macrophages in lymph nodes, bone marrow, spleen, and liver. Virions liberated from these cells augment the viremia and precipitate the prodromal symptoms (fever, chills, headache, muscular aches) which mark the end of the incubation period (usually 3–7 days). Particular target organs may now be infected, commonly muscles (myositis), joints (arthritis), skin (rash), brain (encephalitis), or liver (hepatitis). In the hemorrhagic fevers, including yellow fever, widespread hemorrhages and renal failure may lead to death.

The pathogenesis of different hemorrhagic fevers may be quite distinct. In the case of yellow fever, the massive necrosis of hepatocytes in the mid-zone of the liver lobules leads to a decrease in the rate of formation of the blood coagulation factor, prothrombin, resulting in prolonged coagulation and bleeding times.

The pathogenesis of the dengue hemorrhagic fever/dengue shock syndrome (DHF/DSS) is the subject of considerable controversy. Some believe it merely represents the severe end of the spectrum of clinical presentations of dengue fever, or that it is caused by unusually virulent variants of the virus. However, other workers favor the "immune enhancement" hypothesis, which states that DHF/DSS occurs as a result of enhanced replication of virus in the presence of preexisting antibody against another dengue serotype. Most cases have been noted to occur when dengue type 2 infects either (1) a baby with maternal antibody against dengue, or (2) a child with serological evidence of having been infected during the previous 5 years with a heterologous dengue serotype (1, 3, or 4). Experimentally it can be demonstrated that IgG against say dengue type 1 will bind to cross-reactive determinants on dengue virus 2 but not neutralize it. The virus–antibody complex may then attach to the Fc receptor on a macrophage and be taken up more avidly than is uncomplexed virus. Once inside, the virus is able to multiply in the macrophage which is highly permissive for flaviviruses. Infected monocytes may carry virus throughout the reticuloendothelial system, where further viral multiplication occurs in macrophages and other susceptible cells. In consequence, abnormally high titers of virus are produced in the body, resulting in more severe disease.

This still begs the question of how the virus induces hemorrhage and shock. It has been proposed that the infected macrophages, following activation and or immune cytolysis by cytotoxic T cells, release factors which lead to (1) activation of the complement system, (2) activation of the blood-clotting system, and possibly (3) increase in capillary permeability, all three of which are features of the disease. It is equally possible that such mediators are released from sensitized T lymphocytes, i.e., that delayed hypersensitivity is involved. Immune complexes between virus and nonneutralizing antibody might also play a role in the immunopathology by activation of complement leading to increased vascular permeability.

Immunity

The immunity that follows clinical or subclinical infection with an arbovirus probably lasts for life. In addition, it confers at least partial cross-protection against related agents. For example, infection with Wesselbron, Zika, or dengue viruses provides a significant degree of immunity to related flaviviruses, which may go some way toward explaining why yellow fever is uncommon or mild among the indigenous people in those regions of Africa in which these viruses coexist. Furthermore, when sequential infections with related viruses do occur, there is a broadening of the antibody response to embrace other viruses in the same serogroup and an anamnestic response to the original agent, a phenomenon which limits the reliability of serology as a diagnostic or epidemiological tool.

LABORATORY DIAGNOSIS

Arboviruses are dangerous. Most of the deaths recorded in laboratory personnel over the years have been due to arboviruses and other "exotic" agents. Accordingly, isolation of arboviruses should be attempted only in special reference laboratories with facilities for high-level containment.

Nor is it easy to isolate arboviruses. Viremia declines very early in the illness, so attempts to culture virus are likely to be fruitless unless blood is taken during the first 2 or 3 days and transported to the laboratory without delay, properly refrigerated. Postmortem, virus can sometimes be recovered from appropriate organs, e.g., brain or liver.

Traditionally, arbovirologists have relied upon the newborn mouse for the isolation of arboviruses from man and especially from insects. After intracerebral (ic) inoculation, mice become ill and die. Following passage of the brains of sick mice ic to additional mice or cultured cells, the isolate can be identified serologically.

A more exotic method was developed for the isolation of dengue virus, and could become more generally used, especially for experimental studies of transmission. Mosquitoes can be inoculated intrathoracically, or their larvae intracerebrally; large ones are recommended! Growth of virus is detected by CF or immunofluorescence (IF) on neural tissue.

Cultured cell systems have now been developed for the isolation of arboviruses. The following are the most useful vertebrate cells: primary cultures of chick or duck embryo, the baby hamster cell line BHK-21, and the monkey kidney cell lines Vero and LLC-MK_2. More recently, insect cell lines, which are grown at 28°C, have been developed from *Aedes* mosquitoes and found to be considerably more susceptible than newborn mice to certain arboviruses, e.g., dengue and yellow fever. CPE varies with the virus and cell type: some combinations produce syncytia but many give undramatic cytopathology or none at all. IF identifies viral antigen in the cells.

Characterization of arboviruses is a tricky business which continues to test the ingenuity of arbovirologists, ever on the alert for new agents. Traditionally, isolates of potentially novel agents are put through a Millipore filter to assess their approximate size, and tested for sensitivity to a lipid solvent such as ether or sodium deoxycholate to determine whether the virus is enveloped. Electron microscopy is the most useful method of allocating the virus to its correct family. Then serology is employed to characterize it fully.

More usually, the problem is that of identifying one of a small number of arboviruses known to occur in a particular ecosystem. Clues are provided not only from the source of the isolate, but from observation of clinical signs and death time of infant mice, plaque morphology in cultured cells, and so on. An experienced arbovirologist can then often go straight to plaque reduction neutralization tests to clinch the identification.

Alphaviruses share common antigens, as do flaviviruses. These have traditionally been demonstrated by CF or HI, but IF, EIA, and RIA are coming to be used more frequently. Since togaviruses and flaviviruses agglutinate erythocytes (from geese, at a particular optimal pH between 6 and 8 which varies with the virus) HI has proved to be a convenient procedure for assigning an arbovirus to a particular group. Individual alphaviruses or flaviviruses display much greater cross-reactivity by HI than do viruses belonging to other hemagglutinating families. Neutralization is the appropriate test for characterization at the species level. Even those viruses that induce no CPE in cultured cells usually produce plaques, hence plaque reduction is the most convenient form of neutralization test.

In light of the pronounced cross-reactions between species which render HI and other serological tests less generally useful, there is a need to develop panels of appropriate monoclonal antibodies. For example, one monoclonal antibody might recognize an antigenic determinant common to all flaviviruses, while another recognizes a determinant common to all members of the dengue complex (dengue "types" 1, 2, 3, and 4), yet another recognizes only the type-specific determinant on dengue type 3, and another distinguishes the Caribbean/Pacific subtype from the SE Asian subtype of dengue 3. Such monoclonal antibodies have already proved to be of great value in research and may before long emerge as standard diagnostic reagents, particularly for viruses or geographical regions where serological cross-reactions using polyclonal antisera are a problem.

Oligonucleotide maps of the viral RNA discern even finer differences between isolates from different parts of the world and can be very informative in tracing their origins. For instance, such RNA fingerprints distinguish variants or "topotypes" of dengue type 2 virus from various corners of the globe.

Because attempts to isolate virus from individual patients are often unsuccessful, as virus can usually no longer be recovered from the blood by the time the patient presents, the diagnosis is usually made by demonstrating a rising titer of antibody in the patient's serum. CF can be useful for diagnosing a recent illness, as CF antibody tends to disappear within months. Both CF and HI are cross-reactive within "complexes" (subgenera). They are also subject to the influence of "the doctrine of original antigenic sin": antibodies against a different virus encountered earlier in life are boosted by the current infection with a cross-reacting agent, so confusing the diagnosis. The neutralization test is rather less affected by this phenomenon.

IgM serology is emerging as a relatively simple method for diagnosing arboviral disease. EIAs or RIAs, using anti-human IgM as capture antibody, have been developed for detecting specific IgM in a single acute-phase specimen of serum or, in cases of encephalitis, CSF. However, the reagents used in such tests must be absolutely reliable, and one must also be conscious of the fact that IgM antibodies can persist for many months and/or be elicited by infection with heterologous viruses. IgM serology is ideally suited to rapid diagnosis of individual patients during an epidemic, but is not sufficiently definitive to rely on for identifying a potentially new virus in an "indicator" case prior to such an epidemic.

EPIDEMIOLOGY

Arboviruses may have evolved in invertebrates. Even today many of them appear to be capable of indefinite survival in insects by transovarial transmission from one generation to the next, and in ticks also by transstadial transmission from one developmental stage to the next. These are certainly important mechanisms for surviving long cold winters in temperate climates. Nevertheless vertebrates represent important amplifier hosts, and the typical *maintenance cycle (enzootic cycle)* of an arbovirus is based upon regular alternation between the invertebrate and vertebrate hosts. In contrast to the arthropod, which carries the virus throughout its short life, infected vertebrates usually recover rapidly, eliminate the virus, and develop a lasting immunity to reinfection.

As might be expected in a three-way relationship that has evolved over millions of years, a given arbovirus usually occupies a particular ecological niche. Confined perhaps to the forests of one continent, it infects one or a small number of species of invertebrate that habitually feed off one or a few species of vertebrate. To be an efficient host, the vertebrate must be abundant, must reproduce rapidly, and above all must maintain a high level of viremia for an adequate period. In its turn, the invertebrate, to be an efficient vector, must have a sufficiently low threshold of infection by that virus, must then carry and shed the virus in its salivary secretions for life, but not be adversely affected by it, and must have a distribution, flight range, longevity, and biting habits well adapted to the habits and habitat of its principal vertebrate host. Hence the virus flourishes indefinitely in a state of peaceful coexistence with vertebrate and invertebrate host in which no one gets hurt. The trouble arises when the virus moves beyond this enzootic cycle.

Disease occurs when unnatural hosts become involved. When man lives in regions in which a particular arbovirus is enzootic he is vulnerable to infection and a few of those infected may suffer severe, even lethal, disease. Visitors such as tourists, soldiers, or forest workers are even more vulnerable as, unlike the indigenous population, they will not have acquired immunity from subclinical infection in childhood. In tropical countries where the arthropod vector is plentiful year round, the risk is always present and human disease is endemic (e.g., jungle yellow fever). In regions subject to monsoonal rains, epidemics of mosquito-borne diseases may occur toward the end of the wet season (e.g., Japanese encephalitis in parts of Southeast Asia). In some temperate countries, and particularly in arid areas, human epidemics of mosquito-borne arbovirus disease occur following periods of exceptionally heavy rain. A different species of mosquito, with anthropophilic habits, may become involved in transmission of the virus to humans, and may then maintain the virus in a man–mosquito–man cycle without the intervention of the natural vertebrate or invertebrate host. In fact, dengue and chikungunya (and formerly yellow fever) viruses are maintained indefinitely in a man–*Aedes aegypti*-man cycle in those large towns and cities of the tropics that have not yet succeeded in eradicating this urban vector.

We have discussed the dangers of man intruding into an arbovirus' natural ecosystem. In addition, however, there are circumstances when man so alters his own ecosystem that the arboviruses come to him. Many major dam-building and irrigation projects in arid regions such as Southern California, Australia, and Egypt have so changed the local ecology that mosquitoes and water birds have brought dangerous arboviruses into areas in which they were previously almost unknown.

Unlike mosquitoes, ticks are present throughout the year even in temperate climates and often live through more than a single breeding cycle of their host. The eggs of ticks develop successively through stages (larva to nymph to adult) and a blood meal is generally required at each stage. Arboviruses are passed from one developmental stage (instar) to another (*transstadial transmission*) as well as from one generation of tick to the next (*transovarial transmission*). Some species spend their whole lives on one vertebrate host, whereas others fall off, molt, then find a different host after each meal. The larvae and nymphs generally parasitize birds or small mammals such as rodents, whereas adult ticks prefer larger animals. Domestic and farm animals such as cattle and goats represent important amplifier hosts in the maintenance, and particularly in the spread to man of tick-borne arboviruses.

To illustrate these basic principles we shall now describe the ecology of the major arboviruses affecting man, insofar as we currently understand them. In general, the natural ecosystem of any given virus is probably more complicated than is implied in the basically simple cycles outlined below. Such is the challenge of discovering the basic enzootic cycle of a dangerous arbovirus in the complex fauna of a tropical forest that an epidemiologist is well pleased to discover one important invertebrate and one vertebrate reservoir. In fact, the ecosystem may not be quite so exclusive and, as time goes on, alternative vectors and vertebrate hosts tend to be identified.

It is important to appreciate the difficulties involved in this sort of epidemiological detective work. Arboviruses are generally so difficult to isolate from man that most of them were originally discovered in mosquitoes collected from tropical forests or from "sentinel" mice or chickens left overnight in mosquito-infested areas. Not all of these viruses turn out to be arboviruses, and only a proportion have any relevance to man. The next step is to use the newly isolated virus (ideally having been passaged only in a cultured cell line, not in mice in case an irrelevant passenger is picked up) as antigen in serological tests against reference antisera to ascertain whether it is indeed a new agent. If so, the antigen is used to search for the corresponding antibody in sera collected from man and other vertebrates in the area. The presence of antibodies in significant numbers of a particular species of bird or animal suggests that it may be an important natural reservoir or amplifier host. Attempts are then made to prove this by isolation of virus. Allocating the virus a role in human disease is equally difficult. Most arboviral infections are asymptomatic. Generally one must await an epidemic serious enough to attract attention, before being able to pin the disease on the virus by demonstrating a rising titer of antibodies, or IgM antibody, or virus, in a number of patients with the same illness. Nevertheless, the cycles described below are generally valid enough to provide us with the degree of insight needed to plan rational strategies for control.

Yellow Fever

One of the great plagues throughout history, yellow fever decimated the crews of English sailing ships visiting West Africa and is thought to have been transported to the New World, together with its vector, *Aedes aegypti,* on slave ships some hundreds of years ago. It rapidly became entrenched in the tropical parts of South, Central and North America, including the United States. Thousands of Frenchmen died of yellow fever during the construction of the Panama Canal. The name of Walter Reed, the United States Army doctor who unraveled the epidemiology of this disease in 1900, is now legendary.

In its jungle habitat the virus is maintained in a monkey–mosquito cycle (Fig. 20-1, Plate 20-3). Significantly, Old World monkeys tend to be unharmed by the virus but New World monkeys often die, reflecting the more recent introduction of the virus to the Americas. Various species of jungle canopy-feeding, treehole-breeding mosquitoes serve as vectors—generally of the *Aedes* genus in Africa and the *Haemagogus* genus in the Americas. There is recent evidence that transovarial transmission may enable the virus to be maintained in mosquitoes—and perhaps even in ticks, from which it has been recovered in Africa. Some of these sylvatic mosquitoes, notably *Aedes furcifer-taylori, Aedes simpsoni,* and *Aedes africanus,* can also transmit the virus to man. However, *Aedes aegypti* is the vector responsible for most of the urban epidemics, in which the virus is maintained in a man–mosquito–man cycle.

A concerted campaign to eradicate *Aedes aegypti* from Brazil and the countries surrounding the Amazon basin has dramatically reduced the incidence of yellow fever in Latin America; the few cases still seen occur mainly in adult male forest workers. *Aedes aegypti* is still common

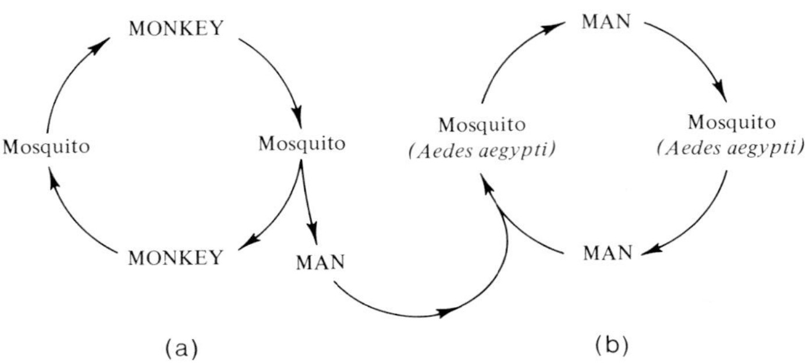

FIG. 20-1. *Jungle (a) and urban (b) cycles of yellow fever.*

throughout Central America and Southern United States. In endemic regions of West Africa infection of children is very common and usually subclinical or mild, most acquiring antibody by adulthood. Major epidemics continue to occur every few years in Africa—Sudan 1940, Ethiopia 1960–1962, Nigeria 1969, Gambia 1978–1979, Ghana/Burkina Faso 1983—an estimated 100,000 human cases with 30,000 deaths having occurred during the Ethiopian epidemic alone. One of the great puzzles of yellow fever is why it has apparently never occurred in Asia. There is some evidence that the Asian strain of *Aedes aegypti* may be a less competent vector for the yellow fever virus.

Dengue

By far the most widely distributed of the arboviruses, dengue, ranks also as the most important cause of morbidity and mortality today. Indeed, the dengue viruses appear to be expanding their range, probably as a consequence of increasing travel and urbanization. For the first time since World War II major epidemics involving hundreds of thousands of people have occurred in the Caribbean (1977–1981), the Pacific (1979), and China (1978–1980) as well as in its habitual strongholds in Southeast Asia and Africa. Furthermore, the disease is increasing in severity as DHF/DSS has emerged as a conspicuous feature of epidemics in Southeast Asia since 1954, and more recently in the 1981 outbreak in Cuba. It has been estimated that 700,000 children have been hospitalized and 20,000 have died from DHF/DSS in the past three decades. Most of these have been dengue type 2 infections in children with serological evidence of recent prior infection with another serotype.

Like urban yellow fever, urban dengue is transmitted from man to man by the domestic mosquito, *Aedes aegypti*. Until recently, no other

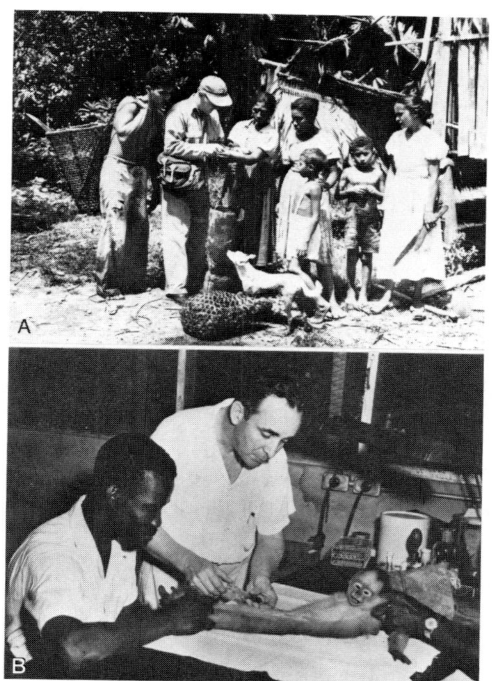

PLATE 20-3. *Epidemiological investigation of yellow fever. (A) Serological survey being conducted among Indians of the Amazon forest. (B) Monkey being bled in Trinidad. (Courtesy WHO and Cambridge University Press. From F. M. Burnet and D. O. White, "Natural History of Infectious Disease." Cambridge University Press, London and New York, 1972.)*

hosts were known but evidence has emerged for the existence of a sylvan cycle analogous to that of yellow fever, involving monkeys and several sylvatic *Aedes* species (Fig. 20-1). Transovarial transmission has been demonstrated experimentally in these forest mosquito species.

Chikungunya

This alphavirus is widespread throughout Africa, India, and Southeast Asia, as far east as the Philippines. It is thought to be maintained in a sylvan cycle with forest species of *Aedes* mosquitoes as vectors, and primates as the natural vertebrate host. Explosive urban outbreaks probably involve a man–*Aedes aegypti*–man cycle, as with dengue and yellow fever.

O'nyong-nyong

This close relative of chikungunya virus is thought to be spread by anthropophilic species of *Anopheles,* making it the only known human arboviral pathogen with anopheline mosquitoes as principal vector. The virus was responsible for an epidemic in East Africa extending from 1959 to 1962 affecting no less than two million people, then mysteriously disappeared.

Ross River

This mosquito-borne alphavirus causes a disease known as epidemic polyarthritis during late summer/early autumn in rural Australia. Explosive epidemics occurred in 1979 and 1980 on several isolated South Pacific islands, affecting up to half the native population, as well as tourists.

Japanese Encephalitis

JE virus is the world's major encephalitogenic arbovirus. Distributed throughout East Asia, its main vector is the mosquito *Culex tritaeniorhynchus,* and its natural host waterbirds such as herons. Domestic pigs are an important amplifier host. Encephalitis in Japan has declined in recent years following a multipronged control program (see below), yet the virus seems to be extending its range. India, Thailand, Vietnam, and China have experienced major epidemics in recent years. Outbreaks tend to occur in late summer/early autumn in temperate zones. In tropical areas by contrast, the virus is endemic, infecting most young children subclinically; outbreaks are occasionally seen at the end of the wet season.

Murray Valley (Australian) Encephalitis

The related flavivirus MVE is enzootic in Papua New Guinea and Northern Australia. Epidemics involving man occur in the Murray River Valley of Southeastern Australia only in occasional summers that follow heavy rainfall with extensive flooding. These conditions encourage explosive increases in the populations of the mosquito vectors and the natural hosts, waterbirds.

West Nile Fever

This flavivirus is believed to be maintained in a *Culex* mosquito–bird cycle. It is endemic in rural Africa, Southern Europe, and Central Asia, east to India.

St. Louis, Western Equine, and Eastern Equine Encephalitis

SLE, WEE, and EEE are all New World viruses, mosquito-borne, with birds as their principal reservoir. The ecology of SLE and WEE viruses was described in Chapter 9 (see also Fig. 9-2). EEE virus is maintained in a bird–*Culiseta melanura* cycle with rare epidemic transmission by other species of mosquito to horses and man, both of which are dead-end hosts. Outbreaks also occur in large caged birds, e.g., pheasants, among which the virus may be transmitted directly by pecking.

Venezuelan Equine Encephalitis

VEE virus is also confined to the Americas. Its enzootic cycle in the rain forests and marshes of South and Central America involves mosquitoes of the genus *Culex* and a number of mammals, principally rodents. During epizootics, however, many mosquito species of other genera are involved in transmitting the virus to horses and man in drier regions. The strains of virus recovered during epizootics tend to be more virulent than those found during interepidemic intervals.

Rocio Encephalitis

This new flavivirus, first isolated in 1975, has been associated with over 1000 cases of human disease including encephalitis in Brazil. A bird–mosquito cycle has been postulated.

Tick-Borne Flaviviruses

The tick-borne encephalitis (TBE) virus(es) (otherwise known as Russian spring-summer encephalitis and Central European encephalitis viruses) were discussed in Chapter 9. As illustrated in Fig. 9-3 the virus is passed transovarially and transstadially in ticks, transmitted to rodents and groundfeeding passerine birds as well as to domestic goats, sheep, and cattle, and acquired by man via tick bite or ingestion of milk. The closely related flavivirus of Northern Britain, louping ill, is also maintained in a tick–rodent cycle and is transmissible to sheep, in which it causes a rapidly lethal disease, or to man, in whom disease is rare. The Omsk hemorrhagic fever (OHF) virus of Siberia and the Kyasanur Forest disease (KFD) virus of India are also closely related to the tick-borne encephalitis viruses but cause hemorrhagic fever instead. In the case of KFD, rodents are again thought to constitute the principal vertebrate reservoir, but monkeys may serve as amplifiers, as do domestic cattle and goats, and the deer and antelopes in nature reserves, hence the concentration of human cases acquired along cattle tracks in Mysore

State forests. Powassan virus infects small mammals in Northern United States, Canada, and USSR and is rarely transmitted to man.

CONTROL

Control of arboviral disease rests upon (1) vector control and (2) vaccination. The tactics of long-term prevention differ from those of short-term containment in the face of an epidemic. Long-term prevention is based upon (1) immunization of the population if a safe effective vaccine is available and the risks of the disease merit such a large-scale program, and (2) elimination of potential mosquito-breeding sites. The reader is referred to Chapter 9, where practical approaches to "source reduction" were discussed. In the emergency posed by an epidemic, immunization is sometimes used but aerial spraying of insecticide is the major additional weapon. It is instructive to observe how successful the Japanese have been in controlling Japanese encephalitis by a combination of these approaches, notably widespread vaccination in infancy, draining of rice paddies at the time when the main vector, *Culex tritaeniorhynchus*, normally breeds, and removal of the principal amplifier host, pigs, from the vicinity of human habitation.

Vaccines

Years before the sophisticated technology of modern virology had been dreamt of, Max Theiler (who completed his medical course but never bothered to collect his degree) developed, on purely empirical grounds, a first-rate vaccine against yellow fever, for which he received the Nobel Prize. He isolated the virus from a monkey, passaged it through mice, then through various primary cell cultures and finally chick embryos, ending up with the avirulent 17D strain that comprises the present-day live vaccine. It is injected subcutaneously, after reconstitution from the lyophilized state. Side effects are minimal (mild headache and malaise in 5–10%). Protection is probably lifelong. Nevertheless, most yellow fever-free countries require that travelers through endemic areas have been vaccinated within the past 10 years. Within endemic areas vaccination policy varies greatly around the world. Some South American nations attempt to vaccinate the whole population as a matter of routine; most African nations do so only as an emergency response to an epidemic.

A formalin-inactivated, mouse brain-derived Japanese encephalitis vaccine is routinely administered to children in Japan and Korea and has dramatically reduced the incidence of the disease. China uses a formalin-killed vaccine derived from virus grown in cultured hamster kidney cells. No encephalitis or other serious side effects have been reported. Live cell culture-derived vaccines are used in horses and in the principal domestic animal reservoir, the pig, and have undergone field trials in man in China.

Formalin-inactivated vaccines are also available for use against tick-borne encephalitis (in Eastern Europe) and again VEE, EEE, and WEE (in the Americas, mainly for laboratory workers and horses).

Clearly, effective vaccines against all the lethal encephalitides and hemorrhagic fevers would be of potential benefit. So far, however, pharmaceutical companies have been reluctant to invest in the costly development of vaccines against those arboviruses that are confined to restricted areas of the tropics and/or that relatively rarely produce lethal or disabling disease.

The desirability of a vaccine against dengue must be considered in the light of the debate about the pathogenesis of DHF/DSS. If "immune enhancement" is a major factor then vaccination could turn out to be disastrous unless protective levels of antibodies against all four serotypes were routinely achieved and maintained for life. On the other hand, if most cases of DHF/DSS are due to superinfection with dengue 2, it may only be necessary to vaccinate against that serotype. An experimental live dengue 2 vaccine based on a *ts*, small-plaque attenuated mutant induces too high an incidence of side effects yet is insufficiently immunogenic (except in those with preexisting yellow fever antibody).

RUBELLA

Though taxonomically a member of the *Togaviridae* family (Plate 20-4A) rubella virus is not an arbovirus. It is conceivable that it once was, but today it infects only man and spreads only by the respiratory route. Rubella, or German measles as it was once known, is a trivial exanthem of childhood. However, in 1941 an Australian ophthalmologist discovered the true significance of this deceptively mild disease. Gregg noticed an unusual concentration of cases of congenital cataract among newborn babies in his practice—an epidemic of blindness. A diligent search of his records revealed that most of the mothers had contracted rubella in the first trimester of pregnancy. Further investigations revealed that these unfortunate children had also suffered a range of other congenital defects including deafness, mental retardation, and cardiac abnormalities.

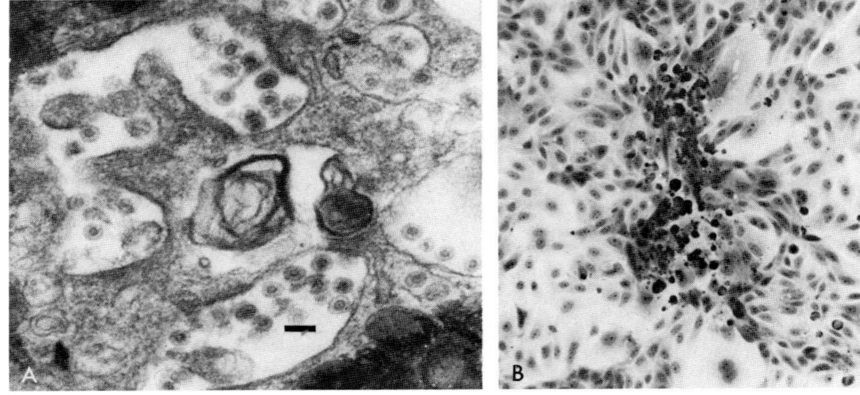

PLATE 20-4. *Rubella. (A) Electron micrograph showing budding of rubella virions into cytoplasmic vesicles. Bar = 100 nm. (B) CPE produced by rubella virus in RK-13 cells (H and E stain, ×60). CPE is minimal, being initially restricted to microscopic "plaques." [Courtesy Dr. I. H. Holmes (A) and I. Jack (B).]*

Clinical Features

Rubella. In children, rubella is such a mild disease that many women are subsequently uncertain about whether they have ever contracted it. The fine, pink, discrete macules of the erythematous ("rubelliform") rash appear first on the face then spread to the trunk and limbs and fade after 48 hours or less. In nearly half of all infections there is no rash at all. Fever is usually inconspicuous and coryza or conjunctivitis may or not be present. A characteristic feature is that postauricular, suboccipital, and posterior cervical lymph nodes are enlarged and tender from very early in the illness. Mild polyarthritis, usually involving the hands is a fairly frequent feature of the disease in adult females; it is usually fleeting but may rarely persist for up to several years. Thrombocytopenic purpura, neuritis, and encephalitis are rare complications. Progressive rubella panencephalitis (PRP) is an even rarer and inevitably fatal complication, developing insidiously in the second decade of life. The analogy with SSPE is striking, but most cases of PRP arise in children with congenital rubella.

Congenital Rubella. At least 20% of all infants infected *in utero* during the first trimester of pregnancy are born with severe congenital abnormalities, usually multiple (Plate 20-5). The commonest are congenital heart disease (especially patent ductus arteriosus, sometimes accompanied by pulmonary stenosis or septal defects), mental retardation (±

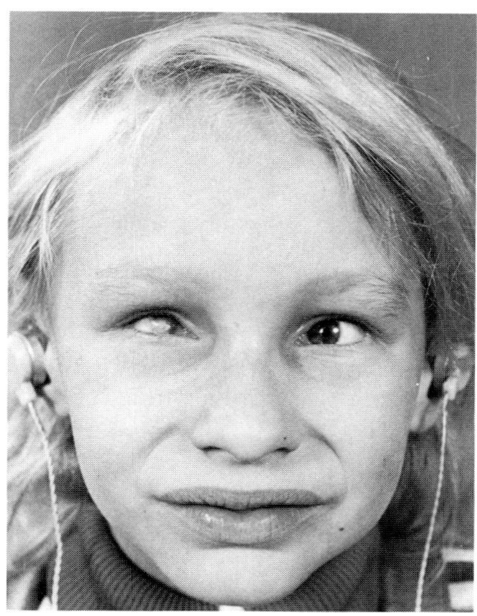

PLATE 20-5. *Rubella syndrome: showing severe bilateral deafness and severe bilateral visual defects (microphthalmia, cataract, corneal opacity, and strabismus). (Courtesy Dr. K. Hayes.)*

microcephaly), deafness (total or partial, due to cochlear degeneration, becoming progressively apparent in the early years after birth), and blindness (total or partial, especially cataracts, but sometimes glaucoma, microphthalmia, or retinopathy). Other common manifestions of the so-called *congenital rubella syndrome* (CRS), not strictly malformations, include hepatomegaly, splenomegaly, osteitis and retardation of growth, and thrombocytopenic purpura. Despite the diversity and severity of this pathology, the rubella syndrome is sometimes missed at birth. This may seriously jeopardize the child's medical, social, and educational prognosis. About 10–20% of babies with CRS die during the first year, and up to 80% of all infected babies develop some evidence of disease within the early years of life.

Pathogenesis and Immunity

The virus enters the body by inhalation to multiply asymptomically in the upper respiratory tract and then spread via local lymph modes to the bloodstream. Not a great deal is known about the events that occupy the 18-day (range: 2–3 weeks) incubation period that precedes the appearance of the rash. Cervical and occipital lymph nodes are swollen and virus is shed from the respiratory tract. Virus has been found in lymphocytes or synovial fluid of adults with rubella arthritis and in the brain of patients with PRP, together with high titers of antibody and immune complexes. Acquired immunity to rubella lasts for many years. Second infections occur occasionally but are usually subclinical and boost immunity further. When rubella virus infects a woman during the first trimester of pregnancy there is a high chance that the baby will suffer congenital abnormalities. Severe damage (blindness, deafness, heart or brain defects) occurs in 20–30% of all infections during the first trimester and in about 5% of those in the fourth month, but rarely thereafter. Minor abnormalities are even more frequent, and following spontaneous abortion or stillbirth, virus can be found in practically every organ.

What makes rubella virus teratogenic, when numerous nonteratogenic viruses are so much more pathogenic postnatally and so much more cytocidal for cultured cells? Paradoxically, this relative lack of pathogenicity may hold the clue to its teratogenicity. More cytocidal viruses may destroy cells and kill the fetus leading to spontaneous abortion. Rubella virus may merely slow down the rate of cell division, as has been demonstrated in cultured human fetal cells, leading to a decrease in overall cell numbers and accounting for the small size of rubella babies. Moreover, death of a small number of cells or slowing of their mitotic rate during the critical period of embryonic differentiation might interfere with the formation of key organs such as the heart, brain, eyes, and ears, all of which are being formed during the first trimester.

There is also clear evidence that neither the mother's nor the baby's immune response is able to clear the virus from the fetus. Although maternal IgG crosses the placenta and the infected fetus manufactures its own IgM antibodies (Fig. 20-2), cell-mediated immune responses are defective and, together with NK responses, remain so postnatally. Clones of infected cells may escape immune cytolysis even though maternal IgG might restrict systemic spread of virus. Whatever the explanation, the rubella syndrome in the fetus is a true persistent infection of the chronic type (see Chapters 5 and 7). Throughout the pregnancy and for several months after birth the baby continues to shed virus in any or all of its secretions.

Laboratory Diagnosis

Rubella virus defied cultivation in the laboratory until 1962 when it was isolated in primary cultures of African green monkey kidney cells,

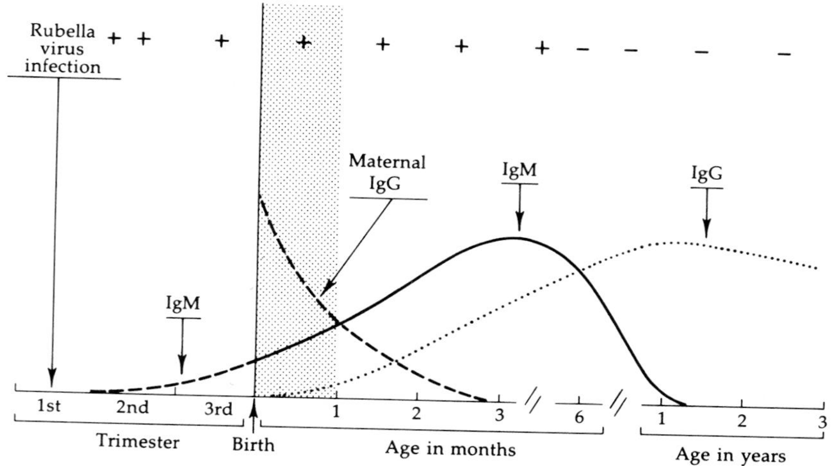

FIG. 20-2. *Schematic illustration of the pattern of viral excretion and antibody response in congenital rubella. (From S. Krugman and R. Ward, "Infectious Diseases of Children and Adults," 5th ed., p. 245. Mosby, St. Louis, 1973.)*

causing no CPE but inducing a state of interference demonstrable by challenge with an unrelated virus. Since then, rubella virus has been found to produce CPE in certain continuous cell lines, notably RK-13 and Vero. Even so, these changes are inconspicuous, commencing as isolated foci, spreading slowly, and rarely involving the whole cell sheet (Plate 20-4B). CPE can be augmented by serial passage, e.g., in BHK-21 cells, and intracellular antigen is identified by immunofluorescence or immunoperoxidase staining.

Because of its lack of sensitivity, isolation in cell culture is not used as a routine method of diagnosing rubella. However, there are two situations in which it does have a place: (1) virus can be isolated from the throat, urine, CSF, leukocytes, or (at autopsy) virtually any organ of an infant with the congenital rubella syndrome; (2) amniocentesis is sometimes used to obtain amniotic fluid from a woman wishing to undergo an abortion only if rubella virus is isolated.

Serology constitutes the standard approach to the laboratory diagnosis of rubella. Fortunately, the few days delay involved does not matter in any of the three common situations in which the clinician requires help from the laboratory:

1. A woman considering vaccination wishes to know whether she has ever had rubella and is now immune.

2. An unimmunized woman develops a rash in the first trimester of pregnancy, or comes into contact with someone with rubella, and wishes to know whether she has contracted the disease and whether she should have an abortion.

3. A baby is born with signs suggestive of the rubella syndrome, or its mother is now believed to have contracted rubella during the first trimester.

The market for rubella serology is such that over 20 diagnostic kits, based on nearly half that number of distinct serological procedures, have been released by various companies in the past few years. Not unnaturally, the family physician feels overwhelmed by the options and cannot be expected to know how to choose between them, let alone what constitutes a meaningful titer. The situation illustrates better than any other the explosion of new serological techniques, and of do-it-yourself kits for clinicians, that has burst upon the market in the 1980s.

Traditionally, hemagglutination-inhibition (HI) has been the preferred procedure for identifying and quantitating rubella antibodies in serum, and it remains today the standard against which newer techniques are compared for sensitivity and specificity. HI (using goose RBC, at 4°C) is sensitive but technically tricky, requiring pretreatment of the patient's serum (with heparin/manganous chloride or dextran sulfate/calcium chloride, followed by heating at 56°C for 30 minutes) to remove nonspecific inhibitors of hemagglutination. Accordingly,HI is being supplanted by a number of procedures with comparable or superior sensitivity which do not require serum pretreatment. The chief competitors are enzyme immunoassay, passive (indirect) hemagglutination, radial hemolysis, immunofluorescence, and latex agglutination. The reader is referred to Chapter 12 for descriptions of the principles, methodology, advantages, and disadvantages of each of these procedures.

As there is but one serotype of rubella virus, and specific IgG continues to be demonstrable in serum for many years after clinical or subclinical infection, detection of antibody by any of these methods is evidence of immunity (see the first of the clinical situations set out above). The second situation—diagnosis of rubella in a pregnant woman—requires the demonstration of either (1) a rising titer of rubella antibody between two specimens of serum taken 10 days or more apart, or (2) rubella antibody of the IgM class in a single specimen.

Several of the serological techniques mentioned above can be adapted to the detection and quantitation of specific IgM. For instance, an appropriate EIA could be set up as follows: mouse monoclonal anti-human IgM (anti-μ) as the capture antibody adherent to the plate; patient's serum (preabsorbed with protein A-Sepharose to remove rheumatoid factor and most of the IgG) as the second layer; rubella antigen as the third; rabbit anti-rubella IgG as the fourth; finally, an enzyme-labeled goat anti-rabbit IgG as the indicator, followed by an appropriate substrate.

Rubella IgM antibody is always demonstrable by 1 week after the appearance of the rash and persists for at least 1 month (occasionally 3 months). IgM serology may be the only method of diagnosis in the case of the woman who first consults her doctor weeks after the rash has gone, when it may no longer be possible to demonstrate a rising titer of (total) rubella antibody.

Diagnosis of the rubella syndrome in a newborn baby is also made by demonstrating rubella IgM in a single specimen of serum—this time the baby's. Any baby will have rubella antibodies of the IgG class, acquired transplacentally from the mother who may have been infected with the virus years earlier. IgM antibodies not only indicate recent infection but do not cross the placenta. Hence rubella IgM detected in umbilical cord blood must have been synthesized by the baby itself *in utero* and is diagnostic of congenital rubella. It continues to be made in detectable amounts for 3–6 months after birth.

Epidemiology

Rubella virus is shed in oropharyngeal secretions and is presumed to be spread by the respiratory route. It is highly transmissible, usually being acquired by school-age children and spread readily within the family. The disease is endemic in all countries of the world.

Prior to the introduction of vaccines, spring epidemics occurred every few years in temperate climates and 80–95% of women had acquired immunity by the time they reached the child-bearing years. Widespread vaccination of infants in the United States commencing in 1969 has significantly reduced the incidence of rubella in that country and raised the average age of acquisition of the disease to late adolescence/early adulthood.

Control

Live attenuated rubella vaccines have been in use since 1969. The two strains most widely used today are Cendehill-51 (derived by 51 serial passages in rabbit kidney cells) and RA27/3 (derived by 27 passages in a human diploid cell line). Either is administered as a single subcutaneous injection. The virus multiplies in the body and is shed in small amounts from the throat but is incapable of spreading to contacts. It induces durable immunity in at least 95% of recipients. Antibody titers are significantly lower than following natural infection and, occasionally, subclinical reinfection occurs several years later, thereby boosting immunity further.

Though substantially attenuated, rubella vaccines commonly induce one significant side effect, namely arthralgia. This is seen mainly in adult females, is usually confined to the small peripheral joints, and only occasionally progresses to frank arthritis, which even more rarely may persist for some years. Lymphadenopathy, a fleeting rash, and low-grade fever are not uncommon.

There is no evidence that the vaccine virus is teratogenic. Hence accidental immunization during pregnancy is not considered to be an indication for termination. Nevertheless, prudence suggests that women not be deliberately immunized during the first trimester and that nonpregnant vaccinees should be advised to avoid conception for the next 2–3 months, and/or be immunized immediately postpartum.

From the outset in 1969, two opposing schools of thought dictated the strategies of rubella vaccine programs in different parts of the world. In the UK and Australia for example, the primary target population has been teenage females; the policy has been to immunize all girls on entry into high school (i.e., usually 12–13 years of age). By contrast, the United States has aimed to immunize all children, male and female, early in the second year of life. An ancillary feature of both policies has been to immunize seronegative females of child-bearing age.

The rationale of the UK approach is that, because naturally acquired immunity is demonstrably more effective and more lasting than that induced by any vaccine, it is desirable not to interfere with the circulation of wild virus in the community but to allow it to infect the majority of children. Immunization of pubescent girls is therefore intended to protect the minority who have not acquired immunity naturally by the time they enter high school. This cohort will be protected during the crucial child-bearing years more efficiently than had they been vaccinated as infants, because immunity is known to decay within the years following immunization. Furthermore, the immunity of these women will be boosted further by subclinical infections they may receive because the wild virus is still circulating.

The logic of the United States policy, on the other hand, is that universal immunization of all children in the second year of life could theoretically eradicate the virus entirely from the community, as has been achieved with smallpox, and very nearly with poliomyelitis and measles, following similar strategies.

Neither strategy can be deemed an unqualified success. As would be expected from the fact that rubella circulates mainly among children of preschool and primary school age, the UK program has had no demonstrable effect on the overall incidence of rubella, which in itself does not matter. Disappointingly, however, it has not yet led to a major decline in the incidence of the congenital rubella syndrome (CRS), probably because the coverage of teenage girls has been inadequate and most women in the 25–45 age group have been missed entirely. The United States program, on the other hand, has led to a substantial decline in the incidence of rubella in children, but not in adolescents or adults, and has had only a modest impact on the occurrence of CRS. This too is not unexpected as the cohort at which the vaccine was targeted during the 1970s is only now moving into the child-bearing years.

Accordingly, the United States Centers for Disease Control have advocated a major effort to increase the rate of vaccination of infants (male and female) to at least 90%, while simultaneously mounting a catch-up campaign to immunize any children previously missed, as well as 15 to 40-year-old women. The latter can most conveniently be vaccinated on entry into educational establishments or workplaces, family planning clinics, or postpartum. Laboratory determination of antibody status is not essential, particularly if the cost of the vaccine is lower than that of the serological test.

Babies with congenital rubella shed substantial amounts of virus from their throats for up to several months after birth and constitute a considerable risk to pregnant women, directly, or via infected nursing staff in maternity hospitals or postnatal clinics. They should be nursed in isolation, preferably by vaccinated or naturally immune personnel.

There remains the difficult problem of what to do with the woman who does in fact contract a laboratory-confirmed rubella infection during the first 3–4 months of pregnancy. There is general agreement that the risk of serious permanent damage to the baby is so substantial that the attending physician should recommend an abortion.

FURTHER READING

Alphaviruses and Flaviviruses

Books

Beran, G. W., ed. (1981). "Viral Zoonoses," CRC Hand. Ser. Zoonoses, Sect. B; Vol. I. CRC Press, Boca Raton, Florida.

Berge, T. O., ed. (1975). "International Catalogue of Arboviruses," DHEW Publ. No. (CDC) 75–8301. U.S. Dept. of Health, Education and Welfare, Washington, D.C.; also see Supplement edited by Karabatsos (1978).

Gibbs, A. J., ed. (1973). "Viruses and Invertebrates." North-Holland Publ., Amsterdam.

Horzinek, M. C. (1981). "Non-Arthropod-Borne Togaviruses." Academic Press, London.

Karabatsos, N., ed. (1978). "Supplement to International Catalogue of Arboviruses Including Certain Other Viruses of Vertebrates." *Am. J. Trop. Med. Hyg.* **27,** 372–440.

Schlesinger, R. W. (1977). "Dengue Viruses," *Virol. Monogr.,* Vol. 16. Springer-Verlag, Berlin and New York.

Schlesinger, R. W., ed. (1980). "The Togaviruses: Biology, Structure, Replication." Academic Press, New York.

Theiler, M., and Downs, W. G. (1973). "The Arthropod-Borne Viruses of Vertebrates: An Account of the Rockefeller Foundation Virus Program 1951–1970." Yale Univ. Press, New Haven, Connecticut.

Walsh, J. A., and Warren, K. S., eds. (1983). "The Great Neglected Diseases of Mankind: Strategies for Control." Univ. of Chicago Press, Chicago, Illinois.

Warren, K. S., and Mahmoud, A. A. F., eds. (1984). "Tropical and Geographic Medicine." McGraw-Hill, New York.

Reviews

Garoff, H., Kondor-Koch, C., and Riedel, H. (1982). Structure and assembly of alphaviruses. *Curr. Top. Microbiol. Immunol.* **99,** 1.

Grimstad, P. R. (1983). Mosquitoes and the incidence of encephalitis. *Adv. Virus. Res.* **28,** 357.

Halstead, S. B. (1981). The pathogenesis of dengue. *Am. J. Epidemiol.* **114,** 632.

Halstead, S. B. (1984). Selective primary health care: Strategies for control of disease in the developing world. XI. Dengue. *Rev. Infect. Dis.* **6,** 251.

Rosen, L. (1982). Dengue—an overview. *In* "Viral Diseases in South-East Asia and the Western Pacific" (J. S. Mackenzie, ed.), pp. 484–493. Academic Press, Sydney, Australia.

Simons, K., Garoff, H., and Helenius, A. (1982). How an animal virus gets into and out of its host cell. *Sci. Am.* **246,** 46.

Strauss, E. G., and Strauss, J. H. (1983). Replication strategies of the single-stranded RNA viruses of eukaryotes. *Curr. Top. Microbiol. Immunol.* **105,** 1.

Tesh, R. B. (1982). Arthritides caused by mosquito-borne viruses. *Annu. Rev. Med.* **33,** 31.

Westaway, E. G. (1980). Replication of flaviviruses. *In* "The Togaviruses: Biology, Structure, Replication" (R. W. Schlesinger, ed.), pp. 531–583. Academic Press, New York.

Woodall, J. P. (1981). Summary of a symposium on yellow fever. *J. Infect. Dis.* **144,** 87.

Rubella

Centers for Disease Control (1984). Rubella prevention: Recommendation of the Immunization Practices Advisory Committee. *Ann. Intern. Med.* **101,** 505.

Lennette, E. H., ed. (1984). "Laboratory Diagnosis of Viral Infections." Dekker, New York.

Preblud, S. R., Serdula, M. K., Frank, J. A., Brandling-Bennett, A. D., and Hinman, A. R. (1980). Rubella vaccination in the United States: A ten-year review. *Epidemiol. Rev.* **2,** 171.

CHAPTER 21

Orthomyxoviruses

Introduction 509
Properties of *Orthomyxoviridae* 510
Clinical Features of Influenza 513
Pathogenesis and Immunity 513
Laboratory Diagnosis 514
Epidemiology 515
Control 516
Further Reading 519

INTRODUCTION

Few viruses have played a more central role in the historical development of virology than that of influenza. The pandemic that swept the world in 1918, just as the Great War ended, killed 20 million people—more than the war itself. The eventual isolation of the virus in ferrets by Smith, Andrewes, and Laidlaw in 1933 was a milestone in the development of virology as a laboratory science. During the ensuing two decades Burnet pioneered technological and conceptual approaches to the study of the virus in embryonated eggs. His system became the accepted laboratory model for the investigation of viral multiplication and genetic interactions until the early 1950s, when newly discovered cell culture techniques transferred the advantage to poliovirus. Hemagglutination, discovered accidentally by Hirst when he tore a blood vessel while harvesting influenza-infected chick allantoic fluid, provided a simple assay method, subsequently extended to many other viruses. The imaginative investigations of Webster and Laver into the continuing evolution of influenza virus by antigenic drift and shift established the discipline of molecular epidemiology. The pièce de résistance was the description by Wiley, Wilson, and Skehel of the location of the antigenic sites on a three-dimensional model of the influenza HA molecule (Fig. 4-5).

PROPERTIES OF *ORTHOMYXOVIRIDAE*

The typical virion is spherical and about 100 nm in diameter, but larger, more pleomorphic forms are commonly seen (Plate 21-1). Filaments up to several microns in length are characteristic of many strains upon first isolation.

The envelope of influenza A and B contains two distinct types of glycoprotein: a hemagglutinin (HA) and a neuraminidase (NA). In influenza type C, both functions are carried by a single glycoprotein, which packs in a regular hexagonal array, giving its envelope a symmetry not seen with A and B.

The single-stranded RNA genome, which has negative polarity, occurs as eight separate molecules (seven for influenza C), most of which code for a single protein. The associated nucleoprotein has loose helical symmetry (Table 21-1).

Classification

The *Orthomyxoviridae* family contains only the viruses of influenza. The three "types," A, B, and C, are properly regarded as species since

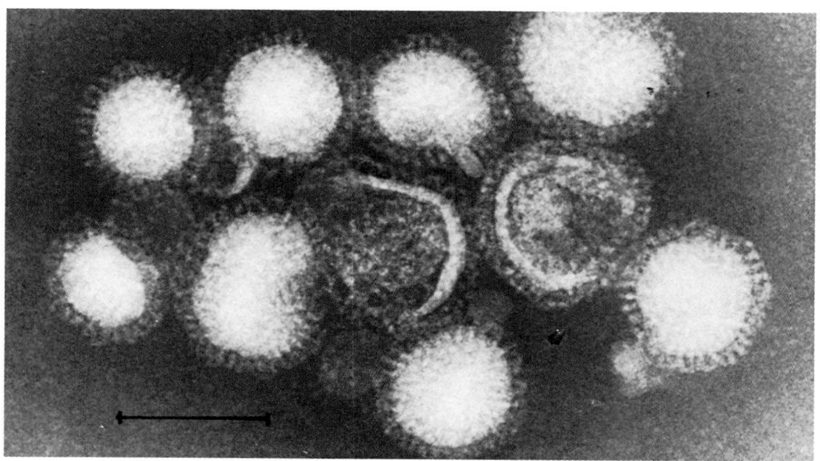

PLATE 21-1. Orthomyxoviridae. *Negatively stained preparation of virions of influenza A virus. Bar = 100 nm. (Courtesy Dr. F. A. Murphy.)*

TABLE 21-1
Properties of Orthomyxoviridae

Pleomorphic spherical or filamentous virion, diameter 80–120 nm
Envelope, containing hemagglutinin and neuraminidase
Nucleoprotein of helical symmetry in segments
ssRNA genome, linear, negative polarity, (7)–8 molecules, total MW 5 million
Capped 5′-termini of cellular RNA cannibalized as primers for mRNA transcription
Nuclear transcription; cytoplasmic maturation by budding from plasma membrane
DI particles and genetic reassortment frequent

they share no common antigens; indeed influenza C is sufficiently distinct to be classified as a separate genus. Influenza A is further divided into subtypes, all of which share a common NP (nucleoprotein) and M_1 (matrix protein) but differ in either their HA or NA. So far, 13 subtypes of HA (H1–H13) and 9 of NA (N1–N9) have been described in birds, animals, or man. Because novel subtypes of influenza A can arise by genetic reassortment, any combination of HA and NA subtypes is theoretically possible. Within each subtype, strains evolve by antigenic drift (Chapter 4). Thus a strain of influenza type A first isolated in Bangkok in 1979 and found to carry HA of subtype 3 and NA of subtype 2 would be designated A/Bangkok/79 (H3N2).

Viral Multiplication

Influenza virus adsorbs via its HA to sialyloligosaccharides in membrane glycoproteins and glycolipids. Following endocytosis of the virion, a conformational change in the HA molecule, occurring only at the pH (5) obtaining in the endosome, facilitates membrane fusion, thus triggering uncoating. The nucleocapsid migrates to the nucleus where viral mRNA is transcribed by a unique mechanism. A viral endonuclease cleaves the capped 5′-terminus from cellular heterologous RNAs and these serve as primers for transcription by the viral transcriptase. Of the 8 cRNA molecules so produced, 6 are monocistronic mRNAs which are translated directly into the proteins representing HA, NA, NP, and the various components of the viral polymerase. The other 2 cRNA molecules undergo splicing, each yielding 2 mRNAs which are translated in different reading frames. In other words, the 8 viral RNA molecules code for 10 proteins: 7 structural and 3 nonstructural. (It will be recalled from Chapter 3 that splicing is otherwise confined to DNA viruses and the retroviruses, which utilize a DNA intermediate.) The sophistication of our understanding of the genes of influenza virus and their protein products is illustrated by the precision of the data in Table 21-2 (see also Plate 3-1).

The HA and NA become glycosylated (Fig. 3-10). In permissive cells, HA is eventually cleaved, but the two resulting chains remain united by disulfide bonds (Fig. 4-5). The virus matures by budding from the apical surface of the cell (Fig. 3-11). It is not known by what mechanism one copy of each of the 8 genes is selected for incorporation in each new virion. DI particles, originally known as "incomplete virus," are quite often produced, especially following infection at high multiplicity (see Chapter 4).

TABLE 21-2

Influenza Virus Genome RNA Segments and Coding Assignments[a]

SEGMENT	LENGTH[b] (NUCLEOTIDES)	ENCODED POLYPEPTIDE[c]	LENGTH[d] (AMINO ACIDS)	MOLECULES PER VIRION	COMMENTS
1	2341	PB2	759	30–60	RNA transcriptase component; host cell RNA cap binding
2	2341	PB1	757	30–60	RNA transcriptase component; initiation of transcription; endonuclease activity?
3	2233	PA	716	30–60	RNA transcriptase component; elongation of mRNA chains?
4	1778	HA	566	500	Hemagglutinin; trimer; envelope glycoprotein; mediates attachment to cells
5	1565	NP	498	1000	Nucleoprotein; associated with RNA; structural component of RNA transcriptase
6	1413	NA	454	100	Neuraminidase; tetramer; envelope glycoprotein
7	1027	M_1	252	3000	Matrix protein; lines inside of envelope
		M_2	96		Nonstructural protein in plasma membrane; spliced mRNA
		?	?9		Peptide predicted by nucleotide sequence only; spliced mRNA
8	890	NS_1	230		Nonstructural protein; function unknown
		NS_2	121		Nonstructural protein; function unknown; spliced mRNA

[a] Adapted from R. A. Lamb and P. W. Choppin (1983). Reproduced, with permission, from the *Annual Review of Biochemistry,* Volume **52,** 467–506, © 1983 by Annual Reviews Inc.

[b] For A/PR/8/34 strain.

[c] Determined by biochemical and genetic approaches.

[d] Determined by nucleotide sequence analysis and protein sequencing.

CLINICAL FEATURES OF INFLUENZA

"Flu" is a term much abused by the layman, and even, one is ashamed to relate, by some of one's medical brethren. Particularly if a few days off work are involved, any sore throat or common cold is likely to be mislabeled influenza. In reality, influenza is a distinct clinical entity with quite prominent systemic manifestations, including fever, chills, headache, generalized muscular aches, even prostration. There is a sore throat, which may be accompanied by hoarseness and cough, but nasal obstruction and discharge are not regular features as with the common cold.

Serious complications are generally confined to vulnerable target groups. Infants quite commonly develop croup. The greater danger however is pneumonia. Uncomplicated viral pneumonitis is unusual; bacterial superinfection by *Staphylococcus aureus, Streptococcus pneumoniae,* or *Haemophilus influenzae* is usually implicated. The elderly and debilitated are the most vulnerable—particularly those with preexisting pulmonary or cardiac disease. During influenza epidemics, epidemiologists note an increase in the overall mortality statistics ("excess deaths") among the aged, particularly in old peoples' homes. Though not usually confirmed by virus isolation, influenza is circumstantially incriminated as the straw that breaks the camel's back.

PATHOGENESIS AND IMMUNITY

The reader is referred back to Chapter 5 for a detailed description of the pathogenesis of influenza.

The genetic basis of virulence in influenza virus has proved to be quite elusive. It has been shown that a single amino acid substitution in the cell-attachment site at the tip of the HA molecule can determine the capacity of avian influenza viruses to replicate in the intestine of ducks. Since virus with uncleaved HA is noninfectious it is not surprising to find also that strains with HA that is insusceptible to the proteases in certain tissues are less pathogenic. As might be expected, however, several other genes can affect virulence in ways we do not yet understand. Why, for example, is fowl plague virus so much more virulent than other avian influenza strains? Why was the 1918 strain of human H1N1 ("Swine flu") far more lethal than the 1957 H2N2 ("Asian flu") and 1968 H3N2 ("Hong Kong flu") both of which infected the majority of people on the face of the globe but killed relatively few? Somewhat alarming in this regard was an apparently unprecedented outbreak of lethal influenza in seals near Boston in 1979–1980. The virus resembled that of fowl plague (H7N7) yet was found to replicate readily in various mammals and caused conjunctivitis in man.

There is debate about which arm of the immune response is most significant in influenza. HI antibodies of the IgA class in the mucus coating the respiratory tract are probably the most important in neutralizing the infectivity of virus, but serum IgG may also make a contribution. Antibodies to the neuraminidase inhibit the release of newly synthesized virions from the cell surface. Cytotoxic T cells bring about recovery from established infection by lysing infected cells. Interferons and alveolar macrophages doubtless also play a part.

LABORATORY DIAGNOSIS

Immunofluorescence may be used for rapid diagnosis of influenza; the specimen would be exfoliated cells from a nasopharyngeal aspirate or sputum, or lung at autopsy.

The developing chick embryo was for two decades the standard host for isolating influenza viruses, and is still for vaccine production. Today however, most laboratories prefer primary monkey kidney (PMK) or Madin–Darby canine kidney (MDCK) cells, because they are generally more sensitive than eggs for primary isolation, especially for influenza type B. Trypsin must be incorporated in the medium of MDCK cultures in order to cleave the viral HA molecule. Growth of influenza virus, at 33–35°C, in cultured cells is recognized by hemadsorption of guinea pig red blood cells after 3–7 days (Plate 2-2D). The isolate is then identified by hemagglutination-inhibition (Plate 12-7).

WHO reference laboratories are constantly on the lookout for novel strains of influenza, particularly at the start of an epidemic. Such laboratories employ eggs as well as PMK because certain subtypes in the past have been recovered more readily in eggs and this may well happen again. Chick embryos are inoculated by both the amniotic and allantoic routes (Fig. 2-1) and the fluids tested for hemagglutination after 3 days. Since a novel subtype arising by antigenic shift may have a new neuraminidase as well as a new hemagglutinin, the isolated virus is analyzed by neuraminidase-inhibition as well as HI tests. Immunodiffusion (Plate 12-2), or a refinement known as single radial immunodiffusion (see Chapter 12), is also able to detect major or minor changes arising by antigenic shift or drift, using reference antisera raised against purified HA and NA.

HI or EIA can also be used to detect rises in serum antibody to a given influenza type or subtype. The problem of "original antigenic sin" (whereby the predominant antibodies elicited by the current strain of virus are against the first strain of influenza virus one experienced years earlier and which are recalled anamnestically) complicates the identification of strain-specific antibodies.

EPIDEMIOLOGY

The influenza A viruses are constantly evolving by the processes discussed at length in Chapter 4. To recapitulate briefly, influenza A occurs naturally in many species of birds and mammals, notably waterbirds, pigs, and horses, as well as man. In ducks, for example, all 13 HA subtypes and 9 NA subtypes have been found in various combinations. Several may cocirculate in the same flock; pondwater often contains several subtypes, and cloacal swabs not uncommonly reveal more than one in a single bird. Clearly then, opportunities abound for genetic reassortment to occur. This is thought to be the mechanism of the antigenic shift which produced "new" subtypes of human influenza virus, causing the pandemics of 1918 (H1N1), 1957 (H2N2) and 1968 (H3N2). The HA of A/Hong Kong/68 (H3N2) differed in nearly 70% of its amino acids from the strain of human H2N2 prevalent immediately prior to 1968 (Fig. 4-6) but was barely distinguishable from the HA of influenza A/Duck/Ukraine/63 (H3N8) or A/Equine/Miami/63 (H3N8). The other 7 genes of human H3N2 were those of human H2N2. It is intriguing that the last three major shifts in human influenza A appear to have originated in China, where the dense, largely rural human population lives in close apposition to ducks and pigs. Genetic reassortment in influenza is readily demonstrable experimentally in eggs, cultured cells, and laboratory and wild animals and birds, and is now occurring naturally in man (see below).

The origin of the latest shift is more of a mystery. The H1N1 subtype which reemerged in man in China in 1977 is virtually identical with the H1N1 strain that was dominant in man until it suddenly disappeared in 1950. Perhaps H1N1 persisted in some animal or avian reservoir for 27 years. Since 1977, H1N1 and H3N2 have cocirculated in human communities, both causing their share of illness (Fig. 21-1). Furthermore, on a number of occasions they have undergone genetic reassortment to yield novel viruses with various constellations of the 8 genes.

Within each subtype, antigenic drift proceeds continuously at a steady rate. Nucleotide sequencing of the HA gene, or amino acid sequencing of the HA protein, reveals that roughly 1% of the amino acids are replaced each year. When substitutions have accumulated in all the major antigenic sites (Fig. 4-5) so that antibody to the previously prevalent strain no longer neutralizes the mutant satisfactorily, a new strain is born. Deletions and insertions occur in addition to point mutations.

Influenza types B and C show much less antigenic diversity than does type A. They have never been observed to cause pandemics, nor to undergo antigenic shift to generate new subtypes. Influenza B tends to circulate in human communities each year, undergoing relatively slow antigenic drift, infecting mainly children. Now and again it may cause quite large outbreaks in school children, with excess deaths in old people as occurred in the United States in 1979–1980. It is not known to infect other animals naturally.

Influenza C virus infects most humans early in life but is an insignificant cause of disease. It is also found in pigs.

Epidemics of influenza A occur every year or two (Fig. 21-1) and influenza B somewhat more sporadically. Most of these epidemics are in winter and run their course in a given city within 2–3 months, having infected more than half the population. Such speed and efficiency of transmission may be attributed to the large numbers of virions shed in droplets discharged by sneezing and coughing, the extremely short incubation period (1–2 days), and the lack of immunity to a novel subtype arising by antigenic shift, or inadequate immunity to strains arising by drift. As up to half the infections are subclinical, and many of those people with symptoms also remain in circulation, the epidemic develops rapidly. The school is the principal marketplace for trading parasites and the children bring the virus home to the family.

CONTROL

No disease illustrates better than influenza the difficulties of control by immunization and the ingenuity required to overcome them. Existing vaccines are continually being rendered obsolete by antigenic shift and drift. WHO, through its extensive network of laboratories around the globe, is ever on the alert for such changes, and supplies the vaccine manufacturers with the latest strain(s) of influenza so that their product may be updated if necessary prior to each flu season.

The current strains of influenza A and B are grown (separately) in the allantois of the chick embryo, then purified by zonal ultracentrifugation, inactivated by formaldehyde or β-propiolactone, and pooled. The resulting polyvalent inactivated vaccine is inoculated intramuscularly each

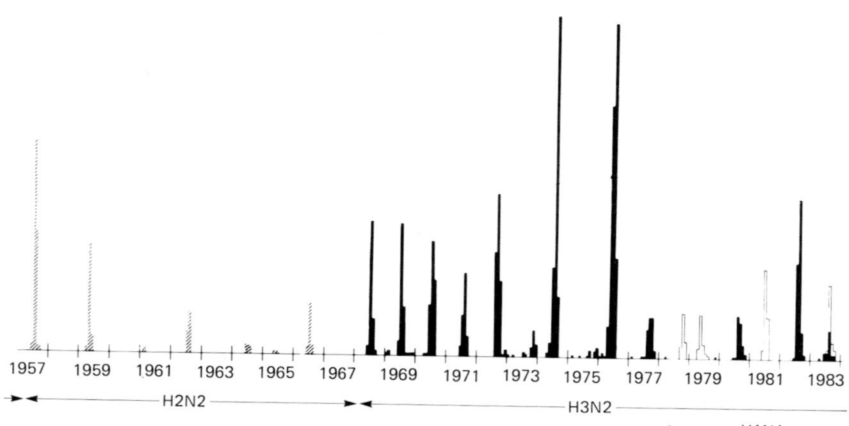

FIG. 21-1. *Epidemic occurrence of influenza A. The histograms show the monthly isolations of influenza A viruses from patients admitted to Fairfield Hospital for Infectious Diseases, Melbourne from 1957 to 1983. Note annual or biennial epidemics in the Australian winter (June to August). The major peak in 1957 represents the country's first experience of the H2N2 subtype which caused the 1957 pandemic of "Asian flu." The next major peak corresponds with the 1968/1969 pandemic of "Hong Kong flu," a new subtype (H3N2). The sporadic winter outbreaks since then, sometimes quite extensive as in 1974 and 1976, have been due to a succession of H3N2 variants arising by genetic drift. However, almost all of the cases in 1978, 1979, and 1981 were of the H1N1 subtype, which reemerged after being totally absent since 1950. Most of the H1N1 cases were in young people, born after 1950. The 1983 outbreak was a mixture of H3N2 and H1N1. (Data courtesy of A. A. Ferris, F. Lewis, N. Lehmann, and I. D. Gust.)*

autumn, or as soon as it becomes clear that an epidemic has begun. It is generally not considered cost effective to immunize the whole community, but only the groups most vulnerable to potentially lethal secondary bacterial pneumonia, namely the elderly, and those with chronic disorders of the pulmonary, cardiovascular or renal systems, metabolic diseases, severe anemia, or compromised immune function. In bad epidemic years there may also be a case for immunizing medical personnel and others providing vital community services, and some countries direct their program at schoolchildren because they are the principal disseminators of virus.

Side effects of this inactivated vaccine, namely redness and induration at the injection site, fever, malaise, and occasionally myalgia, are more prominent in children. The pyrogenicity is somewhat lower with "split-virus" vaccines, derived by disrupting the envelope of the formalin-inactivated virion with detergent. The British "HANA" vaccine which consists of purified hemagglutinin and neuraminidase plus adjuvant is similarly less pyrogenic. Very rarely, hypersensitivity reactions are encountered in people allergic to eggs. A particularly serious problem, the Guillain-Barré syndrome, was encountered in one in every 10^5 Americans vaccinated against "swine flu," i.e., A/New Jersey/76 (H1N1), during a mass campaign in 1976–1977, but no such association has been reported with any previous or subsequent influenza vaccine.

Protection is at best about 70% for a year or so following immunization but is often much lower (around 30%) in the most important target group, the elderly. Annual immunization may merely boost antibodies against irrelevant earlier strains to which the vaccinee has been primed by previous infection or vaccination.

Clearly, current influenza vaccines are not good enough. Research is progressing on a number of fronts to find a better approach. Some of these are directed simply at reducing the cost or increasing the immunogenicity of purified HA (± NA). These molecules have been (1) incorporated in liposomes, (2) synthesized in prokaryotic and eukaryotic cells by recDNA technology, and (3) synthesized *in vivo* by using vaccinia virus as a vector for the influenza HA gene. However, none of these approaches addresses the critical problems, which come back to original antigenic sin and the inability of inactivated vaccines to generate either a local IgA or a cell-mediated immune response. A topically administered live vaccine would appear to be the logical answer.

Cold-adapted (*ca*) mutants of influenza virus with mutations in every gene have now been developed and used as a master strain for genetic reassortment (see Chapter 4) with the prevalent strain of influenza, to produce an attenuated vaccine virus comprising six *ca* genes plus the wild-type HA and NA genes (see Chapter 10). When administered intranasally such *ca* vaccines multiply in the nose but not at the higher temperature prevailing in the lung. They are attenuated in virulence, immunogenic, and reasonably stable genetically. Clinical trials are currently underway to evaluate their efficacy and safety.

Though no substitute for vaccination, chemoprophylaxis with amantadine or rimantadine by mouth has a place in protecting unimmunized old folk during a major epidemic. Recent evidence suggests that they may also be effective in the treatment of established disease when administered as a continuous aerosol, perhaps in conjunction with ribavirin. The use of these agents against influenza A was discussed in Chapter 11.

FURTHER READING

Books

Beare, A. S., ed. (1982). "Basic and Applied Influenza Research." CRC Press, Boca Raton, Florida.

Compans, R. W., and Bishop, D. H. L., eds. (1984). "Segmented Negative Strand Viruses: Arenaviruses, Bunyaviruses, and Orthomyxoviruses." Academic Press, New York.

Kilbourne, E. D., ed. (1975). "The Influenza Viruses and Influenza." Academic Press, New York.

Laver, W. G., ed. (1983). "The Origin of Pandemic Influenza Viruses." Elsevier, Amsterdam.

Nayak, D. P., ed. (1982). "Genetic Variation Among Influenza Viruses," UCLA Symp. Mol. Cell. Biol. Ser., Vol. 21. Academic Press, New York.

Palese, P., and Kingsbury, D. W., eds. (1983). "Genetics of Influenza Viruses." Springer-Verlag, Berlin and New York.

Stuart-Harris, C. H., and Potter, C. W., eds. (1984). "The Molecular Virology and Epidemiology of Influenza." Academic Press, London.

Reviews

Lamb, R. A., and Choppin, P. W. (1983). The gene structure and replication of influenza virus. *Annu. Rev. Biochem.* **52,** 467.

Webster, R. G., Laver, W. G., Air, G. M., and Schild, G. C. (1982). Molecular mechanisms of variation in influenza viruses. *Nature (London)* **296,** 115.

CHAPTER 22

Paramyxoviruses

Introduction ... 521
Properties of *Paramyxoviridae* ... 521
Measles ... 524
Mumps ... 529
Parainfluenza Viruses ... 531
Respiratory Syncytial Virus ... 533
Further Reading ... 538

INTRODUCTION

The paramyxoviruses, as their name implies, morphologically resemble the orthomyxoviruses with which they were originally classified but differ in a number of vital respects including the nature of the genome and strategy of replication. Though few in number, all the human paramyxoviruses are important causes of respiratory disease in children. The parainfluenza and respiratory syncytial viruses are responsible for nearly half of the croup, bronchiolitis, and pneumonia in infants today. Measles and mumps are familiar to every mother, though the first in particular is now receding in the face of widespread immunization.

PROPERTIES OF *PARAMYXOVIRIDAE*

Larger than the orthomyxoviruses and more pleomorphic, the paramyxoviruses range from 150 to 300 nm in diameter (Table 22-1). They are enclosed by a loose lipoprotein envelope which is extremely fragile, rendering the virion vulnerable to destruction by storage, freezing and

TABLE 22-1
Properties of Paramyxoviridae

Pleomorphic spherical virion, 150–300 nm
Envelope containing F and (in most cases) HN glycoprotein
Helical nucleocapsid with associated transcriptase
ssRNA genome, linear, negative polarity, MW 5–7 million
Cytoplasmic replication, budding from plasma membrane

thawing, or even preparation for electron microscopy. Accordingly, particles often appear distorted in electron micrographs and may rupture to reveal their internal nucleocapsid which occurs as a single helix about 1 μm long (Plate 22-1).

In contrast with the genome of orthomyxoviruses, that of paramyxoviruses is not segmented. It consists of a single molecule of single-stranded RNA of negative polarity. Of course a transcriptase is present in the virion.

The envelope contains two types of glycoprotein. One of these (HN) carries both hemagglutinating and neuraminidase activity in the case of the *Paramyxovirus* genus (Table 22-2). The other is known as the fusion protein (F) because it enables the virus to fuse cells together to form the syncytia so characteristic of this family (see Chapter 5). To acquire biological activity the F protein must be cleaved by a cellular protease into two disulfide-linked polypeptides, F_1 and F_2; this fails to occur in certain types of host cell. Since the F protein is essential for viral penetration of the host cell by fusion of the viral envelope with the plasma membrane, and for direct intercellular spread by cell-to-cell fusion, it plays a key role in the pathogenesis of the paramyxoviruses, including persistent infections (see Chapter 7). Paramyxovirus vaccines, to be effective, must elicit antibodies against the F protein (not only against HN) if they are to eliminate the virus by preventing direct cell-to-cell spread. Moreover, oligopeptides analogous to the N-terminus of F_1, adjacent to the cleavage site, are being assessed as potential chemotherapeutic agents (Chapter 11).

Within the family *Paramyxoviridae*, three genera are recognized. Genus *Paramyxovirus* includes the five human parainfluenza viruses and mumps virus (as well as Newcastle disease virus of birds); *Morbillivirus* includes measles virus of man (and its antigenic relatives, canine distemper virus, and rinderpest virus of cattle); *Pneumovirus* contains the human respiratory syncytial virus.

All the *Paramyxoviridae* are labile to the effects of heat or desiccation,

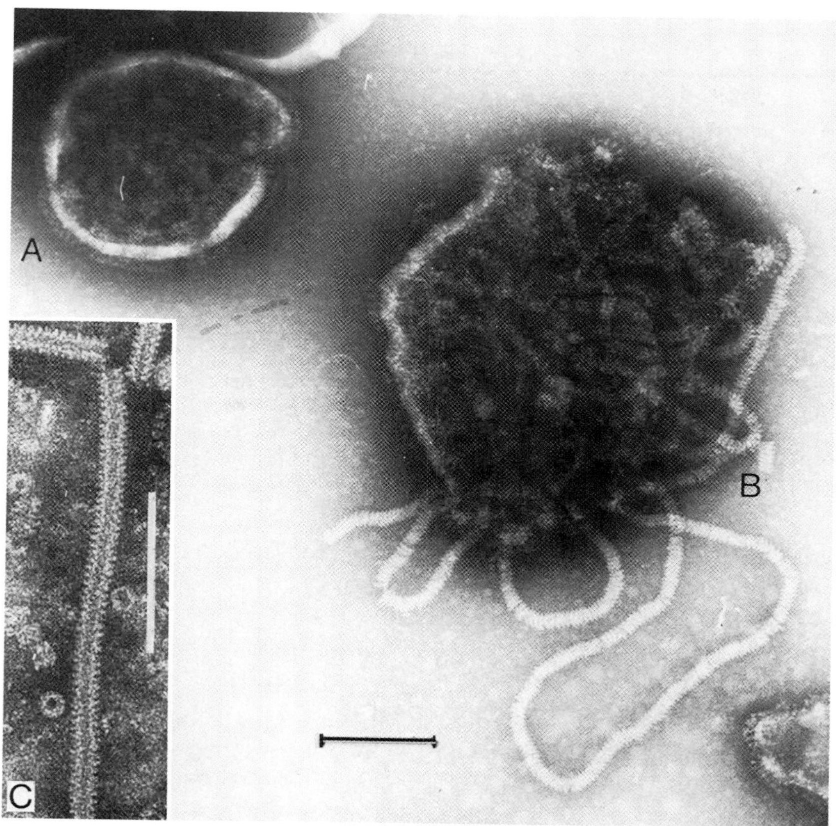

PLATE 22-1. Paramyxoviridae. *Negatively stained virions of mumps virus. (A) Intact virion; peplomers visible at lower edge. (B) Partially disrupted virion, showing nucleocapsid. (C) Enlargement of portion of nucleocapsid, in longitudinal and cross section. Bars = 100 nm. (Courtesy Dr. A. J. Gibbs.)*

hence clinical specimens must be maintained at 4°C and inoculated into cell cultures without delay. They grow best in primary cultures of primate kidney cells. They do not cause extensive cell destruction; indeed carrier cultures readily arise (see Chapter 7). Syncytium formation is a regular feature, with acidophilic inclusions in the cytoplasm (and in the nucleus in the case of *Morbillivirus*) (Plate 22-2). Hemadsorption is readily demonstrable (Plate 2-2), except with *Pneumovirus*.

The strategy of replication of these nonsegmented negative-strand viruses and the detailed sequence of steps involved in glycosylation and

TABLE 22-2
Biological Properties of Human Paramyxoviruses

PROPERTY	MEASLES	MUMPS	PARAINFLUENZA	RSV
Genus	*Morbillivirus*	*Paramyxovirus*	*Paramyxovirus*	*Pneumovirus*
Serotypes	1	1	4 (2 subtypes)	1 (5 subtypes[a])
Fusion (F) protein	+	+	+	+
Hemolysin[b]	+	+	+	−
Hemagglutinin[c]	+[d]	+	+	−
Neuraminidase[c]	−	+	+	−
Hemadsorption	+	+	+	−
Inclusions	N and C[e]	C	C	C

[a] Or "strains," distinguishable by neutralization with sera raised in animals.
[b] Carried by F glycoprotein.
[c] HN glycoprotein carries hemagglutinating and neuraminidase activity.
[d] H glycoprotein, agglutinates monkey RBC only, and has no neuraminidase activity.
[e] N, Nuclear; C, cytoplasmic.

budding were described in Chapter 3 and illustrated in Fig. 3-10 and 3-11 and Plate 3-4.

MEASLES

Until recently measles was perhaps the best known of all the common childhood diseases. The characteristic maculopapular rash, with coryza and conjunctivitis (Plate 22-3), was familiar to every mother. However, as a consequence of the development of an effective live vaccine and vigorous implementation of a policy of immunization of every child, the United States has reduced the incidence of this disease so dramatically that many young doctors have never seen a case. The objective now must be to match this performance in the rest of the world, including the

PLATE 22-2. *Cytopathic effects induced by paramyxoviruses. (A) Respiratory syncytial virus in MA104 cell line (unstained, ×100). Note syncytia (arrows) resulting from fusion of plasma membranes; nuclei accumulate in the center. (B) Respiratory syncytial virus in HEp-2 cells (H and E stain, ×400). Syncytium containing many nuclei and acidophilic cytoplasmic inclusions (arrow). (C) Measles virus in human kidney cells (H and E stain, ×30). Huge syncytium containing hundreds of nuclei. This monolayer was embedded in nitrocellulose before being stripped from the culture tube and stained. (D) Measles virus in human kidney cells (H and E stain, ×400). Note multinucleated giant cell containing acidophilic nuclear (horizontal arrow) and cytoplasmic (vertical arrow) inclusions. (Courtesy I. Jack.)*

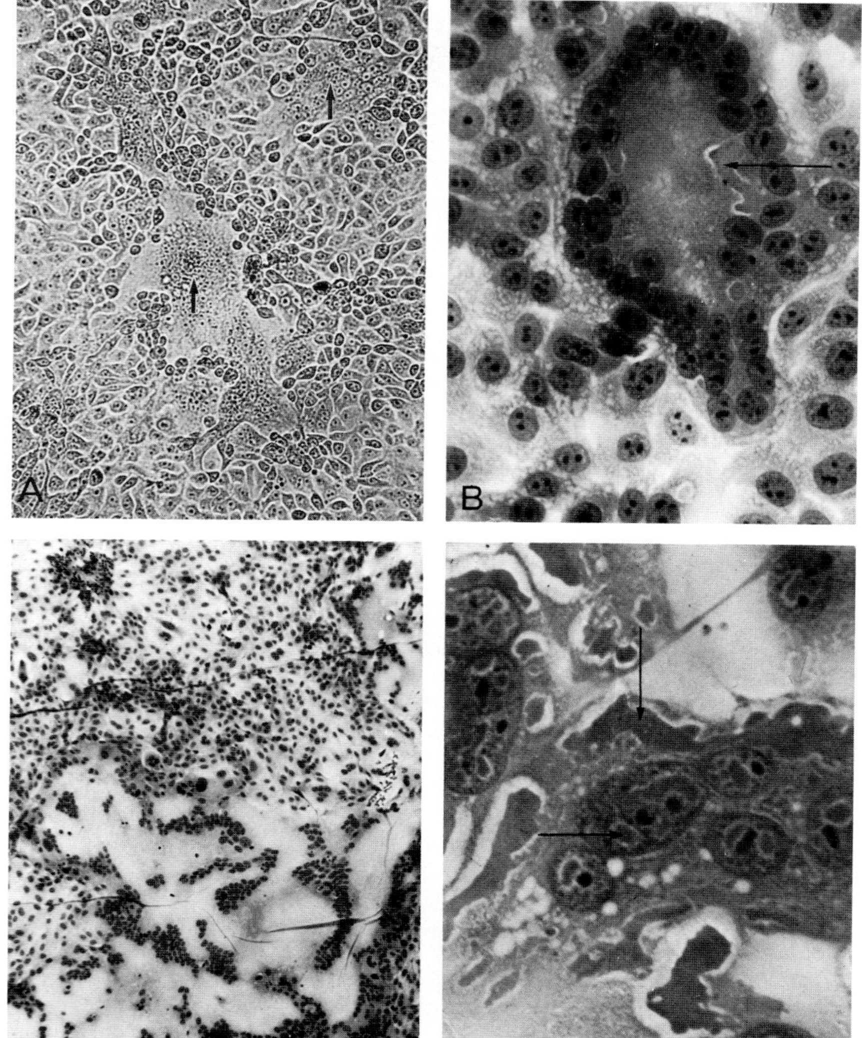
A
B
C
D

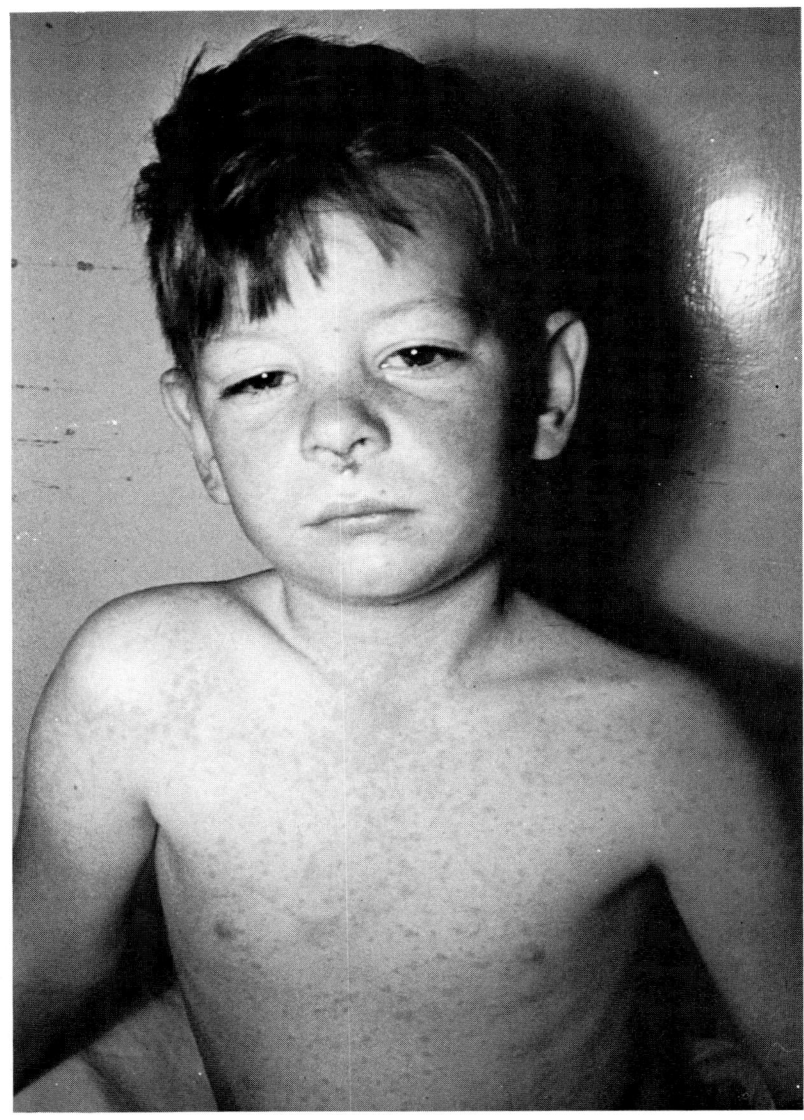

PLATE 22-3. *Measles. Note maculopapular rash and conjunctivitis in young child with upper respiratory infection. (Courtesy Dr. J. Forbes, Fairfield Hospital for Infectious Diseases, Melbourne.)*

developing countries, where the mortality from measles in malnourished infants is such as to make it one of the leading causes of death in children.

Clinical Features

Following a prodrome marked by fever, cough, coryza and conjunctivitis, an exanthem appears on the head and spreads progressively over the chest, trunk then limbs. The rash consists of flat macules that fuse to form blotches rather larger than those of other viral exanthems. They are slow to fade and often leave the skin temporarily stained.

Common complications include otitis media, croup, bronchitis, and bronchopneumonia. Bacterial pneumonia is the usual cause of death when measles kills malnourished children; survivors may go on to bronchiectasis. Middle ear infections may eventually lead to deafness. Immunologically deficient children can die from "giant-cell pneumonia" or "subacute inclusion-body encephalitis", with no sign of a rash. But the most dangerous complication of measles and the principal reason for vaccination programs in the developed nations is encephalomyelitis, which occurs in about one in every thousand cases, and inflicts a mortality of about 15%, with permanent neurological sequelae in many of the survivors. Subacute sclerosing panencephalitis (SSPE) is a very much rarer complication, developing in only one in every million cases some years after apparent recovery from the original infection.

Pathogenesis and Immunity

The pathogenesis of measles was discussed fully in Chapter 5 (page 137 and Fig. 5-5), the importance of T lymphocytes in the generation of the rash being mentioned in Chapter 6. The role of the restriction of full expression of the measles virus genome in the pathogenesis of SSPE was discussed in Chapter 7 (page 198).

Postinfectious encephalomyelitis is a demyelinating disease with a pathogenesis resembling that of experimental allergic encephalomyelitis. Whereas there is no evidence of production of virus or antiviral antibody in the brain, myelin basic protein is found in the CSF and patients' lymphocytes are reactive to myelin.

Essentially all primary measles infections give rise to clinically manifest disease. The resulting immunity is effectively lifelong; second attacks of measles are virtually unknown, even in totally isolated island communities in which subclinical boosts to immunity cannot have occurred for decades (see Chapter 9, pages 255 and 265).

Laboratory Diagnosis

The clinical diagnosis of measles is so straightforward that the laboratory is rarely called upon for help. If necessary, virus can be isolated early in the illness from the nose, throat, conjunctiva, or blood. Primary cultures of human or monkey kidney cells are the most susceptible. Multinucleated giant cells containing numeorus acidophilic inclusions in cytoplasm and nuclei are diagnostic (Plate 22-2C and D). Measles antigen can be identified by immunofluorescence in cultured cells, or in cells aspirated directly from the nasopharynx.

Serology is sometimes used to screen populations for their immune status. HI or EIA is the most convenient.

Epidemiology and Control

The epidemiology of measles in industrialized nations was discussed fully in Chapter 9 (page 255), while in Chapter 6 (page 180) we said something of the very different situation in the developing world where hundreds of thousands of malnourished infants die annually from this eminently preventable disease.

A live measles vaccine was developed by Enders, then further attenuated by Schwartz. In the industrialized countries the vaccine is, or should be, administered subcutaneously to every child at or shortly after 12 months of age. Because maternal antibody has disappeared by this age, seroconversion occurs in over 95% of recipients (cf. only 50–75% at 6 months). The resulting antibody titers are lower than following natural infection but persist for many years at protective levels. Trivial side effects are not uncommon, particularly mild fever (in 5–15%) and transient rashes (5%). Immunocompromised patients, though vulnerable to measles, should not be vaccinated.

Today, indigenous measles has all but disappeared from the United States, where immunization of children prior to school entry is required by law. In the developing world, immunization against measles is a top priority of WHO. There is a case for immunizing as early as 9 (or even 6) months of age in countries with widespread protein malnutrition and high mortality from measles in infants in the first year of life. Maintenance of the "cold chain"is also vital in the tropics. More heat-stable vaccines have recently become available. They are shipped as a powder (freeze-dried) which stores well at 4°C but should be used soon after reconstitution.

Sabin recently reported immunization of children, including 4–6 month olds possessing substantial levels of maternal antibody, by a live attenuated measles vaccine delivered as an aerosol from a nebulizer via a face mask. Further field trials are needed.

Passive immunization still has a place in protecting unvaccinated children, particularly immunocompromised children, following exposure to measles. If administered promptly, pooled normal human immunoglobulin will abort the disease; if given a few days later, the disease may still be modified.

No antiviral agent is effective against measles but bacterial superinfections such as pneumonia or otitis media require vigorous chemotherapy.

MUMPS

Clinical Features

The comical spectacle of the unhappy young man with face distorted by painful edematous enlargement of parotid and other salivary glands, unable to eat or talk without discomfort, is familiar music hall fare, often spiced with the innuendo that the case is complicated by a well-deserved orchitis! Epididymo-orchitis is indeed a painful development which occurs in 20–25% of all cases in postpubertal males and may lead to atrophy of the affected testicle. A wide variety of other glands may be involved, including the pancreas (quite commonly), ovary, thyroid, and breast (more rarely). Benign meningeal signs are detectable in at least 10% of all cases of mumps, and clinical aseptic meningitis, sometimes presenting without parotid involvement, occurs often enough to make mumps the most common single cause of this disease. Fortunately, the prognosis is very much better than with bacterial meningitis and sequelae are rare. Mumps encephalitis on the other hand, though much less frequent, is a more serious development. Chronic mumps encephalomyelitis has also recently been described. Unilateral nerve deafness is an uncommon but important long-term consequence of mumps.

Overall, about one-third of mumps infections are subclinical. In infants the majority are symptomless or present as respiratory infections.

Laboratory Diagnosis

The classic case of mumps (Plate 22-4) can be identified without help from the laboratory but atypical cases and meningoencephalitis present a diagnostic problem. Virus can be isolated from saliva (or from swabs taken from the orifice of Stensen's duct), from urine (mumps being one of the few viruses readily isolated from the latter source), or from CSF in patients with meningitis. Primary cultures of primate kidney cells are the most sensitive and become positive for hemadsorption of chicken or guinea pig erythrocytes within 3-5 days. Immunofluorescence confirms

PLATE 22-4. *Mumps. Note swelling of parotid and other salivary glands (in both patients). (Courtesy Dr. J. Forbes.)*

the diagnosis. Mumps virus also grows well in the amnion of the developing chick embryo, multiplication being detected by testing the amniotic fluid for hemagglutination.

Serological procedures also can be employed to diagnose mumps meningoencephalitis. Traditionally, the CFT has been used for this purpose, but an EIA for IgM antibody may be simpler. Recently, a novel type of "hemadsorption immunoassay" has been developed which uses anti-human IgM as capture antibody, followed by patient's serum or CSF, then mumps hemagglutinin, and finally red blood cells in lieu of indicator antibody. HI is often used to monitor immune status in vaccination studies; care must be taken to remove nonspecific inhibitors from the patient's serum, and problems can be encountered with cross-reactive antibodies to parainfluenza viruses.

Epidemiology

Mumps is transmissible by direct contact with saliva or by droplet spread, from a few days before the onset of symptoms until about a week after. Mumps does not show the dramatic periodicity of the other paramyxoviruses, but tends to cause sporadic cases throughout all seasons, with winter–spring epidemics every few years. A marked decline in the incidence of mumps has been recorded in the United States since the introduction of the vaccine (Fig. 9-5).

Control

An excellent live attenuated vaccine derived by passage in chick fibroblasts is available. It may be used alone in adolescent males, e.g., military recruits, but in the United States is generally mixed with live attenuated measles and rubella for administration to 12- to 15-month-old infants. Protective levels of antibody are conferred by a single subcutaneous injection in at least 95% of recipients and persist for at least a decade. The vaccine is relatively free of side effects, although occasional allergic reactions and rarely parotitis or CNS disturbances have been reported.

PARAINFLUENZA VIRUSES

The human parainfluenza viruses, of which there are four serotypes, are common respiratory pathogens, in the main causing relatively harmless URTI. Nevertheless, they are also the commonest cause of a more serious condition in young children known as "croup."

Clinical Features

Primary infection, typically in a young child, generally manifests itself as coryza and pharyngitis, often with some degree of bronchitis and low fever. However, there are two more serious presentations which are seen in 2–3% of infections (Table 22-3). In an infant, especially under the age of 6 months, parainfluenza type 3 may cause bronchiolitis and/or pneumonia clinically indistinguishable from that more commonly caused by the respiratory syncytial virus (see below). In somewhat older children (6 months to 5 years) parainfluenza type 1, and to a lesser extent type 2, are the major cause of croup (laryngotracheobronchitis). The child presents with fever, cough, stridor, and respiratory distress which may occasionally progress to laryngeal obstruction requiring tracheotomy.

Laboratory Diagnosis

Primary human or monkey kidney cells are significantly more sensitive than human diploid fibroblasts or human epithelial lines for the isolation of parainfluenza viruses. Since, however, the former are no longer generally obtainable and monkey kidneys are often contaminated with the simian parainfluenza virus, SV5, laboratories are turning to the continuous monkey kidney cell line LLC-MK2, which is almost as sensitive provided trypsin is incorporated in the medium to cleave (activate) the F glycoprotein of the virus. Parainfluenza viruses multiply rather slowly and cause very little CPE, except in the case of type 2, which induces syncytia. Viral growth is detected by hemadsorption of guinea pig red cells (Plate 2-2D). Differentiation from other hemadsorbing respiratory viruses is then made by fluorescent antibody staining of the infected monolayer or by hemagglutination-inhibition using virus from the cell culture supernatant.

Rapid diagnosis can be achieved with comparable sensitivity by EIA, RIA, or IF. Adequate solubilization of the respiratory mucus and any accompanying cells, to liberate antigen, is necessary to maximize the sensitivity of EIA or RIA, as discussed in Chapter 12. The specimen for immunofluorescence on the other hand is exfoliated cells.

Epidemiology and Control

Like other respiratory agents, parainfluenza viruses are spread by droplets and by contact with respiratory secretions. The incubation period ranges from 2 to 6 days and shedding continues for about a week. However, a study of an Antarctic party, totally isolated throughout the long winter, provided evidence of chronic infection; parainfluenza viruses were recovered from men involved in an outbreak of coryza occurring months after the last contact with the outside world.

The four types of human parainfluenza virus differ in their epidemiology. Type 3 is endemic throughout the year; infection commonly occurs in the first months of life, about half of all babies becoming infected before the end of their first year, and almost all by the end of their second year. The age of first infection, and the severity of the resulting disease, is influenced somewhat by the titer of maternal antibody in the newborn infant. Parainfluenza 3 ranks second only to respiratory syncytial virus as a cause of bronchiolitis and pneumonia in infants under the age of 6 months. In contrast, types 1 and 2 cause epidemics in the fall, with type 1 (the more prevalent) being prominent in certain years and type 2 in others; primary infection with these types occurs between the ages of 6 months and 5 years. Between them parainfluenza types 1, 2, and 3 are responsible for about 40% of all cases of severe croup. Subtypes 4a and 4b are not rare but produce only trivial illness.

Reinfections with parainfluenza viruses certainly occur, although

TABLE 22-3
Clinical and Epidemiological Features of Parainfluenza and Respiratory Syncytial Virus Infections

VIRUS	MAJOR SYNDROME	AGE	EPIDEMIOLOGY
Parainfluenza 1 and 2	Croup	6 months to 5 years	Autumn epidemic
Parainfluenza 3	Bronchiolitis, pneumonia	First 6 months	Endemic
Parainfluenza 4	URI	Children	Endemic
RSV	Bronchiolitis, pneumonia	First 6 months	Winter epidemic

clinical disease is generally mild and confined to the upper respiratory tract the second time round. Since it may be unrealistic to expect any prospective vaccine to induce a greater degree of immunity than does natural infection, the hope would be to reduce the incidence of severe lower respiratory disease rather than to prevent infection altogether. Experimental "subunit" vaccines comprising liposomes or rosettes of the two surface glycoproteins, F and HN, induce some degree of protection in animals, but are still a long way from being a practical proposition in man.

RESPIRATORY SYNCYTIAL VIRUS

Respiratory syncytial virus (RSV) is the most important respiratory pathogen of childhood, being responsible for about half of all cases of bronchiolitis and a quarter of all pneumonia during the first few months of life.

Clinical Features

The commonest manifestation of RSV infection in all age groups is a febrile rhinitis and/or pharyngitis with limited involvement of bronchi. However, the consequences may be much more serious in certain babies between the second and sixth months of life. Almost 1% of all babies develop an RSV infection severe enough to require admission to hospital and of these about 1% die, particularly those with congenital heart defects. Characteristically, a 1- to 3-month-old infant with rhinorrhea develops fever and cough, progressing to dyspnea, wheezing, and cyanosis (Plate 22-5). Death may occur very rapidly.

About a third of all RSV infections in children involve the middle ear, making this virus the most important causal agent of otitis media. RSV

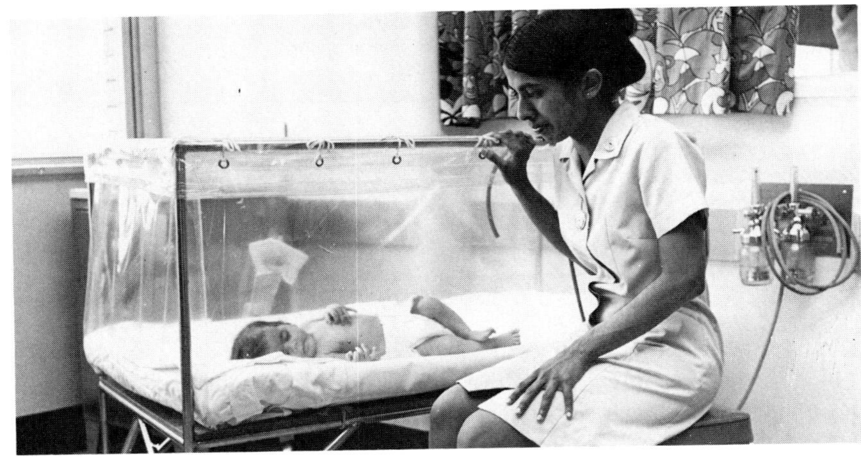

PLATE 22-5. *Respiratory syncytial viral bronchiolitis. Baby maintained in oxygen tent. (Courtesy Drs. H. Williams and P. Phelan, Royal Children's Hospital, Melbourne.)*

can be recovered from the effusion but superinfection with *Streptococcus pneumoniae* or *Haemophilus influenzae* is usual.

In older children and adults, RSV infections are reinfections against a background of partial immunity. The disease resembles a cold, with or without cough and fever. In asthmatic children RSV infection may precipitate an attack of wheezing. In the elderly pneumonia can occur.

Pathogenesis and Immunity

The virus multiplies in the mucous membrane of the nose and throat; in the very young and very old it may involve the trachea, bronchi, bronchioles, and alveoli. The incubation period is 4–5 days. Fatal cases usually show extensive bronchiolitis and pneumonitis with scattered areas of atelectasis and emphysema resulting from bronchiolar obstruction (Plates 22-6, 29-1, and 29-2).

A challenging unanswered question is why severe lower respiratory disease develops only in certain very young infants. Undoubtedly the airways of such young babies, being narrower than those of older children, are much more readily obstructed by inflammation, edema, and mucus. But this does not explain why only a minority of these small infants develop bronchiolitis. There is evidence that the condition may have an immunological basis.

Many years ago, an experimental concentrated formalin-inactivated RSV vaccine was tested in children. When the immunized children en-

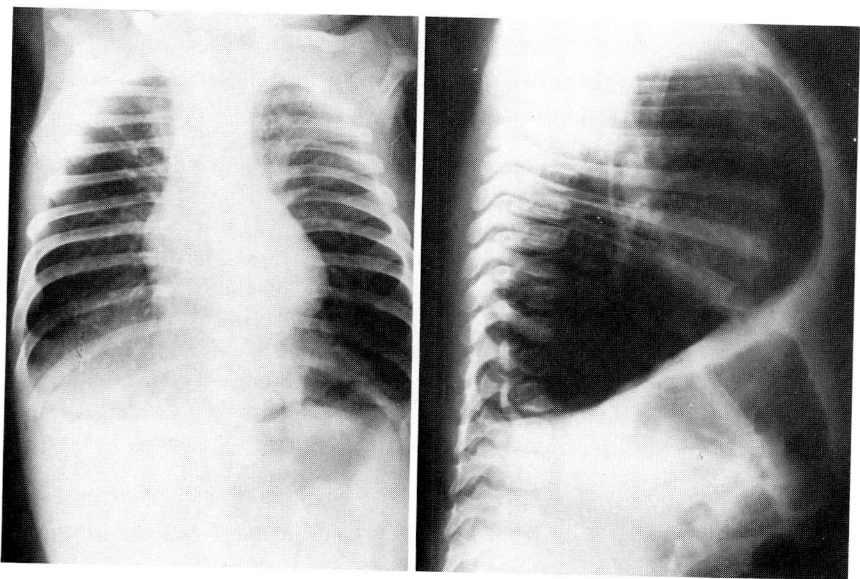

PLATE 22-6. *Radiographs of a baby with respiratory syncytial viral bronchiolitis and pneumonitis. Note grossly overinflated lungfields with depression of the diaphragm and bulging of the anterior chest wall in the lateral view. (Courtesy Drs. H. Williams and P. Phelan.)*

countered RSV during a subsequent epidemic, they actually suffered significantly more serious lower respiratory disease than did unimmunized controls. This alarming occurrence led not only to the abandonment of the killed vaccine but also to speculation about the immunological basis of the episode and its possible relationship to natural RSV bronchiolitis. The first hypothesis was that maternal IgG, present only in young infants, might react with virus multiplying in the lung to produce a hypersensitivity reaction of the Arthus type. However plausible, this idea was abandoned when no correlation could be demonstrated between the titer of maternal antibody and the severity of the RSV-induced illness.

Babies who experience RSV bronchiolitis are more likely than matched controls to develop asthma or exercise-induced bronchial reactivity in later life or to suffer bouts of recurrent wheezing ("allergic bronchitis") associated with subsequent viral infections. Presumably this is not because RSV predisposes to asthma, but because a baby with a hereditary predisposition to allergy has a higher probability of developing some sort of hypersensitivity response if it contracts an RSV infection. Infants whose RSV infections are marked by wheezing and hypoxia have anti-RSV IgE as well as histamine in their nasopharyngeal secretions, and their lymphocytes proliferate more vigorously *in vitro* in response to RSV antigens.

Immunity acquired as a result of RSV infection is inadequate to prevent reinfections with the same virus. During the annual winter epidemics that plague babies' homes the majority of children become reinfected; the severity of the illness is generally diminished the second time, though not necessarily so if the original infection was mild.

It is not known what arm of the immune response is of most importance in RSV infection. Specific antibody of the IgA class is demonstrable, mainly attached to the surface of infected epithelial cells for the first few days, then free in the mucus. However, protection does not correlate well with local IgA titer and there has been some speculation about the possibility of other unidentified factors in respiratory secretions. It has also been demonstrated that RSV is a much less effective inducer of interferon synthesis in normal infants than are influenza and parainfluenza viruses. Children with congenital immune deficiencies described as principally T lymphocyte defects have been shown to excrete RSV (or parainfluenza virus) from their lungs for months.

Laboratory Diagnosis

Three diagnostic methods of approximately equal sensitivity are currently favored by different laboratories: (1) isolation of the virus in cell culture,(2) immunofluorescence (IF) on exfoliated cells, and (3) enzyme immunoassay (EIA) on antigen from nasopharyngeal mucus.

Virus may be recovered from a nasopharyngeal aspirate (Plate 12-4) or nasal wash by inoculation of cultured cells. The extreme lability of RSV makes it mandatory that the specimen be taken early in the illness and that it be added to cultured cells without delay and without preliminary freezing. Some virologists go so far as to recommend that specimens be transferred directly from the patient into the cell culture tubes at the bedside. Human heteroploid cell lines such as HeLa or HEp-2 are the most sensitive, though human embryonic lung fibroblasts (HELF) or human embryonic kidney (HEK) may also be used. A quick answer will not be forthcoming because up to 10 days may elapse before the characteristic syncytia become obvious (Plate 22-2A and B). A trained eye can usually detect early CPE by about the third to fifth day. Fixation and staining generally reveals extensive syncytia containing acidophilic cytoplasmic inclusions, though some strains produce only rounded cells. Absence of hemadsorption distinguishes RSV from all the other paramyxoviruses. Definitive identification can be established by immunofluorescence as soon as early CPE first become apparent. Monoclonal antibody against the viral nucleocapsid protein (NP) highlights the prominent large round inclusions, whereas monoclones directed to the F glycoprotein give uniform diffuse cytoplasmic staining.

Only one serotype of human RSV is currently recognized. Neutralization tests using specific antisera raised in animals, or EIA using monoclonal antibodies, distinguish at least five major strains. Since human sera cross-neutralize all of them there is presumably some degree of cross-protection *in vivo.*

Monoclonal antibody can also be used for indirect IF on exfoliated cells aspirated from the nose and/or oropharynx. This method has the advantage of speed and, in addition, can produce a positive result if the specimen is taken too late to expect viable virus still to be available for culture (because of inactivation by heat or neutralization by IgA).

EIA (or RIA) is as sensitive and as specific as either of the foregoing methods (see Chapter 12). These tests pick up soluble antigens as well as virions and can detect as little as 10–50 ng.

Serum antibodies can be assayed by IF, RIA, EIA, or neutralization (plaque reduction).

Epidemiology

RSV is highly contagious, being shed in respiratory secretions for several days, sometimes weeks, and conveyed by contact to a generally susceptible population, even those with prior experience of the virus having a negligible degree of acquired immunity. Not surprisingly therefore, RSV causes a sharply defined epidemic every winter (Fig. 9-4). Most children become infected at home in their first year or two, then reinfections occur repeatedly throughout life.

Nosocomial infections are also frequent. Outbreaks are occasionally seen in neonatal wards of maternity hospitals, sometimes inflicting high mortality. Moreover, hospital staff and parents of babies with RSV bronchiolitis often develop febrile colds and/or pharyngitis.

Control

Improvements in intensive care facilities have led to a marked reduction in the mortality from RSV pneumonia. Adrenergic bronchodilators such as epinephrine have also helped. Recent trials indicate that ribavirin, administered continuously for 3–6 days as an aerosol in the oxygen tent in which the baby is normally held, reduces the severity and duration of the illness as well as the titer of virus in the lungs. No doubt some or all of the newly available suite of interferons produced by recombinant DNA technology will be tested shortly.

Workers at the United States National Institutes of Health have been trying valiantly for many years to develop a vaccine against RSV. Their early efforts having been thwarted by the problems with inactivated vaccines, they conceived the idea of an intranasal live vaccine, attenuated in virulence by the introduction of mutations that would render the virus temperature-sensitive (*ts*) (see Chapter 10). Intranasal RSV infection of cotton-rats confered only medium-term protection against upper respiratory infection ("nasal immunity") but virtually lifelong protection against pneumonia ("pulmonary immunity"); this was reassuring in view of the frequency of recurrence of upper respiratory infections in man. Apparently suitable *ts* mutants were isolated which multiplied harmlessly in the upper respiratory tract of hamsters and chimpanzees, producing an immune response. Disappointingly, however, the mutants were not completely stable genetically *in vivo* and the project was shelved.

Meanwhile there is some interest in developing a "subunit" or rec-DNA "cloned" F glycoprotein vaccine. In light of the paradoxical experience with formalin-inactivated vaccine, however, the rationale and testing of any RSV vaccine will have to be approached with special caution. The principal target group for a safe RSV vaccine would be very young, especially sickly infants in babies' homes and hospitals. If the putative hereditary defect predisposing to severe lower respiratory tract involvement in RSV could be identified, it might one day be feasible to immunize only those susceptibles. Regardless of this flight of fancy the target group would need to be immunized within the first month of life in order to protect them during the short period of greatest threat. A satisfactory immune response to a vaccine at this very early age, in the presence of maternal antibody, may be exceedingly difficult to achieve.

FURTHER READING

Books

Bishop, D. H. L., and Compans, R. W., eds. (1984). "Non-Segmented Negative Strand Viruses: Paramyxoviruses and Rhabdoviruses." Academic Press, New York.

Reviews

Chanock, R. M., and Murphy, B. R. (1980). Use of temperature-sensitive and cold-adapted mutant viruses in immunoprophylaxis of acute respiratory tract diseases. *Rev. Infect. Dis.* **2,** 421.

Choppin, P. W., and Compans, R. W. (1975). Reproduction of paramyxoviruses. *In* "Comprehensive Virology" (H. Fraenkel-Conrat and R. R. Wagner, eds.), Vol. 4, pp. 95–178. Plenum, New York.

Choppin, P. W., Richardson, C. D., Merz, D. C., and Scheid, A. (1981). Inhibition of functions of paramyxovirus and myxovirus envelope proteins. *Perspect. Virol.* **11,** 57.

Hall, C. B. (1980). Prevention of infections with respiratory syncytial virus: The hopes and hurdles ahead. *Rev. Infect. Dis.* **2,** 384.

Jackson, G. G., and Muldoon, R. L. (1973). Viruses causing common respiratory infections in man. II. Enteroviruses and paramyxoviruses. III. Respiratory syncytial viruses and coronaviruses. *J. Infect. Dis.* **128,** 387 and 674.

Katz, S. L., Krugman, S., and Quinn, T. C., eds. (1983). International Symposium on Measles Immunization. *Rev. Infect. Dis.* **5,** 389–625.

McIntosh, K., and Fishaut, J. M. (1980). Immunopathologic mechanisms in lower respiratory tract disease of infants due to respiratory syncytial virus. *Prog. Med. Virol.* **26,** 94.

Oldstone, M. B. A., and Fujinami, R. S. (1982). Virus persistence and avoidance of immune surveillance: How measles viruses can be induced to persist in cells, escape immune assault and injure tissues. *In* "Virus Persistence" (B. W. J. Mahy, A. C. Minson, and G. K. Darby, eds.), Symp. Soc. Gen. Microbiol., Vol. 33, p. 185. Cambridge Univ. Press, London and New York.

ter Meulen, V., and Carter, M. J. (1982). Morbillivirus persistent infections in animals and man. *In* "Virus Persistence" (B. W. J. Mahy, A. C. Minson, and G. K. Darby, eds.), Symp. Soc. Gen. Microbiol., Vol. 33, p. 97. Cambridge Univ. Press, London and New York.

Tyeryar, F. J., Richardson, L. S., and Belshe, R. B. (1978). Report of a workshop on respiratory syncytial virus and parainfluenza viruses. *J. Infect. Dis.* **137,** 835.

Walsh, J. A. (1983). Selective primary health care: Strategies for control of disease in the developing world. IV. Measles. *Rev. Infect. Dis.* **5,** 330.

CHAPTER 23

Coronaviruses

Introduction 541
Properties of *Coronaviridae* 541
Clinical Features 543
Pathogenesis and Immunity 543
Laboratory Diagnosis 544
Epidemiology 545
Further Reading 545

INTRODUCTION

The *Coronaviridae* family embraces nearly a dozen major pathogens of mammals and birds, displaying tropism for the respiratory tract (avian infectious bronchitis virus), the enteric tract (transmissible gastroenteritis virus of swine; feline infectious peritonitis virus), or the liver (murine hepatitis virus). No convincing evidence has yet been obtained to link human coronaviruses with serious disease affecting any of these systems, but they are an important cause of that trivial but annoying disease, the common cold. In addition, particles morphologically resembling coronaviruses are often seen by electron microscopy in feces, but it has yet to be established whether they ever cause gastroenteritis in man, or even whether they are indeed coronaviruses.

PROPERTIES OF *CORONAVIRIDAE*

The coronaviruses were so named because the unusually large club-shaped peplomers projecting from the envelope give the particle the

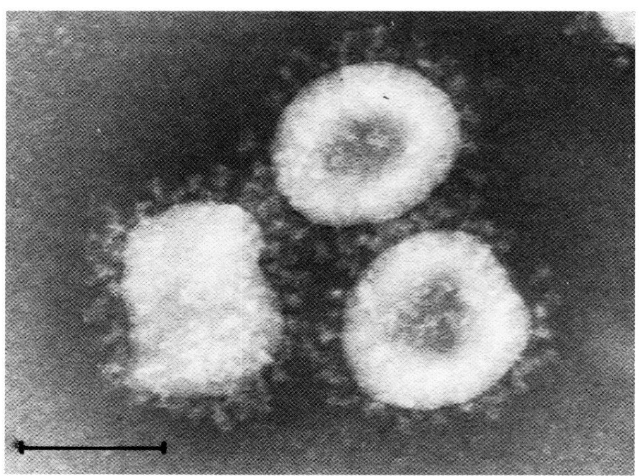

PLATE 23-1. Coronaviridae. *Negatively stained preparation of virions. Bar = 100 nm. (Courtesy Dr. F. A. Murphy.)*

appearance of a solar corona (Plate 23-1). Though typically about 100 nm in diameter, the virion is pleomorphic and can range in size from 60 to 220 nm. The helical ribonucleoprotein, difficult to discern in electron micrographs, seems to be connected directly to an unusual transmembrane glycoprotein that performs the role normally filled by matrix protein in other enveloped viruses.

The genome consists of a single linear molecule of ssRNA of positive polarity, MW about 6 million, which is capped and polyadenylated, and is infectious (Table 23-1).

Viral Multiplication

The strategy of expression of the coronavirus genome is unique. The vRNA molecule is translated directly, one of the products being an RNA polymerase which is then employed to transcribe a full-length cRNA, from which in turn is transcribed a 3′-coterminal *nested set of subgenomic mRNAs*. The nested set comprises half a dozen overlapping species of (+)mRNAs which extend for different lengths from their common 3′ ends. Only the unique sequence toward the 5′ end which is not shared with the next smallest mRNA in the nested set is translated, the product therefore being a unique protein in each case.

The whole of the multiplication cycle is confined to the cytoplasm. The envelope is acquired by budding through those membranes that contain the viral E1 glycoprotein, namely cisternae of the endoplasmic

TABLE 23-1
Properties of Coronaviridae

Pleomorphic spherical virion, 100 (60–220) nm
Envelope with large club-shaped projections
Helical RNP
ssRNA genome, linear, positive polarity, MW 6 million
Multiplies in cytoplasm, budding into ER and Golgi cisternae

reticulum and Golgi apparatus; the virions are then transported in vesicles to the plasma membrane for exit from the cell.

Classification

Antigenically, the mammalian coronaviruses seem to cluster into two groups. Two serotypes (species?) of human coronavirus (HCV), of which the prototype strains are HCV-229E and HCV-OC43, are clearly distinguishable from one another by neutralization or HI, even though each cross-reacts with coronaviruses of other animal species. The status of the "human enteric coronaviruses" (HECV) has yet to be determined.

CLINICAL FEATURES

The typical coronavirus cold is marked by nasal discharge and malaise; cough and sore throat are generally not prominent, and there is little or no fever. The illness lasts about a week, and is of no real consequence. The lower respiratory tract is not involved.

Coronavirus infection may occasionally precipitate attacks of wheezing in asthmatic children, or exacerbations of chronic bronchitis in adults.

PATHOGENESIS AND IMMUNITY

The virus remains localized to the epithelium of the upper respiratory tract and elicits a poor immune response. Serum antibodies are not detectable following about 50% of infections. No doubt local IgA is more important, but this has not yet been formally proven. Immunity to the homologous serotype lasts for a few years at most and there is a high rate of reinfection. There is no cross-immunity between HCV-229E and HCV-OC43.

LABORATORY DIAGNOSIS

Coronaviruses are difficult to grow in cultured cells, hence are rarely recovered from man. HCV-OC43 and related strains were originally isolated in organ cultures of human embryonic trachea or nasal epithelium, then passaged in infant mice, or adapted to human diploid fibroblasts. Organ culture is too tricky and time consuming a technique for a diagnostic laboratory, but some strains can be isolated directly in diploid fibroblastic lines of human embryonic lung, intestine, or kidney (Plate 23-2). Foci of "dirty" or "granular" cells become evident after a week and may progress to reveal some vacuolation before disintegrating; there is no rounding or swelling, but syncytia may form in some cell types. Hemadsorption and hemagglutination are demonstrable with OC43 only.

Direct demonstration of coronavirus antigens in nasopharyngeal aspirates or nasal swabs has more recently been accomplished by enzyme immunoassay and immunofluorescence.

Serological surveys for antibody prevalence can be conducted using coronavirus antigen derived from the human rhabdomyosarcoma (RD) cell line, which produces higher yields of virus than human diploid fetal tonsil or lung fibroblasts, and is cleaner than suckling mouse brain extracts. EIA, CF, and neutralization (plaque reduction) are all satisfactory techniques for titrating antibodies; HI is also available for OC43, and an indirect (passive) HA for 229E.

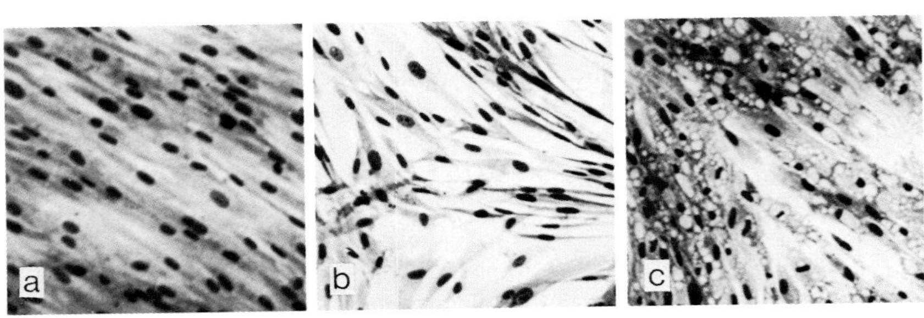

PLATE 23-2. *Cytopathic effect of human coronaviruses 229E and OC43 in diploid human fetal tonsil cells. (a) Uninfected; (b) 229E infected, showing only spindling and detachment of cells; (c) OC43 infected, showing vacuolation. May–Grunwald–Giemsa stain. [From O. W. Schmidt, M. K. Cooney, and G. E. Kenny,* J. Clin. Microbiol. **9,** *722 (1979). Courtesy Dr. O. W. Schmidt.]*

EPIDEMIOLOGY

Colds are readily transmissible to human volunteers by intranasal administration of HCV. After an incubation period of 2–5 (mean 3) days, symptoms develop in about half of those infected, and virus is shed for a further 1–4 days.

Coronavirus colds occur in the winter and early spring. Quite large outbreaks tend to occur with a certain periodicity; every 2–4 years, either 229E or OC43, but rarely both in the same year, triggers an epidemic which may infect up to a third of the population. Adults are affected as well as children because acquired immunity is so ephemeral. Overall, it is estimated that coronaviruses are responsible for 10–30% of all colds.

FURTHER READING

Books

ter Meulen, V., Siddell, S., and Wege, H., eds. (1981). "Biochemistry and Biology of Coronaviruses." Plenum, New York.

Reviews

Macnaughton, M. R., and Davies, H. A. (1981). Human enteric coronaviruses: Brief review. *Arch. Virol.* **70,** 301.

Robb, J. A., and Bond, C. W. (1979). Coronaviridae. *In* "Comprehensive Virology" (H. Fraenkel-Conrat and R. R. Wagner, eds.), Vol. 14, p. 193. Plenum, New York.

Siddell, S., Wege, H., and ter Meulen, V. (1983). The biology of coronaviruses. *J. Gen. Virol.* **64,** 761.

Sturman, L. S., and Holmes, K. V. (1983). The molecular biology of coronaviruses. *Adv. Virus Res.* **28,** 35.

Wege, H., Siddell, S., and ter Meulen, V. (1982). The biology and pathogenesis of coronaviruses. *Curr. Top. Microbiol. Immunol.* **99,** 165.

CHAPTER 24

Arenaviruses

Introduction 547
Properties of *Arenaviridae* 547
Lassa Virus 549
Junin Virus 552
Machupo Virus 552
LCM Virus 553
Further Reading 553

INTRODUCTION

The arenaviruses are largely confined to rodents, in which they produce lifelong inapparent infections transmissible vertically to their offspring. When man becomes infected via rodent excreta the consequence may be a deadly hemorrhagic fever. This is a striking illustration of how viruses which have achieved a state of peaceful coexistence with their natural host over the course of millions of years of evolution may nevertheless be lethal for another host species with which they have no regular contact.

PROPERTIES OF *ARENAVIRIDAE*

The *Arenaviridae* family derives its name from the fact that electron micrographs of sectioned virions (Plate 24-1B) reveal particles resembling grains of sand (*arenosus* = sandy). In fact these are ribosomes picked up fortuitously from the cell in which the virus grew; they have no function in the virion. Arenaviruses are pleomorphic (Plate 24-1A), varying in diameter from about 50 to 300 nm (normally 110–130 nm). Peplomers project from the envelope, but, unusually, no matrix protein lines its inner surface. The internal nucleoprotein forms two loosely helical, closed circular structures with a beaded apearance. This is because the genome comprises two distinct segments of single-stranded RNA, each of which forms a circle as a result of hydrogen bonding between complementary 3′- and 5′-termini.

In general, arenavirus RNA is minus-stranded. However, an extraordinary situation obtains at least with the S (small) segment of the well-studied arenavirus, Pichinde. This molecule is ambisense: the 3′ half is of negative polarity, the 5′ half is positive (Table 24-1). The generality of this situation has yet to be determined.

Viral Multiplication

Arenaviruses replicate in the cytoplasm, although they evidently require some nuclear function. Following uncoating of the genome, the viral transcriptase transcribes (+)mRNA from the negative-sense vRNA. In the case of Pichinde at least, a portion of the vRNA (the 5′ half of the smaller (S) segment) is of positive sense; this is not translated directly, but following transcription of the S segment, the (−) sequence at the 3′

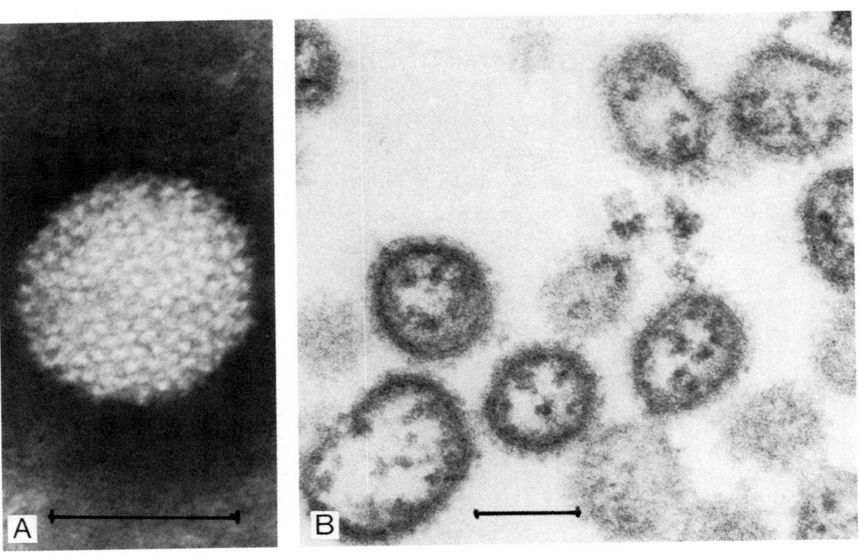

PLATE 24-1. Arenaviridae. *(A) Negative stain, Tacaribe virus. (B) Thin section of cell infected with Lassa virus. Bars = 100 nm. (Courtesy Dr. F. A. Murphy.)*

TABLE 24-1
Properties of Arenaviridae

Enveloped virion, pleomorphic, 120 (50–300) nm
Contains nonfunctional ribosomes
Two closed circular RNPs, plus transcriptase
ssRNA genome, MW 3–5 million, 2 segments, negative polarity or ambisense
Cytoplasmic multiplication, often noncytocidal

end of the transcript is transcribed into (+)mRNA which is then translated. Virions bud from the plasma membrane (Plate 24-1B); aggregates of ribosomes are often seen at the budding site. In some cell culture systems cytopathic effects including cytoplasmic inclusions may occur. In other circumstances there is no inhibition of synthesis of cellular macromolecules and a noncytocidal persistent infection is established. DI particles are plentiful in such carrier cultures. Diploid particles are also common, and reassortment occurs at high frequency when cultures are coinfected with different arenaviruses.

Classification

There are about a dozen known arenaviruses, all but one occurring naturally in rodents. Four of them cause such serious disease when transmitted to man that they are classified as Class-4 pathogens, i.e., may be handled only in maximum security laboratories with facilities meeting the standards known as Class-4 containment (Plate 12-5).

The so-called Tacaribe group, which includes the South American arenaviruses, Junin and Machupo, are antigenically related. The African arenaviruses, Lassa, Mozambique, and 3080, comprise a separate group and confer partial cross-protection against one another (Table 24-2).

All these arenaviruses can be grown in laboratory mice and sometimes in other rodents, e.g., hamsters, guinea pigs, and/or in cultured cells such as Vero or BHK21.

LASSA VIRUS

In 1969 an outbreak of a mysterious "new" disease occurred when a missionary nurse was admitted to a Nigerian hospital and nosocomial spread led to the death of hospital staff. Following the despatch of specimens to Yale University for investigation, a laboratory technician at Yale died too. The chief of the laboratory, one of the world's most famous virologists, also contracted the disease but survived following a

TABLE 24-2
Arenaviruses Causing Disease in Man

VIRUS	RODENT HOST	DISTRIBUTION	HUMAN DISEASE
Lassa	*Mastomys*	West Africa	Lassa fever
Junin	*Calomys*	Argentina	Argentine hemorrhagic fever
Machupo	*Calomys*	Bolivia	Bolivian hemorrhagic fever
LCM	*Mus*	Widespread	Lymphocytic choriomeningitis

massive transfusion of plasma from one of the nurses who had recovered from what came to be known as Lassa fever. A moratorium was placed on the investigation, the putative virus was designated a Class-4 pathogen, and further work was continued only under conditions of maximum containment.

Clinical Features

Lassa fever is extremely variable in its presentation, making it difficult to diagnose clinically. For instance, a traveler returning from West Africa may present with a history of quite insidious development of fever, headache, and malaise, progressing to a very sore throat, abdominal chest and muscular pains, vomiting, and/or diarrhea. However, almost any organ may be involved. In severe cases, pharyngitis, conjunctivitis, pneumonitis, carditis, hepatitis, encephalopathy, proteinuria, facial edema, and hemorrhages are often seen. Death follows cardiovascular collapse. The case fatality rate is around 20% in hospitalized patients, but only about 2% overall. The prognosis is poor if levels of viremia and serum transaminase are both high.

Laboratory Diagnosis

Less than a dozen laboratories in the world are set up with the P-4 facilities necessary to attempt isolation of such dangerous pathogens as Lassa virus. The agent may be recovered from blood (less frequently, throat or urine) by inoculation of Vero cell cultures. Focal necrotic changes become visible after 4 days or so. The diagnosis is established by immunofluorescent staining with monoclonal antibody following fixation and γ irradiation of the infected cells for safety. Alternatively, antigen is detected in the cell culture supernatant by agglutination of antibody-coated RBC (reverse passive hemagglutination).

Another approach is to look for a rising titer of antibody, or antibody of the IgM class, in the patient's serum. Again, indirect immunofluorescence using previously prepared Lassa virus-infected, acetone-fixed, γ-irradiated, Vero (E-6 clone) monolayers is the method of choice.

Rapid diagnosis can sometimes be made by immunofluorescence on infected cells (e.g., conjunctival scrapings) taken directly from the patient.

Epidemiology

It is now recognized that millions of people have been infected with Lassa virus in endemic areas of West Africa. The antibody prevalence rate in some villages is as high as 40%. Clearly, many infections are subclinical.

Infection is acquired from *Mastomys natalensis,* otherwise known as the multimammate mouse. This is a peridomestic rodent in which the virus is enzootic, being transmitted vertically to offspring by transovarial, transuterine, and various postpartum routes. Chronically infected animals shed virus in their urine, which contaminates the environment in village huts, then spreads to man by contact or aerosol. The incubation period is 1–2 weeks.

An antigenically related, less virulent variant known as Mozambique virus is enzootic in the *Mastomys* indigenous to East Africa, while another known as arenavirus 3080 occurs in the rodent *Praomys* in the Central African Republic. Neither of these viruses appears to be as important a cause of human disease as Lassa. No doubt there are other arenaviruses yet to be discovered, each with its own ecological niche in a particular species of rodent with a geographically restricted range.

Control

Rodent control by trapping on a village-wide basis is the simplest approach to the control of arenavirus infections. This is more practicable in the case of Lassa virus, whose reservoir host, *Mastomys natalensis,* transmits the infection to man within the confines of a village, than in the case of, say Junin virus, which is acquired by agricultural workers from its host (*Calomys*) in a rural setting.

Nosocomial spread within hospitals can be minimized by isolation and/or fastidious barrier nursing. Time will tell whether more rigorous precautions are necessary. Some countries, e.g., Australia, have established procedures for transporting suspected cases in plastic isolators to a designated hospital with appropriately secure diagnostic facilities, and nursing them in a Class-4 isolation unit.

Treatment

Ribavirin (Virazole), administered systemically in very high dosage, has been shown to protect monkeys against the lethal effects of Lassa virus. Some degree of protection may also be conferred by administration, at the time of virus challenge, of immunoglobulin with a high neutralizing titer against that particular strain of Lassa virus from the same geographical region. Indeed, ribavirin and specific antibody have been claimed to potentiate one another in monkeys. Clinical trials currently underway have not yet conclusively established whether either or both have a beneficial effect on Lassa fever in man.

JUNIN VIRUS

Argentine hemorrhagic fever (AHF) is an occupational disease of agricultural workers during the maize harvest in Argentina. Outbreaks occur annually but their severity peaks about every 3–5 years with the build up in population density of the rodent *Calomys,* in which Junin virus maintains a lifelong persistent tolerated infection characterized by chronic viremia and viruria.

AHF is a more typical hemorrhagic fever than Lassa. The pathology is largely confined to the circulatory system. There is relatively little inflammation or necrosis; hemorrhage, thrombocytopenia, leukopenia, hemoconcentration, and proteinuria culminate in death from hypovolemic shock. The virus seems to multiply mainly in lymphoid organs and may also damage capillaries directly. The untreated case fatality rate is 15–20%.

Laboratory diagnosis is established by demonstrating a rising titer of antibody in the patient's serum, using indirect immunofluorescence, as for Lassa. Isolation of Junin virus from blood by intracerebral inoculation of newborn mice or guinea pigs, followed by immunofluorescent staining of the rodent's brain, and/or passage in Vero cell cultures, is also feasible but not reliable because viremia in man is limited.

Prompt administration of Junin-immune human plasma has been shown to reduce the mortality from AHF very significantly. An experimental attenuated vaccine is also under trial.

MACHUPO VIRUS

In 1962–1964 the population of *Calomys callosus* built up to plague proportions and these rodents from the vast grassland plain of Eastern

Bolivia invaded the houses in the village of San Joaquin. Bolivian hemorrhagic fever (BHF) broke out among the human population, inflicting a high mortality. In a classical example of epidemiological detective work the reservoir of the virus was first identified then eliminated by a vigorous program of rodent trapping. Half of all the *Calomys callosus* captured were shown to be carrying Machupo virus. The human epidemic of BHF ceased abruptly 2 weeks (one incubation period) after the commencement of trapping. It may be significant that most of the cat population of San Joaquin had recently succumbed to the effects of DDT used for malaria control.

LCM VIRUS

Lymphocytic choriomeningitis (LCM) virus is widely distributed through the common house mouse (*Mus musculus*) populations of Europe and America. This tolerated vertically transmitted lifelong infection has long been the principal experimental model of viral persistence (see Chapter 7). Acquired pre- or perinatally, the virus continues to multiply in all tissues of the mouse throughout its life. It elicits very little antibody and no cell-mediated immune response, hence is never eliminated. The virus is noncytocidal and the animals remain healthy, although immune-complex disease develops in aged mice of certain inbred laboratory strains (see Chapter 7).

Man can become infected with LCM virus through contact with feral or laboratory mice, or pet hamsters. Infection is usually inapparent but there are three possible clinical syndromes: (1) an influenza ("grippe")-like disease, (2) aseptic meningitis, and (3) meningoencephalitis. The latter, which is rare, may occasionally be fatal.

The diagnosis is made by demonstrating a rising titer of antibody in the patient's serum; immunofluorescence is the method of choice. Isolation of virus from CSF or blood can be accomplished with difficulty by intracerebral inoculation of weanling (1-month-old) mice; unlike mice infected *in utero*, adult animals infected in this way will die (see Chapter 6, page 172).

FURTHER READING

Reviews

Bishop, D. H. L. (1985). Replication of arenaviruses and bunyaviruses. *In* "Virology"(B. N. Fields *et al.*, eds.), pp. 1083–1110. Raven Press, New York.

Buchmeier, M. J., Welsh, R. M., Dutko, F., and Oldstone, M. B. A. (1980). The virology and immunobiology of lymphocytic choriomeningitis virus infection. *Adv. Immunol.* **30,** 275.

Buckley, S. M., and Casals, J. (1978). Pathobiology of Lassa fever. *Int. Rev. Exp. Pathol.* **18,** 97.

Casals, J. (1982). Arenaviruses. *In* "Viral Infections of Humans: Epidemiology and Control" (A. S. Evans, ed.), Chapter 6. Plenum, New York.

Compans, R. W., and Bishop, D. H. L (1985). Biochemistry of arenaviruses. *Curr. Top. Microbiol. Immunol.* **114,** 153.

Howard, C. R., and Simpson, D. I. H. (1980). The biology of arenaviruses. *J. Gen. Virol.* **51,** 1.

Johnson, K. M., Webb, P. A., and Justines, G. (1973). Biology of Tacaribe-complex viruses. *In* "Lymphocytic Choriomeningitis Virus and Other Arenaviruses" (F. Lehmann-Grube, ed.), pp. 241–258. Springer-Verlag, Berlin and New York.

Lehmann-Grube, F., Peralta, L. M., Bruns, M., and Löhler, J. (1983). Persistent infection of mice with the lymphocytic choriomeningitis virus. *In* "Comprehensive Virology" (H. Fraenkel-Conrat and R. R. Wagner, eds.), Vol. 18, pp. 43–103. Plenum, New York.

McCormick, J. B., and Johnson, K. M. (1984). Viral hemorrhagic fevers. *In* "Tropical and Geographical Medicine" (K. S. Warren and A. A. F. Mahmoud, eds.), pp. 676–697. McGraw-Hill, New York.

Monath, T. P., ed. (1975). International Symposium on Arenaviral Infections of Public Health Importance. *Bull. W.H.O.* **52,** 381–766.

Rawls, W. E., and Leung, W.-C. (1979). Arenaviruses. *In* "Comprehensive Virology" (H. Fraenkel-Conrat and R. R. Wagner, eds.), Vol. 14, pp. 157–192. Plenum, New York.

CHAPTER 25

Bunyaviruses

Introduction 555
Properties of *Bunyaviridae* 555
Rodent-Borne Nephropathy (Hantaan Virus) 559
Rift Valley Fever 561
Crimean-Congo Hemorrhagic Fever (CCHF) 563
California Encephalitis (La Crosse Virus) 564
Sandfly Fever 564
Further Reading 566

INTRODUCTION

Paradoxically, the largest of the families of mammalian viruses was one of the last to be recognized. Once the dumping ground for relatively poorly characterized arboviruses, the *Bunyaviridae* are now a clearly defined family of over 200 viruses. Most but not all are arthropod-borne. Only a few affect man but some do cause severe disease, e.g., rodent-borne nephropathy, Crimean-Congo hemorrhagic fever, Rift Valley fever, and California encephalitis.

PROPERTIES OF *BUNYAVIRIDAE*

The virion (Plate 25-1) is about the size of influenza virus but lacks a matrix protein beneath the envelope. The negative-sense ssRNA (total MW 4–7 million) occurs as three unique molecules, each in a separate, loosely helical nucleocapsid which has a circular configuration because the 3′ and 5′ ends of each viral RNA segment are hydrogen-bonded together by complementary sequences (Table 25-1).

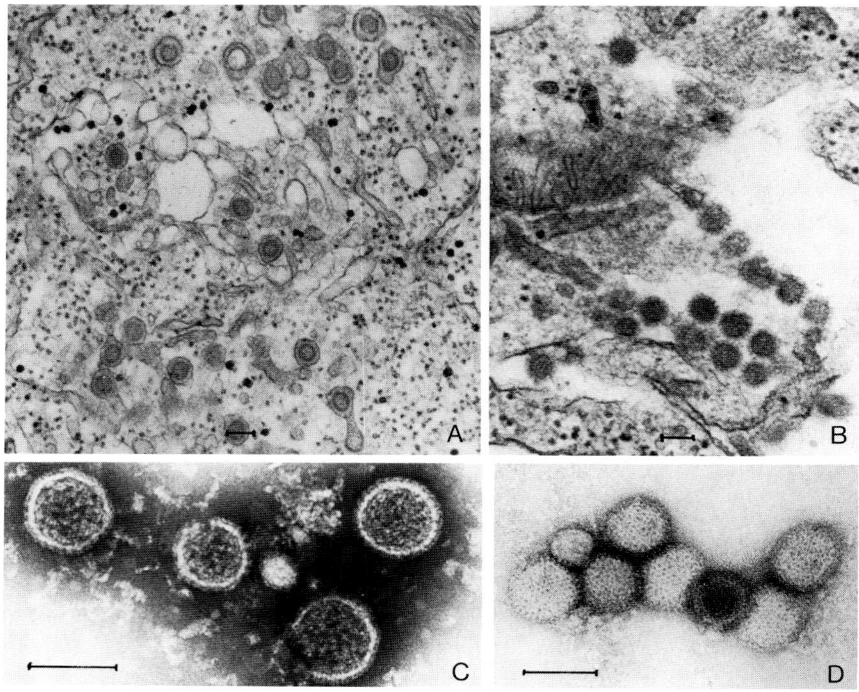

PLATE 25-1. Bunyaviridae. *(A and B) Sections of cultured cells. (A) Virions in Golgi vesicles. (B) Extracellular virions. (C and D) Negatively stained preparations. (C) Hantaan virus. (D) Rift Valley fever virus. Bars = 100 nm. (A and B, courtesy Dr. F. A. Murphy; C, courtesy Drs. J. McCormick and E. L. Palmer; D, courtesy Dr. E. L. Palmer.)*

Classification

Five major genera are differentiated on the basis of serology: species belonging to a given genus share a common CF antigen. Within each genus, species sharing common neutralization and HI determinants are assigned to "serogroups." Members of a given genus tend to display a predilection for particular vectors: members of the *Bunyavirus* genus are transmitted principally (but not exclusively) by mosquitoes, *Phlebovirus* by sandflies, and *Uukuvirus* and *Nairovirus* by ticks. Viruses of the provisional *Hantavirus* genus are not known to be arthropod-borne at all (see Table 25-2).

TABLE 25-1
Properties of Bunyaviridae

Spherical virion, 90–100 nm
Envelope with hemagglutinin but no matrix protein
Three circular nucleocapsids with helical symmetry
ssRNA genome, MW 4–7 million, 3 segments with complementary 3′ and 5′ ends
Negative polarity, but *Phlebovirus* ambisense
Cytoplasmic replication; budding into Golgi vesicles
Genetic reassortment occurs

These arboviruses persist in their arthropod vectors, being transmitted to the next generation transovarially and overwintering in ova, larvae, or nymphs. Within its particular ecological niche each bunyavirus species continues to evolve by "drift" or "shift." For instance, oligonucleotide fingerprints of different isolates of the *Bunyavirus*, La Crosse reveal point mutations, deletions, and duplications in the viral RNA. Genetic reassortment between the three RNA segments leads to much more sudden and major genetic change ("shift"), but though this is readily demonstrable in cultured cells, it seems to occur relatively rarely in nature.

Viral Multiplication. The strategy of replication has recently been unraveled for one member of the *Bunyavirus* genus. The transcriptase associated with each of the three circular nucleocapsids transcribes subgenomic plus-stranded complementary copies (cS, cM, cL) of the three minus-stranded viral RNA molecules (S, M, L). Like influenza mRNAs, bunyavirus mRNAs are found to have a nonviral (cellular?) sequence at the 5′ end. cS is translated from overlapping reading frames into two proteins, N and NS_s; cM into proteins G_1, G_2, and NS_m; and cL into protein L. Bunyaviruses mature by budding into Golgi saccules and vesicles (Plate 25-1A).

The "mixed" transcription strategy of the ambisense S segment of the *Phlebovirus* genome is remarkable. The minus-sense 3′ half of this vRNA segment is transcribed into (+)mRNA, which is translated into the N (nucleocapsid) protein. The + sense 5′ half is not translated directly, but following transcription (replication) of the whole S RNA molecule, the 3′ (−) half of this complementary strand is transcribed into mRNA, which in turn is translated into another protein.

TABLE 25-2
Major Bunyaviruses Affecting Man

VIRUS	GENUS	DISTRIBUTION	ARTHROPOD VECTOR	AMPLIFIER HOST	HUMAN DISEASE
Hantaan	*Hantavirus*	Asia, Europe	Nil	Rodents	Hemorrhagic fever
Rift Valley fever	*Phlebovirus*	Africa	Mosquito	Sheep, cattle, buffalo	Hemorrhagic fever
Crimean-Congo hemorrhagic fever	*Nairovirus*	Central Asia (Africa)	Tick	Sheep, cattle, goats	Hemorrhagic fever
La Crosse	*Bunyavirus*	United States	Mosquito	Chipmunk, squirrel	Encephalitis
Sandfly fever	*Phlebovirus*	Mediterranean, South America	Sandfly	Gerbils?	Conjunctivitis, myositis

RODENT-BORNE NEPHROPATHY (HANTAAN VIRUS)

During the Korean War thousands of United Nations troops developed a severe disease designated Korean hemorrhagic fever, or hemorrhagic fever with renal syndrome (HFRS), but intensive efforts to discover the causal agent proved unsuccessful until 1978 when it was finally isolated from the rodent *Apodemus* and identified as a bunyavirus, named Hantaan. Furthermore, strains of Hantaan virus have been shown to be the cause of a similar but milder condition, nephropathia epidemica, well known in Scandinavia for many years. Hantaan-like viruses are now recognized to be quite widespread and the term rodent-borne nephropathy has been coined to embrace all forms of the clinical syndrome they produce in man.

Properties of Hantaviruses

Though morphologically resembling the *Uukuvirus* genus (Plate 25-1C), hantaviruses are characterized by a unique 3′ terminal nucleotide sequence common to their three RNA molecules. Furthermore, they are serologically unrelated to members of the other four genera and are not arboviruses. For these reasons it has been proposed (but not officially adopted at the time of writing) that they comprise a new genus, *Hantavirus*.

Hantaviruses have diverged in their several rodent hosts into several distinct serotypes, e.g., *Apodemus*-borne (Hantaan), *Clethrionomys*-borne (nephropathia epidemica), *Rattus*-borne, and *Microtus*-borne (Prospect Hill). Within each serotype, strains are distinguishable by oligonucleotide fingerprinting.

Clinical Features

As the descriptive, if cumbersome name, hemorrhagic fever with renal syndrome (HFRS) implies, there is profound renal tubular involvement in this hemorrhagic fever. Lumbar abdominal pain and proteinuria are prominent during the febrile phase, which gives way to a hypotensive hemorrhagic phase, then an oliguric phase with abnormal renal function, and finally a diuretic phase heralding recovery. Hemorrhage can occur from different sites in different patients, e.g., a petechial skin rash, massive gastrointestinal bleeding (sometimes presenting as an acute abdomen), or hemorrhagic pneumonia. Nephropathia epidemica is a nonlethal form of HFRS encountered in Scandinavia and Eastern Europe in which there is no hemorrhage or shock.

Laboratory Diagnosis

The Hantaan virus was originally isolated from wild rodents by passage in laboratory rats, then adapted to Vero-E6 cell cultures in which it grows rather poorly and is noncytocidal. For this reason and because of the need for P-4 facilities, virus isolation is not practicable as a routine diagnostic procedure. However, serum antibodies (IgM, or rising titer) can be demonstrated by immunofluorescence, hemagglutination-inhibition, or neutralization (plaque-reduction). The latter can differentiate the serotypes emanating from different rodent reservoirs.

Epidemiology

Three epidemiological types of rodent-borne virus nephropathy have been recognized: rural, urban, and laboratory associated. The reservoir for the rural type, which is much the commonest form of the disease, accounting for over 50,000 cases annually in China, is *Apodemus* in Korea, China, and eastern USSR, but *Clethrionymus* in Scandinavia and eastern Europe. These are field rodents, hence infections of man occur mainly in rural workers, sometimes as sporadic cases, sometimes as small outbreaks when a number of persons are exposed to a contaminated focus. Outbreaks in military personnel are associated with their exposure to such foci. In China and Korea there is a seasonal peak at harvest time (late fall) when the population of *Apodemus* peaks and much dust is generated. In Scandinavia maximum transmission occurs during the winter, when the reservoir host *Clethrionymus* invades houses.

The urban type of rodent-borne virus nephropathy, first reported in Seoul and Osaka, is transmitted by house rats. The discovery of hantaviruses in *Rattus* alerts us to the possibility of encountering human cases anywhere in the world, especially around ports. Outbreaks of HFRS have also occurred among laboratory animal handlers in several countries. Wistar rats were shown to be the reservoir.

A hantavirus known as Prospect Hill virus has been recovered from *Microtus* and other wild rodents from widely separated regions of North America. The virus has not yet been associated with human disease, although serological evidence of infection has been demonstrated in a small minority of mammalogists.

Experiments with various species of field mice and laboratory mice indicate that infected rodents exhibit no symptoms but excrete virus in their saliva, feces, and especially in their urine for many months. Spread to humans probably occurs by direct or air-borne contact with rodent excreta. Though not arboviruses, the possibility of passive spread by arthropod vectors merits further investigation. There is no evidence of person-to-person spread.

Control

Since rodent control in an agrarian environment is impracticable, there is a clear need for a vaccine against Hantaan virus. Work has commenced on the development of a Vero cell grown, inactivated vaccine, and independently on a recDNA cloned vaccine.

RIFT VALLEY FEVER

Rift Valley fever (RVF) has been known for many years as a devastating disease of ruminants which breaks out every decade or so in East or South Africa, killing the lambs and calves and inducing abortion in the pregnant ewes and cows. At the time of such epizootics occasional cases of a nonlethal dengue-like illness were observed in people who came into contact with sick animals or handled their carcasses. Then suddenly in 1977 a new dimension was added to the problem. For the first time a major epizootic occurred in the delta and valley of the Nile and for the first time large numbers of humans were affected. The disease broke out again in 1978. Over 200,000 Egyptians contracted a particularly virulent form of RVF. At least 600 died.

Clinical Features

The disease began after a very short incubation period (3 days) with a severe version of the previously known dengue-like syndrome—sudden onset of fever, malaise, anorexia, rigor, depression, severe myalgia, headache with retroorbital pain, often marked by leukopenia and hepatitis. However, within a week or so a proportion of the patients progressed to one of three complications: encephalitis (relatively mild and usually without sequelae), retinitis (with diminution of visual acuity sometimes leading to permanent loss of central vision), or hemorrhagic fever (marked by jaundice and widespread hemorrhage, with a mortality of 5–10%).

Laboratory Diagnosis

RVF virus grows in most types of cultured cells and is readily isolated in Vero or BHK-21. However, serology is more commonly employed for diagnosis, as well as for epidemiological surveys. HI is the simplest test, but care must be taken to remove nonspecific inhibitors by adequate pretreatment of the serum and one must also be conscious of partial cross-reactivity with the phlebotomous fever group of viruses.

Epidemiology

Epizootics of RVF break out in East Africa every 5–15 years, usually following exceptionally heavy rains. This points to mosquitoes as the likely vectors and indeed the virus has been isolated from *Culex pipiens.* During such epizootics a wide range of vertebrates may become infected, with sheep and cattle being the principal amplifier hosts. But we do not know what constitutes the natural reservoir that maintains the virus during the long interepidemic intervals. The native fauna is generally seronegative. There is evidence that the virus may be maintained in invertebrates such as mosquitoes (or possibly *Culicoides*) by transovarial transmission.

Laboratory tests so far have not come up with any explanation of how the 1977–1978 Egyptian outbreak was precipitated, nor why it conferred such high morbidity and mortality in man. Unauthorized movement of livestock such as camels from the Sudan has been postulated to have introduced the virus to a new ecosystem, the Nile Valley. But was this a new more virulent strain? Or were the clinical manifestations in man unusually severe because of the prevalence of cross-reacting antibody against phlebotomous fever virus?

Control

The 1977–1978 epizootic raises the specter of future spread of RVF to the Middle East and perhaps beyond. To minimize the chance of this happening it may become necessary to police the international movement of livestock as well as to screen animals serologically from time to time. Spread to man during outbreaks might be reduced by vigorous mosquito control and by implementing safer procedures for the killing and disposal of animals. Live attenuated and formalin-inactivated RVF vaccines are available for immunization of susceptible livestock in enzootic areas. There is also a case for immunizing those people who are particularly at risk, namely laboratory personnel, veterinarians, slaughtermen, etc. in East and South Africa. The present monkey kidney grown inactivated vaccine is in very short supply, but a similar product grown in Vero cells on microcarrier beads in suspension culture should shortly become available. Attempts are also being made to clone the relevant viral gene by recDNA technology, or to produce an avirulent live reassortant.

CRIMEAN-CONGO HEMORRHAGIC FEVER (CCHF)

An exceptionally severe hemorrhagic fever occurs in Central Asia. The causative agent, a bunyavirus of the *Nairovirus* genus, is serologically virtually indistinguishable from the Congo virus which had previously been recovered from ticks, animals, and man in Africa but had been associated only with the nonlethal disease, Congo fever.

Clinical Features

CCHF commences abruptly with fever, headache, severe back and abdominal pain, and progresses to extensive hemorrhages from almost any site—melena, hematemesis, hematuria, and hemorrhagic skin rash (Plate 25-2). Leukopenia, thrombocytopenia, proteinuria, and hepatitis are key findings. Blood loss from internal bleeding leads to shock, pulmonary edema, and death. Case fatality rates range from 5 to 50% depending on the availability of modern medical care.

Laboratory Diagnosis

The virus is lethal for suckling mice and can also be isolated in cultured cells, then identified by immunofluorescent (IF) staining. IF is also employed to demonstrate a rising titer of antibody in the patient's serum. There is a need for an EIA (using inactivated antigen, for safety) to detect specific IgM.

Epidemiology

The vector and principal reservoir of CCHF virus is the tick. The life cycle of ticks was described in Chapter 20 (page 492). Over two dozen species of ixodid (hard) ticks from seven different genera are known to transmit CCHF virus, which is maintained in a tick-to-tick cycle by transovarial and/or transstadial transmission. Vertebrates comprise important amplifier hosts. The larvae and nymphs generally parasitize birds and small animals such as hares, whereas adult ticks are mainly responsible for transmitting CCHF virus to larger animals such as cattle, sheep, goats, camels, and man. The epidemiology of CCHF in different parts of the USSR, the Middle East, and Africa is as variable as is the ecology of the species of tick involved.

Control

Tick control measures in domestic animals and impregnation of clothing with repellent, especially during spring and summer, can re-

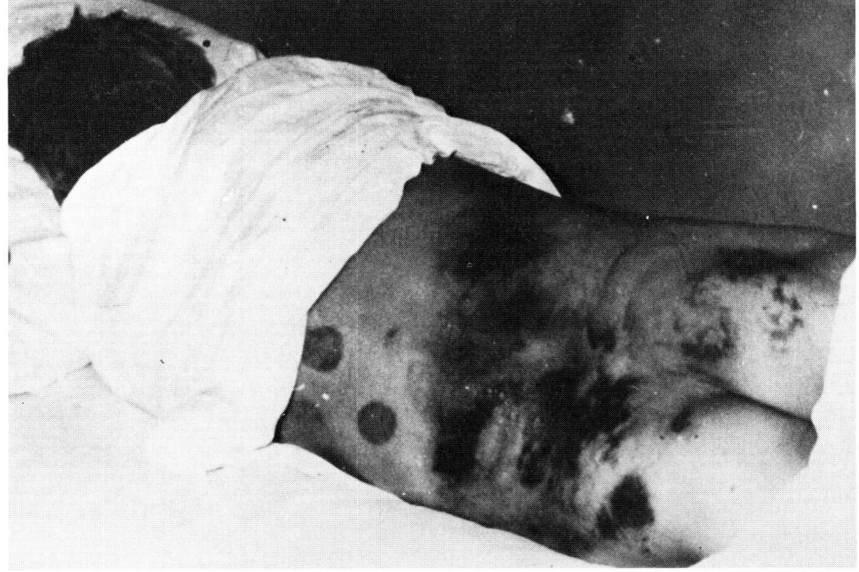

PLATE 25-2. *Crimean-Congo hemorrhagic fever, showing hemorrhagic skin rash. Petechiae progress to extensive ecchymoses. The circular sites on the left were caused by capillary fragility tests. [From J. Casals, B. E. Henderson, H. Hoogstraal, K. M. Johnson, and A. Shelekov,* J. Infect. Dis. **122,** *437 (1970); courtesy Dr. A. Shelekov.]*

duce the incidence of CCHF. There is a need for a safer vaccine than the mouse brain derived version currently avaliable.

CALIFORNIA ENCEPHALITIS (LA CROSSE VIRUS)

The so-called California group within the genus *Bunyavirus* within the family *Bunyaviridae* comprises more than a dozen viruses isolated from mosquitoes or vertebrates in various parts of the world. The best studied and most important human pathogen is La Crosse (LAC) virus, which causes encephalitis in children and forestry workers in wooded areas of the United States. The ecology of the virus is described in the legend to Fig. 25-1.

SANDFLY FEVER

Sandfly (phlebotomus) fever is a common but nonlethal disease transmitted by sandflies (*Phlebotomus papatasii*) to man in countries around the

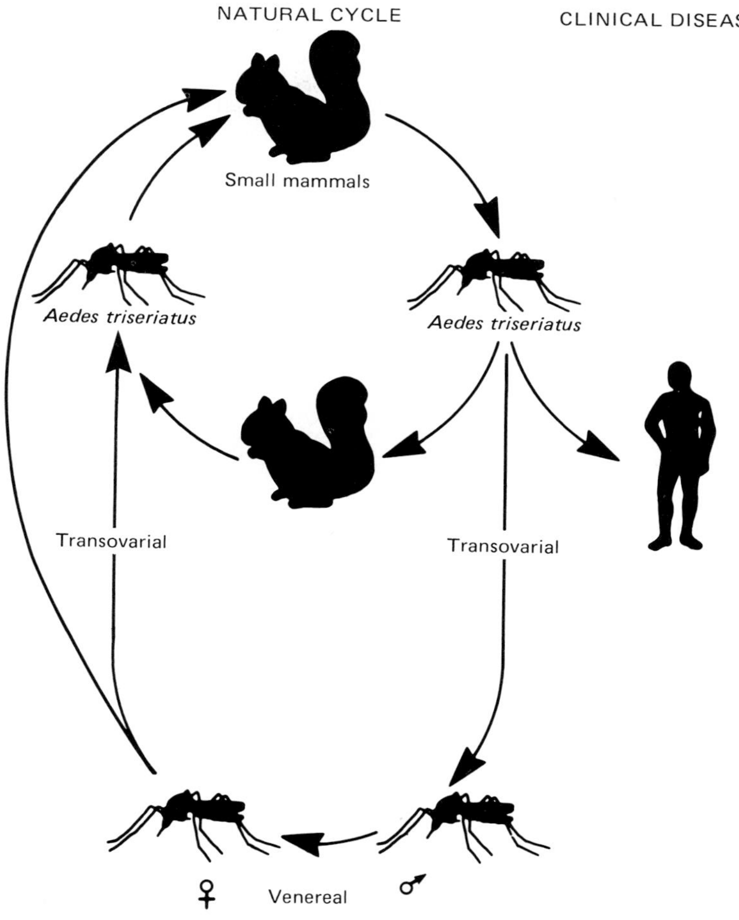

FIG. 25-1. *Cycle of California (La Crosse) encephalitis virus. The "inapparent" cycle of the La Crosse virus is between* Aedes triseriatus, *a woodland mosquito, and chipmunks and tree squirrels. The virus is also maintained indefinitely in mosquitoes by transovarial transmission and is amplified by venereal transmission between the infected male nonbiting mosquito and the uninfected female, which can in turn transmit either to a vertebrate by biting or to the next generation of mosquito by transovarial transmission. Man is the only known host to develop clinical disease and is a dead-end host for the virus. (From R. T. Johnson, "Viral Infections of the Nervous System." Raven Press, New York, 1982.)*

Mediterranean Sea and eastward to the USSR and India. A second major focus is Central and South America where a different variant of this *Phlebovirus* is involved. No vertebrate host other than man has been definitely incriminated but gerbils are among the principal suspects. The disease is a self-limiting dengue-like syndrome marked by fever, myalgia, retroorbital pain, and conjunctivitis.

FURTHER READING

Books

Berge T. O., ed. (1975). "International Catalogue of Arboviruses including certain other Viruses of Vertebrates," DHEW Publ. No. (CDC) 75-8301. U.S. Dept. of Health, Education and Welfare, Washington, D.C.; also see Supplement edited by N. Karabatsos (1978).

Calisher, C. H., and Thompson, W. H., eds. (1983). "California Serogroup Viruses," Prog. Clin. Biol. Res., Vol. 123. Alan R. Liss, Inc., New York.

Compans, R. W., and Bishop, D. H. L., eds. (1984). "Segmented Negative Strand Viruses: Arenaviruses, Bunyaviruses and Orthomyxoviruses." Academic Press, New York.

Karabatsos, N., ed. (1978). "Supplement to International Catalogue of Arboviruses including certain other Viruses of Vertebrates." *Am. J. Trop. Med. Hyg.* **27,** 372–440.

Swartz, T. A., Klingberg, M. A., Goldblum, N., and Papier, C. M., eds. (1981). "Rift Valley Fever." Karger, Basel.

Reviews

Bishop, D. H. L. (1979). Genetic analysis of bunyaviruses. *Curr. Top. Microbiol. Immunol.* **86,** 1.

Bishop, D. H. L. (1985). Replication of arenaviruses and bunyaviruses. *In* "Human Viral Diseases" (B. N. Fields *et al.*, eds.), pp. 1083–1110. Raven Press, New York.

Bishop, D. H. L., and Shope, R. E. (1979). Bunyaviridae. *In* "Comprehensive Virology" (H. Fraenkel-Conrat and R. R. Wagner, eds.), Vol. 14, p. 1. Plenum, New York.

Bishop, D. H. L., Calisher, C., Casals, J., Chumakov, M. P., Gaidamovich, S. Y., Hannoun, C., Lvov, D. K., Marshall I. D., Oker-Blom, N., Pettersson, R. F., Porterfield, J. S., Russell, P. K., Shope, R. E., and Westaway, E. G. (1980). Bunyaviridae. *Intervirology* **14,** 125.

Hoogstraal, H. (1979). The epidemiology of tick-borne Crimean-Congo hemorrhagic fever in Asia, Europe and Africa. *J. Med. Entomol.* **15,** 304–417.

Lee, H. W. (1982). Korean hemorrhagic fever. *Prog. Med. Virol.* **28,** 96.

Meegan, J. M. (1979). The Rift Valley fever epizootic in Egypt 1977–78. I. Description of the epizootic and virological studies. *Trans. R. Soc. Trop. Med. Hyg.* **73,** 618–723.

World Health Organization (1982). Rift Valley fever: An emerging human and animal problem. *WHO Offset Publ.* **63.**

CHAPTER 26

Rhabdoviruses and Filoviruses

Rhabdoviruses 567
Rabies 569
Filoviruses 571
Further Reading 574

RHABDOVIRUSES

The family *Rhabdoviridae* encompasses a fascinating variety of agents parasitizing a great variety of hosts. These distinctive bullet-shaped viruses are to be found not only in vertebrates, both warm blooded and cold blooded, but also in insects and in plants.

One rhabdovirus has the dubious privilege of causing the most lethal of all infectious diseases. Throughout history man has lived in terror of the "mad dog" whose bite transmits rabies (*rabidus* = mad); the agonizing muscular spasms accompanying any attempt to swallow water provided the disease with its evocative pseudonym "hydrophobia." Rabies is also featured in one of the milestones of medical research: in 1884, before the nature of viruses had even begun to be comprehended, Louis Pasteur developed a vaccine and publicly demonstrated its efficacy by saving the life of a peasant boy bitten by a rabid dog.

Properties of *Rhabdoviridae*

Rhabdoviruses (*rhabdos* = rod) are about 75 nm wide and 180 nm long (Plate 26-1). A lipid envelope containing typical glycoprotein peplomers encloses a helical nucleocapsid. The genome consists of a single linear molecule of ssRNA of negative polarity, MW 4×10^6 (Table 26-1).

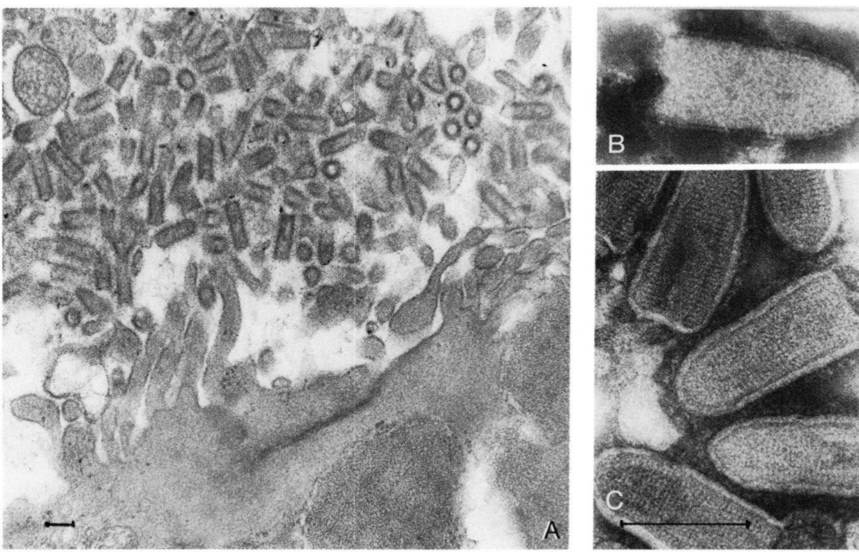

PLATE 26-1. Rhabdoviridae. *(A) Section of fox salivary gland infected with rabies virus. (B and C) Negatively stained preparations. (B) Rabies virus; (C) vesicular stomatitis virus. Bars = 100 nm. (Courtesy Dr. F. A. Murphy.)*

The *Lyssavirus* genus to which rabies virus belongs also includes several African viruses of vertebrates and invertebrates, some of which have been known to infect man. The rabies virus itself, though distributed across every continent except Australia and Antarctica, occurs as a single species with minor geographic variants distinguishable by monoclonal antibodies or oligonucleotide maps.

Viral multiplication takes place in the cytoplasm. Following uncoating, a polymerase associated with the nucleocapsid initiates primary transcription of five subgenomic + strands, representing monocistronic mRNAs to be translated into the five viral proteins. In addition, genome-length copies are transcribed for use as templates for RNA replica-

TABLE 26-1
Properties of Rhabdoviridae

Bullet-shaped enveloped virion, 180 × 75 nm
Helical symmetry
ssRNA genome, linear, negative polarity, MW 4 million
Cytoplasmic replication

tion. Possible mechanisms regulating the transcription and replication modes were discussed on page 74. The steps involved in the glycosylation of the envelope protein, and in the morphogenesis of the virion were set out in Fig. 3-10. Rabies virus buds mainly from intracytoplasmic membranes. Truncated DI particles, commonly produced during rhabdovirus replication, were discussed on page 95.

RABIES

Clinical Features in Man

After a prodomal period of fever, malaise, and often paresthesia around the site of the bite, muscles become hypertonic, and the patient becomes anxious, with episodes of hyperactivity, aggression, and convulsions. Paralysis is often a major feature. Delirium, coma, and death follow inexorably.

Pathogenesis

Man becomes infected when bitten (or licked on an abraded area) by an animal that is secreting virus in its saliva. Overall, the probability of developing rabies following a bite from a rabid dog is about one in six, but is substantially higher in children, especially following a bite to the head. The incubation period varies widely—from a week to a year, with a median of 1–2 months.

The virus infects striated muscle cells around the site of the bite, eventually gaining access to nerve endings then spreading centripetally via both motor and sensory axons to reach the spinal cord. This may result in ascending paralysis. The first neurons in the brain to be infected are those of the limbic system, precipitating the aggressive behavior ("fury") characteristic of the rabid animal and essential to the transmission of the virus by biting. At this time the virus spreads centrifugally via nerves to the salivary glands where acinar cells become infected and virus buds preferentially from the apical side of the plasma membrane directly into the saliva—a remarkable adaptation (Plate 26-1A). Eventually, generalized encephalitis supervenes and the animal dies.

Laboratory Diagnosis

The diagnosis of rabies is best undertaken by experienced reference laboratories. The most common call is to determine whether the animal known to have inflicted the bite is rabid. If clinical observation by a veterinarian suggests rabies, the animal is sacrificed and the hippocampal region of its brain is stained with fluorescent antibody. Rabies antigen is found in neurons in characteristic round cytoplasmic inclusions. These correspond with the classical "Negri bodies" revealed by Sellers or Giemsa stain (Fig. 5-1E).

In man the diagnosis is normally established at autopsy by the same procedure. Confirmation may be made by isolation of virus in cultured cells or by intracerebral inoculation of mice. During life, a corneal impression smear or skin biopsy from the neck is often positive for rabies antigen by immunofluorescence. Antibody becomes demonstrable in the serum or cerebrospinal fluid only later in the illness.

Epidemiology

Man is irrelevant to the ecology of rabies. To the virus he constitutes a blind alley, for humans cannot pass on the infection. In nature, "sylvan" rabies is endemic in a wide variety of carnivores, such as skunks in North America, jackals in Asia, mongooses in South Africa, and wolves in Turkey and Iran. Since World War II rabies has become epizootic in the fox population of Europe, spreading steadily westward through Poland to Germany to Switzerland and France, while raccoon rabies appears to be extending its range in the South-East of the United States. Vampire bats are a major reservoir of rabies in South America; thousands of cattle die annually from rabies transmitted via their bite. A strange outbreak of human rabies in Trinidad in 1929–1930 has been ascribed to vampire bats sucking blood from the toes of sleeping victims. In the United States speleologists have died from rabies acquired by inhaling aerosols created by the secretions of insectivorous bats roosting in caves. However, by far the most important sources of human infection in most parts of the world are the domestic dog and cat. Dogs are the source of thousands of urban cases of human rabies annually in India, China, Indonesia, Thailand, and the Philippines.

Only certain islands such as Australia, Britain, and Japan are free of rabies but even they are vulnerable since rabies virus multiplies in, and is lethal for every animal species in which it has been tested. This raises the puzzling question of how such an uncompromisingly lethal virus perpetuates itself in nature. We have alluded to the survival advantage conferred on the virus by unique features of its pathogenesis. But it is not readily apparent how its host species have survived. A number of factors may be involved: the exceptionally long incubation period of the disease, together with the rapid reproduction rate and limited movements of the host animals.

Prevention and Control

Eradication of sylvatic rabies is of course impossible. However, quarantine, rigidly enforced, continues to exclude the virus from islands such as Australia, aptly described (by an author with other things in mind) as "the lucky country." The strategy for controlling human rabies in most other parts of the world revolves around attempts to eliminate rabies from domestic and feral dogs and cats. Strays need to be destroyed, while pets should be registered and their movements controlled. Immunization of all dogs and cats is a desirable objective.

"Preexposure" immunization of humans is generally confined to those occupationally at risk, such as veterinarians, animal handlers, trappers, park rangers, and long-term visitors to endemic areas. Active immunization is also capable of preventing rabies "postexposure," because of its unusually long incubation period. Indeed, the prompt administration of rabies vaccine is capable of reducing the mortality from this frightening disease from virtually 100% to zero.

Rabies vaccines have come a long way since the days of Pasteur. Today's vaccines are grown from attenuated virus in human diploid fibroblasts or Vero cells, then inactivated with β-propiolactone, or split into subunits with tri-(*n*)-butyl phosphate. Local and mild systemic reactions are not uncommon, but the neurological sequelae and anaphylaxis for which earlier vaccines were notorious do not occur. About 1 in every 1000 recipients develops a mild hypersensitivity reaction following booster doses. A typical postexposure course consists of five or six injections given intramuscularly on days 0, 3, 7, 14, 28, and perhaps 90. A single dose of human rabies hyperimmune globulin is also injected into the wound as soon as it has been thoroughly washed and disinfected.

FILOVIRUSES

In 1967 workers handling kidneys from African monkeys in poliovaccine manufacturing plants in Yugoslavia and in Marburg, Germany contracted a previously unknown disease from which seven people died. Then, in 1976 nosocomial outbreaks of an even more lethal disease now known as Ebola hemorrhagic fever occurred in Zaire and the Sudan. The causative agents resembled rhabdoviruses and are classified as such at the time of writing. However, certain of their physicochemical properties, notably the extraordinary filamentous virions, longer than many bacteria, have led workers in the field to propose their separation off into a new family to be called *Filoviridae* (*filum* = thread).

Properties of Filoviruses

Surely the most bizarre of all viruses, the filoviruses are exceptionally long pleomorphic filaments, constant in diameter (80 nm) but varying in length from about 0.5–14 μm. Branched, circular, and U-shaped forms are also common (Plate 26-2). Maximum infectivity is associated with 790-nm (Marburg) or 970-nm particles (Ebola). The helical nucleocapsid, 50 nm in diameter, is characterized by a 20-nm central axial channel. The genome comprises a single molecule of ssRNA, MW 4×10^6.

Marburg and Ebola viruses are absolutely distinct, differing in genome size, polypeptide profile, and several other properties. The Sudan and Zaire serotypes of Ebola are readily distinguishable from one another by oligonucleotide maps, virulence for suckling mice, and ease of isolation in Vero cells.

Viral multiplication occurs in the cytoplasm. Nucleocapsids accumulate in conspicuous cytoplasmic inclusions. Budding takes place from the plasma membrane.

Clinical Features of Marburg and Ebola Fevers

Marburg and Ebola viruses cause similar types of hemorrhagic fever. Cardinal features in addition to severe hemorrhages and fever are diarrhea, vomiting, abdominal pain, myalgia, pharyngitis, conjunctivitis,

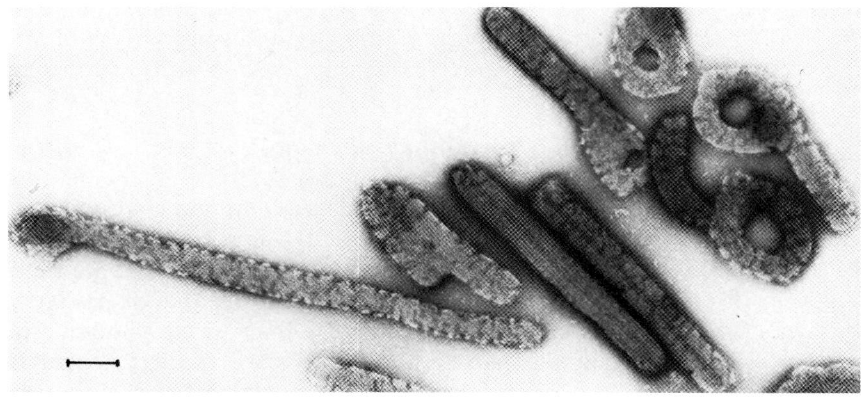

PLATE 26-2. Filoviridae. *Negatively stained preparation of virions of Ebola virus. Bar = 100 nm. (Courtesy Dr. F. A. Murphy.)*

and proteinuria. Onset is sudden, and progress to prostration, profound hypotension, and death is rapid. Mortality rates are extremely high: 25% for Marburg, 60% for the Sudan variant of Ebola, and 90% for the Zaire variant of Ebola.

Laboratory Diagnosis

Because of the inherent danger of these Class-4 pathogens, specimens, typically blood or autopsy material, must be taken with great care and sent, refrigerated and packed according to IATA regulations, to one of the world's half-dozen P4 containment facilities (Plate 12-5). There, filovirus may be isolated in Vero cell cultures and identified by electron microscopy (Plate 26-2) and serologically, by staining the infected cultures (following precautionary γ irradiation) with fluorescent antibody. Specific antigen is also detectable by immunofluorescence in sections or impression smears of liver taken at autopsy. In addition, antibody in the patient's serum is demonstrable by immunofluorescence or radioimmunoassay, using preprepared virus-infected, γ-irradiated Vero cell cultures as antigen.

Epidemiology and Control

Almost certainly these rare lethal African hemorrhagic fevers are zoonoses, but the natural hosts are unknown. Despite the fact that the only known outbreaks of Marburg disease in man were clearly traceable to a consignment of monkeys from Uganda, the deaths of these monkeys and the absence of antibodies in wild African monkeys indicate that this animal, like man, is not the natural host of the Marburg virus. Similarly, seroepidemiological surveys of African wildlife have not yet identified the reservoir of Ebola virus.

Nosocomial transmission, probably via contaminated needles, caused the 1976 Zaire outbreak of Ebola hemorrhagic fever which emanated from a mission hospital and resulted in 318 cases with 280 deaths. Person-to-person spread, by contact with blood or secretions, has also caused many cases of Ebola and Marburg disease in medical personnel and attending family members. However, serological surveys in central Zaire reveal that mild or subclinical Ebola infections are not uncommon among villagers in the tropical rainforests.

As with other exotic hemorrhagic fevers caused by Class-4 agents (see Chapter 29), patients should be maintained in strict isolation, and contacts including hospital staff kept under continuous surveillance.

FURTHER READING

Rhabdoviruses

Books

Baer, G. M., ed. (1975). "The Natural History of Rabies," Vols. 1 and 2. Academic Press, New York.

Bishop, D. H. L., ed. (1979–1980). "Rhabdoviruses," Vols. 1–3. CRC Press, Boca Raton, Florida.

Kaplan, M., and Koprowski, H., eds. (1973). "Laboratory Techniques in Rabies," W. H. O. Monogr. Ser. No. 23. World Health Organ., Geneva.

Kuwert, E., Mérieux, C., Koprowski, H., and Bögel, K., eds. (1985). "Rabies in the Tropics." Springer-Verlag, Berlin.

Reviews

Centers for Disease Control (1980). Rabies prevention. *Morbid. Mortal. Wkly. Rep.* **29,** 265–280.

Johnson, H. N. (1979). Rabies virus. *In* "Diagnostic Procedures for Viral, Rickettsial and Chlamydial Infections" (E. H. Lennette and N. J. Schmidt, eds.), 5th ed., pp. 843–877. Am. Public Health Assoc., Washington D.C.

Plotkin, S. A. (1980). Rabies vaccine prepared in human cell cultures: Progress and perspectives. *Rev. Infect. Dis.* **2,** 433.

World Health Organization (1973). Expert Committee on Rabies, Sixth Report. *W. H. O. Tech. Rep. Ser.* **523.**

Wunner, W. H., Dietzschold, B., Curtis, P. J., and Wiktor, T. J. (1983). Rabies subunit vaccines. *J. Gen. Virol.* **64,** 1649.

Filoviruses

Books

Martini, G. A., and Siegert, R., eds. (1971). "Marburg Virus Disease." Springer-Verlag, Berlin and New York.

Pattyn, S. R., ed. (1978). "Ebola Virus Haemorrhagic Fever." Elsevier/North Holland Biomedical Press, Amsterdam.

Reviews

Kiley, M. P., Bowen, E. T. W., Eddy, G. A., Isaacson, M., Johnson, K. M., McCormick, J. B., Murphy, F. A., Pattyn, S. R., Peters, D., Prozesky, O. W., Regnery, R. L., Simpson, D. I. H., Slenczka, W., Sureau, P., van der Groen, G., Webb, P. A., and Wulff, H. (1982). Filoviridae: A taxonomic home for Marburg and Ebola viruses? *Intervirology* **18,** 24.

McCormick, J. B., and Johnson, K. M. (1984). Viral hemorrhagic fevers. *In* "Tropical and Geographical Medicine" (K. S. Warren and A. A. F. Mahmoud, eds.), pp. 676–697. McGraw-Hill, New York.

Murphy, F. A. (1985). Marburg and Ebola viruses. *In* "Human Viral Diseases" (B. N. Fields *et al.*, eds.), pp. 1111–1118. Raven Press, New York.

World Health Organization International Commission to Sudan (1978). Ebola hemorrhagic fever in Sudan and Zaire, 1976. *Bull. W. H. O.* **56,** 247–293.

CHAPTER 27

Reoviruses

Introduction 575
Properties of *Reoviridae* 576
Rotaviruses 577
Orbiviruses 582
Orthoreoviruses 583
Further Reading 583

INTRODUCTION

The name reovirus is a sigla, short for "*r*espiratory *e*nteric *o*rphan," because the first members of this family to be discovered were found to inhabit both the respiratory and the enteric tract of man and animals, but to be "orphans" in the sense that they are not associated with any disease. The discovery of the human rotaviruses changed all that, for they are recognized to be the most important cause of infantile gastroenteritis throughout the world. In addition, dozens of arboviruses, previously unclassified, have now been allocated to the *Orbivirus* genus of the *Reoviridae* family. Yet other genera infect plants and insects, raising the question of whether these fascinating viruses that cross kingdoms so readily might have evolved in insects.

Furthermore, reoviruses have attracted much attention from molecular biologists because of the unique nature of their genome. Composed of double-stranded RNA, the genome is segmented into 10–12 separate molecules, each representing a different gene. For at least one reovirus, each gene has been cloned and sequenced and its protein product characterized. Moreover, the facility with which these viruses undergo genetic reassortment has been exploited to exchange genes from *ts* mutants and thus determine the role of individual genes in pathogenesis and virulence (see Table 4-4 and Fig. 4-4).

PROPERTIES OF *REOVIRIDAE*

Virions of the *Reoviridae* family possess a striking morphology (Plate 27-1). They have not one but two concentric capsids, each displaying icosahedral symmetry. The genome comprises 10–12 unique molecules of dsRNA, total MW 12–15 million, depending on genus (Fig. 3-6e and Plate 27-2). All five enzyme activities required for the transcription of 5′-capped mRNA are present in the core of the virion (Table 27-1).

The *Reoviridae* family contains six genera, three of which—*Orthoreovirus, Orbivirus,* and *Rotavirus*—contain species that infect man.

Viral Multiplication

Key aspects of the replication of the well-studied *Reovirus* genus have been described in Chapter 3. Briefly, the endocytosed virion is uncoated only partially, by lysosomal hydrolases, to a "subviral particle" (SVP, Plate 3-3). This activates the viral transcriptase and capping enzymes to

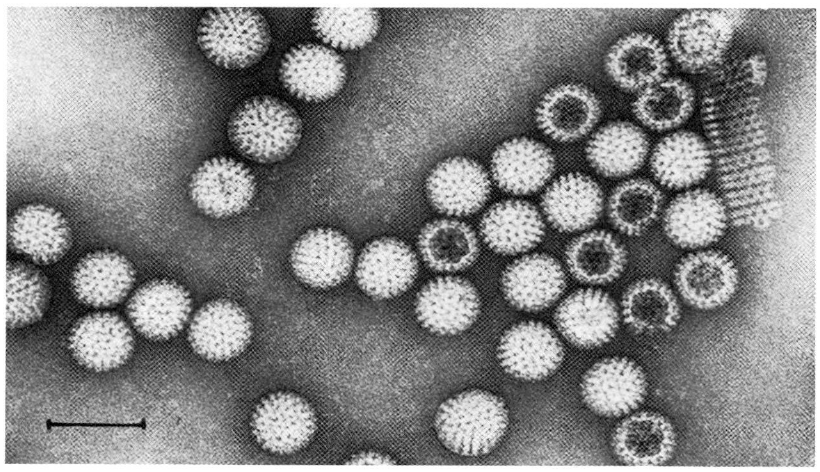

PLATE 27-1. Reoviridae. *Negatively stained preparation of virions of human rotavirus. Bar = 100 nm. (Courtesy Dr. E. L. Palmer.)*

TABLE 27-1
Properties of Reoviridae

Spherical virion, 60–80 nm
Two icosahedral capsids
dsRNA genome, segmented, 10–12 molecules, total MW 12–15 million
Five enzymes in core
Cytoplasmic replication
Genetic reassortment occurs between related species

transcribe 5′-capped mRNA molecules which, uniquely, are not polyadenylated at their 3′ ends. Only certain genes are transcribed initially; the others are derepressed following the synthesis of an early viral protein. The translation of reovirus mRNA and its regulation were described on page 77. Protein associates with each mRNA molecule and a minus RNA strand is synthesized, producing a dsRNA molecule. These in turn serve as templates for the transcription of more mRNA, which this time is uncapped. By a mechanism that is unclear, these uncapped reovirus mRNA molecules are then translated preferentially to yield a large pool of viral structural proteins. Finally, the SVPs associate with additional proteins to complete the maturation of new virions. All this takes place in the cytoplasm.

In the case of the *Rotavirus* genus, morphogenesis involves an unusual type of budding of single-shelled particles into vesicles of rough endoplasmic reticulum; the pseudoenvelopes they thus acquire are subsequently removed and the outer capsid is added in the vesicles, since the major outer capsid protein is glycosylated and its synthesis can be completed only as it traverses the endoplasmic reticulum membrane.

ROTAVIRUSES

In 1973 a new virus was discovered, first in duodenal biopsies then in feces of children with gastroenteritis. Electron micrographs revealed virions with a unique appearance (*rota* = wheel). Quickly it became apparent that rotaviruses are the commonest cause of gastroenteritis in infants and are responsible for countless deaths each year in the Third World.

Properties of Rotaviruses

The rotavirion is particularly photogenic (Plate 27-1). Smooth and round in outline, it is seen to have two concentric icosahedral shells, overall diameter 70 nm. The genome consists of 11 molecules of dsRNA, total MW 12×10^6.

Rotaviruses are ubiquitous. Essentially every species of domestic animal that has been thoroughly searched has its own indigenous rotavirus, causing diarrhea ("scours") in the newborn. All members of the *Rotavirus* genus, they can be separated into half a dozen groups on the basis of the particular type of inner capsid antigen they possess. Most of the typical human and animal rotaviruses can be shown by EIA, IEM, IF, or CF to fall into group A. Others, once called "atypical rotaviruses" or "pararotaviruses," some of which have recently been discovered to cause outbreaks of severe gastroenteritis in adults, belong in newly defined groups together with certain animal rotaviruses.

Differentiation into serotypes is based upon neutralization tests, which recognize the major glycoprotein in the outer capsid. Interestingly, nucleotide sequencing of cloned cDNA copies of the gene for this protein derived from various animal rotaviruses and human rotavirus type 2 demonstrated a remarkable degree of homology (75%). So far, 4 or 5 human serotypes have been identified, some of which also occur naturally in other animals, e.g., dogs and pigs.

Large numbers of *electropherotypes* can be differentiated by polyacrylamide gel electrophoresis of the viral RNA (Plate 27-2). Of course, since these patterns reflect differences in any of the 11 RNA segments, there is no consistent relationship between electropherotypes and serotypes. Nor do the patterns tell us as much as nucleic acid hybridization or sequencing about genetic relatedness or evolutionary history, but they are a valuable tool for tracking the epidemiology of individual strains (see below).

Clinical Features

Asymptomatic infection is the rule in neonates and is quite common in older children and adults. Clinical illness is seen principally between the ages of 6 and 24 months, with the peak around 1 year. After an incubation period of 1–3 days, vomiting generally precedes diarrhea, which lasts for 4–5 days and can lead to severe dehydration. Death is rare in well-nourished children but large numbers die in the poorer tropical countries.

Pathogenesis and Immunity

Rotaviruses multiply selectively in the differentiated columnar epithelial cells found near the apex of the villi of the small intestine. The dividing cells in the crypts of the villi are resistant, as are the similar

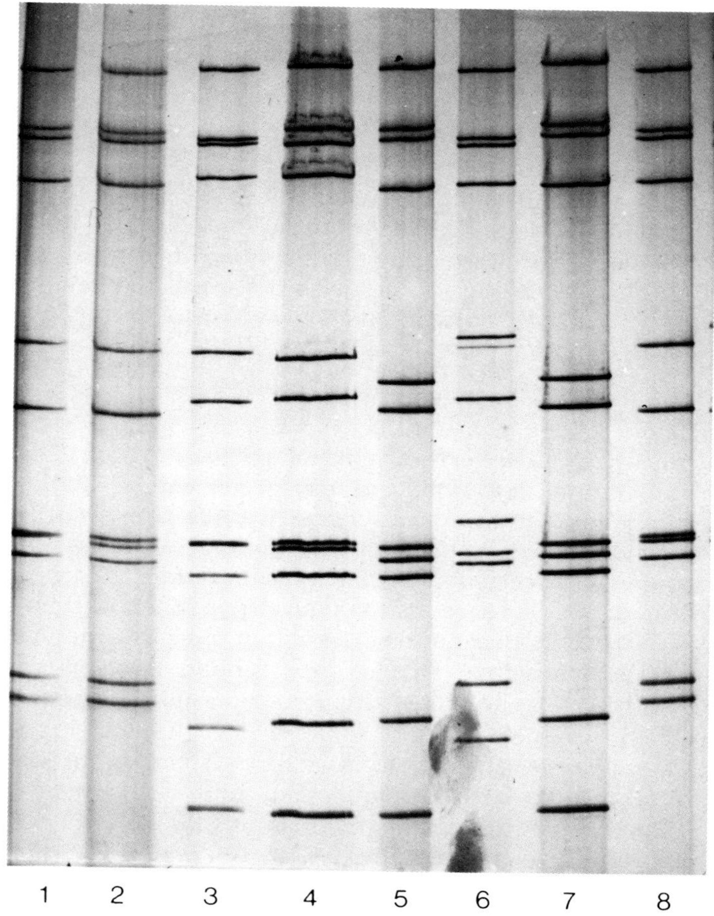

PLATE 27-2. *Electropherotypes of rotavirus. Seven separate isolates of rotavirus obtained from infants in Noumea between 1980 and 1983 were analyzed by polyacrylamide gel electrophoresis of their RNA. Tracks 5 and 7 show two samples of the same electropherotype. (Courtesy Drs. G. Panon and I. H. Holmes.)*

immature cuboidal epithelial cells which move up to cover the shortened tips of virus-damaged villi.

Maternal antibodies of the IgG class transmitted across the placenta do not protect the newborn but IgA antibodies in the colostrum do. Breast-fed infants are significantly less susceptible than bottle-fed infants during the first few days of life. Thereafter, the antibody titer in milk drops off rapidly, but infections acquired during the first few months tend to be inapparent. Neonatal infections themselves do not confer immunity against reinfection, but such reinfections tend to be clinically less severe. Immunity following clinical disease occurring at a later age may be more durable; in a recent trial in volunteers serum antibodies and resistance to challenge were demonstrable 19 months after experimental infection. Coproantibodies are assumed to be the most important. Of interest in this regard is the report that the diarrhea with prolonged rotavirus shedding that occurs in immunocompromised children can be successfully treated with milk containing antibody. All in all, little is known about the duration, mechanism, or type-specificity of postinfection immunity.

Laboratory Diagnosis

Rotaviruses were discovered by electron microscopy and this remains a satisfactory approach to rapid diagnosis (see Chapter 12); the virions are plentiful in feces and are so distinctive that they cannot be mistaken for anything else. However, EIA (or RIA) on fecal extracts is more practicable for the average laboratory, and more sensitive. Reverse passive hemagglutination (RPHA) is also sensitive, specific, and simple (refer Chapter 12 for details). Yet another possible approach is "dot hybridization," using a radioactively labeled or biotin-labeled cDNA probe to anneal with viral RNA extracted from feces and immobilized on nitrocellulose filters.

Only recently has a satisfactory cell culture system for the isolation of human rotaviruses been worked out. Rolled cultures of primary monkey kidney (PMK) or continuous MK cell lines such as MA104 support the growth of group A rotaviruses provided that trypsin is incorporated in (serum-free) medium, to cleave the relevant outer-capsid protein and thus facilitate uncoating. CPE is not striking; immunofluorescence is used to identify rotaviral antigen in the culture.

Neutralization tests (plaque reduction or fluorescent focus reduction) can be used to determine the serotype of the isolate. Polyacrylamide gel electrophoretic analysis of the viral RNA permits even finer discrimination between strains.

Serum antibodies can be measured by EIA, RIA, CF, or HI.

Epidemiology

Rotaviruses are shed in large numbers—up to 10^{10} particles per gram of feces, for 3–5 days, or sometimes longer. Spread presumably occurs mainly by contact with feces and can be decreased by rigorous attention to hygiene, e.g., handwashing, disinfection, proper disposal of contaminated diapers, etc. Nevertheless, nosocomial outbreaks in hospitals and nurseries are so common as to be almost standard. Water-borne epidemics involving adults also occur, the virus being resistant to chlorination. Descriptions of respiratory symptoms in some children and the winter prevalence of rotavirus in temperate climates raise the question of whether transmission can also occur via the respiratory route. The antigenic similarity between certain serotypes of animal and human rotavirus and the facility with which human rotaviruses can be grown in piglets, calves, and dogs raise the possibility of zoonotic infections.

Long-term prospective studies of hospital nurseries and pediatric wards indicate that about half of all newborn infants may become infected. Usually no disease occurs at this early age but a degree of immunity develops. Disease is commoner in the 6-month to 2-year-old group, significant numbers of whom also become infected in these nosocomial outbreaks. A 7-year prospective investigation of the "molecular epidemiology" of rotaviruses in children admitted to a Melbourne hospital revealed that, while one particular strain tends to predominate for months or years before being replaced by another, several electropherotypes are usually found to be cocirculating simultaneously (see also Plate 27-2). Under these circumstances, opportunities abound for genetic reassortment to occur. A logical extension of these studies will be to determine whether or not these genetic changes are accompanied by evidence of antigenic shift or drift analogous to that seen with influenza.

Repeated infections continue to occur throughout life, but most are subclinical. Spread occurs readily within the family unit from children to parents. Rotavirus is also a significant cause of "traveler's diarrhea." The virus is endemic in the tropics throughout the year, but in temperate climates the peak incidence of infections tends to be in winter.

Control

Raising the standard of nutrition and hygiene in the developing world is the long-term answer to the horrific mortality from infantile gastroenteritis. In the meantime many lives can be saved by fluid and electrolyte replacement. A suitable mixture of glucose and electrolytes for administration by mouth has been approved by WHO for use in developing countries where facilities for intravenous rehydration are not widely available. Indeed, such has been the success of oral therapy that intravenous therapy may be necessary only for those infants with shock or severe vomiting.

Patently there is a need for a vaccine and work has begun. A live oral vaccine containing the major human serotypes is the obvious approach. Any vaccine would need to be administered early in life to protect the infant during the all-important first 2 years. It remains to be seen whether the growth of a live vaccine given at that time will be seriously impeded by maternal antibody, how long the resultant immunity will last, and whether it is cross-protective against serotypes not included in the vaccine. In this regard a recent report that live bovine rotavirus given orally to infants induced no disease but protected them during a subsequent epidemic of human rotavirus is provocative. recDNA technology is also being harnessed to clone the gene for the protective outer capsid protein. Thought is being given to immunizing cows with human (or bovine) rotaviruses, or incorporating monoclonal neutralizing antibody in milk or milk substitutes.

ORBIVIRUSES

The *Orbivirus* genus gets its name from the large doughnut- or ring-shaped "capsomers" (*orbis* = ring or circle) that make up the inner capsid. The genome generally consists of 10 molecules of dsRNA, although that of the Colorado tick fever serogroup has 12, hence these viruses may more properly be reclassified as a separate genus. The subdivision of the *Orbivirus* genus into 12 serological groups on the basis of shared CF antigens is in need of revision, because of antigenic cross-reactions that belie the current classification. Subgenera based on more meaningful genetic and evolutionary relationships might be constructed by testing which species can undergo genome reassortment with one another. Almost 100 *Orbivirus* serotypes are distinguishable by neutralization tests. They are all arboviruses, with members of different serogroups being transmitted by different arthropods—ticks, mosquitoes, sandflies, and midges. The most important pathogens are the several serotypes causing bluetongue of sheep and African horse-sickness. Few orbiviruses are known to infect man, the most important being the Colorado tick fever viruses of the United States and the Kemerovo viruses of Siberia, all of which are tick-borne and cause a dengue-like fever in man.

Colorado Tick Fever (CTF)

CTF is a self-limiting fever contracted from ticks by campers and forest workers in the western United States. The virus is maintained in the wood tick, *Dermacentor andersoni,* being transmitted transstadially, and overwintering in hibernating nymphs and adults. Nymphs feed on small mammals such as squirrels and rabbits, which serve as a reservoir for the virus. Adult ticks feed on larger mammals and accidentally infect man during the spring and early summer.

After an incubation period of 3–6 days, the onset of illness is sudden. A characteristic "saddle-back" fever, headache, retroorbital pain, and severe myalgia in the back and legs are the features of this unpleasant but relatively harmless disease. Leukopenia is regularly found and virus can be isolated from red blood cells or detected in them by immunofluorescence for months after symptoms have disappeared. This is a remarkable situation, as erthrocytes of course have no ribosomes and cannot support viral replication. It seems that CTF virus multiplies in erythrocyte precursors in bone marrow, then persists in the mature red cell throughout its life span.

ORTHOREOVIRUSES

Though viruses of the *Orthoreovirus* genus were the first to be discovered and have been most intensively studied by molecular biologists (see above) we have relegated the description of their clinical significance to this postscript because they have never been positively demonstrated to cause disease in man. Yet they must commonly cause inapparent infection because the majority of people acquire serum antibodies by adulthood. The virus is shed in feces for up to several weeks and is assumed to spread mainly by the fecal–oral route. There are three human serotypes, all of which are also widespread in virtually every species of mammal that has been carefully searched.

The viruses were originally isolated in infant mice, but also grow readily in essentially any type of cell culture. CPE is slow to develop and not particularly striking; infected cells acquire a granular appearance, not very different from the nonspecific degeneration seen in uninfected cultures as they deteriorate with time. On staining however, characteristic crescentic perinuclear acidophilic inclusions are seen in the cytoplasm (Plate 5-1B and Fig. 5-1C). The three serotypes share common antigens demonstrable by IF or CF, but can be distinguished by HI (with human RBC).

FURTHER READING

Books

Joklik, W. K., ed. (1983). "The Reoviridae." Plenum, New York.

Tyrrell, D. A. J., and Kapikian, A. Z., eds. (1982). "Virus Infections of the Gastrointestinal Tract." Dekker, New York.

Reviews

Estes, M. K., Palmer, E. L., and Obijeski, J. F. (1983). Rotaviruses: A review. *Curr. Top. Microbiol. Immunol.* **105,** 123.
Fields, B. N. (1982). Molecular basis of reovirus virulence. *Arch. Virol.* **71,** 95.
Fields, B. N., and Greene, M. I. (1982). Genetic and molecular mechanisms of viral pathogenesis: Implications for prevention and treatment. *Nature (London)* **300,** 19.
Holmes, I. H. (1979). Viral gastroenteritis. *Prog. Med. Virol.* **25,** 1.
Joklik, W. K. (1981). Structure and function of the reovirus genome. *Microbiol. Rev.* **45,** 483.
Kapikian, A. Z., Wyatt, R. G., Greenberg, H. B., Kalica, A. R., Kim, K. W., Brandt, C. D., Rodriguez, W. J., Parrott, R. H., and Chanock, R. M. (1980). Approaches to immunization of infants and young children against gastroenteritis due to rotavirus. *Rev. Infect. Dis.* **2,** 459.

CHAPTER 28

Other Viral Diseases

Acquired Immunodeficiency Syndrome (AIDS) 585
Non-A, Non-B Hepatitis 596
Delta (HDV) Hepatitis 599
Spongiform Encephalopathies 601
Further Reading 605

ACQUIRED IMMUNODEFICIENCY SYNDROME (AIDS)

Introduction

In 1981 a new disease was described in the male homosexual populations of New York, San Francisco, and Los Angeles. Acquired immunodeficiency (or immune deficiency) syndrome (AIDS) is a severe immunosuppressive condition of very high mortality, characterized by a marked depletion of helper T cells, leading to death within a year or two from opportunistic infection or cancer. The preponderance of cases among highly promiscuous male homosexuals with a hedonistic life style elicited a predictable backlash against the gay community. The alarm escalated when it became evident that blood donated by homosexuals was transmitting the lethal disease to other members of the community. By 1985 over 10,000 cases of AIDS had been reported in the United States and several thousand in other parts of the world including Europe, Africa, Central and South America, and Australia. Serological evidence of infection was widespread in male homosexuals and in hemophiliacs. It is now clear that a major epidemic of a frightening new disease is underway and that no method of control is in sight. Even if some method of stopping the spread of AIDS were to be discovered today and implemented with 100% efficiency forthwith, hundreds of thousands of the million or more people already infected with the virus would die during the next decade.

The import of the epidemic was quickly appreciated by the Centers for Disease Control in Atlanta and a massive research effort by CDC, the National Institutes of Health, and universities, funded by the United States Government, was mounted. It is impressive testimony to the capability of the public health authorities and the medical research fraternity to respond promptly to such a challenge that within 3 years of the first description of the new disease, its cause had been identified.

Properties of LAV/HTLV-III

In 1983 Montagnier's group at l'Institut Pasteur recovered a novel retrovirus from a male homosexual suffering from lymphadenopathy, or "pre-AIDS." They called the virus LAV, for lymphadenopathy-associated virus, and later showed it to be identical with the virus which they subsequently isolated from patients with typical AIDS. The following year Gallo's team at the United States National Cancer Institute cultured a virus which they called HTLV-III (human T cell lymphotropic virus, type III) from several cases of AIDS, in a malignant T cell line, which supported growth of the virus to high titer. These viruses are now recognized to be essentially the same, and pending adjudication by the ICTV, tend to be known as LAV/HTLV-III.

LAV/HTLV-III is a member of the *Retroviridae* family (see Chapter 8). Morphologically and morphogenetically, it resembles the so-called type-D retroviruses of the *Lentivirinae* subfamily (Plate 28-1). The 100 to 140-nm enveloped virion contains an electron-dense core or "nucleoid" which often appears cylindrical (bar-shaped). The viral RNA genome displays homology with that of the lentivirus visna, but not with the human T cell leukemia viruses, HTLV-I and -II. Like LAV/HTLV-III, members of the *Lentivirinae* subfamily are not oncogenic but are cytocidal for lymphocytes. They are also notorious for their capacity to undergo progressive antigenic drift in their envelope glycoprotein during their prolonged sojourn in a single animal (sheep or goat for visna virus; horse for equine infectious anemia virus). In this connection it is intriguing to note that different isolates of HTLV-III reveal heterogeneity in the *env* region of the genome; about 20% of the nucleotides in the *env* gene of one African strain differ from those in one American strain. This may have repercussions for the use of purified envelope antigen as a vaccine (Table 28-1).

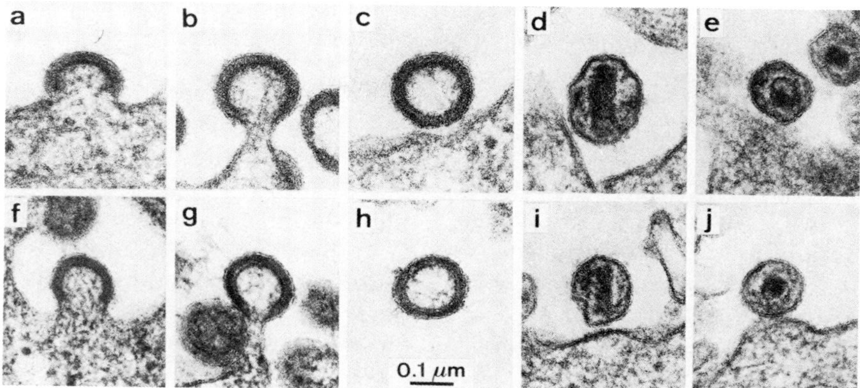

PLATE 28-1. *Electron micrographs of thin sections of cells infected with the retroviruses HTLV-III, the cause of AIDS (a to e), and visna virus, a lentivirus (f to j). [From M. A. Gonda* et al. Science **227,** *173 (1985), copyright 1985 by the AAAS. Courtesy Dr. M. A. Gonda.]*

Clinical and Immunological Features

The spectrum of clinical manifestations of HTLV-III infection has yet to be fully defined. At least half of all infections are asymptomatic, or at most induce a transient fever, lymphadenopathy and rash resembling infectious mononucleosis, after an incubation period of 2–3 weeks. A further 25–35% lead to a nonfatal condition variously known as the lymphadenopathy syndrome (LAS) or AIDS-related complex (ARC), while AIDS itself develops in only about 5–15% (2–20% in various studies). However, because these figures are based on an observation period

TABLE 28-1
Properties of LAV/HTLV-III

Retroviridae family, *Lentivirinae* subfamily
Enveloped "type-D" virion, 100–140 nm, with condensed cylindrical core
Mg^{2+}-dependent, high-MW reverse transcriptase
ssRNA genome, 9.5 kb, homology with *Lentivirinae*
Heterogeneity in *env* gene
Budding from plasma membrane, long stalk
Cytocidal, not transforming, for $T4^+$ lymphocytes
Viral DNA persists in integrated state

of only 2–3 years and it now appears that the average incubation period of AIDS may be longer than this, it is probable that they represent underestimates.

LAS (or ARC) is characterized by generalized enlargement of lymph nodes, fever, diarrhea, and weight loss. These are also prominent features of AIDS, and many clinicians regard LAS as "pre-AIDS" or "prodromal AIDS," but AIDS itself is strictly defined by the clinical and immunological evidence of severe immunosuppression which is set out below.

AIDS is marked by almost total depletion of $OKT4^+$ lymphocytes, the so-called helper/inducer T cells. In contrast, the $T8^+$ lymphocytes, including T_s cells, are not diminished but may actually increase in number. Consequently, the T4 : T8 ratio is inverted, usually falling progressively to <1 as the disease progresses. Cutaneous anergy, i.e., failure to respond with DTH to skin tests against common antigens, is also a regular feature. Other immunological findings are more variable and are influenced by the number and nature of intercurrent infections. They include polyclonal hypergammaglobulinemia and immune complexes, diminished proliferative response of lymphocytes to T cell mitogens *in vitro*, and increased production of acid-labile α-interferon and suppressor factors. The lymph nodes may show involution of follicles and/or disruption of germinal centers (Table 28-2).

The attention of clinicians was first drawn to AIDS by the remarkable frequency of opportunistic infections with unusual parasites such as *Pneumocystis carinii*, and of an otherwise rare malignancy known as Kaposi's sarcoma. Kaposi's sarcoma had been known as a relatively harmless collection of indolent skin tumors in old people of Mediterranean descent, but in young homosexuals with AIDS it presents in a much more invasive form. Other types of malignancy, notably aggressive B cell lymphoma or non-Hodgkin's lymphoma, sometimes involving the brain, and squamous cell carcinomas of anus or mouth, also occur.

In cases with profound immunodeficiency, opportunistic infections are the more common presentation. *Pneumocystis carinii* pneumonia is the most frequent in the United States, being seen in over half of all lethal cases of AIDS. In Africa, cerebral toxoplasmosis or cryptococcosis, and oropharyngeal (or esophageal) candidiasis are especially prominent. Indeed, the clinical picture of AIDS may be complicated by successive intercurrent infections with any of a wide range of opportunistic invaders, including herpesviruses, adenoviruses, and polyomaviruses, as well as bacterial infections of the intestinal and respiratory tracts. The overall case fatality rate of AIDS is of the order of 90% in those diag-

TABLE 28-2
Immunological Features of AIDS

Markedly decreased T4 cells
Normal or increased T8 cells, hence T4 : T8 ratio <1
Cutaneous anergy (negative DTH skin tests)
Decreased proliferation of T cells in response to mitogens *in vitro*[a]
Polyclonal activation of B cells → hypergammaglobulinemia[a]
Immune complexes[a]
Acid-labile IFN-α and β-microglobulin increased[a]
Lymph nodes show follicular involution, germinal center disruption[a]

[a] Irregular, and influenced by intercurrent infections.

nosed 3 years or more earlier. The mean survival time is about a year (Table 28-3).

In the terminal stages of AIDS about 75% of patients develop degenerative changes in the brain, leading to dementia and other neurological problems. The analogy with visna in sheep is noteworthy; this lentivirus also infects brain cells as well as lymphocytes and causes a chronic degenerative neurological disease.

Pathogenesis and Immunity

The evidence that LAV/HTLV-III causes AIDS is now compelling. Antibodies to HTLV-III proteins are found in 90–100% of patients with LAS/ARC or AIDS. The virus can be isolated from about 90% of patients with ARC, and from fewer of those with late AIDS, the lower isolation rate from the latter being explicable by the almost total destruction of $T4^+$ lymphocytes in the terminal stages of the disease. HTLV-III has been demonstrated to replicate in $T4^+$ lymphocytes and to destroy them

TABLE 28-3
Clinical Features of AIDS

Lymphadenopathy syndrome
Fever, lymphadenopathy, diarrhea, weight loss
Malignancy
e.g., Kaposi's sarcoma, lymphoma, squamous cell carcinoma of anus or mouth
Opportunistic infections
e.g., Pneumocystis pneumonia, cerebral toxoplasmosis, oropharyngeal candidiasis
Other parasitic, fungal, viral, and bacterial infections
Neurological involvement
Encephalopathy, dementia

in vitro. However, the strongest evidence for causality has been provided by two types of epidemiological association. Whenever a recipient of a blood transfusion has contracted AIDS and the donor has been traced retrospectively, it has been discovered that the donor is positive for HTLV-III, and quite often he is found also to have died of AIDS or develops AIDS subsequently. Transmission of HTLV-III and AIDS from mother to baby has also been demonstrated many times. In addition, there are many documented instances of prospective studies in which normal individuals from high-risk groups have been observed to undergo seroconversion, reflecting HTLV-III infection, and have then gone on to develop AIDS a year or more later.

We know very little about the pathogenesis of HTLV-III infection. Clearly, the virus can enter the body via transfused blood, and presumably via lymphocytes in the semen of male homosexuals. It replicates in dividing T lymphocytes. Under some circumstances a DNA copy of the viral genome can become integrated into the chromosomes of the T cell and thus remain latent without expression until reactivated.

Seroconversion occurs between 2 and 40 weeks after infection, usually within 3 months. However, most of this antibody appears to be nonneutralizing. Infectious virus continues to be readily demonstrable in the blood of seropositive persons indefinitely. Once infected, even subclinically, every individual must be considered to be a lifelong carrier of the virus and fully capable of transmitting the infection to others.

Another lentivirus, caprine arthritis encephalitis virus, does not induce neutralizing antibodies, while visna and equine infectious anemia viruses undergo antigenic drift *in vivo* to escape neutralization by the antibodies they elicit. Furthermore, antibody titers, against the HTLV-III core protein p24 at least, diminish as the full-blown clinical picture of AIDS develops—presumably due to the progressive destruction of $OKT4^+$ lymphocytes.

Many questions remain to be answered. For example, at the time of writing (early 1985) several commercial kits for the serodiagnosis of HTLV-III infection are about to be licensed for widespread use, but the prognostic value of any given titer of antibody is not yet understood. Moreover, we need to know how the titer of infectious virus detected in various parts or secretions of the body over a period of time correlates with the patient's probability of developing AIDS, or of transmitting the virus to sexual or other types of contact. What cofactors enhance susceptibility to infection with HTLV-III and what affects the chance of an infected individual developing AIDS? Do variants of HTLV-III differ in virulence and are mutants selected *in vivo?* And so on. Answers to some of these questions will come from prospective studies currently underway in the United States.

An animal model would also be of value. The virus has been successfully transmitted to chimpanzees, some of which developed immunological abnormalities and prolonged lymphadenopathy but not AIDS. Some species of African monkeys carry antibody against a virus serologically related to HTLV-III. Another nononcogenic, lymphocytotropic type-D retrovirus related to the Mason–Pfizer monkey virus, but not to HTLVs, has been recovered from macaque monkeys suffering from "simian AIDS"; however, this immunodeficiency seems to be less discriminating than human AIDS, in that there is profound depletion of all lymphocyte populations, plus neutropenia and hypogammaglobulinemia. Another retrovirus again, feline leukemia virus, induces lymphopenia more commonly than it does leukemia; the consequent immunosuppression renders the cat vulnerable to opportunistic infections.

Laboratory Diagnosis

HTLV-III can be isolated from blood, T cells, semen, saliva, or brain. While it has been grown in cultured normal lymphocytes, malignant T (or occasionally B) cell lines are generally employed. In the cloned human T lymphoid leukemia line HT, HTLV-III multiplies to high titer, with virus continuing to be produced indefinitely despite the induction of syncytia. The resulting giant cells are characteristic, the nuclei being arranged in a ring.

At the moment, isolation of virus is a specialized, expensive, and potentially dangerous procedure which is not yet available in routine diagnostic laboratories. Nevertheless, the techniques are likely to be refined for wider application, to the assay as well as the identification of this group of viruses. Serological tests, perhaps employing monoclonal antibodies, will be developed to identify variants, while techniques such as nucleic acid hybridization and sequencing will be harnessed to characterize genomic differences.

Molecular probes are already available for detecting HTLV-III DNA in cells, and assays for HTLV-III reverse transcriptase have also been developed. Immunoassays for HTLV-III antigens in specimens taken directly from the patient are not yet sensitive or reliable enough to have been introduced for routine diagnostic purposes.

Diagnosis in 1985 is based on identification of antibody in serum. Enzyme immunoassay (EIA), immunofluorescence (IF), radioimmunoassay (RIA), radioimmunoprecipitation (RIP), and Western blot (WB) are the serological techniques currently in use. The first test to be developed used immunofluorescence to detect binding of antibody from the patient's serum to HTLV-III infected cells. EIAs have now become standard, and a number of diagnostic ELISA kits compete for the large potential market created by the need to screen all blood donors and to provide a service to high-risk populations. In most of these kits, the capture antigen, adsorbed to beads or microtiter plates, is disrupted virus from infected lymphocyte cultures. Purified or recDNA-cloned antigens may be preferred in the future, but as yet we do not know the diagnostic significance of antibody against the individual antigens. The best of these EIAs are quite sensitive, i.e., pick up the great majority of HTLV-III infections. There can, however, be a problem with specificity; false positives are as frequent as 1% or more with some of the kits that lack a control antigen prepared from uninfected lymphocytes. Because of the serious personal, social, and legal repercussions of false positive diagnosis it is important that, before the subject is informed of the result, any presumptive positive test is repeated. If an EIA that incorporates an uninfected cell control is still positive, another type of immunoassay such as RIP (which detects antibody to the core protein, p24) or Western blot (which detects antibody against all the antigens separately) should be employed for confirmation.

Whole blood or leukocytes from confirmed seropositive individuals can be sent to a reference laboratory for attempted culture of virus. Experience so far indicates that most seropositive people are also found to be carrying infectious virus in their blood.

Epidemiology

At the time of writing, the number of cases of AIDS reported in the United States alone has passed 10,000 and is rising rapidly. An analysis of the first 10,000 United States cases reported to the Centers for Disease Control by April 1985 indicated that 73% of them had occurred in male homosexuals (or bisexuals), 17% in intravenous drug abusers, and smaller numbers in hemophiliacs (receiving factor VIII), recipients of blood transfusions, heterosexual partners of male bisexuals, or newborn babies of infected mothers (Table 28-4). There had been no striking change in the distribution between these groups during the 4 years since the first case was identified and no evidence of spread by casual nonsexual contact. The annual incidence of AIDS was still increasing sharply. In other developed countries the risk categories were the same, although the proportion of cases seen in drug addicts was generally lower.

These data confirm the view that AIDS is spread only by transfusion of blood or certain blood products (factors VIII and IX, leukocytes, platelets), or by sexual intercourse. It is not clear, however, why homosexuality apparently carries a greater risk than heterosexual intercourse.

TABLE 28-4
Populations at Risk of Developing AIDS[a]

Homosexual (or bisexual) males	73%
Intravenous drug users	17%[b]
Female partners of homosexuals or of intravenous drug users	1–2%[c]
Blood transfusion recipients	1–2%
Hemophiliacs (receiving factor VIII)	1%
Babies of infected mothers	1%

[a] Analysis of the first 10,000 United States cases of AIDS (to April 1985).
[b] Not as high in other countries.
[c] Increasing rapidly.

Promiscuity is certainly a major factor; the probability of contracting AIDS is related to the number of different sexual partners (totalling hundreds per year in the case of many of the original homosexual AIDS victims). Yet, so far at least, the incidence of AIDS is not nearly as high in female prostitutes in the United States, suggesting that some aspect(s) of the sexual habits or life style of the promiscuous male homosexual subculture, or the traumatic nature of anal intercourse, is of major importance in the spread of the disease.

The origin of HTLV-III is a mystery. It is conceivable that the virus is a recent mutant or recombinant arising from another human retrovirus. Alternatively, it may be an animal retrovirus such as the African simian HTLV-III-like retrovirus, which may have recently spread to man, with or without genetic change. Intriguing in this regard is the high frequency of AIDS in Haiti and its more recent discovery in Zaire and other parts of Central Africa. It now appears that HTLV-III is much more common in Central Africa than anywhere else, and that its epidemiology there is quite different from that in the more developed nations. AIDS in Africa occurs equally in males and females (cf 15:1 in the United States) and antibodies are commonly found in children. There is evidence that human infection with HTLV-III or a closely related agent was well established in Africa prior to 1981 and that the virus may be originated there. By analogy with hepatitis B, it is not inconceivable that HTLV-III in Africa might be transmitted horizontally among family members by nonsexual routes, e.g., saliva or direct contact with skin sores (for which there is currently no evidence), as well as by heterosexual intercourse, which currently appears to be the principal mode of spread.

A question of major concern for the future is whether the epidemiology of AIDS in the developed countries will shift toward that seen in Central Africa. If, in particular, heterosexual transmission becomes a major mode of spread the epidemic could escalate alarmingly.

Control

The political and social, as well as the medical repercussions of AIDS, complicated as they are by fear and prejudice, have prompted governments in many parts of the world, particularly in the United States, to respond strongly to the challenge. The extent of the progress in only 4 years is testimony to this. The epidemiology of AIDS has been unraveled, the causal agent has been identified, and diagnostic tests have been developed. However, the epidemic has not been controlled. In the long term we need a vaccine and/or a chemotherapeutic agent. In the short term all we can do is to care for the dying, educate the population at large, and implement certain public health measures to slow the spread of the virus.

All seropositive persons require comprehensive clinical and immunological evaluation at regular intervals. Those who develop AIDS over the ensuing few years will spend most of the last year of their lives in hospital. Those who do not—"the worried well"—need psychological support and special counselling about their statistical probability of developing ARC or AIDS, as well as about desirable changes in their life style and many other matters.

Major difficulties are being encountered in balancing the individual's right to privacy against the public's right to protection. At the time of writing, health authorities in most countries are taking the view that confidentiality is the paramount consideration and that, if a carrier of the AIDS virus does not choose to inform his sexual contacts or continues to frequent a homosexual bath house, that no action should be taken. This is inconsistent with the attitude toward blood donors in many countries where legal sanctions are imposed against false declarations regarding seropositive status or even homosexuality. Deep thought needs to be given to this very difficult area of civil rights. Legislation may have to be introduced, not only to protect private citizens against breaches of confidentiality, but also to permit limited disclosure by doctors, laboratory staff, and health authorities in the public interest.

Male homosexuals still constitute by far the most important risk group. Channels of communication with this population are available through various gay clubs and societies, as well as through committees specifically established for this purpose. The level of concern is high and most homosexuals are now reasonably well informed. In some countries the more infamous gay bars and bath houses are being closed down or regulated.Many promiscuous homosexuals have drastically reduced their encounters with multiple anonymous partners, which is the most important single step. A significant minority, however, are not prepared to alter their "fast-lane" life style. The benefits of using condoms and of not sharing items such as toothbrushes and razors are still uncertain.

Intravenous drug users must be informed of the hazards of needle sharing and taught about the importance of proper sterilization of syringes. Pregnancy should be discouraged in seropositive women.

Laboratory workers and health care personnel should implement certain safety procedures to minimize the risk—evidently quite low—of contracting infection from blood or other secretions from known or potential HTLV-III carriers. Self-inflicted pricks from needles or other sharp instruments are potentially hazardous, despite encouragingly low rates of seroconversion in the series of such accidents reported so far. Gloves and gowns should be worn whenever there is a risk of contact with contaminated blood or secretions. Blood spills should be cleaned up promptly with 0.5% sodium hypochlorite, 70% ethanol, or 1% glutaraldehyde. Disposable or heat-resistant equipment should be autoclaved. Biosafety cabinets should be used for processing potentially infected material—and so on. It will be noted that most of these precautions correspond with practices already well established in most hospitals and laboratories for coping with the hazards of hepatitis B.

Irrational fear of AIDS is such that the general public needs to be fully informed of the facts through the media and all other appropriate channels. In particular their concern about the possibility of spread via nonsexual contact must be allayed and faith must be restored in the blood transfusion services. The risk of acquiring AIDS from blood transfusion, artificial insemination, or organ transplantation is already exceedingly low, and will fall sharply, as it has for hepatitis B, now that high-risk groups such as male homosexuals and heroin addicts have been exhorted not to donate blood, sperm, or organs and are required to sign a statutory declaration as well as to subject themselves to a screening test for HTLV-III antibody. The risk to hemophiliacs has also been greatly reduced by the recent introduction of pasteurization (heat treatment at 56°C for 30 minutes) to inactivate HTLV-III in factors VIII and IX—although, unfortunately, a large percentage of hemophiliacs in many countries have already been infected.

Of course, completely different strategies may need to be invoked in Central Africa where the epidemiology is so different. Any heterosexually transmitted disease is notoriously difficult to control, while if it transpires that spread can also occur by saliva or contact in developing countries with low standards of hygiene, control by simple public health measures would be virtually impossible. A vaccine is clearly required.

Vaccine. The highest priority in the control of AIDS is obviously the development of an effective vaccine, initially for the protection of high-risk groups such as male homosexuals and hemophiliacs, and eventually perhaps for universal delivery to infants. A conventional live vaccine against a retrovirus would probably be unacceptably risky, but an inactivated vaccine would be acceptable. recDNA technology could readily be exploited to clone the gene for the appropriate protein in a suitable prokaryotic or eukaryotic expression system, or in vaccinia virus (see Chapter 10). However, there is reason to doubt whether antibodies thus elicited would be protective. The apparent absence of neutralizing antibodies in natural infections is particularly worrying, but protection against initiation of infection may be achievable. The high frequency of mutation in the gene for the envelope glycoprotein might diminish the effectiveness of any vaccine unless a conserved region of the molecule proves to be protective.

Chemotherapy. Specific treatment for AIDS is by no means an impossible objective. The reverse transcriptase (RT) of retrovirus constitutes a tempting target. As discussed in Chapter 11, rifamycins, suramin, and certain other agents have proven efficacy against RT-containing viruses in culture. These and a number of other substances are currently undergoing clinical trials. Interferons, antibodies, and interleukins have been disappointing. If an effective chemotherapeutic agent can be identified, treatment could of course be commenced as soon as laboratory evidence of infection with HTLV-III is obtained, long before AIDS becomes clinically manifest. However, since the viral genome persists indefinitely in integrated form, chemotherapy would have to be continued for life, hence the drug would need to be cheap, nontoxic, and preferably delivered by mouth.

NON-A, NON-B HEPATITIS

Parenteral Form

Following the widespread introduction of sensitive assays for screening blood for hepatitis B it was expected that posttransfusion hepatitis (PTH) would be virtually eradicated. However, that was not to be. Whereas serological investigations indicate that hepatitis B is no longer a major cause of PTH in most developed countries today, there is a large residue of cases with a rather variable incubation period (ranging from 2 weeks to 6 months) and with no markers of hepatitis B (or A). The inescapable conclusion is, therefore, that by far the commonest cause of PTH in the 1980s is a virus, or viruses, yet to be discovered. The putative agent(s) labors under the name of "non-A, non-B" (or NANB) hepatitis virus(es).

Parenteral transmission of NANB hepatitis occurs not only by transfusion but also via any of the other blood-borne routes so well established for hepatitis B. It is very common in drug addicts, and in hemophiliacs who regularly receive factor VIII or IX. In retrospect, the outbreak of hepatitis that conferred a high mortality on staff and patients in an Edinburgh renal dialysis unit in 1969–1970 seems to have been a mixed infection with hepatitis B and non-A, non-B.

The parenteral type(s) of non-A, non-B hepatitis differ clinically from hepatitis B in being generally less severe—shorter preicteric period, milder symptoms, absent or less marked jaundice, and lower serum transaminase levels. However, they lead much more commonly to chronic hepatitis, with moderate elevation of ALT levels persisting for anything up to 1 or 2 years in up to half of all cases; most of these are mild cases of chronic active or chronic persistent hepatitis which spontaneously resolve, but about 10% progress to cirrhosis. The facility with which these agents are transmitted by blood transfusion reflects prolonged carriage of the virus in the bloodstream.

Clearly NANB hepatitis is caused by an infectious agent, presumably a virus. The disease is readily transmissible by blood or blood products not only to man but also serially in chimpanzees following experimental inoculation. Infected animals become immune to reinfection from the same source but remain susceptible to other putative NANB agents. Moreover, multiple episodes of NANB PTH have been reported in hemophiliacs receiving factor VIII from different batches. This suggests the existence of multiple NANB agents. Other epidemiological data support this hypothesis. For example, the incubation periods of NANB PTH not only vary widely but seem to differ strikingly in different series. Furthermore, as will be seen below, the enteral type of NANB hepatitis has a shorter incubation period (10 days–6 weeks), is spread via nonparenteral routes, and induces a more severe acute disease generally without chronic sequelae (see Table 28-5).

A confusing variety of putative NANB viruses have been recovered from infectious batches of factor IX, from the serum and/or hepatocytes of patients with NANB hepatitis, and from chimpanzees to whom the disease had been transmitted, but there is no evidence that any of them causes the disease.

Until the etiological agents are positively identified and cultured, neither diagnosis nor prevention and treatment are likely to advance dramatically. Diagnosis is essentially by exclusion of hepatitis A and B, and other known causes of hepatitis. Several serological tests put forward over the last few years have turned out to be nonspecific. Meanwhile, it

TABLE 28-5
Non-A, Non-B Hepatitis

	PARENTERAL	ENTERAL
Transmission	Blood transfusion Factors VIII and IX Intravenous drug abuse	Water-borne epidemic Sporadic, person to person
Incubation period	2–26 weeks	10–40 days
Clinical disease	Generally mild	Often severe
Chronic hepatitis	Common	Uncommon
Putative agent(s)	?	27-nm virus?

would be possible to reduce the incidence of NANB PTH by moving from paid to volunteer donors, and/or by screening out all blood donors with significantly elevated serum alanine aminotransferase levels. Moreover, the infectivity of factors VIII and IX can be reduced by combined treatment with β-propiolactone (BPL) and ultraviolet light, or BPL and Tween 80; heat treatment, such as is now used to eliminate HTLV-III, may also be effective.

Enteral Form

Another retrospective study recently turned up an even greater surprise. In 1955 a major common source epidemic of presumed hepatitis A resulted from contamination of a drinking water supply by a burst sewerage main in Dehli. Post hoc examination of patients' sera still extant in deep freezers revealed no evidence of hepatitis A IgM—the epidemic was in fact attributable to NANB. Several comparable waterborne epidemics in more recent times have also been shown to be caused by NANB, with an incubation period of 6 weeks or less. Secondary cases, presumably infected by fecal–oral transmission, sometimes occurred within household contacts.

Clearly then, hepatitis NANB has at least two quite distinct modes of spread—parenteral (resembling hepatitis B) and enteral (resembling hepatitis A). In addition, sporadic cases occur, particularly in adolescents and young adults, which may represent person-to-person spread by the enteric route (as occurred within the families of victims of the explosive Indian water-borne outbreaks), or possibly parenteral spread between covert drug addicts.

Clinically, the enteral type of non-A, non-B hepatitis can be severe, with a high incidence of cholestasis, significant mortality from acute fulminating hepatitis, especially in pregnant women and their babies, some persistence, but a relatively low incidence of chronic active hepatitis.

Recently, a 27- to 30-nm virus-like particle (the same size as that of hepatitis A), from feces of patients with water-borne and sporadic NANB hepatitis, has been transmitted to chimpanzees as well as to a hepatitis A immune human volunteer and has been recovered again from their feces. While this agent has not yet been shown to be an etiological agent of NANB hepatitis, nor even to be an *Enterovirus*, it raises the possibility that the enteral form of NANB hepatitis, which so closely resembles hepatitis A both clinically and epidemiologically, is also caused by an *Enterovirus*.

DELTA (HDV) HEPATITIS

In the late 1970s a young Italian physician detected by immunofluorescence a novel antigen, which he called δ, in the nuclei of hepatocytes in some hepatitis B patients. The same δ antigen was later found inside 36-nm virus-like particles, the "δ agent," the capsid of which was serologically indistinguishable from HBsAg. Yet the genome of the δ agent (or hepatitis D virus, HDV, as some now call it) turned out to be a tiny single-stranded RNA molecule, showing no homology with hepatitis B DNA. Since δ is found only in people infected with hepatitis B virus, it appears that it is a defective virus, entirely dependent upon HBV for its replication.

The MW of the δ RNA molecule is only 5×10^5. This is scarcely enough to code for the δ antigen itself. Therefore it is unlikely that there are other undiscovered products of the δ genome. Clearly, the helper virus, HBV, provides the genetic information for the coat protein (HBsAg) of the parasitic δ virus, and it is not inconceivable that it also codes for the enzyme necessary for the replication of δ RNA. At first sight such a proposition is illogical, as HBV itself, being a DNA virus, does not need an RNA-dependent RNA polymerase. However, there is recent evidence that HBV DNA polymerase may carry reverse transcriptase activity as well and it is possible that δ RNA replicates via a DNA intermediate. Another possibility is that δ antigen itself may be the replicase (or a component of it).

The closest analogy to the fascinating δ–HBV relationship is to be found among plant viruses. There are several plant virus "satellites," each totally dependent for its replication on a helper virus, which in some cases codes for the coat protein of the satellite. The genome of these plant satellite viruses is also ssRNA of MW generally about 5×10^5 and with striking secondary structure like that of δ (Table 28-6).

TABLE 28-6
Delta Hepatitis

Delta (δ) virus (HDV)
- Spherical particle, 36 nm, HBsAg coat
- ssRNA genome, MW 5×10^5, much secondary structure
- Defective virus, requiring HBV as helper

Delta hepatitis
- HDV superinfection of HBV carrier leads to
  - Severe δ hepatitis, high mortality
  - Chronic active hepatitis
  - Chronic carrier state (especially in $HBsAg^+$ drug addicts)

When passaged in $HBsAg^+$ carrier chimpanzees, δ virus multiplies to extremely high titers and often induces severe hepatitis, either chronic or fulminant. There appear to be two principal modes of infection in man: (1) simultaneous infection (known as acute *coinfection*, or *coprimary infection*) with δ virus and hepatitis B virus; (2) δ virus *superinfection* of an HBV carrier. Superinfection of a chronic HBV carrier induces an acute hepatitis which seems to be attributable principally to replication of the δ virus in the liver; δ antigen is plentiful in hepatocytes while the markers of HBV infection tend to be depressed. Such δ virus-induced hepatitis may be severe, often fulminating. A recent epidemic ascribed to δ virus superinfection of a tribe of $HBsAg^+$ Indians in a remote part of Venezuela had a case fatality rate of about 20%. Those who recover from this form of δ virus infection often become chronic carriers of the virus and have a high chance of going on to chronic active hepatitis. Indeed, it now appears that a high proportion of the hepatitis B patients who die from fulminant hepatitis or progress to chronic active hepatitis do so because of δ virus superinfection. By contrast with superinfection, coinfection is seen mainly in drug addicts, and may produce either fulminant hepatitis or a disease essentially indistinguishable from hepatitis B alone.

Immunoassays (EIA or RIA) have now been developed to diagnose acute and chronic δ virus infections, the carrier state and immunity. Delta antigen is detectable in serum during acute infection, then is replaced by δ antibody (only transiently in coinfections). Anti-delta IgM is a useful index of active δ infection. Anti-HBc IgM distinguishes simultaneous coinfection from superinfection of an HBV carrier.

About half (20–80% in various studies) of all $HBsAg^+$ intravenous drug users are δ antibody positive. Moreover, in Europe, the United States, and Australia, but not in S.E. Asia, 10–30% of all $HBsAg^+$ carriers with chronic liver disease display markers of δ infection. Antibodies to δ are much less prevalent in healthy HBsAg carriers.

A great deal has yet to be learned about the epidemiology of δ virus. Virtually nothing is known about its normal mode(s) of spread, let alone how such a defective virus has evolved and managed to survive in nature. Clearly, δ virus infection could be prevented by vaccination against hepatitis B.

SPONGIFORM ENCEPHALOPATHIES

In Chapter 7 we discussed the slow infections of the brain that give rise to a group of diseases generally known as the subacute spongiform encephalopathies. The paradigm is scrapie, a disease of sheep, transmissible experimentally to mice. At least two degenerative diseases of the human brain, kuru and Creutzfeldt–Jakob disease, are clinically and histologically virtually indistinguishable from scrapie. Furthermore, preliminary attempts to characterize the etiological agents indicate that they may all be related and that they comprise a completely novel class of infectious agents.

Kuru

Kuru is a disease of the central nervous system characterized by cerebellar degeneration, leading to ataxia, tremors, incoordination, dementia, and certain death within a year or so. It has never been reported outside a single tribe of Melanesians and is confined to an area of 1000 square miles in the South Fore region of the New Guinea highlands (Plate 28-2). According to the recollections of the village elders, kuru was unknown until about 1920. Yet, by 1957, when the disease was first described, kuru was responsible for 80–90% of all deaths in South Fore women. Indeed, it had so decimated the female population that the tribe had come to contain only one-third as many women as men. As most of the deaths were in young women of childbearing age, kuru had left a legacy of orphans, who presented a tragic social problem. Children of both sexes were also afflicted by the disease, but older males were largely spared (except insofar as 20% of all male deaths resulted from ritual murder on suspicion of sorcery in connection with kuru). This led to the hypothesis (now discredited) that kuru was a sex-linked genetic disease, in which the mutant gene had become widely disseminated throughout this isolated tribe by centuries of inbreeding. In order to explain how such a common lethal gene had not exterminated the tribe long ago, it was postulated that, by analogy with diseases such as favism in man or scrapie in sheep, the mutant gene merely predisposed the individual to kuru, and that the actual precipitating cause was perhaps a

PLATE 28-2. *Kuru. A child from the Fore region of the New Guinea highlands showing a disorder of gait due to irreversible cerebellar degeneration. (Courtesy Sir Macfarlane Burnet.)*

plant or a virus first introduced into the area in the early part of this century.

In 1965 Gajdusek and his colleagues working at the United States National Institutes of Health demonstrated that a fatal kuru-like disease could be transmitted to chimpanzees by intracerebral inoculation of brain extracts from patients who had died from the disease. They went on to show that the causative agent is filterable, that it is present in high titer in a variety of human tissues in addition to the brain, and that the infection can be passed to other chimpanzees and monkeys, in which the condition can be more readily studied. Characterization of the etiological agent indicated that it is very similar to that of a clinically similar subacute spongiform encephalopathy, Creutzfeldt–Jakob disease (see below and Chapter 7).

The epidemiological pattern of kuru changed dramatically shortly after Gajdusek's first description of the situation in 1957. The disease gradually declined in incidence, with no new cases at all occurring in children and the average age of onset in adults rising by 1 year every year. The decline coincided in time with the advent of the Australian colonial administration, which first penetrated into this inaccessible area in the 1950s and promptly put an end to the practice of cannibalism. Close interrogation has since revealed that the women of the tribe were introduced to cannibalism around 1910–1920 and had ever since pursued the custom, accompanied by elaborate ritual, of consuming deceased relatives (endo-cannibalism). The men took no part in these ceremonies but the children were involved because tribal custom required that they lived exclusively with the women prior to their initiation. The peculiar sex and age distributions of kuru thus had a social rather than a genetic basis.

Today, the disease is almost extinct. Only the very occasional adult, now in late middle age, dies from kuru following an incubation period of over 20 years after last participating in the ritual cannibalism as a child. The incredible story of kuru is still not fully solved, for the origin of the kuru "virus" remains a mystery. We can surmise, but never know, that the agent of kuru arose from the brain of a relative who died from Creutzfeldt–Jakob disease (see below).

Creutzfeldt–Jakob Disease

This rare subacute presenile dementia, in which the patient becomes uncoordinated and demented and dies within 9–18 months, presents a histological picture so similar to kuru as to suggest that the two may be caused by the same or very similar agents. In both there is a "spongy degeneration" of the brain (status spongiosus), characterized by vacuolation of the axons and dendrites of neurons accompanied by proliferation and hypertrophy of astroglia, and a rather striking electron microscopic picture revealing a tangled web of membranes. The apparent absence of inflammation is consistent with the belief that the etiological agent does not elicit an immunological response. A comparable disease with the same histopathology has been successfully transmitted to chimpanzees, monkeys, and cats by inoculation of filtrates from affected human brain.

Creutzfeldt–Jakob disease (CJD) occurs world-wide at a frequency of about one case per million of population per annum. It tends to strike in middle age. Libyan Jews have an unusually high incidence.

In recent years iatrogenic cases have occurred following corneal transplantation or neurosurgery, e.g., involving implantation of electrodes into the brain. These tragic episodes have alerted surgeons to the necessity of sterilizing instruments according to protocols certain to inactivate the most resistant infectious agents. Autoclaving at 121°C for 60–90 minutes (15 psi) is required to inactivate the CJD agent. Heat-labile equipment may be soaked in sodium hypochlorite, but this oxidizing solution is corrosive. In 1985 it became apparent that a number of people who had been treated years earlier with growth hormone extracted from human pituitary glands had recently died from CJD. The product was withdrawn but it remains to be seen how many others have been infected.

The nature of the spongiform encephalopathy agents remains mysterious. They have never been grown in cell culture and never positively identified by electron microscopy. Animal transmission studies using bacteria-free filtrates established the size of the infectious unit as being about that of a small virus. Experiments to assess their susceptibility to inactivating agents suggested an exceptionally small "target" for various types of irradiation, leading to the conclusion that the genome, if any, is a very small nucleic acid molecule indeed. Furthermore, early attempts to inactivate the infectivity of these agents by heat or chemical treatments seemed to indicate an unprecedented degree of resistance.

More recent work attributes many of these anomalies to the protection conferred on the agents by the crude extracts of brain in which they were suspended. It now appears that the spongiform encephalopathy agents are of a size and a susceptibility to inactivants not inconsistent with the hypothesis that they are small viruses. Long thin filaments have been regularly observed in the brain of scrapie-infected mice, hamsters, and monkeys, kuru-infected monkeys, and CJD-infected humans. They are also found in the spleen of scrapie-infected mice. Called *scrapie-associated fibrils (SAF),* they consist of 2 (or sometimes up to 4) twisted filaments, each filament being 4–6 nm in diameter and of variable length. Overall, the SAF are 11–25 nm in diameter, depending on the method of preparation, staining, and electron microscopy. The protein (MW 27,000–32,000), but not the putative nucleic acid genome, of these rod-shaped structures has been characterized. Despite certain morphological similarities to amyloid, as seen in the amyloid plaques of normal brains from aged people, or in greater numbers in the senile plaques of Alzheimer's disease, the possibility that these structures do indeed represent the infectious entity are being vigorously pursued. Purification procedures that enrich for SAF also enhance the infectivity titer of crude infectious preparations. It is possible that the causal agent of the spongiform encephalopathies has now been identified and that it is the prototype of a new class of filamentous mammalian viruses.

FURTHER READING

AIDS

Broder, S., and Gallo, R. C. (1984). A pathogenic retrovirus (HTLV-III) linked to AIDS. *N. Engl. J. Med.* **311,** 1292.

Gallin, J. I., and Fauci, A. S., eds. (1985). "Acquired Immunodeficiency Syndrome." Raven Press, New York.

Gallo, R. C., Essex, M., and Gross, L., eds. (1984). "Human T-Cell Leukemia Viruses." Cold Spring Harbor Press, Cold Spring Harbor, New York.

Gonda, M. A., Wong-Staal, F., Gallo, R. C., Clements, J. E., Narayan, O., and Gilden, R. V. (1985). Sequence homology and morphological similarity of HTLV-III and visna virus, a pathogenic lentivirus. *Science* **227,** 173 et seq.

Gottlieb, M. S., and Groopman, J. E., eds. (1984). "Acquired Immunodeficiency Syndrome." *UCLA Symp. 16.*

Memorandum from a WHO meeting (1985). AIDS and the WHO Collaborating Centres. *Bull. World Health Organisation,* **63,** in press.

Popovic, M., Sarngadharan, M. G., Read, E., and Gallo, R. C. (1984). Detection, isolation and continuous production of cytopathic retroviruses (HTLV-III) from patients with AIDS and pre-AIDS. *Science* **224,** 497 et seq.

Vilmer, E., Barre-Sinoussi, F., Rouzioux, C., Gazengel, C., Vezinet Brun, F., Dauguet, C., Fischer, A., Manigne, P., Chermann, J. C., Griscelli, C., and Montagnier, L. (1984). Isolation of new lymphotropic retrovirus from two siblings with haemophilia B, one with AIDS. *Lancet* **1,** 753.

deVita, V. T., Hellman, S., and Rosenberg, S. A., eds. (1985). "AIDS." Lippincott, Philadelphia.

NON-A, NON-B HEPATITIS AND DELTA HEPATITIS

Aach, R. D., and Kahn, R. A. (1980). Post-transfusion hepatitis: Current perspectives. *Ann. Intern. Med.* **92,** 539.

Balayan, M. S., Andjaparidze, A. G., Savinskaya, S. S., Ketiladze, E. S., Braginsky, D. M., Savinov, A. P., and Poleschuk, V. F. (1983). Evidence for a virus in non-A, non-B hepatitis transmitted via the fecal-oral route. *Intervirology* **20,** 23.

Deinhardt, F. (1984). The agents of human viral hepatitis, and control of the disease. *Prog. Med. Virol.* **30,** 14.

Deinhardt, F., and Deinhardt, J., eds. (1983). "Viral Hepatitis: Laboratory and Clinical Science." Dekker, New York.

Dienstag, J. L. (1983). Non-A, non-B hepatitis. I. Recognition, epidemiology and clinical features. II. Experimental transmission, putative virus agents and markers, and prevention. *Gastroenterology* **85,** 439–462 and 743–768.

Gerety, R. J., ed. (1982). "Non-A, Non-B Hepatitis." Academic Press, New York.

Hadler, S. C., De Monzon, M., Ponzetto, A., Anzola, E., Rivero, D., Mondolfi, A., Bracho, A., Francis, D. P., Gerber, M. A., Thung, S., Gerin, J., Maynard, J. E., Popper, H., and Purcell, R. H. (1984). Delta virus infection and severe hepatitis. An epidemic in the Yucpa Indians in Venezuela. *Ann. Int. Med.* **100,** 339.

Khuroo, M. S. (1980). Study of an epidemic of non-A, non-B hepatitis. Possibility of another human hepatitis virus distinct from post-transfusion non-A, non-B type. *Am. J. Med.* **68,** 818.

Rizzetto, M., Verme, G., and Bonino, F., eds. (1983). "Viral Hepatitis and Delta Infection." Liss, New York.

Robinson, W. S. (1982). The enigma of non-A, non-B hepatitis. *J. Infect. Dis.* **145,** 387.

Szmuness, W., Alter, H. J., and Maynard, J. E., eds. (1982). "Viral Hepatitis." Franklin Institute Press, Philadelphia, Pennsylvania.

Spongiform Encephalopathies

Farquhar, J., and Gajdusek, D. C., eds. (1981). "Kuru: Early Letters and Field-Notes from the Collection of D. C. Gajdusek." Raven Press, New York.

Gajdusek, D. C. (1977). Unconventional viruses and the origin and disappearance of kuru. *Science* **197,** 943.

Merz, P. A., Rohwer, R. G., Kascsak, R., Wisniewski, H. M., Somerville, R. A., Gibbs, C. J., and Gajdusek, D. C. (1984). Infection-specific particle from the unconventional slow virus diseases. *Science* **225,** 437.

Prusiner, S. B., and Hadlow, W. J., eds. (1979). "Slow Transmissible Diseases of the Nervous System," 2 vols. Academic Press, New York.

CHAPTER 29

Viral Syndromes

Introduction 607
Viral Diseases of the Respiratory Tract 608
Viral Gastroenteritis 613
Viral Diseases of the Central Nervous System 614
Viral Skin Rashes 619
Viral Hemorrhagic Fevers 622
Viral Genitourinary Infections 623
Viral Diseases of the Eye 625
Viral Arthritis 626
Viral Carditis 627
Viral Hepatitis 627
Viral Pancreatitis 629
Congenital and Perinatal Viral Infections 630
Viral Infections in Immunocompromised Patients 631
Further Reading 632

INTRODUCTION

In the preceding 16 chapters we have examined individually the contribution of each of the families of viruses to the spectrum of human disease. In this final chapter we believe it may be helpful to the reader to examine the same scene from the opposite aspect, namely to focus in turn on each of the clinical syndromes. Of course, this is not a textbook of medicine nor even of infectious disease, but of virology. We cannot devote space to detailed clinical descriptions of viral diseases and certainly not to their differential diagnosis from diseases caused by nonviral infectious agents. Appropriate reference works on infectious diseases have been listed under Further Reading. What follows is intended to provide only a bird's eye view of the commoner syndromes so that we may assess the contribution of particular viruses to each one. Insofar as the clinical features as well as the pathogenesis and epidemiology of all the major human viral infections were dealt with in the previous 16 chapters, this chapter should be regarded as little more than an Appendix which brings all these virus–disease associations together in a number of summary tables for ready reference.

VIRAL DISEASES OF THE RESPIRATORY TRACT

Respiratory infections are the most common afflictions of man and most of them are caused by viruses. Children contract on average six respiratory illnesses each year, and adults about two. Admittedly these are mainly trivial colds and sore throats but they account for millions of lost working days and a significant proportion of all visits to family physicians. Furthermore, influenza, parainfluenza, and respiratory syncytial viruses are major causes of serious lower respiratory disease, especially in children.

While some viruses have a predilection for one particular part of the respiratory tract, most are capable of causing disease at any level and the syndromes to be described below overlap somewhat (Fig. 29-1). Nevertheless, for ease of description we will designate five basic diseases of increasing severity as we descend the respiratory tract: rhinitis, pharyngitis, croup, bronchiolitis, and pneumonia (Table 29-1).

Rhinitis (The Common Cold)

The classical common cold (coryza) is marked by copious watery nasal discharge and congestion, sneezing, and perhaps a mild sore throat or cough, but little or no fever. Rhinoviruses are the major cause of rhinitis, several serotypes being prevalent all year round and accounting for about half of all colds. Coronaviruses are responsible for about another 15%, mainly those occurring in the winter months, of shorter duration, without a cough. Certain enteroviruses, particularly coxsackieviruses A21 and A24, and echoviruses 11 and 20 cause febrile colds and sore throats in the summer. In children, respiratory syncytial virus (RSV), parainfluenza viruses, and the lower numbered adenoviruses are between them responsible for up to half of all upper respiratory tract infections ("URTI" or "URI").

Otitis Media or sinusitis sometimes complicate URI. Bacterial superinfection is generally involved, but viruses have also been recovered from

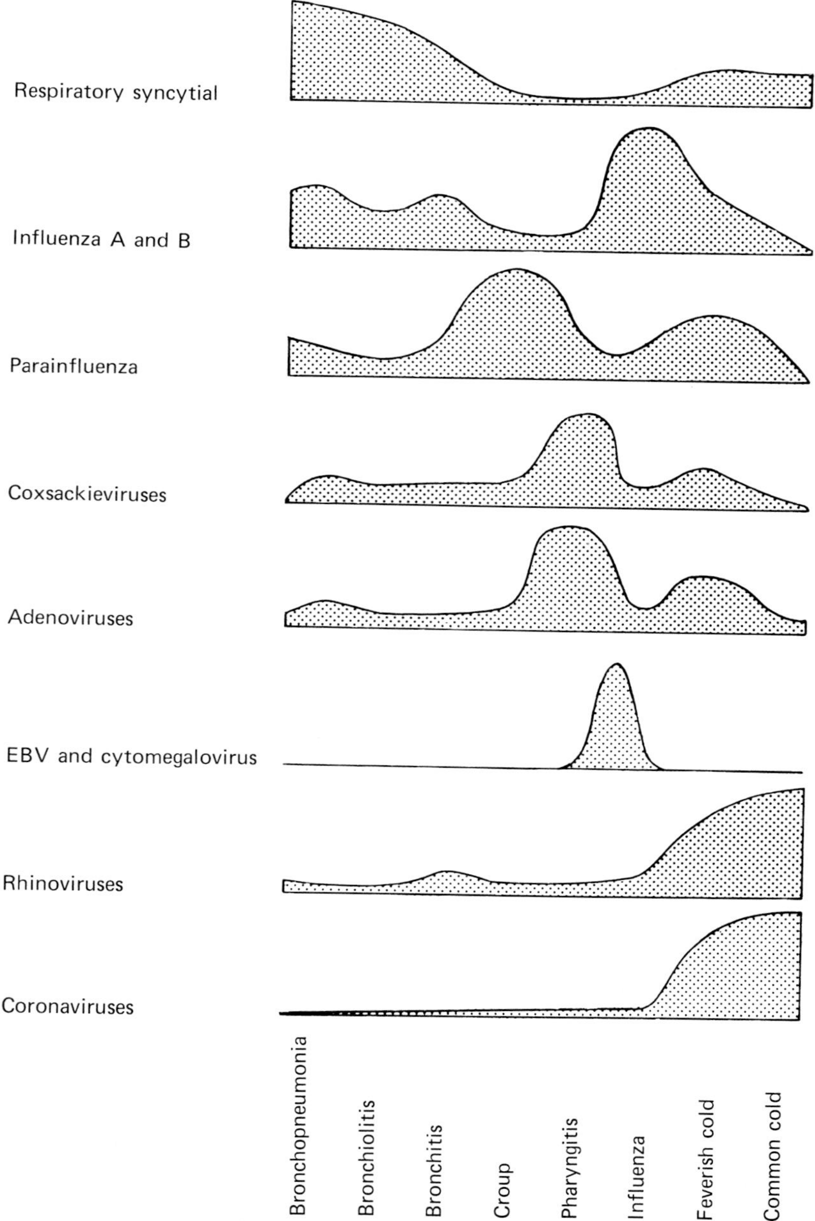

FIG. 29-1. *Diagram showing the frequency with which particular viruses produce disease at various levels of the respiratory tract. (Courtesy Dr. D. A. J. Tyrrell.)*

TABLE 29-1
Viral Diseases of the Respiratory Tract

DISEASE	VIRUS	
	COMMON	LESS COMMON
Rhinitis (common cold)	Rhinoviruses Coronaviruses	RSV, parainfluenza, influenza Adenoviruses Coxsackie A21, A24; echo 11,20
Pharyngitis	Parainfluenza 1–3 Influenza Herpes simplex EB virus Coxsackie A	Rhinoviruses Adenoviruses RSV Cytomegalovirus
Laryngotracheobronchitis (croup)	Parainfluenza 1 > 2 Influenza	RSV
Bronchiolitis	RSV Parainfluenza 3 > 1, 2	Influenza A
Pneumonia	RSV Parainfluenza 3 > 1, 2 Influenza Cytomegalovirus	Adenoviruses 3, 4, 7 Measles Varicella

the effusion. Respiratory infections with RSV, influenza, parainfluenza, adenovirus, or measles viruses predispose to otitis media. Indeed, repeated viral infections can precipitate recurrent middle ear infections, leading to progressive hearing loss.

Pharyngitis

Most pharyngitis is of viral etiology. URI with any of the viruses just described can present as a sore throat, with or without cough, malaise, fever, and/or cervical lymphadenopathy. Influenza, parainfluenza, and rhinoviruses are common causes throughout life, but other agents are prominent in particular age groups: RSV and adenoviruses in young children; herpesviruses in adolescents and young adults. Adenoviruses, though not major pathogens overall, are estimated to be responsible for about 5% of all respiratory illnesses in young children. Pharyngoconjunctival fever is just one particular presentation, which was described in Chapter 15 together with the strange tendency of adenoviruses 4 and 7 to cause outbreaks of "acute respiratory disease" in United States military camps. Primary infection with herpes simplex virus (HSV), if delayed until adolescence, presents as a pharyngitis and/or tonsillitis rather than as the gingivostomatitis seen principally in younger children; the characteristic vesicles, rupturing to form ulcers, can be confused only with herpangina, a common type of vesicular pharyngitis caused by coxsackie A viruses (see Chapters 16 and 19). Infectious mononucleosis (glandular fever) is usually marked by a very severe pharyngitis, often with a membranous exudate, together with cervical lymphadenopathy and fever. This syndrome is generally caused by EB virus in 15–25 year olds, and by cytomegalovirus in older adults.

Laryngotracheobronchitis (Croup)

Croup is one of the serious manifestations of parainfluenza and influenza virus infections. A young child presents with fever, cough, inspiratory stridor, and respiratory distress, sometimes progressing to complete laryngeal obstruction. Parainfluenza viruses are responsible for 40–50% of all cases, type 1 being commoner than type 2, with which it tends to alternate in autumn outbreaks. Influenza and RSV are important causes during winter epidemics.

Bronchitis

Influenza, parainfluenza, and RSV are the main viral causes of acute bronchitis. There is also evidence that chronic bronchitis, which is particularly common in smokers, may be exacerbated by acute episodes of infection with influenza viruses, rhinoviruses, or coronaviruses.

Bronchiolitis

Respiratory syncytial virus is the most important respiratory pathogen during the first year or two of life, being responsible, during winter epidemics, for half of all bronchiolitis in infants. Parainfluenza viruses (especially type 3) and influenza viruses are the other major causes of this syndrome. The disease can develop with remarkable speed. Breathing becomes rapid and labored, and is accompanied by a persistent cough, expiratory wheezing, cyanosis, a variable amount of atelectasis, and marked emphysema visible by X-ray (Plate 29-2A).

Pneumonia

Whereas viruses are relatively uncommon causes of pneumonia in adults, they are the most important in young children. RSV (Plates 29-1A and 29-2A) and parainfluenza (mainly type 3) are between them responsible for 25% of all pneumonitis in infants in the first year of life. Influenza also causes a considerable number of deaths during epidemic years. Adenoviruses 3 and 7 (Plate 29-1B) are rarer but the long-term

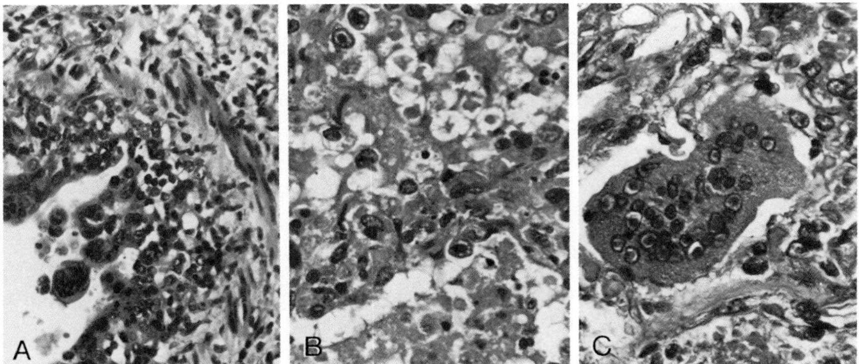

PLATE 29-1. *Viral infections of the lung. Sections from fatal cases of pneumonitis in children. (A) Respiratory syncytial virus, bronchiolitis (H and E stain, ×135). Note multinucleated giant cells being shed into bronchiole, which is surrounded by leukocytes. (B) Adenovirus, pneumonia (H and E stain, × 126). Note characteristic intranuclear inclusions. (C) Measles, giant cell pneumonia (H and E stain, ×216). Note multinucleated giant cell with intranuclear inclusions. (Courtesy I. Jack and A. Williams.)*

sequelae are more serious. Up to 20% of pneumonitis in infants has been ascribed to perinatal infection with cytomegalovirus (see Chapter 16). CMV may also cause potentially lethal pneumonia in immunocompromised patients, as may measles (Plate 29-1C), varicella, RSV, and adenoviruses. Moreover, viral pneumonia not uncommonly develops in adults with varicella (Plate 29-2B), and in military recruits involved in outbreaks of adenovirus 4 or 7, while measles is quite often complicated by bacterial pneumonia, especially in malnourished children. In the elderly, particulary those with underlying pulmonary or cardiac conditions, influenza is a major cause of death (Plate 29-2C and D). Primary influenza pneumonitis occurs but secondary bacterial invasion with *Staphylococcus aureus, Streptococcus pneumoniae,* or *Haemophilus influenzae* is more usual.

Viral pneumonitis often develops insidiously following URI and the clinical picture may be atypical. The patient is generally febrile, with a cough and a degree of dyspnea, and auscultation may reveal some wheezing or moist rales. Unlike typical bacterial lobar pneumonia with its uniform consolidation, or bronchopneumonia with its streaky consolidation, viral pneumonitis is usually confined to diffuse interstitial lesions. The radiological findings are not striking; they often show little more than an increase in hilar shadows or, at most, scattered areas of consolidation (Plate 29-2).

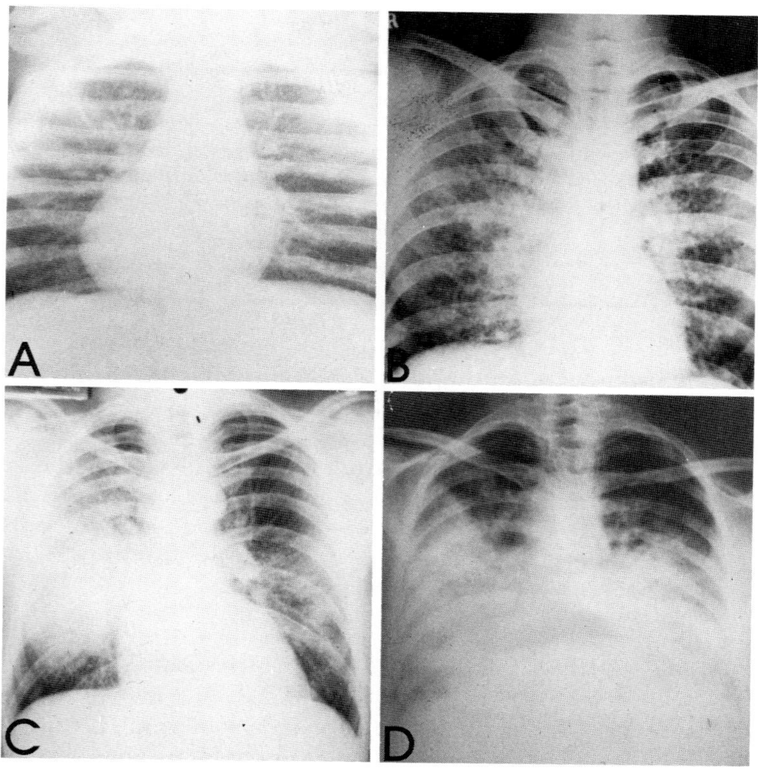

PLATE 29-2. *Radiographs of viral pneumonitis. Note streaky, patchy, or nodular consolidation. (A) Respiratory syncytial virus. (B) Varicella. (C and D) Influenza. (Courtesy Dr. J. Forbes.)*

VIRAL GASTROENTERITIS

Gastroenteritis vies with URI for the mantle of commonest of all infectious diseases and is the greatest cause of death. It has been estimated that 5–10 million children die each year in the Third World from diarrheal diseases, rotavirus infections in malnourished infants being a major contributor. Rotaviruses infect young children, causing severe diarrhea which may last up to a week and lead to dehydration requiring fluid and electrolyte replacement. Most infections are sporadic, but nosocomial outbreaks occur frequently in hospital nurseries.

Caliciviruses, on the other hand, tend to infect older children and adults, often in common-source outbreaks. The illness consists of an

TABLE 29-2
Viral Gastroenteritis

VIRUS	DISEASE
Rotaviruses	Common sporadic cause of diarrhea in infants; potentially lethal
Caliciviruses (including Norwalk)	Outbreaks of vomiting/diarrhea in older children and adults
Adenoviruses (especially 40, 41)	Significant cause of gastroenteritis
Astroviruses	?

explosive episode of vomiting, diarrhea, nausea, malaise, and abdominal cramps, sometimes accompanied by myalgia and/or low-grade fever.

The contribution of other "small round viruses" including the "astroviruses" (Chapter 19) to the global spectrum of gastroenteritis is currently unclear. The recently discovered "fastidious enteric" adenoviruses 40 and 41, and perhaps some others, now also appear to be a significant cause of gastroenteritis (Table 29-2).

VIRAL DISEASES OF THE CENTRAL NERVOUS SYSTEM

Most meningitis and almost all encephalitis is of viral etiology (Table 29-3). Infections of the CNS arise, in the main, as a rare complication of a primary infection established elsewhere in the body which fortuitously spreads to the brain, usually via the bloodstream. Sometimes they occur following reactivation of a latent herpesvirus or papovavirus infection, particularly following immunosuppression. Overwhelming disseminated infections acquired perinatally may also involve the brain.

We know very little about why one individual in every thousand or so infected with a particular arbovirus or herpes simplex virus develops encephalitis while the others escape relatively unscathed. There is clear evidence that certain viruses have a predilection for particular parts of the CNS and the clinical signs of the resulting disease often reflect this, e.g., most enteroviruses do not go beyond the meninges, but polioviruses invade the anterior horn of the spinal cord and the motor cortex of the cerebrum, whereas rabies singles out Ammon's horn, HSV the temporal lobes, and so on. Some viruses lyse neurons directly and there is plenty of evidence of inflammation in the brain (Plate 29-3A); others do their damage in more subtle ways, leading to demyelination of nerves (Plate 29-3B), presumably via immunopathological processes.

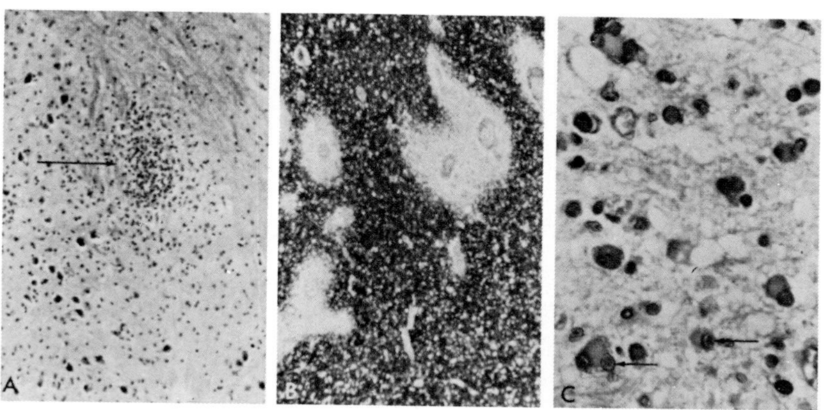

PLATE 29-3. *Viral infections of the brain. Sections from fatal human cases of encephalitis. (A) Arbovirus, Murray Valley (Australian) encephalitis (H and E stain, ×55). Arrow indicates focal concentration of leukocytes. (B) Measles, postinfectious encephalitis (luxol fast blue stain for myelin, ×20). Note areas of demyelination around vessels. (C) Measles, subacute sclerosing panencephalitis (H and E stain, ×220.) Arrows indicate intranuclear inclusions. (Courtesy Drs. I. Jack, R. McD. Anderson, and A. Williams.)*

The many and varied neurological syndromes caused by viruses include meningitis, encephalitis, paralytic poliomyelitis, myelitis, polyneuritis, and several unusual demyelinating and degenerative syndromes.

Meningitis

Viral meningitis is much commoner than bacterial meningitis but is much more benign. Recovery is almost always complete. The patient presents with headache, fever, and slight neck stiffness, with or without vomiting and/or photophobia. Lumbar puncture reveals a clear CSF, perhaps under slightly increased pressure, with near normal protein and glucose concentrations, and only a moderate pleocytosis—the white cell count may range from normal ($<10/mm^3$) to over $1000/mm^3$, but is usually $30–300/mm^3$, with lymphocytes predominating after the first day or so. This is what is generally called "aseptic" meningitis.

By far the most important etiological agents are mumps virus (see Chapter 22) and numerous enteroviruses, including all the coxsackie B types, coxsackie A7 and 9, and many echovirus which were listed in Chapter 19. The herpesviruses, HSV, EBV, and CMV, are rare sporadic causes, while lymphocytic choriomeningitis virus can be acquired from laboratory or pet mice or hamsters.

Meningitis may be the only clinical evidence of infection with these viruses. For example, only half of all cases of mumps meningitis follow typical parotitis. Enteroviral meningitis is only sometimes associated with a rash but often occurs during a summer/autumn epidemic in which others experience rashes, myositis, or other common manifestations of infection with the prevalent agent.

Paralysis

In countries from which polioviruses have not yet been effectively eradicated by vaccination, these viruses remain the major cause of both aseptic meningitis and paralytic poliomyelitis. By contrast, in the United States today wild polioviruses and poliomyelitis are so rare that the oral poliovaccine itself has become the commonest cause of poliomyelitis. Enterovirus 71 and coxsackie A7 are rare causes of a paralytic disease essentially indistinguishable from polio. Other enteroviruses may rarely cause a paralytic syndrome, especially in patients with hypogammaglobulinemia. The radiculomyelitis associated with the enterovirus 70 pandemic in 1969–1971 was generally reversible.

Encephalitis

Encephalitis is the most serious of all common viral diseases. The illness often begins like meningitis with fever, headache, vomiting, and neck stiffness, but alteration in the state of consciousness indicates that the brain parenchyma itself is involved. Initially lethargic, the patient becomes confused then stuporose. Ataxia, seizures, and paralysis may develop before the victim lapses into a coma and dies. Survivors are left with a pathetic legacy of permanent sequelae including mental retardation, epilepsy, paralysis, deafness, or blindness.

Encephalitogenic mosquito-borne or tick-borne togaviruses, flaviviruses, and bunyaviruses, endemic to particular regions of the world, cause epidemics of encephalitis from time to time when the appropriate combination of ecological circumstances develop. The ecology of each of these arboviruses was described in detail in Chapters 20 or 25 (see also Tables 20-3, 20-4; Figs. 9-2, 9-3, 25-1). Encephalitis is also an irregular feature in certain hemorrhagic fevers (see Table 29-5). Rabies causes a uniformly lethal type of encephalitis, described fully in Chapter 26.

In most of the temperate regions of the world however, mumps is the commonest cause of "encephalitis," although this is really a relatively mild meningoencephalitis with only rare sequelae. Herpes simplex virus is the most commonly identified cause of severe sporadic encephalitis. This is a very unpleasant disease indeed, with a 70% case fatality rate

TABLE 29-3
Viral Diseases of the Central Nervous System

DISEASE	VIRUSES[a]
Meningitis	*Enteroviruses* *Mumps* Lymphocytic choriomeningitis Herpes simplex; other herpesviruses rarely
Paralysis	*Polioviruses* Enteroviruses 70, 71; coxsackie A7
Encephalitis	*Herpes simplex* *Mumps* *Arboviruses:* EEE, WEE, VEE, JE, MVE, WN, SLE, Rocio, TBE, louping ill, Powassan, La Crosse, RVF, CrHF, OHF, KFD (see Tables 20-3, 20-4) Arenaviruses and bunyaviruses (see Table 29-5) Rabies Enteroviruses, adenoviruses, other herpesviruses
Postinfectious encephalomyelitis	*Measles,* varicella, rubella, vaccinia, others
Guillain-Barré syndrome	*EB virus,* CMV, others
Reye's syndrome	*Influenza, varicella*
SSPE	*Measles,* rubella
Progressive multifocal leukoencephalopathy	*Polyomavirus JC*
Creutzfeldt–Jakob disease	*Subviral agent?*

[a] The commonest causal agents are italicized.

(see Chapter 16). Enteroviruses, though usually restricting themselves to the meninges, may induce true encephalitis now and again. In neonates or immunocompromised patients the other herpesviruses, and rarely adenoviruses, are also capable of causing encephalitis, generally as part of a widely disseminated and often fatal infection.

Postinfectious Encephalomyelitis

This is a severe demyelinating condition of the brain and spinal cord which occurs as an occasional complication following a few days after any of the common childhood exanthemata—measles, varicella, rubella—and mumps. It also occurred as an occasional complication of vaccination against smallpox, using live vaccinia virus, prior to the eradication of that disease. The pathology of postinfectious encephalomyelitis resembles that of experimental allergic encephalomyelitis, giving rise to the hypothesis that this is an autoimmune disease in which virus infection provokes an immunological attack on myelin. Certainly there is little virus demonstrable in the brain by the time postinfectious encephalomyelitis develops, and the major histological finding is perivenous inflammation and demyelination (Plate 29-3B).

Guillain-Barré Syndrome (GBS)

This is an acute inflammatory demyelinating polyradiculoneuropathy which follows exposure to any one of several viruses. EB virus (which interestingly has also been associated with transverse myelitis and Bell's palsy) has been implicated, GBS appearing 1–4 weeks after infectious mononucleosis, but other agents such as CMV are also involved. Partial or total paralysis develops, usually in more than one limb. Complete recovery occurs within weeks in most cases, but 15% retain residual neurological disability.

An outbreak of GBS in the United States in 1976 was traced to the introduction of a formalin-inactivated vaccine against the so-called Swine strain of influenza. The vaccine was withdrawn and GBS has not been associated with any subsequent flu vaccine. The whole episode remains something of a mystery. It does prove, however, that live virus is not a necessary ingredient in the genesis of GBS. This argues strongly for an immunological basis of the demyelination.

Reye's Syndrome

Reye's syndrome is a postinfectious encephalopathy with a 25% case fatality rate which follows influenza (B > A) or chickenpox in children. There is cerebral edema but no evidence of inflammation. Fatty infiltration of the liver is the other major feature. An epidemiological association with the administration of aspirin during the original fever has been noted.

Chronic Demyelinating Diseases

Certain of the rarer demyelinating diseases are known to be due to viruses. *Subacute sclerosing panencephalitis (SSPE)* is a rare late sequel to measles while *progressive rubella panencephalitis* is even rarer. The pathogenesis of SSPE was discussed in Chapter 7. *Progressive multifocal leukoencephalopathy (PML)* is a different type of demyelination seen when immunosuppression for renal transplantation or malignancy reactivates infection with the human polyomavirus JC (see Chapters 7 and 14). These associations have quickened interest in the possibility that more common demyelinating diseases of unknown etiology, notably multiple sclerosis, might also be caused by viruses. However, despite suggestive epidemiological evidence and many false alarms, no virus has yet been incriminated.

Chronic Degenerative Diseases of the CNS

In Chapter 28 we discussed the role of putative viral or "subviral" infectious agents in *kuru* and *Creutzfeldt–Jakob disease,* two degenerative diseases of the brain generally classified as *subacute spongiform viral encephalopathies,* of which scrapie in sheep is the paradigm (see Chapter 7). The search is now on for viruses as possible etiological agents of much more important degenerative diseases of the CNS, such as amyotrophic lateral sclerosis, Parkinson's disease, and Alzheimer's disease (presenile dementia).

VIRAL SKIN RASHES

Many viruses involve the skin in one way or another (Table 29-4). Some, such as papillomaviruses, poxviruses, and recurrent herpes simplex produce relatively localized crops of lesions and few if any systemic symptoms. Others, such as those causing the childhood exanthemata, produce a generalized rash as part of a wider clinical syndrome that follows a systemic infection. These rashes vary greatly in their anatomical distribution and in the morphology of the individual lesions. They are classified for convenience into maculopapular, vesicular, nodular, and hemorrhagic rashes (Plate 29-4).

Macules are flat, colored spots; papules are slightly raised from the surface of the skin but contain no expressible fluid. Virus is not shed from the lesions of maculopapular rashes. Many of them may in fact result from a hypersensitivity response to the virus growing in cells of the skin or capillary endothelium.

The differential diagnosis of maculopapular rashes is difficult—not only because many rashes are of toxic, allergic, or psychogenic origin, but also because they are a common feature of countless infectious diseases caused by bacteria, rickettsiae, fungi, protozoa, and metazoa as well as viruses! The rash itself is rarely pathognomonic; the whole clinical syndrome must be taken carefully into account.

The classical standards of reference against which other rashes are compared are the so-called "morbilliform" rash of measles and the "rubelliform" rash of rubella. The exanthem of measles (Plate 22-3) consists of flat reddish-brown macules which coalesce to form rather large blotches; after the rash fades on day 5 or 6 the skin retains a brownish stain for a time then undergoes desquamation. In contrast, the exanthem of rubella consists of much smaller (pinpoint) pink macules which tend to remain discrete, giving the rash a fine or erythematous appearance; it usually disappears after 2–3 days.

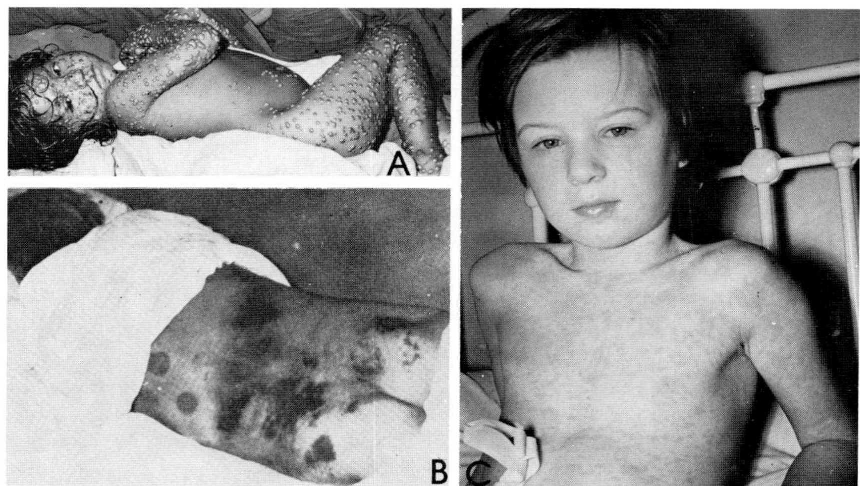

Plate 29-4. *Three distinct types of viral rash. (A) Vesiculopustular rash of smallpox. (B) Hemorrhagic rash of hemorrhagic fever. (C) Maculopapular rash of measles. (A, courtesy Prof. A. W. Downie; B, courtesy Dr. A. Shelekov; C, courtesy Dr. J. Forbes.)*

Numerous unrelated viruses produce rashes almost indistinguishable from one or other of these two prototypes. Infections with literally dozens of different enteroviruses can present as a maculopapular rash, generally in children, and often during late summer epidemics. These exanthems are usually ephemeral and nonpruritic. They are mainly "rubelliform" or "morbilliform," but can be erythematous, petechial, urticarial, or vesicular in character. Space does not allow description of the syndromes associated with each of the 30-plus enteroviruses involved. Sufficient to note that the serotypes most frequently responsible for cutaneous eruptions are echoviruses 4, 9, 16 and coxsackieviruses A9, A16, B5.

Erythema infectiosum, or fifth disease, now known to be caused by a parvovirus, B19, is recognized for its unique rash: the child first develops flushed red cheeks, contrasting with pallor around the mouth, then a rubelliform eruption on the limbs which develops a lace-like appearance as its fades. Some 5% of cases of infectious mononucleosis, whether caused by EB virus or cytomegalovirus, have a maculopapular rash, especially if treated with ampicillin. Many arthropod-borne alphaviruses and flaviviruses, including dengue, chikungunya, Sindbis, and Ross River viruses, also produce a maculopapular or scarlatiniform rash lasting 2–3 days. Finally, mention should be made of the urticarial rash

TABLE 29-4
Viral Skin Rashes

RASH	VIRUSES
Maculopapular	Measles
	Rubella
	Parvovirus B19
	Echoviruses 4, 9, 16, many others
	Coxsackie A9, A16, B5, many others
	EB virus, cytomegalovirus
	Dengue, chikungunya, Ross River
	Hepatitis B
Vesicular	Varicella-zoster
	Herpes simplex 1, 2
	Coxsackie A9, A16; enterovirus 71
	Vaccinia, cowpox, orf
Nodular	Papillomaviruses
	Molluscum contagiosum
	Milker's nodes

that forms part of the serum sickness syndrome seen fleetingly in the prodromal phase of 10–20% of cases of hepatitis B.

Vesicles are blisters, containing clear fluid from which virus can readily be isolated. Vesicular rashes do not present a great diagnostic problem, particularly now that smallpox (Plate 29-4) has disappeared. A generalized vesicular rash in a febrile child today is almost certainly chickenpox (varicella). The lesions occur in crops, initially concentrated on the trunk, then spreading centrifugally. Each vesicle progresses to a pustule and a scab which then falls off. In herpes zoster the lesions are largely (but not necessarily exclusively) confined to a particular dermatome (Plate 16-5) as is also the case with the recurrent form of herpes simplex (Plate 16-2B). However, in the case of disseminated herpes simplex or zoster, as seen in newborn infants or immunocompromised patients, the lesions may be widespread throughout the body. Something of a curiosity is the condition known as hand-foot-and-mouth disease caused by certain coxsackieviruses, in which vesicles or even bullae occur on the palms, soles, and buccal mucosa. Coxsackie A viruses also produce a similar type of vesicular enanthem on the mucous membrane of the throat and palate ("herpangina").

Poxviruses preferentially infect the skin, producing multiple nodular lesions, as in molluscum contagiosum and milker's nodes, or large single ulcerating lesions such as orf and cowpox. These and other zoonotic poxvirus diseases were described in Chapter 17 and illustrated in Plate 17-4.

Papillomavirus infections were described in Chapter 14. The papilloma, or wart, is a benign hyperplastic growth, usually multiple, occurring in crops on skin or mucous membranes. Dermatologists classify them in various ways but generally recognize common warts, flat warts, plantar and palmar warts, epidermodysplasia verruciformis, and genital warts—all of which are clinically distinct and tend to be caused by different human papillomavirus types (see Table 14-2).

VIRAL HEMORRHAGIC FEVERS

Though not an entirely homogeneous group of diseases, the hemorrhagic fevers share the common characteristic of widespread hemorrhage from the body's epithelial surfaces, including internal mucosae such as the gastrointestinal tract as well as the skin. The skin "rash" is often a mixture of pinpoint hemorrhages (petechiae) and massive "bruising" (ecchymoses), as depicted in Plate 29-4. The pathogenesis of these important diseases is generally not well understood. Thrombocytopenia and leukopenia are almost always present but no general mechanism has been discovered to explain the hypovolemic shock without major blood loss which may lead to death within hours in dengue or Lassa fever, for example. Severe liver damage, extensive bleeding, and disseminated intravascular coagulation may be the key to the high mortality in the African hemorrhagic fevers, Crimean hemorrhagic fever, and the hemorrhagic form of Rift Valley fever. Renal tubular necrosis and severe oliguria are distinctive features of Hantaan virus infection.

Detailed descriptions of the epidemiology and clinical features of each of the 11 major hemorrhagic fevers (plus yellow fever, which could also be so regarded) were given in the chapters on flaviviruses (20), arenaviruses (24), bunyaviruses (25), and filoviruses (26). Suffice it to add here that the "typical" florid clinical syndromes, though protean in their presentation, are distinguishable by criteria such as those set out in Table 29-5. Dengue hemorrhagic fever, Hantaan hemorrhagic nephrosonephritis, Rift Valley fever, and Lassa fever are the most prevalent on a world scale. The problem in Western countries is that the disease is likely to be completely outside the experience of the clinician who first sees it, and may also be a mild or atypical case with little to show other than an undifferentiated fever, possibly acquired abroad. The alternatives are limited by the traveler's recent itinerary—with Africa providing the most options. False alarms are frequent, especially in countries

TABLE 29-5
Viral Hemorrhagic Fevers[a,b]

DISEASE	VIRUS	HEMORRHAGE	SHOCK	HEMOCONCENTRATION	PROTEINURIA	OTHER FEATURES
DHF	Dengue 1–4	+	++	++	−	P
HNN	Hantaan	+	++	++	++	N > P
RVF	RVF	++[c]	++	±	+	H > E
LF	Lassa	+	±	+	++	H > P, E
AFHF	Marburg	++[c]	++	+	++	H > P
AFHF	Ebola	++[c]	++	+	++	H > P
CrHF	CrHF	++[c]	++	+	+	H > E
OHF	OHF	±	±	±	+	P > E
KFD	KFD	±	±	±	+	E
AHF	Junin	+	++	+	+	E
BHF	Machupo	+	++	+	+	E

[a] Modified from K. M. Johnson, *In* "Viral Diseases in S.E. Asia and the Western Pacific" (J. S. Mackenzie, ed.), Academic Press, Sydney, 1982; and K. M. Johnson, *In* "Infectious Diseases" (P. D. Hoeprich, ed.), Harper & Row, Philadelphia, 1983.

[b] Abbreviations: DHF, dengue hemorrhagic fever; HNN, hemorrhagic nephrosonephritis; RVF, Rift Valley fever; LF, Lassa fever; AFHF, African hemorrhagic fever; CrHF, Crimean-Congo hemorrhagic fever; OHF, Omsk hemorrhagic fever; KFD, Kyasanur Forest disease; AHF, Argentine hemorrhagic fever; BHF, Bolivian hemorrhagic fever; ++, very frequent and/or severe; P, pneumonia; N, nephritis; H, hepatitis; E, encephalitis; >, more prominent than.

[c] Disseminated intravascular coagulation demonstrated or highly suspect.

in which expensive facilities for transporting, nursing, and diagnosing Class-4 pathogens have been established, but "discretion is the better part of valor" in such circumstances. Isolation of the patient and laboratory identification of the etiological agent are essential.

VIRAL GENITOURINARY INFECTIONS

There are two major viral sexually transmitted diseases (STD), *genital herpes* and *genital warts,* both of which have dramatically increased in frequency in recent years. The painful itchy lesions of genital herpes (Plate 16-3) and the accompanying local and systemic symptoms were described in Chapter 16. Dozens or even hundreds of recurrences, mainly attributable to HSV-2, may dominate the life of the hapless carrier. Genital warts (condyloma acuminata, Plate 14-2), caused by the human papillomaviruses HPV-6, 11, 16, and 18, were described in Chapter 14, and their association with cervical carcinoma in Chapter 8. Adenovirus 37 has recently been associated with cervicitis. Molluscum contagiosum is also occasionally an STD. Other viruses are shed in semen and in female genital secretions and are transmitted venereally but cause no disease in the genital tract itself. These include HTLV-III, hepatitis B, and cytomegalovirus.

Many more viruses are regularly conveyed between male homosexuals. These include enteric viruses such as hepatitis A and polioviruses, as well as HTLV-III. Most highly promiscuous male homosexuals reveal serological evidence of infection with cytomegalovirus, EB virus, and hepatitis B.

Viruses rarely infect the urinary tract (Table 29-6). *Urethritis* can complicate infections with HSV and has been reported with adenovirus 37. *Acute hemorrhagic cystitis,* an unusual disease of young children, has been associated principally with adenoviruses 11 and 21.

Glomerulonephritis is sometimes observed as a manifestation of immune-complex disease in chronic hepatitis B (see Chapter 13). It is safe to predict that future research may reveal that some cases of "idiopathic" glomerulonephritis are also caused by chronic persistent infections with other viruses yet to be identified (see Chapter 7).

Cytomegalovirus persists asymptomatically in renal tubules (Plate 16-7A), from which cytomegalic cells as well as virus are shed into the urine. When primary infection or reactivation occurs during renal transplantation, rejection of the graft may be accelerated by *CMV-induced nephropathy* (see Chapter 16). The human polyomaviruses, BK and JC (Chapter 14) also persist in the urinary tract, are reactivated by renal transplantation and may or may not play a role in rejection of the graft.

TABLE 29-6
Viral Infections of the Genitourinary Tract

DISEASE	VIRUS
Genital	
Genital herpes	Herpes simplex (HSV-2 > HSV-1)
Genital warts	Human papillomaviruses 6, 11, 16, 18
Cervicitis	Adenovirus 37
Molluscum contagiosum	Molluscum contagiosum
Urinary	
Urethritis	HSV, adenovirus 37
Acute hemorrhagic cystitis	Adenoviruses 11, 21, others
Glomerulonephritis	Hepatitis B
Nephropathy	Cytomegalovirus; polyomaviruses BK and JC?
	Hantaan
Hemolytic-uremic syndrome	Enteroviruses

Clearly there is profound malfunction of the kidneys in *hemorrhagic fever with renal syndrome,* caused by the bunyavirus, Hantaan. The clinical developments were described in Chapter 25 but the pathology is still unclear.

Hemolytic-uremic syndrome is characterized by acute microangiopathic hemolytic anemia, intravascular coagulopathy, and impaired renal function. The patient often presents with bloody diarrhea. Various enteroviruses (especially coxsackie B viruses, as well as coxsackie A4 and echo 11) have been isolated from family clusters of HUS. One study indicated that all cases had evidence of recent infection with two of these viruses, giving rise to the hypothesis that HUS may be a Shwartzman-type reaction provoked by concurrent infection with two different viruses.

VIRAL DISEASES OF THE EYE

It is not generally appreciated how frequently viruses can involve the eyes (Table 29-7). *Conjunctivitis* is a transient feature of a number of common childhood exanthemata such as measles (Plate 22-3), rubella, and certain enteroviral infections, and is an important component of the dengue-like syndromes caused by many arboviruses, e.g., phlebotomus (sandfly) fever. Infections with adenoviruses, notably Ad3, 4a, and 7 in children, present as a bilateral follicular conjunctivitis, or as pharyngoconjunctival fever.

Keratoconjunctivitis, on the other hand, is potentially more dangerous as it involves the cornea. Adenoviruses 8, 19a, and 37 are major causes of epidemic keratoconjunctivitis, which spreads readily by contact to adults, and usually involves only one eye, but may take up to a year to resolve. The main cause of sporadic keratoconjunctivitis—indeed the commonest infectious cause of blindness in the Western World—is herpes simplex virus. Pathognomonic "dendritic" or "geographic" ulcers develop on the cornea (Plate 16-3) and, if infection progresses to involve the stroma beneath, the immunological reaction may lead to disciform keratitis, scarring, and loss of vision. Recurrent attacks are particularly damaging, as can be the application of corticosteroids. When herpes zoster involves the fifth cranial nerve, ophthalmic zoster (Plate 16-5) can cause lasting damage to the eye.

Acute hemorrhagic conjunctivitis exploded on the world in 1969 and has since infected millions of people in a succession of epidemics particularly involving the Third World. Subconjunctival hemorrhages, keratitis, and uveitis are quite common features; neurological complications are rare. The etiological agents are enterovirus 70 and coxsackie A24.

TABLE 29-7
Viral Infections of the Eye

DISEASE	VIRUS	FEATURES
Conjunctivitis	Adenoviruses 3, 4a, 7, others	Pharyngoconjunctival fever
	Phlebotomus fever Dengue fever	Dengue-like syndrome
	Measles Rubella	Exanthemata
	Marburg, Ebola	Hemorrhagic fever
Keratoconjunctivitis	Adenoviruses 8, 19a, 37, others	Epidemic
	Herpes simplex	Primary or recurrent
	Herpes zoster	Ophthalmic zoster
Acute hemorrhagic conjunctivitis	Enterovirus 70; coxsackie A24	Pandemics; ± radiculomyelitis
Chorioretinitis	Cytomegalovirus	Immunosuppressed or congenital
	Rift Valley fever	
Cataracts Glaucoma Microphthalmia Retinopathy	Rubella	Congenital rubella syndrome

Retinitis, sometimes leading to permanent loss of central vision, was a feature of the 1977 epidemic of Rift Valley fever in the Nile Valley. Chorioretinitis is also a feature of cytomegalovirus infections in immunocompromised (e.g., organ transplant) patients, as well as in congenitally infected babies with cytomegalic inclusion disease. Retinopathy, glaucoma, microphthalmia, and especially cataracts are the major eye abnormalities encountered in the congenital rubella syndrome (Plate 20-5); total or partial blindness may result.

Finally, a rare accidental cause of eye infection is autoinoculation with certain animal viruses, including Newcastle disease virus of chickens, seal influenza, or vaccinia virus.

VIRAL ARTHRITIS

Arthritis, usually accompanied by myositis and a fleeting rash, is a common presentation of infections with togaviruses of the *Alphavirus* genus (Table 29-8). Chikungunya, o'nyong-nyong, and Ross River viruses have caused huge epidemics of polyarthritis in Africa, Asia and the Pacific Islands (see Chapter 20). Arthritis is a somewhat less prominent feature of rubella but is common in adult females following either natural infection or rubella vaccine. In all these diseases the polyarthritis

TABLE 29-8
Viral Arthritis

VIRUS	FEATURES
Ross River Chikungunya O'nyong-nyong	Arboviral fevers + polyarthritis
Rubella	Especially in adult females
Mumps Varicella Coxsackieviruses	Rare
Hepatitis B	Immunologically mediated during prodrome
RA-1 (parvovirus?)	?Role in rheumatoid arthritis

tends to flit from one joint to another, involving principally the extremities such as the hands; only rarely does it persist for more than a few weeks. Much less frequently, ephemeral arthritis is seen in mumps, varicella, and coxsackievirus infection. The arthralgia sometimes observed in the prodromal stages of hepatitis B is immunologically mediated. Many people sense that rheumatoid arthritis may be of viral origin; the latest in a long succession of candidates to come under scrutiny is the putative parvovirus RA-1 (Chapter 18).

VIRAL CARDITIS

Coxsackie B viruses are now recognized to be the most important single cause of carditis (Plate 19-4C). Most cases of myocarditis or pericarditis in the newborn, and at least one-third of those at any age can be shown by EIA to have IgM against a coxsackie B virus. The syndrome is characterized by fever, chest pain, dyspnea, tachycardia, cardiac enlargement, abnormal heart sounds, and ECG changes. Progressive autoimmune damage may supervene and lead to dilated congestive cardiomyopathy. From time to time the heart may be infected in the course of systemic infections caused by many viruses, such as other enteroviruses, influenza viruses, cytomegalovirus, EB virus. Moreover, congenital infection with rubella commonly damages the heart, while prenatal mumps has been associated with endocardial fibroelastosis.

VIRAL HEPATITIS

Our knowledge of viral hepatitis has developed remarkably over the past decade. Accordingly we have devoted a good deal of space else-

TABLE 29-9

Viral Hepatitis

	HEPATITIS A	HEPATITIS B	NON-A, NON-B		DELTA
			PARENTERAL	ENTERAL	
Virus	Enterovirus 72	Hepadnavirus	?	27 nm?	Defective 36 nm RNA (Hepatitis B helper)
Transmission	Enteric	Parenteral Perinatal, venereal	Parenteral	Enteric	Parenteral?
Incubation period	2–6 weeks	6–26 weeks	2–26 weeks	1.5–6 weeks	?
Clinical disease	Mild or moderate	Moderate	Mild	Severe	Severe
Chronic carrier state	No	Yes	Yes	No	Yes

TABLE 29-10
Other Viruses Causing Hepatitis

VIRUS	DISEASE
EB virus	Infectious mononucleosis
Cytomegalovirus	Cytomegalic inclusion disease; immunocompromised
Herpes simplex	Neonatal; immunocompromised
Varicella	Neonatal, immunocompromised
Rubella	Congenital rubella syndrome
Yellow fever Marburg Ebola Lassa Rift Valley fever Crimean-Congo HF	Hemorrhagic fevers

where in this volume to hepatitis A (Chapter 19), hepatitis B (Chapter 13), non-A, non-B hepatitis, and the delta agent (Chapter 28). Here, we simply produce a summary Table (29-9) which brings together for easy comparison some of the main clinical and epidemiological features of the five "hepatitis viruses."

It should be stressed that hepatitis is an occasional feature of the clinical syndromes induced by several other viruses as well, e.g., infectious mononucleosis (Table 29-10). Yellow fever, caused by a flavivirus (Chapter 20), is characterized by a very severe hepatitis—hence its name. Indeed, hepatitis is prominent in many of the hemorrhagic fevers, particularly Marburg, Ebola, Lassa, Rift Valley fever, and Crimean-Congo hemorrhagic fever. It is also a major feature of most of the disseminated viral infections which overwhelm neonates (neonatal herpes simplex or varicella, cytomegalic inclusion disease, congenital rubella syndrome) or immunocompromised patients (herpes simplex, varicella, cytomegalovirus).

VIRAL PANCREATITIS

Several viruses occasionally infect the pancreas in man. Mumps, for example, can be complicated by severe pancreatitis. Of greater research interest is the question of whether viruses may trigger diabetes mellitus. A "diabetogenic variant" of coxsackievirus B4 isolated from the pancreas of a child who died from acute-onset juvenile diabetes mellitus of the insulin-dependent type (IDDM) was shown to destroy β cells and cause diabetes in mice. The association of IDDM with particular HLA types encouraged a hypothesis that pancreatic infection with any virus may trigger autoimmune destruction of β cells, but proof is lacking.

CONGENITAL AND PERINATAL VIRAL INFECTIONS

Numerous viruses, e.g., measles, mumps, varicella, enteroviruses, can cross the placenta, infect the fetus, and may precipitate a miscarriage. However, only two commonly cause serious *congenital abnormalities.* Rubella during the first 3–4 months of pregnancy causes severe teratogenic effects; the congenital rubella syndrome (Plate 29-5A) and its pathogenesis was described in considerable detail in Chapter 20. Prenatal infection with cytomegalovirus (Chapter 16) induces cytomegalic inclusion disease (Plate 29-5B). The congenital varicella syndrome is extremely rare.

In contrast to these congenital (*prenatal*) infections, several other viruses may infect the infant during or shortly after birth (Table 29-11). Such *perinatal* (alias *neonatal* or *natal*) infections may be acquired during passage of the baby through an infected birth canal (herpes simplex,

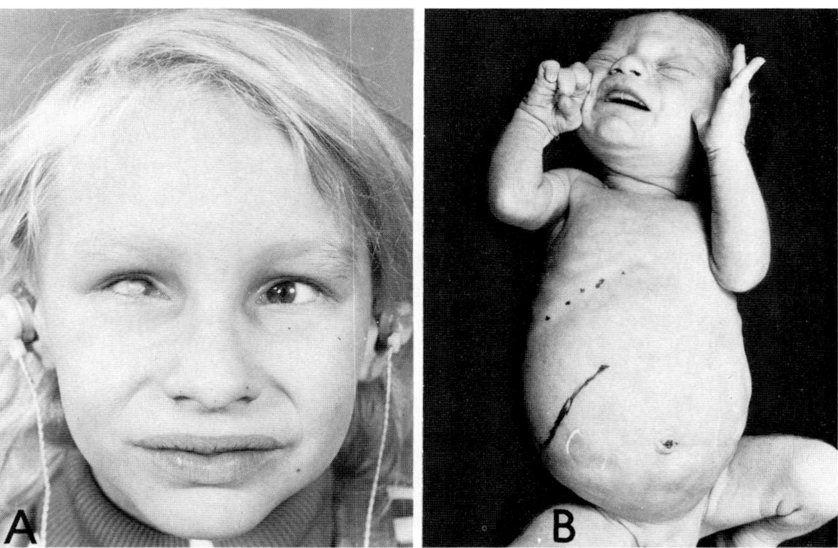

PLATE 29-5. *Viral embryopathies. (A) Rubella syndrome. (B) Cytomegalic inclusion disease. (Courtesy Dr. K. Hayes.)*

TABLE 29-11
Congenital and Perinatal Viral Infections

TIME OF INFECTION	DISEASE	VIRUS
Prenatal	Congenital rubella syndrome	Rubella
	Cytomegalic inclusion disease	Cytomegalovirus
	Congenital varicella syndrome	Varicella
Natal	Herpes neonatorum	Herpes simplex
	Myocarditis of the newborn	Coxsackie B
	Disseminated varicella-zoster	Varicella
	Subclinical or pneumonia	Cytomegalovirus
Postnatal	Hepatitis B carrier state	Hepatitis B

cytomegalovirus), or by contamination with feces (coxsackie B, echovirus 11). Disseminated neonatal herpes, disseminated varicella-zoster, and myocarditis of the newborn (Plate 19-4C) are all overwhelming generalized infections with high mortality. Since primary maternal infection occurs around the time of parturition, the baby is not protected by maternal antibody. These various congenital syndromes may be difficult to distinguish without recourse to virus isolation and IgM serology (see Chapter 12). Unequivocal laboratory confirmation is important because progressive damage can occur after birth and long-term follow-up is essential.

Still other viruses are commonly acquired within the first few weeks of life (*postnatal* infections). Newborn babies of $HBeAg^+$ mothers almost always themselves become lifelong HBsAg carriers within the first few months of life; possible routes of infection were discussed in Chapter 13. It is probable that the transmission of HTLV-III from mother to baby occurs via similar mechanisms. At risk of stretching the definition of postnatal infection, we should for completeness remind the reader that cytomegalovirus and HSV-1 tend to be acquired "vertically" from the mother via milk (CMV) or saliva (CMV, HSV-1) very early in life, particularly in the Third World, whereas rotavirus and respiratory syncytial virus may be picked up by "horizontal" spread, e.g., nosocomially, in hospital nurseries.

VIRAL INFECTIONS IN IMMUNOCOMPROMISED PATIENTS

In Chapter 6 we discussed the role of the various arms of the immune response in controlling viral infections, and in Chapter 7 the reactivation of persistent infections by immunosuppression. Table 7-7 listed the viruses commonly reactivated in immunocompromised patients. In the context of this final overview of viral disease it is sufficient merely to add that reactivation of latent herpesviruses, in particular, has become commonplace as a result of three advances in modern medicine: (1) organ transplantation, requiring as it does, profound immunosuppression to prevent rejection of the allograft, (2) chemotherapy of cancer using highly cytotoxic drugs as well as radiotherapy, and (3) survival of children with profound congenital immunodeficiency syndromes who would not previously have lived. The dangers are enhanced by the fact that the massive blood transfusions (or hemodialysis), so often an integral part of the life-saving therapeutic regimen, carry the risk of iatrogenic transmission of exogenous viruses such as non-A, non-B hepatitis, cytomegalovirus, or HTLV-III, which may overwhelm a patient already desperately ill. Severely burned patients are also highly vulnerable to invasion by HSV. The magnitude of the problem is dramatically illustrated by the observation that it is quite standard for more than one, and sometimes all four, of the herpesviruses, HSV, VZV, CMV, and EBV, to be reactivated in bone marrow transplant recipients; up to 25% of all such patients die of CMV pneumonia alone! The pathogenesis, clinical manifestations, and management of such patients by antiviral chemotherapy, active or passive immunization, and appropriate virological and immunological screening were discussed for each of these viruses individually in Chapter 16. Reactivation of human polyomaviruses was addressed in Chapter 14, and of adenoviruses in Chapter 15.

FURTHER READING

Books

Beran, G. W., ed. (1981). "Viral Zoonoses," CRC Handb. Ser. Zoonoses, Sect. B; Vol. I. CRC Press, Boca Raton, Florida.

Bolte, H.-D., ed. (1984). "Viral Heart Disease." Springer-Verlag, Berlin and New York.

Braude, A. I., ed. (1981). "Medical Microbiology and Infectious Diseases." Saunders, Philadelphia, Pennsylvania.

Deinhardt, F., and Deinhardt, J. B., eds. (1983). "Viral Hepatitis: Laboratory and Clinical Science." Dekker, New York.

Feigin, R. D., and Cherry, J. D. (1981). "Textbook of Pediatric Infectious Diseases," Vol. II. Saunders, Philadelphia, Pennslyvania.

Fox, J. P., and Hall, C. E. (1980). "Viruses in Families." PSG Publishing, Littleton, Massachusetts.

Hanshaw, J. B., and Dudgeon, J. A. (1978). "Viral Diseases of the Fetus and Newborn." Saunders, Philadelphia, Pennsylvania.

Hoeprich, P. D., ed. (1983). Infectious Diseases," 3rd ed. Harper & Row, Philadelphia, Pennsylvania.

Jackson, G. G., and Muldoon, R. L. (1975). "Viruses Causing Common Respiratory Infections in Man." Univ. of Chicago Press, Chicago, Illinois.

Johnson, R. T. (1982). "Viral Infections of the Nervous System." Raven Press, New York.

Katz, S., and Krugman, S. (1981). "Infectious Diseases of Children," 7th ed., Mosby, St. Louis, Missouri.

Mandell, G. L., Douglas, R. G., and Bennett, J. E., eds. (1985). "Principles and Practice of Infectious Diseases." 2nd ed. Wiley, New York.

Mims, C. A., Cuzner, M. L., and Kelly, R. E., eds. (1984). "Viruses and Demyelinating Diseases." Academic Press, London.

Naseman, T. (1977). "Viral Diseases of the Skin, Mucous Membranes and Genitals." Saunders, Philadelphia, Pennsylvania.

Remington, J. S., and Klein, J. O., eds. (1983). "Infectious Diseases of the Fetus and Newborn Infant." Saunders, Philadelphia, Pennsylvania.

Szmuness, W., Alter, H. J., and Maynard, J. E., eds. (1982). "Viral Hepatitis." Franklin Institute Press, Philadelphia.

Tyrrell, D. A. J., and Kapikian, A. Z., eds. (1982). "Virus Infections of the Gastrointestinal Tract." Dekker, New York.

Warren, K. S., and Mahmoud, A. A. F., eds. (1984). "Tropical and Geographical Medicine." McGraw-Hill, New York.

REVIEWS

Blacklow, N. R., and Cukor, G. (1981). Viral gastroenteritis. *N. Engl. J. Med.* **304,** 397.

Gajdusek, D. C. (1977). Unconventional viruses and the origin and disappearance of kuru. *Science* **197,** 943.

Grist, N. R., Bell, E. J., and Assaad, F. (1978). Enteroviruses in human disease. *Prog. Med. Virol.* **24,** 114.

Holmes, I. H. (1979). Viral gastroenteritis. *Prog. Med. Virol.* **25,** 1.

Jenson, A. B., and Rosenberg, H. S. (1984). Multiple viruses in diabetes mellitus. *Prog. Med. Virol.* **29,** 197.

McCormick, J. B., and Johnson, K. M. (1984). Viral hemorrhagic fevers. *In* "Tropical and Geographical Medicine" (K. S. Warren and A. A. F. Mahmoud, eds.), pp. 676–697. McGraw-Hill, New York.

Pan American Heath Organization (1982). "Acute Respiratory Infections in Children." Pan Am. Health Organ., Washington, D.C.

Pavan-Langston, D. (1984). Ocular viral diseases. *In* "Antiviral Agents and Viral Diseases of Man" (G. J. Galasso, T. C. Merigan, and R. A. Buchanan, eds.), 2nd ed., pp. 207–246. Raven Press, New York.

Server, A. C., and Johnson, R. T. (1982). Guillain-Barré syndrome. *Curr. Clin. Top. Infect. Dis.* **3,** 74.

Tesh, R. B. (1982). Arthritides caused by mosquito-borne viruses. *Annu. Rev. Med.* **33,** 31.

Wolinsky, J. S., and Johnson, R. T. (1980). Role of viruses in chronic neurological diseases. *In* "Comprehensive Virology" (H. Fraenkel-Conrat and R. R. Wagner, eds.), Vol. 16, pp. 257–296. Plenum, New York.

World Health Organization (1980). Viral respiratory diseases. *W. H. O. Tech. Rep. Ser.* **642.**

Index

A

Abortion, 141–142, 346, 502, 507, 561, 630
Abortive infection, 98
Abortive transformation, 239
Acquired immunodeficiency syndrome, *see* AIDS
Acronym, 22
Acupuncture, 375–376
Acute hemorrhagic conjunctivitis, 463, 467, 625
Acute hemorrhagic cystitis, 394, 624
Acute respiratory disease (ARD), 393
Acute transforming retroviruses, 223
Acycloguanosine (Acyclovir), 317–319, 413–414, 418
Acylation, 78, 83–84
Adaptation, of viruses, to new hosts, 97, 285
ADCC, 164
Adenine arabinoside (Ara-A), 315, 418
Adeno-associated viruses, 98, 446–447
Adenoviruses, 26, 389–400
  classification, 391–392
  clinical features, 393–394, 608–612, 623–625
  epidemiology and control, 398–399
  genetics, 100
  laboratory diagnosis, 395–398
  multiplication, 62–63, 66, 72–73, 79, 82, 87, 121, 393
  oncogenesis, 238–239
  pathogenesis and immunity, 394–395
  properties, 8, 13, 389–391
  transformation, 238–239
  vaccines, 399
Adenosine deaminase, 307
S-Adenosylhomocysteine analogs, 320
Adjuvants, 292, 378
Adsorption, *see* Attachment of viruses
Aerosol, 344
African horse-sickness, 582
Age, effect on infection, 180–181, 201, 213, 255, 267, 269, 376, 386, 415–416, 426, 457, 470, 506, 513, 534, 610-613, 631
Aggregation of virions, 354, 460, 465
AIDS, 585–596
  chemotherapy, 322, 596
  classification and properties, 586–587
  clinical features, 587–589, 631–632
  control, 594–596
  epidemiology, 592–593
  immunity, 587–591
  laboratory diagnosis, 591–592
  pathogenesis, 589–591
  vaccine, 596
AIDS-related complex (ARC), 587–589
Air travel, 274, 277
Aleutian disease of mink, 174, 205
Alimentary, *see* Enteric infections
Alleles, 367
Allergic bronchitis, 535
Alphaviruses, 483–484, *see also* Togaviruses
Alzheimer's disease, 604, 619

Amantadine, 321–322
Ambisense, 15–16, 74, 548, 557
Amphotropic retroviruses, 223
Amplifier hosts, 492
Amyloid plaques, 604
Amyotrophic lateral sclerosis, 619
Anamnestic response, 170
Anchorage independence, 126, 239
Animals, *see also* individual animals and viruses
  bite, transmission of viruses by, 129–130, 344, 569–571
  cells for recDNA cloning, 108
  as experimental models, 109–117, 168–181, 204–210, 223–224, 239–241, 591, 604
  as reservoirs, 252, 257–262, 440–443, 484–485, 490–498, 551–553, 560–574, 582
  for testing antivirals, 308
  for vaccine testing, 285, 297
  for virus isolation, 44–45, 180, 348–349, 465–466, 488, 583
  viruses of, 114, 289, 404, 578, 593
Antibiotics, 4, 303, 343, 529, 620
Antibodies, 154–156, 161–164, 170–174, 180–181
  anti-idiotype, 293–294
  enhancing, 487
  fluorescent, *see* Immunofluorescence
  IgA, 154–156, 171, 180, 214, 295, 459, 579
  IgE, 155, 173–174, 536
  IgG, 154–156, 171, 301–302
  IgM, 154–156, 358–360, *see also* IgM serology
  immune complexes, 173–174, 205–207, 214, 371
  maternal transfer of, 155, 579, 631
  measurement of, 356–360
  monoclonal, 115, 162, 290, 331–332, 490
  neutralization of viruses by, 114–116, 161–164, 291–292, 351–354, 465
  non-neutralizing, 487–488
  passive immunization, 301–302, 377–378, 471, 552, 571
Antigens,
  antigen–antibody complexes, 173–174, 205–207, 214, 371
  cross-reactions, 349–350, 391, 483, 488–490, 582
  detection methods, 328–339, 349–356
  in hepatitis B, 367, 371–374
  as vaccines, 291–294
Antigenic determinants, 162–163, 291–292, 483, 490
Antigenic drift, 114–118, 210–211, 515, 586, 590
Antigenic shift, 114–118, 515–517
Antigenic variants, 354
Anti-idiotypic antibodies, 294
Anti-oncogenes, 234
Aplastic crises, 448
Ara-AMP, 315, 418
Arboviruses,
  bunyaviruses, 561–566
  definitions, 479
  epidemiology and control, 258–262, 479–480, 490–499
  orbiviruses, 582–583
  pathogenesis, 129–130, 251, 487–488
  togaviruses and flaviviruses, 479–499
Arenaviruses, 30, 547–554, 622
  multiplication, 62–67, 548–549
Argentine hemorrhagic fever, 552, 623
Arildone, 306
Arteritis, 174, 205–207
Arthritis, 626–627
  parvovirus, 449
  rubella, 500, 502, 506
  togaviruses, 485
Arthropod transmission, of viruses, 129–130, 251, 258–262, 276, 484–498, 562–565
Artificial insemination, 595
Aseptic technique, 376–377
Assays
  hemagglutination, 48–50
  infectivity, 45–50
Assembly, of virions, 82–87
  inhibition of, 322
Asthma, 472, 534–535, 543
Asymptomatic, *see* Inapparent infections
Astroviruses, 475
Attachment, of viruses, 65
Attenuation, of viruses, 113, 284–288, 296–297, 458
Atypical lymphocytes, 420, 427–428
Australia antigen, 365
Autoclaving, 595, 604
Autoimmune disease, 174–175, 617, 630
Autopsy, *see* Necropsy

Autoradiography, 57, 76, 108, 340, 355
Avian infectious bronchitis virus, 541
Avian leukosis, 207–208, 232
Avidin, 333, 337, 340

B

Bacteria, 4
  for gene cloning, 104–108, 287–288, 290–291, 310–311, 340
  secondary infection by, 179, 181, 303, 415, 473, 513, 588, 608, 612
Bacteriophages, 45, 53, 104–108, 328
Barrier nursing, 551
Bats, 259, 570
B cells, 148, 153–154
  growth of viruses in, 427–429
B virus, 295
Bell's palsy, 618
BHK-21 cells, 38, 346, 489, 503
Biopsy, 337, 342, 570
Biotin, 333, 337, 340
Birds
  as reservoirs, 259–261, 484–485, 490–498, 515
  viruses of, 207–208, 259–261, 289, 404, 511, 522, 541
BK virus, 386–388
Blind passage, 347
Blindness, 501, 616, 625
Blood banks, 375–376
Blood transfusion, 144, 202–203, 251, 375–378, 420, 423, 590, 592, 595, 597, 632
Bluetongue, 582
Bolivian hemorrhagic fever, 553, 623
Bornholm disease, 462
Boston exanthem, 462
Bovine papillomavirus, 241
Brain, *see* Central nervous system, viral diseases of
Breast feeding, 422, 459, 579
Bromovinyldeoxyuridine (BVDU), 319–320, 409
Bronchiolitis, 531, 534, 609, 611
Bronchitis, 472, 543, 609, 611
BS-C-1, cell line, 346, 468
Budding, of virions, 84–87, 403, 569, 587
Buffalo (African) green monkey kidney cell line (BGM), 465
Bunyaviruses, 31, 555–566
  clinical features, 559–566, 616, 622
  multiplication, 62–67, 557
Buoyant density, of virions, 452, 474
Burkitt's lymphoma, 233, 235–236, 273
BVaraU, 319

C

Caliciviruses, 7, 29, 474–476
California encephalitis, 564–565
Cancer, 217–246
  adenoviruses, 239–240
  hepadnaviruses, 241–243
  herpesviruses, 234–239
  human, 218, 243–245
  papovaviruses, 238–241, 383–388
  retroviruses (oncoviruses), 218–234
  transformation, 125–127, 233, 238–239
  vaccines, 235, 244
Canine distemper, 522
Cannibalism, 210, 603
Cap, 5′, of RNA, 15, 73
Caprine arthritis encephalitis virus, 590
Capsid, 5–8, 109–111
Capsomer, 5–8, 17
Capture antibody, 330
Carditis, 462, 627
Carnivores, in rabies, 570
Carrier cultures, 125, 197
Carrier state, 590, 597, 600
Case fatality rate, 267
Cat,
  as reservoir, 570
  viruses of, 474, 541, 591
Cattle, viruses of, 241, 441–442, 475, 483, 497, 522, 562, 582
Cell(s)
  effects on viruses on, 120–128, 347–348
  fusion, 122–124
  receptors, 176–177
  shutdown, 87, 120–121
  transformation, 125–127, 233, 238–239
Cell culture, 35–48, 97, 194–197, 290, 297–298, 346–348, *see also* individual viruses, Persistent infections
  transformation, 125–127, 233, 238–239
Cell lines, 297–298, 346, *see also* individual cell lines
Cellular oncogenes, 228
Cendehill-51 vaccine, 505
Central nervous system (CNS) infections, *see also* Encephalitis, Meningitis, Po-

CNS infections (*cont.*)
liomyelitis, individual viruses and diseases
chronic degenerative diseases, 619
clinical features, 614–619
latent infections, 198–202
pathogenesis, 138–141
slow infections, 198, 207–211, 387
spongiform encephalopathies, 601–605
Centrifugation, 10, 56, 290, 328, 344–345, 378, 452, 460, 516
Cervical carcinoma, 239, 241
Cervicitis, 394, 623
Caesarean section, 413
Chang conjunctival cell line, 396
Chemical composition, of viruses, 9–32, 58, 512, *see also* individual families
Chemotherapy, antiviral, 303–324, *see also* individual viruses
Chick embryo, 43–44, 349, 411, 436, 514, 516, 530
Chickenpox, *see* Varicella
Chickens, viruses of, 224, 235
Chikungunya virus, 495, 620, 626
Chimpanzee, *see* Monkey
Chinese hamster ovary (CHO) cell line, 108, 291
Chlamydia, 4, 303
Chlorination, 251, 275, 399, 581
Chorioretinitis, 626
Chromatography, 10, 56, 290, 355, 460
Chromosomal abnormalities, 126
translocations, 232–233, 236
Chronic infections, 194–195, 203–207
IgM serology, 358
Class-4 pathogens, 344–345, 549–551, 560, 573, 623
Classification, of viruses, 20–33, 61–67, 354, 391–392, *see also* individual families
bunyaviruses, 556–558
orbiviruses, 582
picornaviruses, 452–453
togaviruses and flaviviruses, 483, 489–490
Cleavage,
of proteins, 64, 66, 75, 78, 82, 84, 322, 512–514, 522, 580
of mRNA, 64, 66
Climate, *see* Seasons
Clinical syndromes, 607–633, *see also* individual viruses and diseases
Cloning,
of cells, 126
of genes, 104–108, 287–288, 290–291, 310–311, 340
of interferons, 182, 188, 310–311
of viruses, 45–48
Coated pits and vesicles, 68, 84
Cocultivation, 244, 427, 449
Coinfection, 98–104, 600
Cold-adapted (*ca*) mutants, 285–286, 518
Cold chain, 298, 460, 528
Cold sores, 405
Colorado tick fever, 582
Common cold, 265–266, 270, 463, 608–610
chemotherapy, 312, 323
coronaviruses, 543–545
rhinoviruses, 472–473
Common-source outbreaks, 275, 470, 476
Communicability, 253–254, 267
Complement, 156–157, 163–165, 336, 353, 488
Complement fixation, 350–351
Complementation, 102–103
Complex virion, 434
Computer modelling, 269
Concentration of viruses, 290
Concurrent infections, 355, *see also* Bacteria
Condyloma acuminata, 623
Confidentiality, 594
Congeners, 305
Congenital immunity, 171, 181, 289
Congenital infections, 141–143, 257, 419–424, 630–631
IgM serology, 358
malformations, 419–420, 449, 500–502, 533, 630
Congenital rubella syndrome, 142, 501, 626, 630
Congenital transmission, 207–208
Congenital varicella syndrome, 415, 630
Conjunctivitis, 129, 463, 526, 566, 625
Consensus sequences, 14, 59, 74
Contact infection, 248–249, 254, 258
Control of viral diseases, 273–280, *see also* individual viruses
Copro-antibodies, 459, 580
Copy DNA, 105, 107

Core, of virion, 5, 434, 586
Coronaviruses, 30, 541–545
  clinical features, 543, 608–609
  multiplication, 62–67, 542
Corticosteroids, 181–182, 625
Coryza, *see* Common cold
COS cell line, 291
Coughing, 250
Counterimmunoelectrophoresis, 334
Cowpox, 441, 621
Coxsackieviruses, 452–453, 461–468, *see also* Picornaviruses
  clinical features, 461–464, 608–631
  laboratory diagnosis, 349, 464–467
Creutzfeldt–Jakob disease, 210, 603–605, 619
Crimean-Congo hemorrhagic fever, 563, 622
Cross-reactivation, 102
Cross-reactivity, of antigens, 296, 349–350, 379, 391, 483, 488–490, 582
Croup, 531, 609, 611
Cryoenzymology, 305
Cryptogram, 23–32
C-type particle, 219
Cultivation, of viruses, 35–51, 341–349
Cytocidal infections, 120–124, 126
Cytomegalic cells, 422
Cytomegalic inclusion disease, 419–420, 630
Cytomegalovirus (CMV), 419–426, *see also* Herpesviruses
  chemotherapy, 318, 425
  clinical features, 419–421, 609, 611–612, 620, 624, 626, 631–632
  epidemiology and control, 421–425
  laboratory diagnosis, 424–425
  pathogenesis and immunity, 202, 421–423
  vaccine, 426
Cytopathic effects (CPE), 40–42, 120–124, 347–348, *see also* individual viruses
  adenoviruses, 396–397
  enteroviruses, 455–456
  herpesviruses, 410
  paramyxoviruses, 525
Cytotoxic T cell, 159–161, 168–169, 173

D

Dane particle, 366
Deafness, 501, 529, 616
Defective retroviruses, 224
Defective viruses, 98, 103, 599
Deletion mutants, 287
Delta hepatitis (HDV), 599–601, 628
Dementia, 589, 601, 603, 619
Demyelination, 141, 198, 208, 387, 527, 615, 618
Dendritic cells, 158
Dendritic ulcers, 625
Dengue fever, 494–495
  clinical features, 484–486, 620–623
  hemorrhagic fever/shock syndrome, 487–488, 622–623
  immunity, 488
  laboratory diagnosis, 488–490
  vaccine, 499
Dentistry, 375–376, 412, 595
Deoxyacyclovir, 319
Deoxycholate, inactivation of viruses, 489
Dependovirus, 446
Dermatitis, 407
Detergents, inactivation of viruses, 11, 18–19, 350, 489
DHPA, 320
DHPG, 319
Diabetes mellitus, 464, 629
Diagnostic kits, 327, 358, 504, 590–592
Diagnostic procedures, 325–361, *see also* individual viruses
  AIDS, 591–592
  adenoviruses, 395–398
  arenaviruses, 550–553
  bunyaviruses, 560–563
  coronaviruses, 544
  enteroviruses, 464–467
  hepatitis B, 371–374
  herpesviruses, 409–412, 416, 424–425, 428–429
  influenza, 514
  paramyxoviruses, 528–532, 536–537
  parvoviruses, 449
  poxviruses, 436–437
  rhabdoviruses, 569–570, 573
  rotaviruses, 580
  togaviruses and flaviviruses, 488–490
Diarrhea, *see* Gastroenteritis
Diarylamidines, 306–307
4′-6-dichloroflavan, 323
DI particles, 95–96

Diploid cell strains, *see* Human diploid fetal fibroblasts
Diploid genome, 219
Diseases, *see* Virus-disease associations and Chapters 13–29
Directed mutagenesis, 96
Disinfection, by chemicals, 17–20, 275, 344, 376, 471, 581, 595, 604
Disseminated intravascular coagulation, 174, 622
DNA
  cDNA, 105
  probes, 57, 240, 242, 269, 339–341, 382, 580
  transcription, 70–74
  replication, 79–80
  of viruses, 10–14, 16, 56–57, 66–67
DNA polymerase, 317–321
Dodecahedron, 82
Dog, in rabies, 570
Dot hybridization, 580
Droplet nuclei, 250
Drug addicts, 375, 592, 595, 597, 600
Dry ice, 343
Ducks, viruses in, 368, 515

E

Early genes, 403
Eastern equine encephalitis, 486, 497
Ebola, 571–574, 622–623
EB virus, 426–430
  chemotherapy, 318
  clinical features, 426–427, 609, 611, 618, 620, 624, 632
  epidemiology and control, 429–430
  laboratory diagnosis, 428–429
  oncogenesis, 235–238
  pathogenesis and immunity, 202, 427–428
  vaccines, 430
Echoviruses, 453, 461–468, *see also* Picornaviruses
  clinical features, 461–464, 608, 615, 620, 625, 631
Eclipse period, 55
Ecology, *see* Epidemiology
Ecotropic retroviruses, 223
Eczema vaccinatum, 439
Edema, 127, 172, 534, 563
Efficiency of plating, *see* Plating efficiency
Electron microscopy, 5–8, 49–50, 70, 350, 434, 576, 580, *see also* individual viruses
  for diagnosis, 328–329
Electrophoresis, polyacrylamide gel, 16, 56–59, 76, 354, 579
Electropherotypes, 578–579, 581
ELISA, *see* Enzyme immunoassay
Encephalitis, 616–618, 623
  arboviruses, 259–262, 484–486, 496–498
  bunyavirus, 564
  herpes simplex, 408
  measles, 527
  mumps, 529
  post-infectious, 141, 617–618
  rabies, 569
  vaccinia, 439
  varicella, 415
Endemic disease, 255
Endocardial fibroelastosis, 627
Endocytosis, 65–69
Endothelial cells, 127, 136, 173
Enhancer, in retrovirus LTR, 225, 232
Enhancing antibody, 163
Enteric infections, *see also* individual viruses
  adenoviruses, 394, 398–399
  caliciviruses, 474–476
  control, 275–276, 470–471
  enteroviruses, 451–471
  epidemiology, 250, 262–263
  gastroenteritis, 613–614
  hepatitis, 468–471, 598–599
  laboratory diagnosis, 328–334
  pathogenesis, 133–134, 144
  rotaviruses, 577–582
Enteroviruses, 21, 451–471, *see also* Hepatitis A, Polioviruses
  clinical features, 461–464, 608, 615–617, 620, 625, 627, 630
  epidemiology and control, 275–276, 467–471
  laboratory diagnosis, 353, 464–467
  pathogenesis and immunity, 464
Entry, of viruses
  into body, 128–134
  into cell, 65–70
Enucleate cells, 435
Envelope, viral, 5–9, 85, 402
Enviroxime, 323, 473

Enzyme immunoassay (EIA), 332–334, 340, 591–592
Enzymes, viral, 17, 305–307
Epidemiology, 112–118, 247–281
  see also Epidemiology sections of Chapters 13–28
  definitions, 266–267
  methods, 266–273, 355
  shedding, 143–144, 249–257
Epidermodysplasia verruciformis, 241, 384
Epilepsy, 616
Episomes, 211, 235–237, 240, 340, 427
Epitopes, 162–163, 291–292, 483, 490
Epizootics, 562, 570
Equine infectious anemia, 210, 568, 590
Eradication, of viral diseases, 278–280, 457
Erythema infectiosum, 448, 620
Erythema nodosum, 173
Erythrocytes, *see also* Hemagglutination
  viruses in, 136, 448, 583
Ethanol, 595
Ethidium bromide, 355
Ethylenimines, 290
Eukaryotic cells, for gene cloning, 108, 291, 310
Evolution, of viruses, 112–118, 233, 252–259, 269, 404, 490–491, 582, 593
Exanthemata, 619–623, *see also* Rashes
Excretion, of viruses, 143–144, 247–257, 341
Extrinsic incubation period, 252
Extinct diseases, 603
Eye infections, 129, 625–626
  adenoviruses, 392–394, 398–399
  arboviruses, 501, 561, 566
  chemotherapy, 313–318
  enteroviruses, 463
  herpesviruses, 417, 420

F

Factors VIII and IX, 592, 595, 597
False positives, in serology, 592
Families, of viruses, 21–32, 61–67, *see also* Chapters 13–28
Fc receptors, 337, 402, 487
Feline infectious peritonitis virus, 541
Feline leukemia virus, 591
Fetus, 141–143, *see also* Congenital infections
Fever, 128, 182, 199, 290, 312
Fever blisters, 199–201, 405
Fibers, of virions, 390
Fifth disease, 448, 620
Filoviruses, 32, 571–574, 622–623
Filtration of viruses, 4, 344, 354, 489, 604
Flavans and flavones, 473
Flaviviruses, 29, 479–499, *see also* Arboviruses
  classification and properties, 479–485
  clinical features, 484–486, 616, 620, 622
  epidemiology, 260–262, 479–480, 490–499
  laboratory diagnosis, 488–490
  multiplication, 483
  pathogenesis and immunity, 487–488
  vaccines, 498–499
Fluorescent antibody, *see* Immunofluorescence
Fluorocarbons, 10
Fluoroiodoaracytosine (FIAC), 320
Flushed cheek syndrome, 448
FMAU, 329
Fomites, 250
Food-borne viruses, 275, 470–471, 476
Foot and mouth disease, 4, 298
Formaldehyde, 289–290, 378, 460, 516
Forssman antibody, 428
Fowl plague, 513
Fragment rescue, 102
Freeze-drying, of viruses, 528
Freeze-storage, of viruses, 10, 18, 343–344, 357, 521
FRhK-4 and FRhK-6 cell lines, 468
Fungi, 588
Fusion
  of cells by virus, 122–124
  of cells, for virus rescue, 239
  in entry/uncoating, 69
  protein, of paramyxoviruses, 522
  protein, in recDNA cloning, 108

G

Gastroenteritis, 613–614
  adenoviruses, 394–395, 398
  caliciviruses, 474–476
  control, 275
  rotaviruses, 577–582
G + C content, 435
Gel diffusion, *see* Immunodiffusion
Genera, of viruses, 21–32

Gene amplification, 233
Genes, of viruses, 13–16, 58–67, 70–77, 511–512, 579
Gene cloning, 104–108, 287–288, 290–291, 310–311, 340
Genetic engineering, 104–108, 287–288, 290–291, 310–311, 340
Genetic interactions, 98–104
Genetic reassortment, 286, 515, 557, 581
Genetic recombination, 225
Genetic resistance, to viruses, 176–177, 255
Genetics, of viruses, 91–118
Genital herpes, 406–414, 623
Genital infections, *see* Sexually transmitted diseases
Genital warts (*condyloma acuminata*), 241, 384–386, 623
German measles, *see* Rubella
Giant cell, 122–124, 422, 525
Giant cell pneumonia, 527
Gingivostomatitis, 404, 610
Glandular fever, 420, 426, 620
Glomerulonephritis, 174, 205–207, 371, 624
Glutaraldehyde, 344, 376, 595
Glycosylation, of proteins, 83–84
Golgi vesicles, 62, 78, 87, 436, 543, 556
Graham (293) cell line, 396
Guillain-Barré syndrome, 518, 618

H

Hamster, 45, 239, 386–387, 538, 553
Hand, foot and mouth disease, 463
Hand washing, 471, 581
Hantaan virus, 559–561, 622
Harvey murine sarcoma virus, 229
HBcAg, 366, 372
HBeAg, 366, 372
HBsAg, 366, 372, 599
Heart, infections, *see* Carditis
HeLa cells, 38, 346, 536
Helical symmetry, 5–9, 523
Helper T cells, 160
Helper virus, 103, 224, 447, 599
Hemadsorption, 42, 124, 524, 530
Hemagglutination, 42, 48–50, 509
  adenoviruses, 392
  arboviruses, 489
  paramyxoviruses, 522, 524
Hemagglutination inhibition (HI), 350–352, 504
Hemagglutinin, 8, 109, 114–118, 162, 335, 512, 518
Hemoconcentration, 623
Hemodialysis, 375–378, 597, 632
Hemolysin, 524
Hemolytic anemia, 448, 625
Hemolytic-uremic syndrome, 464, 625
Hemophiliacs, 375–378, 592, 597
Hemorrhagic fever, 174, 622–623
  arenaviruses, 550–553
  bunyaviruses, 559–564
  filoviruses, 572–573
  togaviruses and flaviviruses, 486
Hemorrhagic fever with renal syndrome (HFRS), 559, 625
HEp-2 cells, 38, 346, 536
Hepadnaviruses, 24, 365–380
  carrier state, 242–243, 371–376
  chemotherapy, 379–380
  classification and properties, 365–367
  clinical features, 368–369, 621, 624, 627–628, 631
  control, 376–380
  epidemiology, 374–376
  immunity, 371
  laboratory diagnosis, 371–374
  multiplication, 368
  oncogenesis, 241–243, 369
  pathogenesis, 203, 369–371
  vaccines, 287, 294, 377–379
Hepatitis, 623, 627–629
  hepatitis A, 468–471
  hepatitis B, 203, 241–243, 365–380
  non-A, non-B hepatitis, 203, 596–599
Hepatitis A, 468–471, 628
Hepatitis B, *see* Hepadnaviruses
Hepatocellular carcinoma (hepatoma), 242–243
Herd immunity, 265, 269
Herpangina, 463, 611
Herpes simplex, 404–414, *see also* Herpesviruses
  chemotherapy, 313–321, 413–414
  clinical features, 404–408, 610, 616, 621, 630–632
  epidemiology, 355, 412–413
  immunity, 408–409
  laboratory diagnosis, 409–412

oncogenesis, 238
pathogenesis, 111, 199–201, 408–409
vaccines, 414
Herpesviruses, 26, 401–431, *see also* individual viruses
chemotherapy, 313–321, 413–414, 425
classification and properties, 13, 402–404
clinical features, 404–408, 415–416, 419–421, 426–427
epidemiology and control, 412–413, 418–419, 421–425, 429–430
laboratory diagnosis, 409–412, 416, 424–425, 428–429
immunity, 409, 416, 421–423, 427–428
multiplication, 62–67, 403
oncogenesis, 234–239
pathogenesis, 111, 408–409, 416, 421–423, 427–428
vaccines, 414, 419, 426, 430
Herpes zoster, *see* Varicella-zoster
Heteroduplex mapping, 382
Heterokaryon, 122
Heterophil(e) agglutinins, 427–428
Hexamer, 402
Hexon, 389–391
Histones, 79, 402
Histopathology, 120–124, 370, 423, 466, 612, 615
Homogeneous assays, 333
Homology, of nucleic acids, 391–392
see also Nucleic acid hybridization
Homosexuals, 277, 376–378, 406, 422, 429, 470, 585, 592–595, 624
Horizontal transmission, 257
Hormones, 125, 128, 181–182, 185, 409, 604
Horses
as reservoirs, 114, 260–261, 483, 497, 515
viruses of, 582, 586
Horse RBC agglutinin test, 428
HT cell line, 591
HTLV-III/LAV, *see* AIDS
Human diploid fetal fibroblasts (HDF or HEF), 37–39, 297, 346, 460, 544
Human embryonic kidney (HEK), 346
Human embryonic lung (HEL), 37–39, 346
Human rhabdomyosarcoma (RD) cell line, 544
Human T cell lymphotropic viruses (HTLV), 227
HTLV-III, *see* AIDS
Human volunteers, 270–272, 285, 449, 475, 545, 599
Humidity, 263
Hybridoma, 123
Hydrophobia, 567
Hygiene, 275–276, 457, 470–471, 581
Hypersensitivity, delayed, 160, 169, 172–174, 535, 571, 619
Hypogammaglobulinemia, 170, 177

I

Iatrogenic infection, 130, 144, 375–377, 398, 422, 425, 573, 590, 604, 632, *see also* Nosocomial infection
Icosahedral symmetry, 5–8
Icosahedron, 389
ID_{50}, 48, 175
Idiotype, 153, 294
IgA, 154–156, 171, 180, 214, 295, 459, 579, *see also* Antibodies
IgE, 155, 173–174, 536, *see also* Antibodies
IgG, 154–156, 171, 301–302, *see also* Antibodies
IgM, 154–156, *see also* Antibodies
serology, 358–360, 425, 467, 470, 490, 504–505, 530, 631
Immortalization of cells, 234, 236, 240, 427
Immune adherence hemagglutination, 350
Immune complexes, 173–174, 205–207, 214, 371
Immune cytolysis, 159–160, 163–164, 173
Immune enhancement, 487
Immune globulin, 301
Immunity, against viruses, 147–175, 186–188
see also Immunity sections of Chapters 13–29
epidemiological significance, 264–266
persistent infections, 211–215
vaccines, 294–296
Immunization, 283–302, *see also* Vaccines
Immunocompromised patients, 631–632, *see also* Immunosuppression
Immunodeficiency, 168–170, 177–179, 189, 415, 426, 536, 588, 632
Immunodiffusion, 334–335
Immunoelectron microscopy, 329, 469, 475

Immunofluorescence, 43, 336–339, 350, 570
Immunogenicity, 289
Immunological tolerance, 205, 214
Immunopathology, 172–175, 189
Immunoperoxidase staining, 338, 340
Immunopotentiation, 292–294
Immunosuppression, 169, 177–179, 189, 198, 201–202, 213–214, 345, 387, 395, 408, 415–416, 419, 423, 427, 459, 588–591, 631–632
IMP dehydrogenase, 307
Inactivated poliovaccine (IPV), 458–460
Inactivation, of serum, 350–353, 357, 504
Inactivation, of viruses, 17–20, 275–276, 289–290
  by chemicals, 17–20, 275, 289, 301, 303, 376, 460, 489, 598, 604
  by freeze-thawing, 521–523, 536
  by heat, 17–18, 276, 298, 343, 376, 458, 460, 468, 522, 528, 536, 595, 598, 604
  by irradiation, 17, 19–20, 276
Inapparent infections, 126, 181, 252–254, 491
Incidence, 266, 268
Inclusion bodies, 121–123, 347–348, 410, 525
  see also individual viruses
Incomplete virus, *see* Defective virus
Incubation period, 207, 210, 250, 254, 569–571, 588, 597
Index case, 267
Infantile paralysis, 457
Infectious dose (ID_{50}), 48, 175
Infectiousness, of viruses, 253–254, 267
Inflammation, 172–174, 182, 534
Influenza, *see* Orthomyxoviruses
Ingestion, of viruses, 250, 275, 344, 470, 476
Inhalation, of viruses, 249–250
Inhibitors, of viral replication, 57, *see also* Chemotherapy
Injection, transmission of viruses by, 129–130, *see also* Iatrogenic infection, Nosocomial infection
Insecticides, 276–277
Insects, *see* Arthropod transmission
Insect viruses, 575
Insertional mutagenesis, 232
In situ hybridization, 341
Interference, 43, 182, 298, 458, 503, *see also* Interferons
Interferons, 182–190, 310–314, 414, 418, 473
Interleukins, 152
Intersecting serum pools, 352
Intestinal, *see* Enteric infections
Intron, 13, 72–73, 188, 228
Inverted repeat (IR) sequences, 13, 221
Iododeoxyuridine, 314
Ir (immune response) genes, 177
Irrigation, 492
Island communities, 255
Isolation of virus, 35–51, 341–356
  see also Laboratory Diagnosis section of Chapters 13–28
Isomer, 403

J

Japanese encephalitis, 496, 498
Jaundice, 368, 372, *see also* Hepatitis
JC virus, 386–388
Joints, infections, *see* Arthritis
Junin virus, 552

K

Kaposi's sarcoma, 588
K cell, 164, 188
Kemerovo viruses, 582
Keratoconjunctivitis, 394, 407, 625
Kilobase, 13
Kinetic neutralization, 354
Kirsten murine sarcoma virus, 229
Koch's postulates, 244, 272–274, 355, 589–590
  see also Virus-disease association
Koplik's spots, 137–138
Korean hemorrhagic fever, 559
Kuru, 210, 601–603, 619
Kyasanur Forest disease, 497, 623

L

Laboratory safety, 344–345, 357, 488, 591, 595
La Crosse virus, 564–565
Lactic dehydrogenase virus (LDV), 205
Langerhans cells, 158
Laryngotracheobronchitis, 531, 609, 611
Laryngeal papilloma, 384

Lassa virus, 549–552, 622
Latent infections, 194–195, 199–202. *see also* individual viruses
Latent period, 55
Lateral bodies, 435
Latex agglutination, 358, 504
LAV/HTLV-III, *see* AIDS
LD_{50}, 175
Lentiviruses, 208–210, 585–596
Lethal viruses, 570, 573, 588, 601
Leukemia, in man, 227
Leukopenia, 622
Lipids of viruses, 9, 18, 85, 231, 290, 435
Liposomes, 159, 293, 309, 430
LLC-MK2 cell line, 489, 532
Long terminal repeat (LTR), 221, 225
Louping-ill, 497
Lucké frog virus, 234
Lumbar puncture, 615
Luxury functions, of cells, 128
Lymphadenopathy, 426–427, 500, 587–589
Lymphadenopathy syndrome (LAS), 587–589
Lymph nodes, 134, 172
Lymphoblastoid cell lines, 427
Lymphocytes, 147–154, 159–161, 167–174, 186–187, 212, 310
  growth of viruses in, 588–591
Lymphocytic choriomeningitis (LCM), 172–174, 205, 553, 615
Lymphokines, 152, 187
Lymphotropic polyomavirus, 387–388
Lyophilization, 18, 298, 438
Lysosomes, 68, 321
Lyssavirus, 568

M

MA104 cell line, 580
Machupo virus, 552
Macrophages, 134, 136, 156–158, 165–168, 176, 179, 186–188, 214, 487
Macules, 619
Maintenance cycle (enzootic cycle), 491
Malnutrition, 179–180
Mammary tumor viruses, 207
Mapping, genetic, 71–72
  heteroduplex, 57, 340
  R loop, 57, 340
  marker rescue by transfection, 107
  oligonucleotide, 57, 340, 354–355
  peptide, 57
  transcription and translation, 59
Marburg virus, 571–574, 623
Marek's disease, 235, 404
Marker rescue, 102, 107
Mason–Pfizer monkey virus, 591
Master strains, 286
Maternal immunity, 155, 579, 631
Mathematical modeling, 269
Matrix (M) protein, 6, 9, 85, 198, 512
Maturation, of virions, 82–87
MDCK cell line, 36, 38, 514
Measles, 524–529, *see also* Paramyxoviruses
  clinical features, 527, 612, 617, 619, 625, 630
  epidemiology and control, 180, 255–256, 265, 528–529
  eradication, 279
  immunity, 178, 255, 265, 527
  laboratory diagnosis, 528
  malnutrition, 180
  pathogenesis, 137–138, 527
  SSPE, 198
  vaccine, 528
Mechanical transmission, by arthropods, 251
Media, for cell culture, 36, 347
Meningitis, 408, 462, 529, 553, 615–617, *see also* individual viruses
Mental retardation, 420, 500–501, 616
Mesenteric adenitis, 394
Messenger RNA, *see* RNA
Metabolic inhibition test, 353
Methylation, of RNA, 73, 320, 408
MHC antigens, 148–153, 159–160, 187
Microcarrier beads, 290, 310, 460
Micro-methods, for diagnosis, 327, 353, 358
Midge, 582
Milk, 144, 171, 180, 261–262, 422, 459, 580, 631
Milker's nodes, 442, 621
Mimotope, 292
Mink encephalopathy, 210
Minus-stranded RNA, 15–16, 28, 62–66, 74–76, 79
Mitotic cycle, 446
Modulation, of antigen by antibody, 164
Molecular epidemiology, 268–269, 581
Molluscum contagiosum, 441, 621, 623

Monkey
as experimental animal, 45, 211, 236, 349, 367, 597, 599–603
as reservoir, 176, 573
kidney cell culture, 297, 346, 531
for vaccine testing, 285, 287, 458, 471, 538
for virus isolation, 468, 603
viruses of, 196, 203, 224, 234, 258, 297, 346, 386–387, 404, 440, 443, 532, 591
Monkeypox, 440
Monocistronic mRNA, 66
Monoclonal antibodies, 115, 162, 290, 331–332, 490
Monokines, 158
Mononucleosis, 420, 426, 620
"Morbilliform" rash, 619
Morphogenesis, of virions, 82–87
Mortality rate, 267
Mosquito
for arbovirus isolation, 489
cell culture, 489
control, 276–277
Mosquito-borne viruses, 129–130, 251–252, 258–262, 276, 487, 491–492
bunyaviruses, 562, 565
orbiviruses, 582
poxviruses, 112–114, 253, 443
togaviruses and flaviviruses, 484–498
Mouse,
as experimental animal, 44, 210, 168–172
immunology, 149, 151, 165, 168–170, 172
oncogenic viruses in, 241
as reservoir, 551–553, 560
for virus isolation, 349, 465–466, 488, 552–553, 563, 583
viruses of, 205, 224, 386, 541
Mozambique virus, 551
MRC-5 cells, 346
Multiple puncture, 439
Multiplication, of viruses, 53–89
adenoviruses, 72–73, 79, 82, 87, 238–239, 393
arenaviruses, 63, 548–549
bunyaviruses, 63, 557
caliciviruses, 75
coronaviruses, 542–543
flaviviruses, 75, 483
hepadnaviruses, 368
herpesviruses, 403
orthomyxoviruses, 58, 74, 511–512
papovaviruses, 71–72, 79–80, 238–240
paramyxoviruses, 58, 69, 74, 76, 83–86
parvoviruses, 446
picornaviruses, 75, 78, 81, 82
poxviruses, 435–436
reoviruses, 69–70, 76–77, 576–577
rhabdoviruses, 58, 74, 76, 83–86, 568–569
strategy for chemotherapy, 305–307, 322
togaviruses, 68, 75, 482
Multiplicity, of infection, 55
Multiplicity reactivation, 100
Mumps, 529, 615, 617, 627, 629–630
Murine hepatitis virus, 541
Murray Valley (Australian) encephalitis, 496
Mutagenesis, 96
Mutants, 92–97, 108–118, 162, 284–287, 586
cold-adapted, 285–286, 518
conditional lethal, 95
defective interfering (DI), 95–96
deletion, 287
drug-resistant, 309, 319
oncogenes, 229, 233
revertants, 285–287, 297, 459
suppressor, 93
temperature-sensitive, 95, 102–103, 228, 285–286, 538
Myalgia, epidemic, 462
Mycoplasma, 4, 303
Myositis, 462, 626
Myxomatosis, 112–114, 253
Myxovirus, *see* Orthomyxoviruses

N

Nairovirus, 558
Nasopharyngeal carcinoma (NPC), 237
Natal, *see* Perinatal infections
Natural killer (NK) cells, 158, 166, 170, 186–188
2′-NDG, 319
Necropsy, 337, 342, 570, 573
Negative straining, 5, 329, 576
Negative-stranded RNA, 15–16, 28, 62–66, 74–76, 79
Negri body, 570

Neonatal, *see* Perinatal infections
Nephritis, 174, 205–207, 371, 623–624
Nephropathia epidemica, 559
Neplanocin A, 320
Nerves, 138–141, *see also* CNS infections
Nested set, of subgenomic mRNAs, 67, 542
Neuraminidase, 8, 49, 114, 163, 335, 512, 518, 522
Neurological disease, *see* Central nervous system infections
Neutralization, of viruses by antibody, 114–116, 161–164, 291–292, 351–354, 465
Newcastle disease virus (NDV), 522, 626
Nomenclature, of viruses, 22–33
Non-A, non-B hepatitis, 203, 596–599, 628
Non-neutralizing antibody, 590
Nonproductive infection, 125
Northern blotting, 340
Norwalk agent, 329, 475
Nosocomial infection, 393, 399, 412, 419, 422, 464, 468, 507, 537, 551, 573, 581, 631–632, *see also* Iatrogenic infection
Nucleic acid, *see also* individual viral families
  hybridization, 57, 240, 243, 269, 339–341, 382, 391–392, 580
  infectious, 11
  replication, 79–81
  transcription, 70–76
  of viruses, 10–16, 56–57, 66, 579
Nucleocapsid, 5–9, 74, 523, 555
Nucleoside analogs, as antiviral agents, 307, 314–321
Nucleoprotein (NP), 523

O

Okazaki fragments, 79
Oligodendrocytes, 387
Oligonucleotide
  as antiviral agents, 322
  mapping, 269, 340, 354–355, 456, 490, 557
Oligopeptides, as antiviral agents, 322
Oligosaccharides, 83
Omsk hemorrhagic fever, 497, 623
Oncogenes, 227–234
Oncogenic viruses, 217–246
Oncoviruses, 220–234, *see also* Retroviruses
O'nyong-nyong virus, 496, 626
One-step growth curve, 54
Oophoritis, 529
Opportunistic infections, 588
  see also Immunosuppression, Nosocomial infection
Opsonization, 165–166
Oral poliovaccine (OPV), 458–460
Orbiviruses, 582–583
Orchitis, 529
Orf, 442, 621
Organ culture, 36, 348, 473, 544
Organ transplantation, 421, 423, 595, 604, 632
Original antigenic sin, 172, 360, 490, 515
Orphan viruses, 453, 575
Orthomyxoviruses, 29, 509–519
  chemotherapy, 321–322
  classification and properties, 8, 58, 162–163, 510–511
  clinical features, 513
  epidemiology and control, 114–118, 515–518
  genetics, 109, 114–118, 188
  immunity, 162–163, 169, 171–173, 188, 513–514
  laboratory diagnosis, 335, 349, 352, 514–515
  multiplication, 58, 62–68, 74, 83–85, 511–512
  pathogenesis, 130–133, 513–514
  vaccines, 516–518
Otitis media, 472, 527, 608
Overwintering of arboviruses, 259, 491
Ox red cell hemolysin test, 428
Ozone, 275

P

P4 containment laboratories, 344–345, 550–551, 560, 573, 623
Packing sequence, 82
Paired sera, 356
Palindrome, 79, 448
Pancreatitis, 529, 629
Pandemics, 114, 463, 515
Papanicolaou smear, 411
Papillomaviruses, 381–386
  chemotherapy, 313
  oncogenesis, 240–241
Papovaviruses, 25, 381–388
  chemotherapy, 313

Papovaviruses (*cont.*)
multiplication, 71–72, 79–80, 100
oncogenesis, 127, 238–241
properties, 12, 381–382
Papules, 619
Parainfluenza, 531–533, 608–611, *see also* Paramyxoviruses
Paralysis, 454–455, 462, 616, 618
Paramyxoviruses, 30, 521–539, *see also* Measles, Mumps, Parainfluenza, RSV
chemotherapy, 537
classification and properties, 521–525
clinical features, 527, 529, 531, 533–534
epidemiology and control, 180, 255, 528, 530–533, 537–538
genetics, 104
laboratory diagnosis, 528–532, 536–537
multiplication, 62–67, 69, 74, 76, 83–86
pathogenesis and immunity, 137–138, 255, 265, 527, 534–536
steady-state infections, 196
vaccines, 528, 531, 533, 538
Pararotaviruses, 578
Parasites, 179
Parkinson's disease, 619
Paronychia, 408
Parvoviruses, 7, 27, 445–450
Passive immunity, 171, 181, 289
Passive immunization, 301–302, 377, 471, 529, 552, 571
Pasteurization, 595
Pathogenesis, of viruses, 119–145
see also Pathogenesis section of Chapters 13–28
persistent infections, 211–215
Pathogenicity, of viruses, 125–126
Pathology, 119–145, 172–175
Paul–Bunnell test, 428
Penetration, of cells by viruses, 65–70
Pentakis dodecahedron, 468
Pentamer, 402
Penton, 390–391
Peplomer, 6, 9, 85, 542
Peracetic acid, 344, 376
Periarteritis nodosa, 173
Perinatal infections, 257, 630–631
cytomegalovirus, 421–424
enteroviruses, 462
hepatitis B, 376
herpes simplex, 408
rotavirus, 579–581
RSV, 537
varicella, 415
Persistent infections, 124, 193–215, 256–257, 286
adenoviruses, 395
arenaviruses, 551–553
congenital rubella, 502
delta virus, 600
hepatitis B, 242–243, 371–376
herpesviruses, 408, 416, 421, 427
LAV/HTLV-III, 590
measles (SSPE), 198, 286
non-A, non-B hepatitis, 597
polyomaviruses, 386–388
rubella syndrome, 142, 271, 500–507
spongiform encephalopathies, 210, 601–605, 619
Petechiae, 622
pH, 18, 68, 133, 163, 321, 396, 453, 472, 511
Phage, 45, 53, 104–108, 328
Phagocytosis, 65–69, 157, 166, *see also* Macrophages
Pharmacology, of antiviral agents, 308
Pharyngitis, 393, 405, 426, 463, 531, 609–611, *see also* individual viruses
Pharyngoconjunctival fever, 393
Phenotypic mixing, 103–104, 226
Phlebotomus fever, 564–566, 625
Phlebovirus, 558
Phosphonoformic acid (PFA), 321
Phosphorylation of protein, 230–231
Photodynamic inactivation, 357
Picornaviruses, 27, 451–473, *see also* Hepatitis A, Poliovirus, Rhinovirus
chemotherapy, 473
classification and properties, 7, 451–453, 468–469
clinical features, 454, 461–464, 469–470, 472, 608–631
epidemiology and control, 456–460, 467–468, 470–471, 473
genetics, 104
immunity, 454–455, 464, 469–470, 472
laboratory diagnosis, 349, 353, 455–456, 464–467, 470, 472–473
multiplication, 62–67, 69, 75, 78, 81–82
pathogenesis, 454–455, 464, 469–470, 472
vaccines, 458–460, 471, 473

Pigs
as reservoirs, 114, 496, 515
viruses of, 474, 483, 541
Plant viruses, 9, 575, 599
Plaque assay, 45–50, 92, 489
Plaque-reduction test, 353, 465, 489
Plasma cell, 154
Plasmapheresis, 377–378
Plasmid, 104–108, 287
Plating efficiency, 50
Pleiotropism, 97
Pleomorphic viruses, 572
Pleurodynia, 462
PMK cells, 346
Pneumocystis carinii, 588
Pneumonia, 609, 611–613, 623
adenovirus, 393
AIDS, 588
cytomegalovirus, 422–423
influenza, 513
measles, 527
parainfluenza, 531
RSV, 534–536
varicella, 415
Pocks, 43, 47, 349, 411
Polarity, of RNA, 15–16, 28, 62–66, 74–76, 79, 446
Poliovirus, 454–460
clinical features, 454, 616, 624
epidemiology and control, 456–460
eradication, 278–279
genetics, 109, 285
laboratory diagnosis, 455–456
pathogenesis and immunity, 454–455
vaccines, 93, 109, 285, 295, 298, 458–460
Polyadenylation, 15–16, 71, 73
Polyarthritis, 496, 626
Polyclonal B cell activation, 427
Poly(I)·poly(C), 310
Polykaryocytosis, 122
Polyomavirus
human, 198, 386–388, 624, 632
mouse, 127, 238–239
Polyploidy, 104
Polyprotein, 15, 66, 75
Post-infectious encephalomyelitis, 141, 617–618
Post-mortem, *see* Necropsy
Powassan virus, 498
Poxviruses, 26, 433–443
classification and properties, 13, 434–435
clinical features, 440–443, 620
genetic interactions, 98–103
laboratory diagnosis, 47, 436–437
multiplication, 62–67, 123, 435–436
myxomatosis, 112–114, 253
smallpox, epidemiology, 253, 262, 265
smallpox, eradication, 274, 278–280
smallpox, vaccine, 437–439
vaccinia, vector for recombinant vaccines, 287–288
Pregnancy, 181, 213, 271, 345, 387, 415, 421, 595, 598
Prenatal infections, 257, 630–631, *see also* Perinatal infections
cytomegalovirus, 421–424
rubella, 142, 271, 500–507
varicella, 415
Prevalence, 267–268
Prime strains, 354
Primer, 79, 81, 107, 219, 368, 511
Procapsid, 82
Processing, of RNA, 73, *see also* Polyadenylation, Splicing
Prodrug, 317
Productive infection, 98, 125
Progressive lymphoproliferative disease, 427
Progressive multifocal leukoencephalopathy (PML), 198, 386–387, 618
Progressive rubella panencephalitis (PRP), 500, 618
Promoter, 225, 232
Propagative transmission, by insects, 258–262
β-Propiolactone, 290, 516, 571, 598
Prospect Hill virus, 560
Prospective studies, 271, 590
Proteases, 482, 513, 522
inhibitors of, 322
Protein A, 330, 360
Protein
"early", 77
fusion, 108
glycosylation, 83–84
nonstructural, 77
by recDNA cloning, 104–108, 291, 357
synthesis, 77–78, 83–84, 87

Protein (*cont.*)
- transport, 78, 83–84
- as vaccines, 291–294
- of viruses, 16–17, 57, 61, 77–78, 87, 120–121

Protein kinase, 230–231
Proteinuria, 623
Protooncogenes, 228, 232–233
Protozoa, 588
Provirion, 82
Provirus, 207–208, 211, 221–222, 225, 340
Psoralens, 290, 357
Purification of viruses, 9–10, 290
Pyknosis, 124

Q

Quarantine, 274, 279–280, 562, 571

R

RA27/3 vaccine, 505
Rabbit kidney cell cultures, 409
Rabies, 123, 569–571
- pathogenesis, 138–140, 569

Radiation, *see* Inactivation of viruses
Radiculomyelitis, 463, 616
Radioimmunoassay (RIA), 330–332
Radioimmunoprecipitation (RIP), 591–592
Radioisotopes, 57–58, 330–332, 339–341
Rapid diagnostic methods, 327–341, 358–366
Rashes, 136–138, 143, 173, 619–623
- arboviruses, 484–486
- arenaviruses, 552
- bunyaviruses, 559–563
- enteroviruses, 462–463
- herpes simplex, 404–408
- measles, 526–527
- papillomaviruses, 383–386
- parvovirus, 448
- poxviruses, 437–443
- rubella, 500
- varicella, 415, 417

Rat transmission, 551, 560
RD line of human rhabdomyosarcoma, 346
Reaction of identity, 334
Reactivation,
- genetic, 100–102
- of latent infection, 178, 198–202, 213, 387, 421, 631–632

Reading frame, 72
Reassortment, of genes, 100–101
Receptors, for virus on cell, 65
Recombinant DNA technology, 104–108, 287–288, 290–291, 310–311, 340
Recombinant vaccines, 286–288, 290–291
Recombination, genetic, 98–109
Recovery, from infection, 167–170, 189, 211–215
Recurrent infections, 404–405
Regulation
- of cell, by virus, 87
- of viral transcription, 72–76
- of viral translation, 77–78

Reiterated sequences, 11–13, 16, 402
Reoviruses, 32, 575–584, 613
- classification and properties, 109–111, 575–577
- genetics and pathogenesis, 109–111
- multiplication, 62–67, 69–70, 76–77, 122–123, 576–577
- vaccines, 289, 581–582

Replicase, 81
"Replication complex", 81
Replication, of viral nucleic acid, 79–81
- inhibition by antiviral agents, 307, 314–321

Replication of viruses, *see* Multiplication of viruses
Replicative intermediate (RI), 81
Reptiles, 252, 404
Rescue
- by complementation, 102–103
- by defective virus, 98
- fragment rescue, 102
- of intracellular viral genome, 239
- marker rescue (reactivation), 99–102

Resistance, to viral infection, 175–190
Respiratory infections, 21, 608–613, *see also* individual viruses
- epidemiology, 249–250, 262–264, 276
- laboratory diagnosis, 337–338, 341
- pathogenesis, 130–133, 143

Respiratory syncytial virus (RSV), 533–538, *see also* Paramyxoviruses
- chemotherapy, 321, 537
- clinical features, 533–534, 608–611, 631
- epidemiology and control, 264, 537–538
- laboratory diagnosis, 536–537

pathogenesis and immunity, 174, 534–536
vaccines, 538
Restriction endonuclease, 71, 96, 104–108, 341, 355
mapping, 382, 392
Reticulocytes, 448
Retinitis, 626
Retroviruses, 31, 218–234, 585–596
chemotherapy, 322
HTLV-III/LAV and AIDS, 585–596
lentiviruses, 208–210, 585–596
multiplication, 224–226
oncogenesis, 207–208, 223–234
Reverse passive hemagglutination, 336, 376, 550, 580
Reverse transcriptase, 219, 225, 368, 587
inhibitors of, 322, 596
Reye's syndrome, 618
Rhabdomyosarcoma line (RD), 465
Rhabdoviruses, 32, 567–574, *see also* Rabies
DI particles, 95
multiplication, 58, 62–67, 69, 74, 76, 83–86, 568–569
Rheumatoid arthritis, 449, 627
Rheumatoid factor (RF), 163, 359
Rhinitis, 608–610, *see also* Common cold
Rhinoviruses, 472–473, 608–610, *see also* Common cold
chemotherapy, 313, 323, 473
epidemiology, 266, 473
Rhombic triacontahedron, 468
Ribavirin, 320–321, 537, 552
Ribosomes, in virions, 547
Rickettsiae, 4, 303
Rifamycins, 596
Rift Valley fever, 561–562, 622, 626
Rimantadine, 321–322
Rinderpest, 176
Risk groups, 594
RK-13 cell line, 38, 346, 503
RNA,
cRNA, 79
mRNA, 15, 62–67, 70–76
mRNA, 5′-capping, 15, 73, 306, 320
mRNA, 3′-polyadenylation, 15, 71, 73, 306
polarity, 15–16, 28, 62–66, 74–76, 79
replication, 79–81
transcription, 73–76
translation, 77–78
tRNA in retroviruses, 219, 225
of viruses, 10, 14–16, 56–57, 66–67, 579
RNA-dependent RNA polymerase, 15, 63, 66, 70, 79–81, 511–512
Ro 09-0410, 323
Rocio virus, 497
Rodents, 259, 261
control, 551–553, 561
Rodent-borne nephropathy, 559–561
Ross River virus, 496, 620, 626
Rotavirus, 577–582, 613
Rous sarcoma virus, 177, 224, 228, 322
Rubella, 499–507
clinical features, 500, 617, 619, 625–627
congenital rubella syndrome, 142, 271, 500–507

S

Sabin vaccine, 285, 458–460
Safety
in laboratory, 344–345, 357, 488, 591, 595
of vaccines, 296–297
SAH hydrolase, 320
Salivary transmission, 249, 256, 376, 412, 422, 429, 530, 569, 631
Salk vaccine, 290, 458–460
Sandfly, 262, 582
fever, 564–566
Sanitation, 275–276, 457, 470–471
Satellite viruses, 599
Scarlatiniform rash, 620
Schwarz vaccine, 528
Scrapie, 210, 601, 604
Scrapie-associated fibrils (SAF), 604
Seals, 114, 626
Seasons, effect on virus infection, 113, 262–264, 266, 276, 473, 491, 516, 545, 560, 562, 583, 608, 620
Secondary attack rate, 267
Semen, 144, 251, 376, 421, 590–595, 624
Sendai virus, 123, 310
Sense, of nucleic acids, *see* Polarity
Sentinel animal, 258, 492
Sequencing, of nucleic acids, 11, 22, 57, 59, 105, 115–117, 269, 354, 382, 512
Seroconversion, 268, 590
Seroepidemiology, 267–269

Serology, 328–339, 349–360, *see also* individual families
cross-reactions, 349–350, 379, 391, 483, 488–490, 582
in hepatitis B, 372–374
IgM, 358–360, *see also* IgM serology
surveys, 267–269, 357, 493, 507, 590–592
in taxonomy, 22, 349–354, 391
Serum, for antibody assay, 350–353, 356–357, 504
Serum sickness, 368, 621
Sexually transmitted diseases (STD), 129, 623–624
adenoviruses, 394
AIDS, 585–596
cytomegalovirus, 422
genital herpes, 406–414
hepatitis B, 376
papillomavirus, 241, 384–386
poxviruses, 441
Sewage, 250, 275, 457, 470–471, 476
Shedding, *see* Excretion of viruses
Shellfish, 251, 471, 476
Sheep, viruses of, 208–210, 442, 475, 497, 562, 582, 586, 601
Shingles, 199, *see also* Herpes zoster
Shock, 487–488, 552, 563, 573, 622–623
Shwartzman reaction, 464
Sickle cell anemia, 448
Sigla, 575
Signal sequence, 83
Simian viruses, 297
Sindbis virus, 484, 620
Single radial (immuno)diffusion, 334, 514
Sinusitis, 472, 608
Site-directed mutagenesis, 286
Skin, 129, 136–138, 143, 173, 344, 619–623, *see also* Rashes
Slow infections, 194–195, 207–211, *see also* individual viruses
Slow transforming retroviruses, 223
Smallpox, 437, 620
epidemiology, 253, 262, 265
eradication, 274, 278–280
vaccine, 287–288, 437–439
Snakes, 259, 366
Sneezing, 250
Sodium deoxycholate, 354
Sodium hypochlorite, 344, 376, 595, 604
Southern blotting, 340
Species, of viruses, 21–22, 391–392, 452, 483, 511
Specimens, for virus isolation, 341–345
Sperm, 595
Splicing, of RNA, 66, 71, 73, 511
Spicules, 436
Split-virus vaccines, 290, 518, 571
Spongiform encephalopathies, 210, 601–605, 619
Spread of viruses
in the body, 134–145, *see also* Pathogenesis
in the community, 247–281, *see also* Epidemiology
Spumavirinae, 221, *see also* Retroviruses
SSPE, 198, 286
St. Louis encephalitis, 260–261, 497
Steady-state infections, 124, 196
Sterilization, 17–20, 376, 399
Stomatitis, 404, 610
Storage, of viruses, 17–18
Strategy, of expression of the viral genome, 61–67
Structure, of viruses, 5–17, 21–32, 58, 109–111, 512, *see also* individual families
Subclinical, *see* Inapparent infections
Subgenomic mRNA, 66, 74–75, 482–483
Subunit vaccines, 290, 414, 533
"Subviral particle", of reovirus, 69–70
Summer grippe, 463
Superhelix, 11, 80
Suppressor mutation, 286
Suppressor T cells, 161, 173–174, 212
Suramin, 322, 596
Surveillance, in control of viruses, 279, 327
Susceptibility, to viral infection, 175–190, 255, 267
Suspension culture, 290, 298, 310
SV5, 196, 532
SV40, 297, 386
Swimming pools, 275, 398–399, 441, 471, 476
Symmetry, in viral structure, 5–9
Syncytium, 122–124, 347–348, 525
Syndromes, viral, 607–633
see also Clinical Features section of Chapters 13–29

Synergism, in chemotherapy, 309, 314
Synthetic peptides, 291–292, 379, 414
Systems, body, diseases of, 607–633

T

T antigens, 79, 126, 239, 387
T cells, 148–153, 159–161, 168–169, 172–174, 186–187, 214, 295
  growth of HTLV-III in, 588–591
T particles, 96
Tacaribe viruses, 548–549
Tanapox, 443
Target organs, 487
Tattooing, 375–376
Taxonomy, *see* Classification of viruses
Tegument, 402
Temperature
  body, 182, 285
  environmental, 18, 113, 251, 259–260, 276, 298, 343, 347, 473, 491
  non-permissive, 286
Teratogenic viruses, 419–422, 500–502, 630
Terminal repeats, 11–13, 16, 219, 403
Thalassemia, 448
Therapeutic index, 308
Thrombocytopenia, 622
Thymidine kinase, 112, 287, 317–320, 408, 436
Thymidylate synthase, 307
Tick-borne viruses, 492
  bunyaviruses, 563
  flaviviruses, 497, 259–262, 485, 499
  orbiviruses, 582
Time-resolved fluoroimmunoassay, 332
Tissue culture, *see* Cell culture
Togaviruses and Flaviviruses, 29, 479–508, *see also* Rubella, Arboviruses
  classification and properties, 479–485
  clinical features, 484–486, 616, 620, 626
  epidemiology and control, 260–261, 479–480, 490–499
  laboratory diagnosis, 488–490
  multiplication, 62–63, 66–68, 75, 83, 86, 482–483
  pathogenesis and immunity, 487–488
  rubella, 499–507
  vaccines, 498–499
Tolerance, immunological, 142
Tonsillitis, 405, 427
Trans-acting transcriptional regulation, 227
Transcapsidation, 104
Transcriptase, 81, *see also* RNA-dependent RNA polymerase
Transcription, 66, 70–76, 403
Transduction, 228, 231
Transfection, 102, 107, 228–229, 232–233, 240
Transformation, of cells, 46, 125–127
  by oncogenes, 233
  by retroviruses, 223–234
Translation
  cellular, viral inhibition of, 87
  viral, 77–78
Translocation
  of genes on chromosomes, 232
  of virus through plasma membrane, 69
Transmissibility of viruses, 253–254, 267
Transmissible gastroenteritis of swine, 541
Transmission of viruses, 247–281, *see also* Epidemiology
Transovarial transmission, in arthropods, 259, 261, 492–494, 563, 565
Transplantation, of organs, 387, 395, 421–423, 631–632
Transstadial transmission, in ticks, 492, 563
Transverse myelitis, 618
Traveler's diarrhea, 581
Trifluorothymidine, 315
Trisodium phosphonoformate (foscarnet), 321
Tumor-specific transplantation antigens (TSTA), 124, 126
Tumor suppressors, 234
Typing, of viruses, 22, 391, 452, 483, 510
  electropherotyping, 578
  nucleic acid homology, 391
  oligonucleotide fingerprinting, 354–355
  restriction endonuclease mapping, 392
  serotyping, 349–356, 391
Tyrosine (-specific protein) kinase, 230–231

U

Ultraviolet radiation, 59, 96, 123, 199, 276, 357, 598

Uncoating, of viruses, 69–70
  inhibition of, 306, 321
Urethritis, 394, 624
Urine, viruses in, 251, 256, 387, 421, 551, 624
Urogenital tract, 129, 144, 251, 623–625
  see also Sexually-transmitted diseases
URTI, 608

V

Vaccines, 277–280, 283–302
  adenoviruses, 399
  AIDS, 594–595
  arboviruses, 498–499
  bunyaviruses, 562, 564
  cytomegalovirus, 426
  EB virus, 430
  hepatitis A, 471
  hepatitis B, 377–379
  herpes simplex, 414
  influenza, 516–516, 618
  measles, 528
  mumps, 531
  parainfluenza, 533
  poliomyelitis, 458–460
  rabies, 571
  respiratory syncytial virus, 538
  rotaviruses, 581–582
  rubella, 505–507
  schedules, 300–301
  smallpox, 287–288, 437–439
  varicella, 419
  yellow fever, 498
Vaccine trials, 271
Vaccinia virus
  clinical features, 439, 617, 626
  in smallpox eradication, 278–280
  vaccination, 437–439
  as vector for recombinant vaccines, 287–288
Varicella-zoster, 415–419, *see also* Herpesviruses
  chemotherapy, 313–321, 418
  clinical features, 415–416, 612, 617–618, 621, 625, 631
  epidemiology and control, 256, 418–419
  laboratory diagnosis, 416
  pathogenesis and immunity, 199–202, 416
  vaccines, 419
Vector control, 276–277
Vectors, for gene cloning, 104–108, 287–288, 291, 383
Venereal transmission, 251, *see also* Sexually-transmitted diseases
Venezuelan equine encephalitis, 497
Vero cell line, 38, 346, 460, 489, 503, 550
Verrucae, 383–386
Vertical transmission, 207–208, 222, 257, 551–553, 590, 631
Vesicles, 406, 439, 621
Vesicular stomatitis virus, 568
Vidarabine, 315
Virazole, 320–321, 537, 552
Viremia, 135–141, 144, 251, 487
Virgin-soil epidemics, 255, 265, 457
Virion, 5–8
Virolysis, 164
Virosomes, 293
Virulence, of viruses, 108–114, 125–126, 253–254, 285, 287, 297, 513
Virus-disease association, 244, 252, 272–274, 327, 355, 493, 589–590, 599, 604, 607–633
Virus transport medium, 343
Visna, 208–210, 586–587, 589

W

Warts, 383–386, 622
Water transmission, of viruses, 251, 275–276, 470–471, 476, 581, 598
Weather, *see* Seasons
Wesselbron virus, 488
Western blotting, 57, 591–592
Western (equine) encephalitis, 260–261, 497
West Nile fever, 496
WI-38 cells, 37, 346

X

Xenotropic retroviruses, 223
X-linked lymphoproliferative (XLP) syndrome, 426
X-ray crystallography, 5, 115, 305, 323
X-ray of lungs, 535, 611, 613

Y

Yabapox, 443
Yeasts, 105, 108, 291, 310, 379, 588

Yellow fever, 493–495, 622
  clinical features, 486, 629
  epidemiology, 493–495
  laboratory diagnosis, 489
  immunity, 171, 176
  vaccine, 498

Z

Zika virus, 488
Zoonoses, 257–262
  arenaviruses, 551–553
  bunyaviruses, 560–566
  filoviruses, 571–574
  poxviruses, 440–443
  rhabdoviruses, 567–571
  rotaviruses, 581
Zoster immune globulin (ZIG), 418